# METHODS *in* PLANT MOLECULAR BIOLOGY *and* BIOTECHNOLOGY

*Edited by*

Bernard R. Glick
John E. Thompson

582.087
M585
1993

CRC Press
Boca Raton Ann Arbor London Tokyo

**Library of Congress Cataloging-in-Publication Data**

Methods in plant molecular biology and biotechnology / editors,
Bernard R. Glick, John E. Thompson.
p. cm.
Includes bibliographical references and index.
ISBN 0-8493-5164-2
1. Plant molecular biology—Methodology. 2. Plant biotechnology—Methodology. I. Glick, Bernard R. II. Thompson, John E.
QK728.M483 1993
582′.087328—dc20

92-45589
CIP

This book represents information obtained from authentic and highly regarded sources. Reprinted material is quoted with permission, and sources are indicated. A wide variety of references are listed. Every reasonable effort has been made to give reliable data and information, but the author and the publisher cannot assume responsibility for the validity of all materials or for the consequences of their use.

Neither this book nor any part may be reproduced or transmitted in any form or by any means, electronic or mechanical, including photocopying, microfilming, and recording, or by any information storage and retrieval system, without permission in writing from the publisher.

All rights reserved. Authorization to photocopy items for internal or personal use, or the personal or internal use of specific clients, is granted by CRC Press, Inc., provided that $.50 per page photocopied is paid directly to Copyright Clearance Center, 27 Congress Street, Salem, MA, 01970 USA. The fee code for users of the Transactional Reporting Service is ISBN 0-8493-5164-2/93 $0.00 + $.50. The fee is subject to change without notice. For organizations that have been granted a photocopy license by the CCC, a separate system of payment has been arranged.

The copyright owner's consent does not extend to copying for general distribution, for promotion, for creating new works, or for resale. Specific permission must be obtained from CRC Press for such copying.

Direct all inquiries to CRC Press, Inc., 2000 Corporate Blvd., N. W., Boca Raton, Florida, 33431.

© 1993 by CRC Press, Inc.

International Standard Book Number 0-8493-5164-2

Library of Congress Card Number 92-45589
Printed in the United States 2 3 4 5 6 7 8 9 0
Printed on acid-free paper

# Preface

Within the last decade, advances in molecular biology have resulted in the development of a host of powerful techniques that have application beyond the horizons of the discipline itself. Indeed, many of the tools of molecular biology are being used increasingly by those who are practitioners of cell biology, biochemistry, and physiology. The reasons for this are numerous and compelling. In particular, these techniques are very powerful and can be used to provide new information about metabolism, development, regulation, and subcellular structure that was not previously possible. As well, the techniques are becoming easier to deploy as various kinds of commercial kits become available.

The motivation to prepare a comprehensive treatise on techniques in plant molecular biology and biotechnology arose, therefore, from a growing awareness that plant biologists of many stripes are now using these techniques and that there was a need to set down detailed protocols for the techniques for use by the specialist and nonspecialist alike. The book covers a wide range of methods with emphasis on well-tested approaches. There are several chapters dealing with the various dimensions of recombinant DNA technology, including the isolation and characterization of plant nucleic acids, the construction of vectors and libraries, and the introduction of foreign DNA into plants. There are also chapters that describe methods for visualizing and assessing gene expression in plants, methods for the production and analysis of plant mutants, methods for assessing and using relevant computer software, and methods for DNA mapping and analysis of DNA polymorphism. Finally, there are chapters that deal with the detection and characterization of plant pathogens, plant-bacterial symbioses, and the isolation and characterization of plant growth promoting rhizobacteria.

The authors are all experienced practitioners of the technology they describe and were encouraged to share with the reader nuances in the protocols that are usually not recorded in research papers and that can only be gleaned through working with the techniques over a period of time. Many of the chapters also provide an overview of the methodology under consideration, including discussion of the relative merits of the various methods available, detailed protocols for the methods being described, and an accounting of their limitations.

It is our hope that the book will be used by the experienced molecular biologists as well as by those in other fields who can achieve heightened elucidation of biological phenomena and mechanisms in general through the application of these techniques. In particular, we hope that the development of these techniques will be made easier by the availability of a comprehensive, user-friendly treatise.

B. R. Glick
J. E. Thompson

N. C. WESLEYAN COLLEGE
ELIZABETH BRASWELL PEARSALL LIBRARY

# The Editors

**Bernard R. Glick,** Ph.D. is a Professor in the Department of Biology at the University of Waterloo in Waterloo, Ontario, Canada. Dr. Glick received his B.S. degree in Chemistry at the City College of New York in 1969, and an M.Sc. and Ph.D. in Chemistry and Biochemistry at the University of Waterloo in 1972 and 1974, respectively. Dr. Glick's post-doctoral training included four years at the University of Toronto and one year at the National Research Council of Canada in Ottawa. In 1979 Dr. Glick joined a fledgling biotechnology company located in Toronto as the Head of the Molecular Genetics and Biochemistry Group. In 1982 he was appointed Associate Professor in the Department of Biology at the University of Waterloo. He was promoted to full Professor in 1988.

Dr. Glick has presented over 40 papers at scientific meetings and has published over 70 research papers. He is currently an Editor of the review journal Biotechnology Advances. Dr. Glick was recently awarded a Lady Davis Fellowship to work and teach a biotechnology course by which certain plant growth-promoting bacteria stimulate the growth of plants and the development of integrated systems for the large scale production of proteins from recombinant microorganisms.

**John E. Thompson**, Ph.D., is a Professor in the Department of Biology at the University of Waterloo in Waterloo, Ontario, Canada. Dr. Thompson received his B.S.A. degree from the University of Toronto in 1963 and a Ph.D. in Plant Biochemistry from the University of Alberta in 1966. He spent a post-doctoral period in the Department of Medical Biochemistry at the University of Birmingham, England before joining the Department of Biology at Waterloo as an Assistant Professor in 1968. He was promoted to Associate Professor in 1972 and to Full Professor in 1976.

Dr. Thompson has presented numerous papers at scientific meetings, has published over 100 research papers, and is a Fellow of the Royal Society of Canada. He is currently Dean of Science at the University of Waterloo.

# Contributors

**Richard M. Amasino**
Department of Biochemistry
University of Wisconsin
Madison, Wisconsin

**Gynheung An**
Institute of Biological Chemistry
Washington State University
Pullman, Washington

**Alain Asselin**
Département de Phytologie
Faculté des Sciences de l'Agriculture et de l'Alimentation
Université Laval
Sainte-Foy, Québec, Canada

**Yoav Bashan**
Department of Microbiology
Division of Experimental Biology
The Center for Biological Research (CIB)
La Paz, B.C.S., Mexico

**Nicole Benhamou**
Département de Phytologie
Faculté des Sciences de l'Agriculture et de l'Alimentation
Université Laval
Sainte-Foy, Québec, Canada

**P. J. Charest**
Petawawa National Forestry Institute
Forestry Canada
Chalk River, Ontario, Canada

**Paula P. Chee**
Molecular Biology Unit-7242
The Upjohn Company
Kalamazoo, Michigan

**William L. Crosby**
Molecular Genetics Section
Plant Biotechnology Institute
National Research Council of Canada
Saskatoon, Saskatchewan, Canada

**Roger F. Drong**
Molecular Biology Unit-7242
The Upjohn Company
Kalamazoo, Michigan

**Erwin B. Dumbroff**
Department of Biology
University of Waterloo
Waterloo, Ontario, Canada

**P. F. Fobert**
John Innes Institute
John Innes Centre for Plant Research
Norwich, United Kingdom

**Shimon Gepstein**
Department of Biology
The Technion-Israel Institute of Technology
Haifa, Israel

**Bernard R. Glick**
Department of Biology
University of Waterloo
Waterloo, Ontario, Canada

**Paul H. Goodwin**
Department of Environmental Biology
University of Guelph
Guelph, Ontario, Canada

**Bruce M. Greenberg**
Department of Biology
University of Waterloo
Waterloo, Ontario, Canada

**Robert Gregerson**
Department of Agronomy and Plant Genetics
University of Minnesota
St. Paul, Minnesota

**Margaret Y. Gruber**
Forage Section
Agriculture Canada Research Station
Saskatoon, Saskatchewan, Canada

**John J. Heikkila**
Department of Biology
University of Waterloo
Waterloo, Ontario, Canada

**Ann M. Hirsch**
Department of Biology
University of California, Los Angeles
Los Angeles, California

**Gina Holguin**
Department of Microbiology
Division of Experimental Biology
The Center for Biological Research (CIB)
La Paz, B.C.S., Mexico

**David H. Hubbell**
Department of Soil and Water Science
University of Florida
Gainesville, Florida

**V. N. Iyer**
Department of Biology
Carleton University
Ottawa, Ontario, Canada

**Younghee Kim**
Institute of Biological Chemistry
Washington State University
Pullman, Washington

**Benoit S. Landry**
Agriculture Canada Research
St-Jean-sur-Richelieu
Québec, Canada

**Ran Lifshitz**
Beak Consultants Limited
Brampton, Ontario, Canada

**Russell L. Malmberg**
Botany Department
University of Georgia
Athens, Georgia

**James R. Manhart**
Department of Biology
Texas A&M University
College Station, Texas

**Heather I. McKhann**
Department of Biology
University of California, Los Angeles
Los Angeles, California

**B. L. Miki**
Agriculture Canada
Plant Research Center
Central Experimental Farm
Ottawa, Ontario, Canada

**Annette Nassuth**
Department of Botany
University of Guelph
Guelph, Ontario, Canada

**A. N. Nunberg**
Department of Biology
Texas A&M University
College Station, Texas

**J. J. Pasternak**
Department of Biology
University of Waterloo
Waterloo, Ontario, Canada

**K. Peter Pauls**
Department of Crop Science
University of Guelph
Guelph, Ontario, Canada

**Jerry L. Slightom**
Molecular Biology Unit-7242
The Upjohn Company
Kalamazoo, Michigan

**Judith Strommer**
Department of Horticultural Science
Department of Molecular Biology and Genetics
University of Guelph
Guelph, Ontario, Canada

**Brian H. Taylor**
Department of Biology
Texas A&M University
College Station, Texas

**T. L. Thomas**
Department of Biology
Texas A&M University
College Station, Texas

**Allen Van Deynze**
Department of Crop Science
University of Guelph
Guelph, Ontario, Canada

**Michael Vayda**
Department of Biochemistry,
Microbiology and Molecular Biology
University of Maine
Orono, Maine

**Kang Fu Yu**
Department of Molecular Biology
and Genetics
University of Guelph
Guelph, Ontario, Canada

# Contents

# 1

# The Use of Recombinant DNA Technology to Produce Genetically Modified Plants

*Bruce M. Greenberg and Bernard R. Glick*

## I. INTRODUCTION

The relatively recent development of techniques to isolate and manipulate specific DNA fragments, more or less at will, has led to advances in the ability to selectively alter the genetic make-up of an organism. Unlike random mutagenesis, a simple but largely uncontrollable process with unpredictable results, genetic alteration using recombinant DNA technology is highly selective. This notwithstanding, the isolation of a specific gene and its introduction and expression in a new host are often technically challenging.

In the early 1980s, it became clear that plants were amenable to genetic engineering.[1,2] This is because foreign genes can be stably introduced into plant chromosomal DNA by a variety of techniques.[2,3] If a given foreign gene contains the appropriate regulatory sequences, the gene product will be synthesized by the transformed plant.[4] Furthermore, most plant cells are totipotent,[2] thereby allowing regeneration of a fertile "transgenic" plant from a single transformed cell. If the transgenic plant flowers and produces viable seed, the acquired trait will be preserved in the progeny.

Conceptually, there are two reasons for transforming a plant. One is to confer a trait to a crop plant that will improve its agricultural, horticultural, or ornamental value. The second is to facilitate an understanding of a particular biological process. Some of the traits that can be "engineered" into plants by the addition of a single gene (or a small cluster of genes) include insecticidal activity,[5] tolerance to viral infection,[6] herbicide resistance,[7] altered fruit ripening,[8] altered flower pigmentation,[9] improved nutritional quality of seed proteins,[10] and self-incompatibility.[11] To date, numerous transgenic plants have been generated, including crop species such as tomato, potato, lettuce, canola, cotton, soybean, corn, and rice.[2,3] With genetic manipulation of plants now well established, plant biotechnology will begin to have an impact on agricultural practice over the next decade.

## II. PLANT TRANSFORMATION

### A. METHODS OF DNA TRANSFER

Although straightforward in principle, in practice plant transformation can be quite complex. Once a gene of interest is isolated, which in itself can be elusive and arduous, it is introduced into a plant transformation vector. The vector DNA facilitates manipulation of the gene in *Escherichia coli* prior to plant transformation, as well as transfer of the gene to the host plant (often via the plant pathogenic bacterium, *Agrobacterium tumefaciens*). An idealized vector would contain a multiple cloning site, an antibiotic resistance gene allowing

0-8493-5164-2/93/$0.00+$.50
© 1993 by CRC Press, Inc.

for selection in both *E. coli* and *A. tumefaciens* (e.g., a gene encoding ampicillin resistance), a broad-host bacterial origin of replication, and an antibiotic resistance gene (e.g. a gene encoding kanamycin resistance) for selection of the foreign DNA in transformed plants. If the *Agrobacterium* system is used for transformation, the vector may contain the Ti plasmid virulence genes and T-DNA borders as well.

There are many approaches for introducing DNA into plants, of which the most commonly used are the *A. tumefaciens* "agro-infection" system, electroporation, and the particle gun.[3] The wild type form of *A. tumefaciens* contains a Ti plasmid that directs the production of tumorigenic crown gall growth on host plants. The crown gall is produced following the transfer of the tumor inducing T-DNA region from the Ti plasmid into the genome of an infected plant. This DNA fragment encodes genes for auxin and cytokinin biosynthesis, and it is these hormones in high concentration that promote growth of undifferentiated cells in the crown gall. Transfer of the T-DNA to the plant genome requires the Ti plasmid-encoded virulence genes as well as the T-DNA borders, a set of direct DNA repeats that delineate the region to be transferred. The tumor inducing genes can be removed from Ti plasmid vectors, disarming the pathogenic nature of the system, without affecting the transfer of DNA fragments between the T-DNA borders. Therefore, the tumor inducing genes are generally replaced with a gene encoding resistance to kanamycin, to allow for selection of transformants, and a gene encoding the desired trait.

The *Agrobacterium* containing the engineered plasmid is co-cultivated with cultured plant cells or wounded tissue. The de-differentiated plant cells are then propagated on selective media, and a transgenic plant is subsequently regenerated from the transformed cells by altering the levels of auxin and cytokinin in the growth medium. One of the disadvantages of this technique is that monocots are not a natural host for the bacterium and, for the most part, do not have the proper wound responses to be agro-infected. They are therefore generally excluded from *A. tumefaciens* transformation.[3,12]

Current protocols for *Agrobacterium* mediated transformation often employ binary vector systems,[13,14] which divide the Ti-plasmid into two components, a shuttle vector and a helper plasmid. The helper plasmid (~100 kb), which is permanently placed in the *Agrobacterium* host, carries the virulence genes. However, a much smaller shuttle vector contains T-DNA borders, a broad-host range bacterial origin of replication, antibiotic resistance markers, and a multiple cloning site for incorporation of the foreign gene. In this way the shuttle vector can be easily manipulated and engineered in *E. coli* prior to introduction into *A. tumefaciens* and subsequent co-cultivation with plant tissue. A similar strategy employs co-integrating Ti plasmid vectors.[14,15] In this system, an intermediate plasmid containing antibiotic resistance, the gene to be transferred and one T-DNA border are used to transform *A. tumefaciens* containing a disarmed Ti plasmid possessing the virulence genes and one T-DNA border. The two plasmids homologously recombine *in vivo* at the T-DNA borders, placing the antibiotic resistance gene and the gene of interest between two T-DNA borders, one from each plasmid. The genes are then transferred into plant tissue upon co-cultivation.

Electroporation is another effective means of introducing foreign DNA into plant cells.[3,16] This technique involves the electrophoretic transfer of naked DNA into cells. Since any DNA fragment can be delivered to the cell, this technique has the advantage of allowing assimilation of a gene without having to clone the DNA into an *A. tumefaciens* vector. One disadvantage of this method is that during incorporation into the host nuclear genome, DNA rearrangement sometimes occurs. Furthermore, only plants from which protoplasts can be isolated may be transformed by this technique, and not all protoplast systems can be used to regenerate flowering plants. However, even if a plant cannot be regenerated, this method provides an extremely useful means for studying transient gene expression.

A new and promising technique for delivering DNA to intact plant tissue or protoplasts is ballistic projection.[3,17,18] With this technique, microprojectile particles (e.g., 1.2 μm Tungsten beads) coated with DNA are accelerated at high speeds into plant tissue. This approach has the distinct advantage of applicability to any intact plant tissue or region of the plant,[12] and has even been used to transform organellar DNA.[18,19] With this technique, any piece of DNA can be used for the transformation process, but again the disadvantage exists that regions of DNA may be randomly lost during incorporation. This technique is especially useful for plants that are otherwise recalcitrant with respect to transformation and/or regeneration. For instance, ballistic methods have been successfully used to produce transgenic monocot cereal plants.[3,12,20]

### B. VARIETY OF TRANSFORMED PLANTS

Since plant cells are totipotent, if a cell can have foreign DNA incorporated into its genome and survive, then in principle one can regenerate a transgenic plant from that cell. However, in practice not all plant species have proven to be amenable to regeneration from dedifferentiated tissue.[3] Also, as indicated above, not all plants are acceptable hosts for *A. tumefaciens* and transformable by this technique.

In general, dicot plants are easier to transform and regenerate from protoplasts or explant sections than monocot plants.[3] Furthermore, the natural hosts for *A. tumefaciens* are dicots. Consequently, this approach has worked extremely well with many dicot crop plants, including tomato, cotton, tobacco, lettuce, sunflower, canola, sugar beet, alfalfa, flax, soybean, potato, celery, cucumber, carrot, cauliflower, and grapes.[2,3] Two examples of noncrop plants that have been genetically transformed and subsequently used for basic biological studies are *Arabidopsis* and petunia.[2]

It has been difficult to generate transgenic monocot plants. One reason is that this group of plants, which includes the economically important grain crops such as rice, corn, sorghum, and wheat, is not susceptible to infection by *A. tumefaciens*. While it is not a problem to deliver exogenous DNA to protoplasts derived from monocots, it is difficult to regenerate viable plants from the transformed cells.[3,12] However, there has been progress in the area of regeneration, especially in the cases of rye, maize, and rice.[2,12,20] Also, with the growing success of ballistic methods for plant transformation, if the DNA can be stably introduced prior to or during embryogenesis and the embryos are not destroyed, one can obtain transgenic monocots. For instance, maize and rice embryos bombarded with DNA coated microprojectiles have been transformed by this method.[12,20,21]

## III. TRAITS CONFERRED TO TRANSGENIC PLANTS

### A. INSECT RESISTANCE

If functional insecticides can be genetically engineered into plants, it may be feasible to develop crops that are intrinsically tolerant of insect predators. Such plants would not have to be sprayed with costly and hazardous chemical pesticides. In addition, as is the case with many biologically generated insect pathogens, such insecticides should be required at much lower concentrations then exogenously applied synthetic pesticides. Moreover, since biological insecticides are usually specific, they are generally not hazardous to the intended consumers of the food.

Two main strategies have been used to confer resistance against insect predators. The first involves transforming plants with a gene for an insecticidal protoxin produced by one of several subspecies of *Bacillus thuringiensis*.[5,22,23] The protoxin does not persist in the environment nor is it hazardous to mammals, and thus, it is a safe means of protecting plants. The gene for the toxin has been introduced and selectively expressed in a number of plant species including tomato, tobacco, potato, and cotton.[2,5] The prospects for the successful application of this approach to crop protection appear to be quite good, provided that the overuse of *Bacillus thuringiensis* insecticidal toxins does not facilitate the development of resistant insects.

The trypsin inhibitor protein from cowpea has been shown to be effective against a wide variety of insects;[24] its presence restricts the ability of insects to digest food (i.e., plants) by interfering with hydrolysis of plant proteins by the insect. Since the cowpea trypsin inhibitor is a natural plant protein, it can be expressed in plants without adversely affecting the physiology of the host. Similarly, the gene encoding a plant proteinase inhibitor can be introduced readily by transformation, with the expectation that its expression will not be problematic. One drawback, however, is that relative to the *B. thuringiensis* toxin, high concentrations of cowpea trypsin inhibitor are required for effectiveness.[5] Since most proteinase inhibitors are inactivated by heating, a transformed plant expressing this polypeptide should not be hazardous to humans if the produce is cooked. Furthermore, if expression is limited to tissues that an insect is likely to attack (e.g., roots and leaves) and not present in the tissues intended for the end-user (e.g., fruit), then transgenic plants which produce a proteinase inhibitor should be safe even without cooking.

### B. VIRAL RESISTANCE

For many transgenic plants, following the expression of a viral coat protein, the ability of the virus to subsequently infect the plant and spread systemically is greatly diminished.[6] Excess coat protein may interfere with virus assembly and/or inhibit expression of the viral genome.[25] A number of different viral coat protein genes have been introduced into plant genomes, expressed, and subsequently found to confer viral tolerance, including tobacco mosaic virus, cucumber mosaic virus, alfalfa mosaic virus, tobacco streak virus, tobacco rattle virus, potato viruses X and Y, and tobacco etch virus.[2,6,25-27] Plants that have been successfully engineered with viral resistance include tobacco, tomato, and potato. Although total immunity has not been achieved, reasonably high levels of viral tolerance have been reported.[6] This approach has great potential as the introduction of a coat protein from one virus can often provide tolerance to a spectrum of seemingly unrelated viruses.[6] In addition, transgenic plants carrying the coat protein have been shown to be virally resistant in the field as well as in the laboratory.[28]

Another strategy that has been successful for introducing viral tolerance into plants is to transform the plant genome with an antisense segment of a viral genome.[6,25,29,30] This approach may provide viral tolerance by interfering with either the translation of viral mRNAs or the replication of the viral genome. However, antisense RNA appears to provide less resistance than overexpression of viral coat proteins. Moreover, this approach may only provide protection against the specific virus from which the antisense RNA is derived, since strong hybridization may be required for antisense RNA to be effective.

### C. HERBICIDE RESISTANCE

The development of herbicide resistant plants is of economic importance to agriculture and is an area where genetic engineering is making significant contributions.[31] If one can engineer a crop to be herbicide resistant, the entire field where the crop is grown can be put under an effective program of weed control.

In the early 1980s, the photosystem II herbicides (e.g., atrazine and DCMU) were considered to be very promising targets for conferring herbicide resistance. The site of action for these chemicals was known to be the D1 photosystem II reaction center protein, and its gene (*psb*A, which is chloroplast encoded) from several species had been sequenced.[32] It had also been found that a point mutation in the *psb*A gene, converting Ser-264 to a Gly, yielded plants that were resistant to Atrazine.[33,34] However, due to the inherent difficulty of transforming chloroplast DNA, this gene did not become a target for bioengineering. Now, with the advent of ballistic methods of transformation, chloroplast DNA can be transformed and atrazine tolerance can be selectively placed into a plant.[18,35] Unfortunately, atrazine is a particularly noxious herbicide that is both toxic to animals and persistent in the soil, so there may not be a good rationale for pursuing the development of atrazine resistant plants. This notwithstanding, molecular biological methods applied to atrazine and DCMU action have contributed greatly to our understanding both of herbicide action and photosystem II function.[32]

In contrast to atrazine, Glyphosate is an environmentally "friendly" herbicide that is degraded to nontoxic compounds in the soil.[7] Glyphosate inhibits 5-*enol*pyruvylshikimate-3-phosphate synthase (EPSPS), an enzyme in the shikimate pathway preventing synthesis of aromatic amino acids. The gene for the enzyme is nuclear encoded and thus can be manipulated readily by the established methods of plant transformation.[7,36] An EPSPS gene has been isolated from glyphosate resistant *E. coli*, given a plant promoter, and introduced into plants, yielding glyphosate tolerant plants. Transgenic species carrying resistance to glyphosate have been developed in tobacco, petunia, tomato, potato, cotton, and *Arabidopsis*.[2,7,36]

Sulfonylurea herbicides (e.g., Glean and Oust) inhibit the enzyme acetolactate synthase, thereby blocking the biosynthesis of branched chain amino acids.[37,38] Site specific mutagenesis of the gene for this enzyme yields variants that are no longer inhibited by sulfonylurea herbicides. In addition, since this gene is nuclear encoded, it is possible to generate transgenic plants containing an acetolactate synthase gene that confers tolerance to these herbicides. Resistance to sulfonylurea herbicides has been transferred to several plant species including tobacco, *Brassica*, and *Arabidopsis*.[37-40]

Bromoxynil is a phenolic-type herbicide that acts by inhibiting photosystem II.[31,41] Rather than attempting to modify the chloroplast encoded gene (*psb*A) specifying the protein site of action of this herbicide (D1 protein), the strategy used to confer tolerance has been to introduce an enzyme into plants which can detoxify this compound before it comes in contact with the D1 protein. The gene for a bacterial nitrylase, which can inactivate bromoxynil, was transferred to tobacco, and the resultant transgenic plants were found to be resistant to the otherwise toxic effects of the herbicide.[41]

### D. STRESS TOLERANCE

Due to their immobility, plants cannot avoid stresses such as high light, UV irradiation, heat, and drought. At the molecular level, stress often induces the production of oxygen radicals. Thus, tolerance of active oxygen species could increase the resistance of plants to many forms of environmental stress. A common form of active oxygen is the superoxide anion, which may be detoxified via conversion to hydrogen peroxide by superoxide dismutase (SOD) and then to water by any of a number of peroxidases and/or catalases.[42] There was an attempt to increase tolerance to active oxygen by transformation of plants with the gene for a Cu/Zn-SOD under the control of a very active promoter (the promoter for a ribulose-1,5-bisphosphate carboxylase/oxygenase gene).[43] This approach met with only limited success, possibly because superoxide anion is not completely consumed by the Cu/Zn-SOD and could react with hydrogen peroxide to form the highly toxic hydroxyl radical via the Mehlor reaction. Recently, however, Van Montagu and coworkers[44] achieved tolerance to active oxygen by transforming tobacco with a Mn-SOD gene under the control of the cauliflower mosaic virus 35S promoter. It is possible that this protein is effective at alleviating oxidative stress because it quantitatively converts superoxide anion to hydrogen peroxide, eliminating potential contributions from the Mehlor reaction. The addition of a functional Mn-SOD gene has also been demonstrated to confer resistance from oxidative stress to both *E. coli* and cyanobacteria.[45,46] An additional benefit of increasing the Mn-SOD concentration is tolerance to the herbicide methyl viologen.[44-46]

A natural mechanism of plant protection from UV-B stress is the synthesis of UV screening flavonoids in the epidermal cell layer.[47] In leaves, the synthesis of enzymes involved in the flavonoid pathway is under the control of a UV-B responsive promoter.[48] Many of these genes (e.g., phenylalanine ammonia lyase and chalcone synthase) have been isolated along with their promoters,[49] so it should be feasible to transfer them into plants that are especially sensitive to UV-B irradiation, along with stronger promoters, in an effort to provide the plant with greater UV protection. One problem is that flavonoid synthesis involves an extensive biosynthetic pathway, so it may require the transfer of too many genes to be feasible. It is possible, however, that overproduction of one enzyme which is at a key juncture in the pathway (e.g., chalcone synthase) might be enough to significantly increase flavonoid synthesis.

One of the functions of chaperonins appears to be prevention of denaturation of cellular proteins during stress.[50] In particular, the HSP60 and HSP70 classes of proteins may play this role during heat, drought, and light stress. Since the genes for many of these proteins have been isolated, it is possible that they could be placed into plants under the control of strong promoters that are induced during a particular form of stress.

## E. IMPROVED SEED STORAGE PROTEINS

Seed storage proteins are generally made up of a limited number of amino acids organized into repetitive structural units.[51] They are often missing one or more of the amino acids that are essential to human diets, thus limiting the potential food value of a given grain crop. Recently, progress has been made on altering the amino acid composition of seed proteins so that they contain a better spectrum of amino acids (especially those with sulfur or nitrogen containing residues), consequently raising their food value.[10,52,53] This can be done in a few ways. First, foreign seed storage protein genes that encode a broad spectrum of amino acids can be transferred into plants with poor quality seed proteins.[54] This has been shown to be a feasible approach in model studies with tobacco; a phaseolin gene from a legume was faithfully expressed, and the protein product was correctly compartmentalized in transgenic tobacco.[52] Second, it is possible to alter native seed storage protein genes to produce a protein with an improved amino acid composition (i.e., more lysine and/or methionine residues) and then to reintroduce them into the host plant.[53,55] Unfortunately, these altered proteins are not always properly compartmentalized into protein bodies in the seeds. However, recent work on the transport and assembly of storage proteins may greatly expand the utility of this approach.[54] A different approach to this problem is to insert the desired amino acid sequences at the hypervariable regions near the carboxy termini of storage proteins.[56] This seems to have less of a perturbing impact on protein assembly into protein bodies.

Another area in which the quality of seed proteins is of importance is in the production of bread.[57] High molecular weight glutenin subunits confer different degrees of elasticity to bread doughs. In particular, the subunit 10 glutenin has longer and more regular β-turn structures and, as a result, confers better elasticity to the bread.[58] Perhaps the genes for these proteins could be modified and reintroduced into bread wheats.

## F. ALTERED FRUIT RIPENING

A major problem in the fruit industry is premature ripening and softening, either prior to or during transport to the market place. Several ripening specific genes have been isolated (e.g., those coding for cellulase and polygalacturonase),[8,59] and by altering the expression of such genes, it is possible to alter fruit ripening patterns in tomato.[59] In particular, an antisense polygalacturonase gene introduced into tomato slows the softening process during fruit maturation while allowing ripening to proceed.[60] Another means of slowing postharvest ripening and softening is to interfere with ethylene production. This has been done in two ways. First, a gene coding for the enzyme that degrades amino cyclopropane carboxylic acid

(ACC), the immediate precursor to ethylene, was used to transform target plants.[61] Second, transgenic plants with an antisense gene for either ACC synthase or the ethylene forming enzyme were generated.[62] Such transgenic plants produce tomatoes that are likely to have a significantly longer shelf-life. These sorts of genetic manipulations may result in vine ripened tomatoes that have good transportation qualities. The work with tomato ripening has also demonstrated the utility of expression of antisense genes in the regulation of plant development.

### G. ALTERED FLOWERS AND ORNAMENTALS

The flower industry, a billion dollar a year business, is continually improving flower appearance and postharvest lifetime. It has been shown that manipulating the genes for enzymes in the flavonoid biosynthesis pathway can result in novel coloring patterns in flower petals.[9,63] Thus, understanding the regulation of this pathway may ultimately provide the means to specifically design unique and preferred flower patterns.

As is the case for fruit, preservation of cut flowers during shipping is of great importance. Recently, there has been progress in the understanding of some of the mechanisms of flower senescence at the molecular level. It is clear that oxygen radicals play a key role in causing cut flowers to deteriorate.[64] Similar to stress tolerance, if one could increase the endogenous oxygen scavenging capacity in cut flowers by overexpression of a Mn-SOD gene, under the control of a flowering specific promoter, their shelf-life might increase.

### H. CYTOPLASMIC MALE STERILITY AND SELF-INCOMPATIBILITY

When a plant cannot self-fertilize, production of hybrid seed by cross-fertilization from specific male and female parental lines is greatly facilitated. Thus, the traits of cytoplasmic male sterility and self-incompatibility both eliminate the need for the labor intensive removal of the stamin from a female host in order to prevent self-fertilization.[11] In some plant species, self-incompatibility is dictated by the so-called S-locus, which codes for a group of glycoproteins that interfere with pollen tube development when the pollen of the self-plant binds to the stigma, thus forcing cross-fertilization.[65]

Recently, Goldberg and coworkers[66] reported that they were able to selectively confer male sterility to tobacco and *Brassica napus* by bioengineering. This was achieved by placing genes for RNases from *Aspergillus oryzae* (RNase T1) and *Bacillus amyloliquefaciens* (barnase) under the control of a tobacco anther specific promoter. This manipulation prevents pollen sac formation but does not interfere with embryogenesis if the pollen comes from another parent. The anther specific promoter appears to be conserved in a number of different plant species, so this technique may have wide applicability.

## IV. FUTURE DIRECTIONS

The above are just a few examples of how plant molecular techniques can be used to alter the phenotypes of agriculturally important plants. At this juncture, it is clear that bioengineering has great potential for crop improvement, and the possibilities are just beginning to be realized. As techniques of transformation improve and other useful genes are isolated and characterized, it will be possible to add new and different traits to plants. For example, as the technology improves and our knowledge of gene regulation increases, it should be possible to transform plants with more than one gene in order to engineer more complex traits. We envision the manipulation of entire biosynthetic pathways, thereby providing the ability to produce a wider range of specific bioactive compounds in plants. Furthermore, over the next decade many transgenic crop plants should come into commercial use (over 200 field studies have already been performed), many more plants will become regeneratable from tissue culture, and a much better understanding of the biochemistry of a whole range of plant specific processes is likely to be realized. Thus, the future promises to be very exciting in the areas of plant biotechnology and molecular biology.

## REFERENCES


1. **Goldberg, R. B.,** Plants: novel developmental processes, *Science*, 240, 1460, 1988.
2. **Gasser, C. S. and Fraley, R. T.,** Genetically engineering plants for crop improvement, *Science*, 244, 1293, 1989.
3. **Potrykus, I.,** Gene transfer to plants: assessment of published approaches and results, *Annu. Rev. Plant Physiol. Plant Mol. Biol.*, 42, 205, 1991.
4. **Benfey, P. N. and Chua, N.-H.,** Regulated genes in transgenic plants, *Science*, 244, 174, 1989.
5. **Brunke, K. J. and Meeusen, R. L.,** Insect control with genetically engineered crops, *Trends Biotechnol.*, 9, 197, 1991.
6. **Beachy, R. N.,** Plant transformation to confer resistance against virus infection, in *Gene Manipulation in Plant Improvement,* Vol. 2, Gustafson, J. P., Ed., Plenum Press, New York, 1990, 305.
7. **Shah, D. M., Horsch, R. B., Klee, H. J., Kishore, G. M., Winter, J. A., Tumer, N. E., Hironaka, C. M., Sanders, P. R., Gasser, C. S., Aykent, S., Siegel, N. R., Rogers, S. G., and Fraley, R. T.,** Engineering herbicide resistance in transgenic plants, *Science*, 233, 478, 1986.
8. **Kramer, M., Sheehy, R. E., and Hiatt, W. R.,** Progress towards the genetic engineering of tomato fruit softening, *Trends Biotechnol.*, 7, 191, 1989.
9. **Mol, J., Stuitje, A., Gerats, A., van der Krol, A., and Jorgenson, R.,** Saying it with genes: molecular flower breeding, *Trends Biotechnol.*, 7, 148, 1989.
10. **Hall, T. C., Bustos, M. M., Anthony, J. L., Yang, L. J., Domoney, C., and Casey, R.,** Opportunities for bioactive compounds in transgenic plants, in *Bioactive Compounds from Plants,* Ciba Found. Symp. No. 154, Chadwick, D. J. and Marsh, J., Eds., John Wiley & Sons, Chichester, England, 1990, 177.
11. **Haring, V., Gray, J. E., McClure, B. A., Anderson, M. A., and Clarke, A. E.,** Self-incompatibility: a self-recognition system in plants, *Science*, 250, 937, 1990.
12. **Wu, R., Kemmerer, E., and McElroy, D.,** Transformation and regeneration of important crop plants: rice as the model system for monocots, in *Gene Manipulation in Plant Improvement,* Vol. 2, Gustafson, J. P., Ed., Plenum Press, New York, 1990, 251.
13. **An, G., Watson, B. D., Stachel, S., Gordon, M. P., and Nester, E. W.,** New cloning vehicles for transformation of higher plants, *EMBO J.*, 4, 277, 1985.
14. **Pasternak, J. J. and Glick, B. R.,** Utilization of Ti plasmid for plant genetic engineering, in *Frontiers in Applied Microbiology,* Vol. 3, Mukerji, K. G., Singh, V. P., and Garg, K. L., Eds., Rastogi and Co., Meerut, India, 1989, 1.
15. **Rogers, S. G., Klee, H., Horsch, R. B., and Fraley, R. T.,** Use of co-integrating Ti plasmid vectors, in *Plant Molecular Biology Manual*, Vol. A2, Kluwer Academic Publishers, Dordrecht, 1988, 1.
16. **Saul, M. W., Shillito, R. D., and Negrutiu, I.,** Direct DNA transfer to protoplasts with and without electroporation, in *Plant Molecular Biology Manual*, Vol. A1, Kluwer Academic Publishers, Dordrecht, 1988, 1.
17. **Tomes, D. T., Weissinger, A. K., Ross, M., Higgins, R., Drummond, B. J., Schaaf, S., Malone-Schoneberg, J., Staebell, M., Flynn, P., Anderson, J., and Howard, J.,** Transgenic tobacco plants and their progeny derived by microprojectile bombardment of tobacco leaves, *Plant Mol. Biol.*, 14, 261, 1990.
18. **Svab, Z., Hajdukiewicz, P., and Maliga, P.,** Stable transformation of plastids of higher plants, *Proc. Natl. Acad. Sci. U.S.A.*, 87, 8526, 1990.
19. **Boynton, J. E., Gillham, N. W., Harris, E. H., Hosler, J. P., Johnson, A. M., Jones, A. R., Randolph-Anderson, B. L., Robertson, D., Klein, T. M., Shark, K. B., and Sanford, J. C.,** Chloroplast transformation in *Chlamydomonas* with high velocity microprojectiles, *Science*, 240, 1534, 1988.
20. **Gordon-Kamm, W. J., Spencer, T. M., Mangano, M. L., Adams, T. R., Daines, R. J., Start, W. G., O'Brien, J. V., Chambers, S. A., Adams, W. R., Jr., Willetts, N. G., Rice, T. B., Mackey, C. J., Krueger, R. W., Kausch, A. P., and Lemaux, P. G.,** Transformation of maize cells and regeneration of fertile transgenic plants, *Plant Cell*, 2, 603, 1990.
21. **Christou, P., Ford, T. L., and Kofron, M.,** Production of transgenic rice (*Oryza sativa* L.) plants from agronomically important Indica and Japonica varieties via electric discharge particle acceleration of exogenous DNA into immature zygotic embryos, *Bio/Technology*, 9, 957, 1991.

22. **Vaeck, M., Reynaerts, A., Hofte, H., Jansens, S., De Beuckeleer, M., Dean, C., Zabeau, M., Van Montagu, M., and Leemans, J.,** Transgenic plants protected from insect attack, *Nature*, 328, 33, 1987.
23. **Perlak, F. J., Fuchs, R. L., Dean, D. A., McPherson, S. L., and Fischhoff, D. A.,** Modification of the coding sequence enhances plant expression of insect control protein genes, *Proc. Natl. Acad. Sci. U.S.A.*, 88, 3324, 1991.
24. **Hilder, V. A., Gatehouse, A. M. R., Sheerman, S. E., Barker, R. F., and Boulter, D.,** A novel mechanism of insect resistance engineered into tobacco, *Nature*, 330, 160, 1987.
25. **Hemenway, C., Fang, R.-X., Kaniewski, W. K., Chua, N.-H., Tumer, N. E.,** Analysis of the mechanism of protection in transgenic plants expressing the potato virus X coat protein or its antisense RNA, *EMBO J.*, 7, 1273, 1988.
26. **Hill, K. K., Jarvis-Eagan, N., Halk, E. L., Krahn, K. J., Liao, L. W., Mathewson, R. S., Merlo, D. J., Nelson, S. E., Rashka, K. E., and Loesch-Fries, L. S.,** The development of virus-resistant alfalfa, *Medicago sativa* L., *Bio/Technology*, 9, 373, 1991.
27. **Golemboski, D. B., Lomonossoff, G. P., and Zaitlin, M.,** Plants transformed with a tobacco mosaic virus nonstructural gene sequence are resistant to the virus, *Proc. Natl. Acad. Sci. U.S.A.*, 87, 6311, 1990.
28. **Nelson, R. S., McCormick, S. M., Delannay, X., Dube, P., Layton, J., Anderson, E. J., Kaniewska, M., Proksch, R. K., Horsch, R. B., Rogers, S. G., Fraley, R. T., and Beachy, R. N.,** Virus tolerance, plant growth, and field performance of transgenic tomato plants expressing coat protein from tobacco mosaic virus, *Bio/Technology*, 6, 403, 1988.
29. **Day, A. G., Bejarano, E. R., Buck, K. W., Burrell, M., and Lichtenstein, C. P.,** Expression of an antisense viral gene in transgenic tobacco confers resistance to the DNA virus tomato golden mosaic virus, *Proc. Natl. Acad. Sci. U.S.A.*, 88, 6721, 1991.
30. **Rezaian, M. A., Skene, K. G. M., and Ellis, J. G.,** Anti-sense RNAs of cucumber mosaic virus in transgenic plants assessed for control of the virus, *Plant Mol. Biol.*, 11, 463, 1988.
31. **Mazur, B. J. and Falco, S. C.,** The development of herbicide resistant crops, *Annu. Rev. Plant Physiol. Plant Mol. Biol.*, 40, 441, 1989.
32. **Trebst, A.,** The three dimensional structure of the herbicide binding niche on the reaction center polypeptides of photosystem II, *Z. Naturforsch.*, 42c, 742, 1987.
33. **Hirschberg, J. and McIntosh, L.,** Molecular basis of herbicide resistance in *Amaeranthus hybridus*, *Science*, 222, 1346, 1983.
34. **Goloubinoff, P., Edelman, M., and Hallick, R. B.,** Chloroplast-coded atrazine resistance in *Solanum nigrum: psb*A loci from susceptible and resistant biotypes are isogenic except for a single codon change, *Nucl. Acids Res.*, 24, 9489, 1984.
35. **Przibilla, E., Heiss, S., Johanningmeier, U., and Trebst, A.,** Site-specific mutagenesis of the D1 subunit of photosystem II in wild type *Chlamydomonas*, *Plant Cell*, 3, 169, 1991.
36. **della-Cioppa, G., Bauer, S. C., Taylor, M. L., Rochester, D. E., Klein, B. K., Shah, D. M., Fraley, R. T., and Kishore G. M.,** Targeting a herbicide resistant enzyme from *Escherichia coli* to chloroplasts of higher plants, *Bio/Technology*, 5, 579, 1987.
37. **Lee, K. Y., Townsend, J., Tepperman, J., Black, M., Chui, C. F., Mazur, B., Dunsmuir, P., and Bedbrook, J.,** The molecular basis of sulfonylurea herbicide resistance in higher plants, *EMBO J.*, 7, 1241, 1988.
38. **Haughn, G. W., Smith, J., Mazur, B. J., and Somerville, C.,** Transformation with a mutant *Arabidopsis* acetolactate synthase gene renders tobacco resistant to sulfonylurea herbicides, *Mol. Gen. Genet.*, 211, 266, 1988.
39. **Odell, J. T., Caimi, P. G., Yadav, N. S., and Mauvais, C. J.,** Comparison of increased expression of wild type and herbicide resistant acetolactate synthase genes in transgenic plants, and indication of posttranscriptional limitation on enzyme activity, *Plant Physiol.*, 94, 1647, 1990.
40. **Miki, B. L., Labbe, H., Hattori, J., Ouellet, T., Gabard, J., Sunohara, G., Charest, P. J., and Iyer, V. N.,** Transformation of *Brassica napus* canola cultivars with *Arabidopsis thaliana* acetohydroxyacid synthase gene and analysis of herbicide resistance, *Theor. Appl. Genet.*, 80, 449, 1990.
41. **Stalker, D. M., McBride, K. E., and Malyj, L. D.,** Herbicide resistance in transgenic plants expressing a bacterial detoxification gene, *Science*, 242, 419, 1988.
42. **Asada, K. and Takahashi, M.,** Production and scavenging of active oxygen in photosynthesis, in *Photoinhibition*, Kyle, D. J., Osmond, C. B., and Arntzen, C. J., Eds., Elsevier, Amsterdam, 1987, 227.
43. **Tepperman, J. M. and Dunsmuir, P.,** Transformed plants with elevated levels of chloroplastic SOD are not more resistant to superoxide toxicity, *Plant Mol. Biol.*, 14, 501, 1990.
44. **Bowler, C., Slooten, L., Vandenbranden, S., De Rycke, R., Botterman, J., Sybesma, C., Van Montagu, M., and Inze, D.,** Manganese superoxide dismutase can reduce cellular damage mediated by oxygen radicals in transgenic plants, *EMBO J.*, 10, 1723, 1991.
45. **Bloch, C. A. and Ausubel, F. M.,** Paraquat-mediated selection for mutations in the manganese-superoxide dismutase gene *sodA*, *J. Bacteriol.*, 168, 795, 1986.

46. **Gruber, M. Y., Glick, B. R., and Thompson, J. E.,** Cloned manganese superoxide dismutase reduces oxidative stress in *Escherichia coli* and *Anacystis nidulans, Proc. Natl. Acad. Sci. U.S.A.*, 87, 2608, 1990.
47. **Caldwell, M. M., Robberecht, R., and Flint, S. D.,** Internal filters: Prospects for UV-acclimation in higher plants, *Physiol. Plant.*, 58, 445, 1983.
48. **Schmelzer, E., Jahnen, W., and Hahlbrock, K.,** *In situ* localization of light-induced chalcone synthase mRNA, chalcone synthase, and flavonoid end products in the epidermal cells of parsley leaves, *Proc. Natl. Acad. Sci. U.S.A.*, 85, 2989, 1988.
49. **Wingender, R., Rohrig, H., Horicke, C., and Schell, J.,** cis-Regulatory elements involved in ultraviolet light regulation and plant defence, *Plant Cell*, 2, 1019, 1990.
50. **Ellis, R. J.,** Molecular chaperones: the plant connection, *Science*, 250, 954, 1990.
51. **Shutov, A. D. and Vaintraub, A. I.,** Degradation of storage proteins in germinating seeds, *Phytochemistry*, 26, 1557, 1987.
52. **Altenbach, S. B., Pearson, K. W., Meeker, G., Staraci, L. C., and Sun, S. S. M.,** Enhancement of the methionine content of seed proteins by the expression of a chimeric gene encoding a methionine-rich protein in transgenic plants, *Plant Mol. Biol.*, 13, 513, 1989.
53. **Shaul, O. and Galili, G.,** Increased lysine synthesis in tobacco plants that express high levels of bacterial dihydrodipicolinate synthase in their chloroplasts, *Plant J.*, in press.
54. **Lee, W. S., Tzen, J. T. C., Kridl, J. C., Radke, S. E., and Huang, A. H. C.,** Maize oleosin is correctly targeted to seed oil bodies in *Brassica napus* transformed with the maize oleosin gene, *Proc. Natl. Acad. Sci. U.S.A.*, 88, 6181, 1991.
55. **Hoffman, L. M., Donaldson, D. D., and Herman, E. M.,** A modified storage protein is synthesized, processed, and degraded in the seed of transgenic plants, *Plant Mol. Biol.*, 11, 717, 1988.
56. **Dickinson, C. D, Scott, M. P., Hussein, E. H. A., Argos, P., and Nielsen, N. C.,** Effect of structural modifications on the assembly of a glycinin subunit, *Plant Cell*, 2, 403, 1990.
57. **Payne, P. I.,** Genetics of wheat storage proteins and the effect of allelic variation on bread-making quality. *Annu. Rev. Plant Physiol.*, 38, 141, 1987.
58. **Flavell, R. B., Goldsbrough, A. P., Robert, L. S., Schnick, D., and Thompson, R. D.,** Genetic variation in wheat HMW glutenin subunits and the molecular basis of breadmaking quality, *Bio/Technology*, 7, 1281, 1989.
59. **Grierson, D., Slater, A., Spiers, J., and Tucker, G. A.,** The appearance of polygalacturonase mRNA in tomatoes: one of a series of changes in gene expression during development and ripening, *Planta*, 163, 263, 1985.
60. **Schuch, W., Bird, C. R., Ray, J., Smith, C. J. S., Watson, C. F., Morris, P. C., Gray, J. E., Arnold, C., Seymour, G. B., Tucker, G. A., and Grierson, D.,** Control and manipulation of gene expression during tomato fruit ripening, *Plant Mol. Biol.*, 13, 303, 1989.
61. **Klee, H. J., Hayford, M. B., Kretzmer, K. A., Barry, G. F., and Kishore, G. M.,** Control of ethylene synthesis by expression of a bacterial enzyme in transgenic tomato plants, *Plant Cell*, 3, 1187, 1991.
62. **Hamilton, A. J., Lycett, G. W., and Grierson, D.,** Antisense gene that inhibits synthesis of the hormone ethylene in transgenic plants, *Nature*, 346, 284, 1990.
63. **van der Krol, A. R., Mur, L. A., de Lange, P., Gerats, A. G. M., Mol, J. N. M., Stuitje, A. R.,** Antisense chalcone synthase genes in petunia: visualization of variable transgene expression, *Mol. Gen. Genet.*, 220, 204, 1990.
64. **Thompson, J. E.** Physical changes in the membranes of senescing and environmentally stressed plant tissues, in *Physiology of Membrane Fluidity*, Vol. 2, Shinitzky, M., Ed., CRC Press, Boca Raton, FL, 1984, 85.
65. **Clarke, A. E., Anderson, M. A., Atkinson, A., Bacic, A., Ebert, P. R., Jahnen, W., Lush, W. M., Mau, S.-L., and Woodward, J. R.,** Recent developments in the molecular genetics and biology of self-incompatibility, *Plant Mol. Biol.*, 13, 267, 1989.
66. **Mariani, C., De Beuckeleer, M., Truettner, J., Leemans, J., and Goldberg, R. B.,** Induction of male sterility in plants by a chimaeric ribonuclease gene, *Nature*, 347, 737, 1990.

# 2

# Production and Analysis of Plant Mutants, Emphasizing *Arabidopsis thaliana*

*Russell L. Malmberg*

## I. INTRODUCTION

Mutational analysis is a powerful tool for dissecting complex biological processes. In *Arabidopsis thaliana* and other plant species, mutants are currently being used as tools for studying plant development and for examining metabolic pathways. The fundamental assumption of a mutational analysis, sometimes not explicitly stated, is that an examination of the phenotype of a mutant will shed light on the normal functioning of the nonmutant, wild type versions of that gene. This underlying assumption has consequences for the optimum design of mutagenesis, for mutant identification, and for mutant characterization experiments. By way of contrast, a different set of usefulness criteria is appropriate when a new mutant marker is needed for a transmission genetic experiment, such as analysis of recombination within a chromosomal region.

The goal of this chapter is to summarize some practical and theoretical aspects of mutant isolation specifically for mutational analyses of complex processes. The focus will be strongly on *Arabidopsis thaliana* (mouse ear cress), but occasional mention will be made of other outstanding model genetic systems, such as *Zea mays* (maize). Questions to be asked include: What experimental designs have worked previously? How are mutants identified? How should a new mutant be characterized?

Although there are highly tedious aspects to mutant screening, the entire process can be one of the most rewarding endeavors in biology. Successful isolation and characterization of a new mutant can totally change the interpretation of a biological phenomenon. The analysis of mutants permits one to develop testable hypotheses about complex processes and also to apply readily the appropriate critical tests.

## II. TYPES OF MUTANTS

In the discussion that follows, a variety of terms derived from Muller[1] are used that define the different kinds of mutant alleles:

- amorph — A phenotype resulting from complete absence of a gene product or function.
- hypomorph — A phenotype resulting from an amount of a gene product less than the wild type, but not from complete absenceof this product.
- hypermorph — A phenotype resulting from an excess of gene function, more than the wild type level.
- antimorph — A phenotype in the opposite direction from that of the wild type; a more extreme phenotype than an amorph.

0-8493-5164-2/93/$0.00+$.50
© 1993 by CRC Press, Inc.

neomorph — A phenotype that is in a new direction, a completely different property than the wild type.

If a wild type allele could be assigned a value on an axis of +1, an amorph would be 0, a hypomorph would be between 0 and 1, a hypermorph would be greater than 1, an antimorph would be less than 0, and a neomorph would be on a different axis. It is easy to imagine that neomorphs and hypermorphs might be dominant mutants, whereas hypomorphs and amorphs would be expected to be recessive.

These terms are not only useful, but they also permit certain behavioral predictions with respect to gene dosage. If a mutant allele is a true hypomorph, increasing the gene dosage should restore the phenotype to nearly normal. An amorph, in contrast, should be unaffected by gene dosage. If a mutant is really a hypermorph, increasing the gene dosage of the wild type allele should mimic the mutant phenotype. In maize, it is possible to change gene dosage readily via AB translocations, where portions of the normal, well-behaved chromosomes (the A set) have been translocated onto accessory chromosomes (the B set) that do not obey the same rules of centromere behavior. This allows the maize geneticist to vary the dosage of a gene by transposing it to a B chromosome, where it will sometimes nondisjoin during cell divisions. These kinds of gene dosage manipulations can be used to help characterize and classify a newly isolated mutation. Similar chromosomal mechanics cannot yet be performed readily in *Arabidopsis*.

The kind of mutation that is most useful for analysis of a complex process is an amorph, a recessive knockout of a gene in which none of the gene products are made, the equivalent of a deletion. This should allow the clearest analysis of gene function, whether it governs altered development or the absence or surplus of a metabolite.

In practice, it may not be easy to decide if a given new phenotype is due to a gene knockout or to partial gene function. Some genes that control development can so derange the organs of a plant that it can initially be difficult to determine what each structure is related to or derived from. Both of these problems can be alleviated if several mutant alleles with different degrees of severity exist at a locus. The most severe allele phenotype is presumably closest to being an amorph. The intermediate or weaker alleles may also help one to interpret the distorted phenotype of the severe mutant alleles, by providing structures intermediate between the wild type and the severe alterations for comparison. An excellent example of an allelic series in *Arabidopsis thaliana* is present at the *ap2* locus,[2] where at least seven different alleles have been characterized. The basic tendency of these mutants is to replace sepals and petals with carpels and stamens, but the extent to which these transformations occur varies greatly with each allele. In some alleles the transformation is nearly complete; in others only patches of cells on otherwise normal sepals and petals may reveal the phenotype. The most extreme alleles are presumably nearly complete amorphs.

Neomorph mutations may be difficult to interpret. A neomorph results in the creation of a structure or event in some place or time at which it has never been located before. The *Knotted* mutation of maize is an excellent example.[3] This dominant mutation results in lumps, bumps, twists, and other alterations to the maize leaf. It is the result of expression of a transcriptional regulatory protein in a tissue where it is normally not expressed, thereby triggering cell divisions and other morphologies that would normally be found in other contexts. However, this leaf morphology mutation does not tell one anything about leaves per se. Mutations that reveal something about normal leaf development would be expected to be recessive amorphs, knocking out a portion of the normal leaf developmental controls. The *Knotted* neomorph is revealing something about the developmental controls of completely different cells and tissues, perhaps how those cells and tissues behave when transplanted to an abnormal environment.

It is also useful to distinguish between developmental arrest and developmental control mutations, as defined by Kimble and Schedl.[4] A developmental arrest is a mutation in which development of a structure stops at a defined point and then does not continue further. A developmental control mutation transforms one structure into another structure that is normally made at a different place or time. Developmental control mutations that are primarily spatial, replacing one structure with an alternative that is normally made at a different location, can be classified as homeotic. Developmental control mutations that are best interpreted as changing the developmental timing of a structure can be considered heterochronic. Key points in the analysis of developmental arrest and developmental control mutations are the characterization and identification of abnormal structures being made, and a growth comparison with wild type to determine the stage at which mutant development veers from wild type.

A very useful category of mutations are those that are temperature sensitive (*ts*), which usually designates mutants that cannot grow at an elevated temperature at which the wild type can. Cold sensitive mutations (*cs*) are equally useful but less commonly found. As has been shown many times in many organisms,[5,6] temperature sensitive mutations can allow the investigator to identify the developmental times at which a gene product is required. If a particular gene product is needed only at stage x, then the plant will develop normally up to stage x and after the end of stage x, even when grown at the elevated (restrictive) temperature. The sensitive time period can be identified by a combination of shift-up and shift-down experiments, where the *ts* individual is switched from permissive to restrictive, or restrictive to permissive, at different developmental stages. A nice example of the use of *ts* mutants with *Arabidopsis* is the characterization of the floral homeotic mutant allele *ap2-1* by Bowman et al.[7]

Most temperature sensitive mutations are discovered accidentally. A mutant is identified; then a characterization of growth over a range of temperatures reveals that the phenotype is nearly normal at one temperature and aberrant at another temperature. The probability of finding a temperature sensitive mutant can be increased by doing the mutant screening or selection at an elevated temperature. In practice this is not done as much as it might be because it requires extensive growth chamber space.

Finally, one needs to consider the possibility that a mutation may be cytoplasmic, i.e., chloroplastically or mitochondrially encoded, as opposed to nuclear encoded. In this article we will focus on nuclear mutations as the ones most commonly encountered. If it is possible that a cytoplasmic mutation might have occurred, the genetic test to distinguish it from nuclear mutations is simple. Since the cytoplasm is normally inherited maternally, 100% of the offspring of the cross should have the phenotype of the female parent. Two crosses, one in which the mutant is the pollen parent and one in which the mutant is the ovule parent, should reveal this pattern of inheritance clearly for a cytoplasmically encoded trait.

## III. MUTAGENESIS

Mutagenesis of *Arabidopsis thaliana* is usually performed by treating the seed with the mutagen, letting the surviving seeds germinate, and then recovering the progeny for analysis. The plant generation that grows from the treated seeds is referred to as the M1; it should contain heterozygote chimeras for any given mutation. Progeny collected after selfing are referred to as the M2 generation, and should be segregating both heterozygotes and homozygotes for a given mutation. The existence of heterozygotes in the M2 is particularly important if the mutation of interest turns out to be sterile or lethal when homozygous. This will be discussed from the standpoint of setting up M1 seed pools.

Treating seeds with a mutagen is not equivalent to treating a single cell. The seed after mutagenesis will be a chimeric individual, containing some nonmutant cells and a variety of cells that have mutations in their DNA. The seed will be chimeric not only for wild type versus mutant, but will also be chimeric for multiple mutations. The cells that matter are the ones that will contribute to the next generation, the M2. Mutations that are in cell lineages that do not lead to the germ line will be lost. Li and Rédei[8] have estimated that there are two precursor cells to the germ line in the *Arabidopsis* seed, based on patterns of genetic segregation in the progeny of just mutagenized seeds. Studies in tobacco and maize[9,10,35] of the shoot apex using chimeral/clonal-analysis/fate-mapping suggest that the number of cells in the seed that give rise to the upper body of the shoot and the reproductive organs is quite small, with three cells being a reasonable approximation. In other words, the goal of seed mutagenesis is to target the roughly two to three cells that will give rise to the reproductive tissues later in development. A second consequence is that a given M1 plant may have several independent mutant cell lineages. Seeds collected from siliques on one portion of the M1 shoot may be genetically different from those collected at another position on the same shoot.

The variety of *Arabidopsis* chosen as the wild type starting material deserves some consideration. The varieties Landsberg (with the *erecta* dwarf allele) and Columbia have been chosen more than others. Some varieties, such as Nossen-0 or RLD, are more readily transformable with Agrobacterium T-DNA than Landsberg *erecta* and Columbia.[11-13] It is worth knowing the lineage of the seeds obtained; i.e., where did the investigator's source get the seeds, how many times have they been propagated, etc.? It is also worth pointing out that the taxonomy of the *Arabidopsis* genus has only been lightly investigated.[14] A thorough phylogeny of the genus is sorely needed.

The discussion that follows is directed towards identification of recessive mutants, since they are the most common and since recessive amorphs are ideal for mutational analyses. There may be occasion to screen for dominant mutants as well. In theory this may be possible in the M1 generation, in addition to the M2. A dominant morphological mutation should be visible in some cell lineages derived from the just treated seeds. Such a mutation would thus give rise to a visible chimera consisting of mutant and wild type tissues, and recovery of the mutant would only be in seeds derived from that particular cell lineage. In practice, dominant mutants are on average at least ten-fold rarer than recessive, so that a screen for visible dominant mutants might prove inordinately tedious.

Lehle Seeds (Table 4) is a commercial supplier of *Arabidopsis* seeds, including some mutagenized with EMS and fast neutron bombardment. If one intends to use this supplier, it would be wise to review at least Section D below, to understand the issues of seed pool size and mutant independence.

## A. SELECTION OF A MUTAGENIC AGENT

EMS (ethylmethane sulfonate, a base alkylating agent) works well (see sample protocol), whether it is used in seed mutagenesis of *Arabidopsis*[15] or pollen mutagenesis of maize.[16] Expressed in terms of killing versus mutagenesis, as described below, EMS has a high induced mutation rate versus its toxicity. At optimium doses for total mutant recovery, more mutants are likely to be obtained from EMS than other mutagens; however, there is considerable merit to other mutagens, in spite of the efficacy of EMS. As already discussed, for proper interpretation of a mutational analysis, one would prefer to have a gene knockout/amorph; EMS yields some of these, but it also gives many hypomorphs. EMS induced mutations cannot be rapidly characterized at the DNA level; one usually must start with RFLPs and then walk to the locus in order to clone the gene.

Among other mutagens, diepoxyoctane, diepoxybutane, and gamma rays have been reported to cause deletion mutations in other eucaryotic systems.[17] The efficiency of these mutagens in generating deletions in plants is still being investigated. Deletion mutations have the advantage that they are, almost by definition, amorphs. In addition, subtractive cloning methods have advanced to the point that it is possible to think about cloning a deletion mutation by direct comparison of the DNA with wild type. At least one gene in *Arabidopsis* has already been cloned by subtractive hybridization, using a deletion mutant generated by fast neutron bombardment mutagenesis.[18]

Feldmann and colleagues[19] have generated a large collection of *Arabidopsis* lines, each independently transformed with the T-DNA from the *Agrobacterium* Ti plasmid. Since the T-DNA integrates randomly, it can sometimes insert into the middle of an important gene. Mutants created by T-DNA insertion into a gene are likely to be functional knockouts by disruption, and the T-DNA provides a convenient handle for molecular cloning. A number of interesting mutants have been identified and their respective genes cloned by examining this collection of transformed lines. A disadvantage to screening this collection has been that the frequency of the T-DNA mutagenesis has been low enough that some of the mutant phenotypes obtained were in fact due to spontaneous nucleotide changes, not insertions. A thorough description of the transformation process and the analysis of the mutants has recently been published.[20]

Several laboratories are attempting to develop the maize *Ac/Ds* transposon system into a mutation-inducing system in *Arabidopsis thaliana*, by introducing these elements into the *Arabidopsis* genome. *Ac* and *Ds* certainly transpose in *Arabidopsis*,[21] but their propensity for preferential local transposition (at least in maize) will need to be overcome before *Arabidopsis* geneticists can start bombarding the genome with *Ds* elements. This system, or use of other naturally occurring transposable elements, seems very promising. As noted with T-DNA, transposable elements could provide both gene knockouts and a convenient strategy for subsequent molecular cloning. Within a short time, various constructs may be developed that would make transposon-tagging the mutagenesis of choice.

### B. DOSE OF MUTAGEN

How does one decide what dose of a chemical or physical mutagen to use? Usually mutagenesis is a balance between killing the treated cells versus increasing the yield of mutants with a higher dosage. After mutagenesis one needs also to have readily scorable markers to assess the induced mutant frequency. An additional consideration is the frequency with which multiple mutants can be expected to occur, and hence the probability that the phenomenally interesting phenotype just discovered is the product of twelve interacting loci, not one.

#### 1. Killing Seeds vs. Creating Mutants

One rational way to choose a dose of a mutagen is to select the dose that will optimize the total yield of mutants. A very simple calculation suggests that the optimum mutagen dose can be calibrated initially based on the percent seed survival.

$M$ = number of mutants
$N$ = number of seeds, alive or dead
$l$ = fraction of seeds alive after mutagenesis
$m$ = fraction of seeds mutant among those alive
$d$ = mutagen dose
$k$ = exponential survival constant
$j$ = linear dose response constant

The number of mutant seeds after mutagenesis will be

$$M = mlN \tag{1}$$

Assume exponentially declining survival of seeds and linearly increasing fraction of mutants with dose of mutagen:

$$l = e^{-kd} \quad \text{and} \quad m = jd \tag{2}$$

Substitute these assumptions into the basic equation:

$$M = jde^{-kd}N \tag{3}$$

To optimize the total yield of mutants, take the derivative of $M$ with respect to $d$, set equal to 0, and then solve for $d$:

$$dM/dd = 0 \quad \text{and} \quad d = l/k \tag{4}$$

Substituting this dose back into $l$ gives:

$$l = e^{-k/k} = e^{-1} = 0.368 \tag{5}$$

This simple calculation thus suggests that the optimum yield of mutants, when considering both increasing seed death and increasing mutant yield as a function of mutagen dose, should occur when the mutagenized seed survival, relative to wild type, is about 37%.

### 2. Scoring the Increase in Mutant Frequency

The ultimate measure of success is the production of mutants. A relatively simple and quick method for doing this is the use of embryonic-lethals, as pioneered by Meinke.[22] Embryonic-lethal mutations are, by definition, expressed very quickly after fertilization. They can be scored directly in the siliques of the M1 plants, appearing as a 3:1 segregation of pale white embryos instead of the normal green. Since many loci can yield embryonic-lethals, the frequency of this phenotype can be relatively high. After a successful EMS mutagenesis, the percentage of M1 plants segregating embryonic lethals is typically 5 to 10% (David Meinke, personal communication). Many investigators will do a quick dose response curve for each new batch of mutagen, to make sure it has the desired efficacy. It is also possible to score the frequency of albinos in the M2, or the frequency of albino chimeras in the M1, as another measure of general mutagenesis. These frequencies will be on the order of 1 in 1000 to 1 in 250 after a successful EMS mutagenesis (Ruth Wilson, personal communication).

In the terms of the previous theoretical section, scoring mutant frequency as a function of surviving plants is a measurement of the dose-response curve, while scoring mutant frequency as a function of treated seeds measures the combination of both dose-response and seed kill.

## C. SAMPLE PROTOCOL

An overview of mutagenesis and a sample EMS mutagenesis protocol are given in Tables 1 and 2, respectively. The basic strategies can be used with a variety of chemical mutagens, provided that the dose of mutagen is calibrated against seed survival and/or increase in mutant frequency.

## D. USE OF M1 SEED POOLS : INDEPENDENCE AND STERILE OR LETHAL MUTANTS

How does one collect the seeds from the M1 plants? Two extreme methods are to harvest the plants and put all of the M2 seeds in one bag, or, alternatively, to collect the seeds from each M1 plant individually. The seeds are then planted in soil or agar as appropriate for

## TABLE 1
## Outline of *Arabidopsis thaliana* Seed Mutagenesis Strategy

Mutagenesis:

1. Estimate needed M1 and M2 sizes based upon the desired number of mutants to isolate and the expected mutant frequency.
2. Choose mutagen based on efficacy and need for specific types of mutations, such as deletions, for further experiments. Find out how to neutralize or detoxify the mutagen if it is a chemical.
3. Do a survival curve for the mutagen, comparing survival and germination frequency after treatment at a variety of doses. Pick a dose that gives 35 to 40% survival compared to the zero dose control.
4. Choose the pool size of M1 plants. A useful number is the square root of the projected total M1 population.
5. Begin mutagenizing the seeds and plant the M1 generation. It may be convenient to plant all the M1 plants for a given pool in one flat, for ease of collection.
6. Collect pools of M1 seeds.
7. Plant sufficient M2 seeds to get a 95% representation of each M1 pool.
8. Select or screen mutants as appropriate.
9. If the phenotype of the mutant is a recessive lethal or sterile, replant the critical M1 pool seeds in an organized manner, so that heterozygotes segregating for the mutant can be identified.

Mutant Characterization:

1. If more than one mutant of similar phenotypes develops, choose only one mutant from each M1 pool to characterize.
2. Backcross the mutant to wild type at least five times. This process can be accelerated by crossing presumptive heterozygotes and then using progeny analysis to check the phenotype of the heterozygotes.
3. Begin mapping the mutant to chromosomes using standard visible, RFLP, selectable, or other markers.
4. Cross independent mutants from different M1 pools with each other for a complementation test to determine the number of loci identified or the number of alleles per locus.
5. Cross new mutants with other similar mutants to test for interactions between loci.
6. Determine the developmental stage at which the mutant phenotype diverges from the wild type phenotype.

Secondary Mutagenesis:

1. If multiple alleles per locus are needed, either repeat the original mutagenesis to acquire more independent lines or attempt the suggested pollen mutagenesis, with lack of complementation as an assay.
2. Plan mutagenesis for isolation of suppressors or revertants.

## TABLE 2
## Typical EMS Mutagenesis Protocol

Use 0.2% v/v EMS (methanesulfonic acid ethyl ester)

Safety Precautions:

- Wear gloves
- Work in a fume hood
- Neutralize EMS solutions, glassware, gloves, etc. with 1 *M* NaOH

0. Before doing this protocol on a large scale, estimate dose versus the percent survival and germination under the given conditions and current batch of EMS. Also, check for the frequency of embryonic lethals or albinos induced by each dose of the mutagen.
1. Place dry seeds in a beaker that is large relative to the volume of seeds. Add the EMS solution, and incubate 16 to 20 h with occasional stirring.
2. Wash seeds with distilled water for 2 to 3 h. Change the wash every 10 to 15 min. About halfway through the washes, transfer seeds to a fresh beaker.
3. Plant seeds in soil. Thoroughly soak the soil ahead of time and then sow the seeds in an even distribution across the soil. *Arabidopsis thaliana* seeds may be pipeted directly if one moves quickly, or they can be suspended in 0.1% gelatin for a more leisurely delivery. Cover the soil with some protective coating to keep moisture in, and protect delicate seedlings during their first week. We have used cheese cloth, saran wrap, and pine straw. Water the plants with a fine mist.
4. Keep careful track of the survival rate and actual number of plants in the M1 generation. Collect seeds from the M1 by harvesting either siliques or the aerial portions of the plant, assembling them in pools of appropriate sizes. Brown paper bags work well as a temporary dessication/storage holder.
5. Let the collected plant parts dry for several days and then collect the seeds. Seed germination of the M2 may be enhanced by several days storage in the cold.
6. Plant sufficient M2 seeds to obtain a 99% representation of each M1 pool. Screen or select for mutants. Keep good records of the origin of each batch of seeds, so that it can be determined whether similar mutants are independent or not.

selection or screening. The issue of pooling M1 seeds together is almost unique to *Arabidopsis thaliana*. Mutagenized seeds from large seeded plants are collected on a plant by plant basis or, as is the case for maize, as the product of a single fruit on a cob. The small seed size and ease of mass collection of *Arabidopsis thaliana* seeds compels the use of a probabilistic solution.

Obviously, collecting seeds in one pool is easier. The process of collecting seeds from each individual plant and then planting individually has multiple advantages. First, if two mutants are identified with the same phenotype, one can be sure they are of independent origin; if they fail to complement each other, then two alleles have been isolated at the same locus. Second, if the phenotype entails sterility or lethality, then the mutation can be recovered from its heterozygous sibling seeds. The sectors of the M1 plant containing the mutation are heterozygous, so their selfed progeny are naturally segregating heterozygotes for the mutation as well as homozygotes. The phenotypically normal progeny from the same plant will contain some +/m heterozygotes, which can be identified in turn based on analysis of their progeny. Careful bookkeeping and collecting of seeds from each M1 plant individually thus allow one to recover sterile or lethal mutants, as well as to be absolutely sure that all mutants of the same phenotype are independent.

In practice, collecting seeds from each plant separately can be very tedious and not necessarily the most efficient method. The alternative is to collect seeds from a pool of M1 plants. For example, if 50 M1 seeds are sown on a single flat, the M2 seeds from these might be collected together as a single pool. If the goal is to collect seeds from 10,000 M1 plants, then one would end up with 200 separate pools. If two mutants with similar phenotypes are identified as originating from separate pools, then one can be sure they are of independent origin. If two mutants originate from the same pool and if they are in the same complementation group, then there is a good chance that they are identical to each other. Since mutants segregate as 25% of the seeds from the M1, mutant homozygotes will likely be present multiple times in any one seed pool. The best procedure is to save only one of the mutants of the same phenotype that come out of a given M1 seed pool.

Recovering the sibling heterozygotes for a sterile or lethal mutant can also be done, provided the M1 seed pools are not too large. Additional M2 seeds from the same pool that yielded the mutant are planted, and then seeds from the nonmutant segregants are collected individually from each plant. The seeds from each plant are then sown separately, and the mutant heterozygotes can be identified by the presence of mutants in their progeny at the expected 25% rate. Thus, to recover the heterozygotes for a sterile or lethal mutant, it is eventually necessary to recover seeds and score progeny on an individual plant by plant basis. This task is made easier by reducing the number of plants per M1 seed pool; on the other hand, the initial mutant seed collection and screening is made much easier the larger the M1 seed pools are. A good suggestion, as a balance point, is to take the square root of the total M1 plant population and then collect seeds in pools of that size. This should provide the most efficient two generation search and recovery for a sterile or lethal mutant.

If M1 seeds are collected in pools, how many plants should be examined from each M1 pool to be sure of an adequate representation of the total mutants in the pool? Since one may wish to recover a sterile or lethal mutant, one certainly does not want to plant all the seeds from a pool. If the total number of plants in an M1 seed pool is $N$, and there are two to three independent cell lineages contributing to the next generation in each M1 plant, then there are at most $2N$ to $3N$ possible independent mutant lineages. Further, only 25% of the M2 seeds will be homozygous for the mutation. Sampling from $2N$ to $3N$ lineages, and with 25% segregants, is thus a problem in binomial probability. The probability of not obtaining a representative of each mutant lineage in a sample of $k$ seeds will be

$$\left(1-\frac{1}{3N\cdot 4}\right)^{k} \tag{6}$$

If we specify that we wish a 95% probability of sampling all mutant lineages, then

$$\left(1-\frac{1}{3N\cdot 4}\right)^{k} = 0.05 \tag{7}$$

which is equivalent to

$$k = \frac{ln 0.05}{ln\left(1-\frac{1}{3N\cdot 4}\right)} \tag{8}$$

For an M1 pool of 50 plants, the number of M2 plants one should examine per pool is 1796, in order to have a 95% probability of screening all present mutants. This calculation also suggests that even if seeds are collected completely on an individual plant basis, it is advisable to screen at least 34 M2 plants per M1 plant, to assure a 95% probability of discovering any mutants present.

The choice of 95% probability is arbitrary. The balance is between how sure the investigator wishes to be that all the mutant alleles in a given pool have been sampled versus the likelihood of seeing the same mutants over and over again.

Our recommendation is thus to project the size of the M1 that will be used, then take the square root of this and use the square root as the size of the M1 seed pools. For a projected M1 of 10,000 mutagenized plants, the seeds should be collected in 100 pools of seeds from 100 M1 plants. During subsequent selection or screening, only enough seeds should be used from each pool to ensure an adequate representation of the M1 seed pool, leaving the rest of the seeds unplanted as a resource for recovery of mutant heterozygotes. In our example with a pool size of 100, solving the probability statement given above indicates that a sample of 3593 M2 plants from each pool of 100 will be sufficient to ensure a 95% probability that all mutant cell lineages will be examined. If more than one mutant in the same complementation group is obtained from a single M1 seed pool, only one of the mutants should be retained in order to assure independence.

## IV. MUTANT SCREENING AND SELECTIONS

Mutants can be identified either by screening or by selection. In the sections that follow, we describe a few examples of both approaches.

### A. BIOCHEMICAL MUTANTS

The metabolic pathways of plants are not as well defined as those in bacteria, yeast, or mammalian systems. Sometimes plants have different pathways than the textbook paradigms; sometimes plants have alternative pathways. In addition, plants produce a large array of secondary products, many with medicinal value. The metabolic diversity of plants is extraordinary. For these reasons, a mutational analysis of a plant metabolic pathway will often help define the biochemical steps in that pathway. Characterization of a biochemical mutant will usually be more complex than simply looking up the pathway and rapidly deducing the afflicted gene product.

The thiamine auxotrophs of Rédei[23] are a classic example of identifying a biochemical pathway and mutants deficient in that pathway by screening. Rédei looked for *Arabidopsis* plants that were unhealthy on minimal medium but could be recovered when grown on enriched medium. Analysis of growth on media supplemented with various substances (auxanography) was then used to discover the precise metabolic steps affected. He uncovered mutants in a number of steps in the thiamine biosynthetic pathway in this manner. Recently, a biotin auxotroph was identified by Meinke and coworkers[24] in a slightly different fashion. Meinke initially screened for embryonic lethal mutations and then looked among these for embryos that could be recovered when cultured on supplemented media. Logical continuations of this approach, including measurements of enzyme activity, precursor accumulation, and product decline, can then be used to characterize the mutation. In spite of the successes with biotin and thiamine, the search for auxotrophs by screening has often been less than productive for reasons that are still obscure.

An alternative approach is to develop a selection method for absence of a function. Two examples are the use of chlorate resistance to identify nitrate reductase mutants[25] and the use of 5-methylanthranilate to identify tryptophan auxotrophs.[26] The principle of both of these methods is the same — a target enzyme converts a nontoxic substrate analog into a toxic product analog. Mutants that lack the enzyme will thereby be resistant to the chlorate or 5-methylanthranilate. This positive selection for absence of a function should work in other pathways for which similar toxic product analogs can be identified.

Growing *Arabidopsis* seedlings on supplemented media can prove difficult if a given seed batch is heavily contaminated with fungi or bacteria. One observes that standard seed sterilization protocols work well when the seeds are subsequently germinated on minimal media, but not necessarily when germinated on enriched media. Presumably the rich media allow growth of contaminants that cannot grow on minimal media. A protocol is included in Table 3, developed by George Jen of CIBA-GEIGY, that has proven successful in seed sterilization where other protocols have failed.

Selection of dominant resistant mutants (neomorphs, hypermorphs, acquisition-of-a-function), should be straightforward, provided the relative rarity of dominant mutants is kept in mind. The selection can occur either by plating sterilized seeds on agar medium containing the agent or by spraying seedlings directly. Usually a green survivor will stand out against the dying, yellowing background of seeds.

The use of enriched media is an indirect screen for a biochemical mutation. An example of a direct screen is the mutational identification of the fatty acid biosynthesis pathway by Somerville and Browse.[27] This group screened mutagenized M2 seedlings by directly injecting extracts into a gas chromatograph and recording the levels of various fatty acids and lipids. Mutants were identified based on the absence or superabundance of a particular chromatography peak. The success of this mutational analysis allowed the investigators to define the fatty acid biosynthetic pathway in *Arabidopsis*. Direct biochemical screening such as this can be successful if the assay is relatively quick and if a labor force exists to perform several thousand assays. Since multiple fatty acids could be monitored on one chromatogram, the screening method simultaneously searches for mutants in more than one gene, thereby increasing the probability of finding any one mutation. The delineation of fatty acid biosynthesis in *Arabidopsis* this way is a splendid achievement that can serve as a useful model for other biochemical genetic projects.

**TABLE 3**
**Seed Sterilization for Aseptic Seedling Screening or Selections**

Work in an aseptic laminar flow hood as much as possible

1. Measure up to 2 g of M2 seeds into 50-ml plastic screw-cap centrifuge tubes.
2. Add 30 ml of 70% ethanol, shaking briefly to wet the seeds. Allow the seeds to settle, and decant the floating debris.
3. Add 25 ml of 30% bleach (which is 5.25% sodium hypochlorite) plus 0.02% Triton X-100. Shake at 350 rpm, to keep the seeds suspended, for 20 min.
4. Allow the seeds to settle and decant any floating debris.
5. Wash seeds three times with distilled water. Spin seeds briefly to pellet them.
6. Suspend the seeds in 25 ml of benlate solution (62 mg/100 ml, prewarmed to 45°C). Incubate at 45°C with shaking for 1 h.
7. Melt 100 ml of 0.2% agarose in water. Add 21 mg of benlate per 100 ml. Let cool to 42°C.
8. Let the seeds settle, or briefly spin them to a pellet.
9. Resuspend seeds in agarose-benlate solution, and pipet on to Petri plates at 4 ml per plate. The plating density will depend upon whether the medium is selective (high density plating) or nonselective (low-density plating). A typical high density plating would be 100-mg starting material per plate; low density might be 10-mg starting material per plate.
10. Wrap plates with parafilm or micropore paper tape. Incubate in a lighted growth chamber or on a well-lit bench top.
11. Differences between healthy and dying seedlings can usually be seen in 1 to 4 weeks.

From Jen, George C., CIBA-GEIGY Corp., Research Triangle Park, NC, unpublished results. With permission.

## B. DEVELOPMENTAL MUTANTS

The ability to identify amusing and valuable developmental mutants of *Arabidopsis* is generally limited only by the cleverness with which an appropriate visual screen can be devised. We will describe two different examples to illustrate this point.

Chory et al.[28] developed a screen for mutants in the signal-transduction pathway after light irradiation. Dark grown plants are etiolated — elongated and pale; hence, one effect of light is to affect the growth rate and morphology of seedlings. Chory's method consisted of screening for mutants that did not etiolate properly in the dark. They identified *det* (de-etiolated) mutants by visually screening 100,000 M2 seeds for lack of the typical seedling elongation response to darkness. These mutants did not turn green, indicating that there are separate signal transduction pathways for the elongation and greening responses to light. Given the difficulties of analyzing a signal-transduction chain biochemically, defined mutants that lack a step should prove invaluable. The key to the success of this study is simply anticipating the probable phenotype of the desired mutants, and then constructing a screen in such a way that the mutant could be found and subsequently rescued.

Mutants that affect root growth patterns and obstacle-avoidance behavior have been identified by Okada and Shimura.[29] They found that they could grow *Arabidopsis* plants on agar set vertically, allowing the growth of the root to be visualized. Thus, they were able to examine M2 seeds and detect altered root growth. One of the cleverest screens used a very high concentration of agar in the medium to simulate impenetrable obstacles. The *Arabidopsis* seeds were placed on a vertical agar plate, which was later shifted to a 45° angle. Wild type roots attempted to penetrate the stiff agar, failing which, the roots grew in characteristic oscillating spirals and then tried to penetrate again. Repetition produced a characteristic wavy growth pattern of wild type. Mutants that could not perform the oscillating spiral were easily identified by their lack of a wavy growth pattern.

One of the most powerful uses of *Arabidopsis thaliana* mutagenesis is precisely to find new approaches for studying these physiological and developmental problems. Maize is superior for a number of genetic studies because of its immense genetic resources and well-developed chromosome mechanics. The advantage of *Arabidopsis* lies in the development of novel mutational analysis approaches to previously messy problems in plant biology.

## V. SECONDARY MUTAGENESIS

### A. ADDITIONAL ALLELES AT A SINGLE LOCUS

As discussed previously, it is extremely useful (almost essential) to develop an allelic series for any important morphological locus. How does one obtain additional alleles? The obvious method is simply to keep plugging away at the original mutagenesis strategy, perhaps trying a different mutagen, and to eventually identify additional mutants with a similar phenotype. These are then crossed with the original mutant, and complementation groups are assigned.

We suggest an alternative, highly directed strategy, which has also been tried by Estelle and Wilson (personal communication). The concept is to mutagenize wild type pollen rather than seeds and then to apply that pollen to pistils of the original mutant (named *m1* here). If no mutation at the same locus occurs in the treated pollen, then the genotype of the next generation will be heterozygous (+/*m1*). If a new mutation occurs at the same locus (*m2*), then the seed from this fertilization will be *m2*/*m1* heterozygotes. These plants should display either the original phenotype or a modification of it, based on the dominance relations between *m2* and *m1* alleles. The new allele, *m2*, could then be recovered from a self of the *m2*/*m1* heterozygote or by subsequent crossing and selfings with wild type.

Several genetic tricks would greatly facilitate this proposed strategy of lack-of-complementation. First, the recipient mutant plant (*m1*/*m1*) should be either male sterile or self-incompatible. Second, the original mutant should have been genetically mapped and linked to a readily scorable marker. This facilitates distinguishing the *m1* and (hoped-for) *m2* alleles by tagging the *m1* chromosome with the scorable marker. To summarize, with the goal of isolating additional alleles at the *M* locus, currently defined by *m*1,

1. Mutagenize wild type pollen with EMS or by irradiation,
2. Fertilize a homozygous, male sterile, *m1 x* plant, where *x* is a linked marker,
3. Collect seeds,
4. Examine next generation for *m*/*m* phenotypes,
5. Identify new *m* allele by its lack of linkage to the *x* marker.

This strategy is awkward with *Arabidopsis* because of the need for a male sterile or incompatibility marker. It should be quite easy with maize for the following reasons: maize chemical mutagenesis is most commonly performed on the pollen;[16] plenty of useful markers are available to be linked to the mutant of interest; the standard crossing methods prevent selfings in any case.

### B. REVERTANTS AND/OR SUPPRESSORS

One of the best ways of finding additional loci that affect the same developmental process as the original mutation is to isolate suppressors or second site revertants of the mutant. This requires generating a large seed supply of the original mutant and then mutagenizing, generating an M1 and M2, and screening for a modified or restored mutant phenotype. Any mutants so generated are then mapped to determine if they are at a different genetic locus than the primary mutant. Second site revertants or suppressors may be dominant mutations. This has two consequences. The bad news is that the mutation rate may be much lower than that to obtain a recessive mutant and therefore may require a larger screen. The potentially good news is that it may be possible to identify a dominant revertant as a sector on an M1 plant. If a reversion of the mutant could be detected when chimeric, then one may be able to screen directly in the M1 generation. Of course, dominant mutations will be present as entire plants in the M2. Recessive suppressor or interacting mutations will require M2 screening as usual. To avoid accidental contamination of the putative suppressor/revertants with wild type seeds that can float around the greenhouse or growth chamber, it is often useful to put additional linked mutants (markers) in the same strain with the original mutant. A revertant that contains these markers is thus truly a revertant and not a contaminant.

## VI. MUTANT ANALYSIS

### A. MULTIPLE BACKCROSSES

How often will one obtain multiple mutants in an individual M2 plant? Always. The per locus mutant frequency in the M2, after successful EMS treatment, is probably in the range of 1 in 2000 to 1 in 5000. We can reinterpret this as the probability for a given locus to be mutated in a single seed and then estimate the number of mutants per seed by multiplying by an estimate of the total gene number. Kamalay and Goldberg[30] have estimated that the total number of mRNAs expressed in all organs throughout the life of the tobacco plant is 60,000. Using this figure produces a guess of 12 to 30 mutations per treated M2 seed. If you believe there are as few as 10,000 active genes in *Arabidopsis*, then the number of mutants per M2 seed could be as small as two to five. This calculation ignores the possibility of significant variance from seed to seed in effective mutagen treatment. Mutagenesis may produce jackpots of mutants in some nuclei and very low frequencies in other nuclei, to produce the 1 in 2000 to 1 in 5000 average. If this is so, then identifying a mutant plant in the M2 will certainly mean that one has found one of the jackpots, and the total number of mutants per seed could be higher than 12 to 30.

The consequences of the multiple mutants per seed are simple. Any new mutant must be backcrossed to wild type repetitively until the genetic background is cleaned up. Until this is done, one cannot be sure that a given phenotype is due to a single locus or multiple interacting loci. The number of nonlinked extraneous background mutations is halved with every backcross generation. Starting with the original M2 seed mutation, it would take five cycles of backcrossing followed by selfing to identify the mutant homozygotes to obtain a 32-fold reduction in background mutant frequency. Even with the accelerated *Arabidopsis* life cycle, ten generations is a long time.

The time taken by the suggested five cycles of backcrossing and selfing can be reduced by a combination of progeny analysis and good recordkeeping. The initial M2 mutant plant (*m*/*m*) is crossed to wild type (+/+), and then the resulting F1 (+/*m*) is selfed, yielding standard 3:1 segregation. Among the portion of the F2 that has a wild type phenotype, 67% will be heterozygotes (+/*m*). The heterozygotes can be identified retroactively by progeny analysis — collecting seeds and scoring which plants segregate the mutant. While the progeny analysis is occurring, the +/+ and +/*m* plants can be immediately crossed with wild type again. When the progeny analysis is completed and the progenitor heterozygotes are identified, the pattern of +/+ and +/*m* backcrossing can be continued. This process is diagramed in Figure 1. Essentially, the mutant heterozygotes are identified in retrospect by analyzing the segregation of the mutation in their selfed progeny. One is always doing some +/+ x +/+ crosses inadvertantly, but this is a small price for the great acceleration in time allowed. In practice it never hurts to keep backup crosses going at the same time. The *m*/*m* segregants in the progeny analysis can be crossed with wild type, just in case the +/*m* cross fails for some reason.

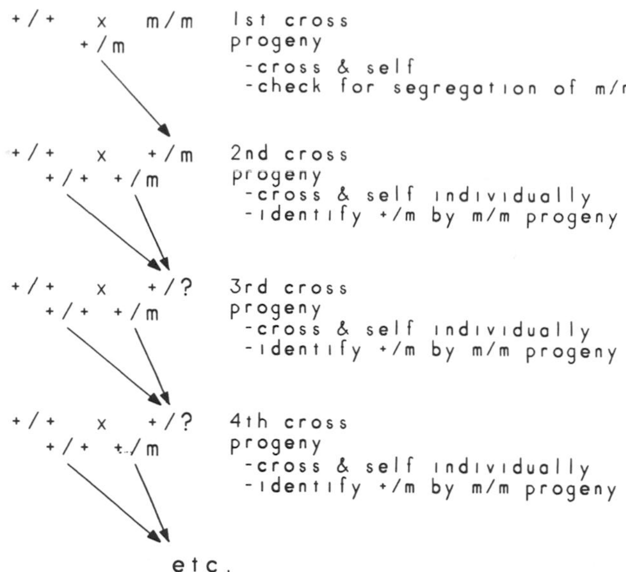


**FIGURE 1.** Accelerated backcrossing scheme using progeny analysis to identify heterozygotes carrying the mutation.

## B. MAPPING

Always map mutants or transformants. A variety of kinds of mapping strains are now available, including visible markers,[31] RFLP markers,[32, 33] and T-DNA transformants.[20] Some of these strains may be obtained from the various stock centers (see Table 4). Visual markers are usually used in two sequential steps. First, a stock with two or three genetic markers on each chromosome is used to assign the gene to a particular chromosome; then a second stock is used that is multiply mutant within that chromosome to put the mutant on the genetic map. If one's lab is routinely performing RFLP analysis, crossing the mutant with an appropriate other strain may allow relatively rapid and easy mapping. T-DNA transformant stocks that are already mapped to all chromosomes can provide a nice, dominant marker, kanamycin resistance that is very useful for mapping other mutants.

## C. DOSAGE ANALYSIS

As discussed in the section on types of mutants, one should be able to distinguish types of mutants by a dosage analysis. In *Arabidopsis* this is difficult at present. In maize, a wonderful series of chromosome translocations exist, in which portions of the regular (A) chromosomes are translocated to extra supernumerary (B) chromosomes that have less regular patterns of replication and segregation. One can therefore categorize mutants as amorphs, hypomorphs, etc. by varying the dosage of the mutant or wild type alleles of the gene using these A/B translocation stocks. A good example of the use of this method is Dudley and Poethig's analysis of the *Teapod* mutation.[34] This paper illustrates another useful genetic technique — crossing the mutant allele into a variety of genetic backgrounds to check for the effects of polygenic modifier loci; *Teapod* mutants had different morphologies and severities depending upon the genetic background.

## D. CHIMERISM AND CELL AUTONOMY

If your mutation is closely linked to a color marker, a chorophyll or anthocyanin gene for example, you can test for cell autonomy of the trait by constructing chimeras. This is

**TABLE 4**
***Arabidopsis* Genetic Stock Centers and Other Addresses**

**Centers for Mutants, RFLPs, and Other Genetic Stocks**

Arabidopsis Biological Resource Center at Ohio State
Ohio State University
1735 Neil Avenue
Columbus, OH 43210, U.S.A.
*fax*: 614-292-0603
*E-Mail*: arabidopsis+@osu.edu

Nottingham Arabidopsis Stock Centre
School of Biological Sciences
University of Nottingham
University Park
Nottingham, NG7 2RD, United Kingdom
*fax*: 55-603-424270
*E-Mail*: pbzmlh@vax.ccc.nothingham.ac.uk
*or*: pbzmlh@uk.ac.notthingham.ccc.vax

RFLP, Contig, & Mutation Database
*E-Mail*: curator@weeds.mgh.harvard.edu

Arabidopsis Electronic Mail Bulletin Board
*E-Mail subscriptions*: biosci@net.bio.net
*Usenet newsgroup*: bionet.genome.arabidopsis

**Commercial Source of EMS and Fast Neutron Irradiated Mutagenized Seeds**

Lehle Seeds
6531 North Camino Katrina
Tucson, AZ 85718, U.S.A.
*fax*: 602-797-9009

usually done using gamma ray treatment at a defined time during development on *cis* double heterozygotes of the mutant and color marker. Sometimes the gamma rays will knock out the wild type chromosomal sector, leading to production of a color and mutant morphology sector on the plant. If either the mutant or wild type morphology extends across the sector boundary, then the trait must be noncell autonomous, carried by a diffusible factor. This is a very powerful method of developmental analysis. Poethig recently demonstrated that one of the *Teapod* loci[9, 35] encoded a diffusible factor regulating maize shoot development.

## E. MULTIPLE LOCI AND EPISTASIS

When studying a given biological process, one would always like to isolate mutants in all the genes that affect that process, i.e. saturation mutagenesis. Complementation tests allow one to distinguish whether mutants with similar phenotypes are in the same gene or not. An analysis of double mutants using two different loci that affect the same process can tell whether the mutants are acting independently of each other, or whether there is a significant interaction. In some cases, the analysis of a double mutant can allow construction of a dependent pathway.

Epistasis is a term with both a broad and narrow definition. In the broadest sense it refers to unexpected (nonadditive) interactions between specific alleles at different loci. In the narrow sense it can mean complete suppression of the phenotype of one locus by a specific

N.C. WESLEYAN COLLEGE
ELIZABETH BRASWELL PEARSALL LIBR

allele at another locus. This is most commonly illustrated with an example like flower color. Suppose there were several biochemical steps required to achieve a certain red color:

$$\text{compound 1 (colorless)} \xrightarrow{\text{enzyme 1}} \text{compound 2 (colorless)} \xrightarrow{\text{enzyme 2}} \text{red compound}$$

If compound 1 and compound 2 are colorless, then a mutation in either gene 1 or gene 2 will render the flower colorless. If gene 1 is defective, it does not matter what the genotype of gene 2 is, and vice versa. This is mutual epistasis in the narrow sense.

This form of epistasis is partially a product of only scoring the final phenotype, the accumulation, or lack thereof, of the red compound. Suppose one could also score the levels of compound 1 and compound 2 readily, perhaps by HPLC or GC. A mutation in gene 1 would then accumulate compound 1 and have none of compound 2. A mutation in gene 2 would accumulate compound 2 and have none of the red stuff. A double mutant would accumulate compound 1 and have none of compound 2 or the red stuff; it would resemble mutant 1 alone. Thus, when the intermediate products are assayable, it is possible to distinguish mutant 1 from mutant 2. The double mutant demonstrates epistatic suppression of mutant 2 by mutant 1, but not the reverse. If it was not known which step in the pathway came first, this could be deduced by analysis of the double mutant and the accumulation of compound 1. Double mutant analysis makes it possible to deduce dependency: gene 2 works after gene 1 because it depends upon gene 1 for expression.

The analysis of double mutants for morphological traits can be similar, but it also can be more complicated. In addition to independence of action and epistasis, double mutants can give quite unexpected, interactive phenotypes as well. Sometimes it may be difficult to decide whether the altered structure is due to a simple addition of the two mutant phenotypes, or if an unexpected interaction has occurred. Deducing a developmental pathway, or order of gene action, based on the concepts of dependency and epistasis can lead to a significantly deeper understanding of the morphological trait. An example is the recent study of the *leafy* mutation of *Arabidopsis*, and the way it acts before other floral development homeotic loci.[36] This is also an example of how valuable it is to determine the time at which the pattern of mutant development diverges from that of wild type.

## VII. CONCLUSIONS

There probably is no perfect mutagenesis and mutant isolation experiment. Experimental needs vary from time to time and as one gains more experience with the system at hand. Rather than trying to design a single massive-scale experiment that will satisfy all future needs, it is more useful to expect to do a variety of different treatments, with different genetic stocks and under varying circumstances. It can also be difficult to guess precisely what the phenotype of interest will be.

A successful *Arabidopsis thaliana* mutagenesis poses another, more philosophical problem. Usually so many interesting, weird looking plants show up in the M2 that it is easy to lose the initial focus. One suddenly discovers that all of plant growth is interesting. There is no cure for this. Most of the time, a quick search of the literature will reveal that someone else has already reported the phenotype. Once in a while, however, one may be in the fortunate but difficult situation of deciding if one's original goals are more important than the weird plant that the mutation has produced.

## ACKNOWLEDGMENTS

The author would like to thank Michael Cellino, Sue Gibson (Michigan State University), Debra Mohnen, Barry Palevitz, Wayne Parrot, and Ruth Wilson (Michigan State University) for reading the manuscript and making helpful suggestions. Research in the author's laboratory has been supported in recent years by N.S.F. grant DCB8715799, U.S.D.A.-C.R.G.O. grant GAM8901056, and D.O.E. Energy BioSciences grant DEFG0991ER20034.

## REFERENCES

1. **Muller, H. J.,** Further studies on the nature and causes of gene mutations, in *Proc. 6th Int. Congr. Gen.*, 1, 213, 1932.
2. **Kunst, J., Klenz, J. E., Martinez-Zapater, J., and Haughn, G. W.,** *Ap2* gene determines the identity of perianth organs in flowers of *Arabidopsis thaliana, Plant Cell,* 1, 1195, 1989.
3. **Hake, S., Vollbrecht, E., and Freeling, M.,** Cloning *Knotted,* the dominant morphological mutant in maize, using *Ds2* as a transposon tag, *EMBO J.*, 8, 15, 1989.
4. **Kimble, J. and Schedl, T.,** Developmental genetics of *Caenorhabditis elegans,* in *Developmental Genetics of Higher Organisms,* Malacinski, G., Ed., Macmillan, New York, 1988, ch. 8.
5. **Edgar, R. S. and Lielausis, A.,** Temperature-sensitive mutants of bacteriophage T4D: their isolation and genetic characterization, *Genetics,* 49, 649, 1964.
6. **Suzuki, D. T.,** Temperature sensitive mutants in *Drosophila melanogaster, Science,* 170, 695, 1970.
7. **Bowman, J. L., Smyth, D. R., and Meyerowitz, E. M.,** Genes directing flower development in *Arabidopsis, Plant Cell,* 1, 37, 1989.
8. **Li, S. and Rédei, G. P.,** Estimation of mutation rates in autogamous diploids, *Radiat. Bot.*, 9, 125, 1969.
9. **Poethig, R. S.,** Clonal analysis of cell lineage patterns in plant development, *Am. J. Bot.*, 74, 5814, 1987.
10. **McDaniel, C. N. and Poethig, R. S.,** Cell lineage patterns in the shoot apical meristem of the germinating maize embryo, *Planta,* 175, 13, 1988.
11. **Chaudhury, A. M. and Signer, E. R.,** Relative regeneration proficiency of *Arabidopsis thaliana* ecotypes, *Plant Cell Rep.*, 8, 368, 1989.
12. **Damm, B. and Willmitzer, L.,** Regeneration of fertile plants from protoplasts of different *Arabidopsis thaliana* genotypes, *Mol. Gen. Genet.*, 213, 15, 1988.
13. **Valvekens, D., Van Montagu, M., and Van Lijesbettens, M.,** *Agrobacterium tumefaciens* mediated transformation of *Arabidopsis* root explants using kanamycin selection, *Proc. Natl. Acad. Sci. U.S.A.*, 85, 5536, 1988.
14. **Al-Shehbaz, I.,** The genera of sysmbrieae (cruciferae, brassicaceae) in the southeastern United States, *Arabidopsis, J. Arnold Arboretum,* 69, 213, 1988.
15. **Rédei, G. P. and Li, S. L.,** Effects of X-rays and ethyl methanesulfonate on the chlorophyll b locus in the soma and on the thiamine loci in the germ line of *Arabidopsis, Genetics,* 61, 453, 1969.
16. **Neuffer, M. G. and Coe, E. H., Jr.,** Paraffin oil technique for treating corn pollen with chemical mutagens, *Maydica,* 22, 21, 1977.
17. **Brockman, H. E., deSerres, F. J., Ong, T., DeMarini, D. M., Katz, A. J., Griffiths, A. J. F., and Stafford, R. S.,** Mutation tests in *Neurospora crassa, Mut. Res.*, 133, 87, 1984.
18. **Ping,T. and Ausbuel, F.,** *Science,* in press.
19. **Feldmann, K. A., Marks, M. D., Christianson, M. L., and Quatrano, R. S.,** A dwarf mutant of *Arabidopsis* generated by T-DNA insertion mutagenesis, *Science,* 243, 1351, 1989.
20. **Feldmann, K. A.,** T-DNA insertion mutagenesis in *Arabidopsis*: mutational spectrum, *Plant J.*, 1, 71, 1991.
21. **Van Sluys, M. A., Tempe, J., and Fedoroff, N.,** Studies on the introduction and mobility of the maize Activator element in *Arabidopsis thaliana* and *Daucus carota, EMBO J.*, 6, 3881, 1987.
22. **Meinke, D.W.,** Embryo-lethal mutants of *Arabidopsis thaliana:* analysis of mutants with a wide range of lethal phases, *Theor. Appl. Genet.*, 69, 543, 1985.
23. **Li, S. and Rédei, G. P.,** Thiamine mutants of the crucifer *Arabidopsis, Biochem. Genet.*, 3, 163, 1969.
24. **Franzmann, L., Patton, D. A., and Meinke, D. W.,** In vitro morphogenesis of arrested embryos from lethal mutants of *Arabidopsis thaliana, Theor. Appl. Genet.*, 77, 609, 1989.

25. **Oostindier-Braaksma, F. J. and Feenstra, W. J.,** Isolation and characterization of chlorate resistant mutants of *Arabidopsis thaliana, Mut. Res.,* 19, 175, 1973.
26. **Last, R. L. and Fink, G. R.,** Tryptophan requiring mutants of the plant *Arabidopsis thaliana, Science,* 240, 305, 1988.
27. **Somerville, C. and Browse, J.,** Plant lipids: metabolism, mutants, and membranes, *Science,* 252, 80, 1991.
28. **Chory, J., Peto, C., Feinbaum, R., Pratt, L., and Ausubel, F.,** *Arabidopsis thaliana* mutant that develops as a light grown plant in the absence of light, *Cell,* 58, 991, 1989.
29. **Okada, K. and Shimura, Y.,** Reversible root tip rotation in *Arabidopsis* seedlings induced by obstacle touching stimulus, *Science,* 250, 274, 1990.
30. **Kamalay, J. C. and Goldberg, R. B.,** Regulation of structural gene expression in tobacco, *Cell,* 19, 935, 1980.
31. **Patton, D. A., Franzmann, L. H., and Meinke, D. W.,** Mapping genes essential for embryo development in *Arabidopsis thaliana, Mol. Gen. Genet.,* 3, 337, 1991.
32. **Chang, C., Bowman, J. L., DeJohn, A. W., Lander, E. S., and Meyerowitz, E. M.,** Restriction fragment length polymorphism linkage map for *Arabidopsis thaliana, Proc. Natl. Acad. Sci. U.S.A.,* 85, 6856, 1988.
33. **Nam, H. G., Giraudaut, J., denBoer, B., Moonan, F., Loos, W. D. B., Hauge, B., and Goodman, H.,** Restriction fragment length polymorphism linkage map of *Arabidopsis thaliana, Plant Cell,* 7, 699, 1989.
34. **Dudley, M. and Poethig, R. S.,** The effect of a heterochronic mutation, *Teapod2,* on the cell lineage of the maize shoot, *Development,* 111, 733, 1991.
35. **Poethig, R. S.,** A non-cell autonomous mutation regulating juvenility in maize, *Nature,* 336, 82, 1988.
36. **Schultz, E. A. and Haughn, G. W.,** *Leafy,* a homeotic gene that regulates inflorescence development in *Arabidopsis, Plant Cell,* 3, 771, 1991.

# 3

# DNA Sequence Organization and Gene Copy Number Determination

*J. J. Pasternak*

## I. INTRODUCTION

Currently, almost all molecular biology research that is concerned with the genome is focused on isolating and characterizing individual genes. However, it is often important to generate information about the overall genomic organization of a particular organism. A number of DNA reassociation studies have been carried out on plant species[1-4] using the principles and methodologies established by Britten, Davidson, Kohne, Wetmur, and others.[5-12] Among other things, these analyses reveal the proportions of the genomic DNA that consist of repetitive and single copy DNA sequences, the fraction of the genome that "folds-back", i.e., contains inverted repeats, and the pattern of interspersion between repetitive and single copy DNA elements.

There is considerable diversity in the genomic organization of repetitive and single copy DNA in plants, with the patterns ranging from short period interspersion where segments of repetitive DNA that are about 0.15 to 1.0 kilobases (kb) in length alternate with single copy sequences that are ≥1.5 kb in length[1] to at least one case where there is no evidence of interspersion.[13] As well, among plant taxa, angiosperms are distinctive because of the high levels (50 to 80%) of repetitive DNA that occur within the genomic DNA. In the first part of this chapter, procedures for characterizing plant genomes by DNA reassociation kinetics and thermal denaturation are presented. In the second part, strategies for determining the copy number of specific DNA sequences are outlined.

## II. DETERMINATION OF GENOMIC DNA SEQUENCE ORGANIZATION

When double-stranded DNA is denatured and then reassociated (renatured), the kinetics of the overall reaction can be represented as a Cot curve. A Cot curve is usually a plot where the y-axis is the fraction of reassociated DNA ($1-c/C_0$) and the x-axis is the log of the Cot values. A Cot value is the original concentration ($C_0$) of denatured DNA in nucleotides (mol•litre$^{-1}$) multiplied by the time (sec) of the reaction. The term c denotes the concentration of single-stranded DNA (mol•litre$^{-1}$) at a particular time, i.e., Cot point, during the reassociation reaction.

### A. DNA EXTRACTION

For DNA reassociation kinetics experiments, the starting DNA must be both very pure and of high molecular weight. Any procedure that produces good quality DNA is acceptable.

0-8493-5164-2/93/$0.00+$.50
© 1993 by CRC Press, Inc.

DNA samples and buffers that are used for generating reassociation kinetics curves should be treated with Chelex-100 (available from either Sigma or BioRad) or an equivalent chelating resin to remove excess divalent cations. The Milli-Q system or an equivalent process should be used to produce highly purified water for the buffer solutions.

### B. DNA SHEARING AND SIZING

Native DNA can be sheared by treatment with an ultrasonic probe for various durations at certain intensities. Trials will have to be run to determine precisely the settings that are needed for a particular sonicator. For example, DNA fragments with an average length of 550 nucleotide pairs (np) can be created by sonication for 3 min with a 0.05-in. probe using a BIOSONIC III unit at full intensity.[13] With a 0.25-in. probe for 1 min at setting 20, fragments of ≅1500 np can be produced. Larger fragments (≅4900 np) with reasonable size homogeneity can be generated by homogenizing the original DNA sample in a multi-blade mixer at 25,000 rpm for 6 min. During the shearing treatment, it is important to keep the DNA sample immersed in an ice bath. Agarose gel electrophoresis can be used to estimate the apparent mean molecular mass of a sheared DNA sample.

### C. DNA REASSOCIATION

Sonicated DNA samples (approximately 2 ml) with an average size of ≅500 np at concentrations from 0.25 to 1750 µg/ml in 0.12 *M* sodium phosphate, pH 6.8, are sealed in Pasteur pipettes, denatured by boiling for 5 min and then maintained at 60°C until a specified Cot value is reached. Roughly, a particular Cot value is equivalent to 0.5 $A_{260}$ at zero time of the DNA sample, multiplied by the time of the reaction in hours. Although 60°C is the standard temperature for reassociation kinetics, optimal reassociation occurs at about 25°C below the median melting temperature (*Tm*) of the genomic DNA. After reassociation to a particular Cot value, each sample is quickly frozen by immersion in a solid $CO_2$/ethanol mixture. To determine the fraction of nonreassociated (i.e., single-stranded) and reassociated DNA (i.e., duplex fraction), each frozen DNA sample is thawed and slowly applied (100 µg of DNA per 2.5 ml) to reagent grade hydroxyapatite (0.5 g) in a 5 ml water-jacketed column that has been equilibrated with 0.12 *M* phosphate buffer, pH 6.8, and maintained at 60°C. The column is then washed three times with 2.5 ml 0.12 *M* phosphate buffer, pH 6.8, to elute single-stranded DNA. To elute double-stranded (renatured) DNA, the column temperature is raised to 97°C and washed four times with 2.5 ml aliquots of 0.12 *M* sodium phosphate, pH 6.8. Alternatively, double-stranded DNA can be eluted from the hydroxyapatite column at 60°C with four 2.5 ml aliquots of 0.5 *M* sodium phosphate buffer. Recoveries from hydroxyapatite columns usually are greater than 95%. To reduce nonspecific binding and light scattering, boil and wash the hydroxyapatite in 0.12 *M* sodium phosphate buffer before packing the column.

The absorbance of each eluted sample is read at both 260 and 320 nm. Assume that one absorbance unit at 260 nm ($A_{260}$) is equivalent to approximately 50 µg DNA. The reading at 320 nm is used to correct for light scattering effects. The fraction of DNA that is single-stranded at each Cot value is determined from the relationship

$$\frac{A_{260}\text{ss - DNA}}{A_{260}\text{ss - DNA} + A_{260}\text{ds - DNA}} \tag{1}$$

where ss-DNA is the single-stranded DNA fraction and ds-DNA is the double-stranded DNA fraction. For plant genomic DNA, the x-axis, i.e., log Cot, of a Cot curve can range over seven log intervals; therefore, about 30 to 50 or more Cot points are often required to generate a reliable Cot curve.

For Cot points with high values (>200), a high salt buffer system can be used to decrease the incubation time. For these runs, native, size-fragmented DNA is suspended in 0.2 *M* NaCl buffer (0.2 *M* NaCl, 2 m*M* phosphate buffer, 0.2 *M* EDTA, (pH 6.8)), denatured by boiling and chilled, and then the solution is brought to a final concentration of 1 *M* NaCl, 0.01 *M* phosphate pH 6.8, and 1 m*M* EDTA with 5 *M* NaCl buffer. The samples are maintained at 70°C to prescribed Cot values. Prior to hydroxyapatite chromatography, the sample solution is adjusted to 0.12 *M* phosphate with 1.2 *M* phosphate buffer, pH 6.8.

To standardize a Cot plot of a plant genome and to facilitate calculating genome size and other parameters, *E. coli* DNA should be run as a reference. The *E. coli* Cot plot can be generated from independent reassociation kinetics experiments.

## D. DATA ANALYSIS

Although Cot curves can be analyzed empirically, it is advisable to use least squares analysis[14] and other computer programs[7,13,15] to extract as much reliable information as possible from the data. These programs will determine the relative proportion of the genomic DNA that is single copy, moderately repetitive, highly repetitive, and fold-back (i.e., 'zero-time' binding fraction), the rate constants for each class of DNA, and the relative number of copies of sequence elements that comprise the moderately and highly repetitive fractions.

## E. DETERMINING THE PATTERN OF INTERSPERSION

The nature of the interspersion of repetitive and single copy DNA elements can be ascertained by carrying out separate Cot plot determinations with two DNA samples that differ in the size of the sheared fragments. The smaller fragments should be about 300 np, and the larger fragments should be greater than 3000 np. If there is no interspersion, then the observed rate constants for single copy DNA of the two Cot curves should vary as a function of the square root of the ratio of the molecular weights of the two DNA samples, i.e.,

$$\frac{K1}{K2} = \sqrt{\frac{L1}{L2}} \tag{2}$$

where $K1$ and $K2$ are the rate constants for the reassociation of the short ($L1$) and long ($L2$) fragment lengths in nucleotides, respectively.[12] A value for $K1/K2$ that is greater than 1.0 is indicative of interspersion of repeated and single copy DNA sequences.

## F. SIZE DISTRIBUTION OF REPETITIVE DNA SEQUENCES

To determine the average length of repetitive DNA sequences, renature a genomic DNA sample to a Cot value at which only repeated sequences are fully renatured, e.g., Cot 10. Size-fragmented DNA samples with either short or long lengths can be used for these experiments. After reassociation, digest the sample with Sl nuclease to remove single-stranded regions and then isolate Sl-resistant duplex DNA by hydroxyapatite chromatography. The length of DNA can be determined by agarose gel electrophoresis.

Sl nuclease digestion entails treating the 'Cot 10' DNA fraction in 0.15 *M* NaCl, 0.05 PIPES, 0.025 *M* Na acetate, and 0.1 m*M* $ZnSO_4$, pH 4.55 with 100 units Sl nuclease per microgram DNA for 45 min at 37°C. The reaction is terminated by chilling and bringing the reaction mixture to 0.1 *M* phosphate buffer, pH 6.8.

## G. DETERMINING GENOMIC GC CONTENT

Native, sheared DNA (25 to 50 µg/ml) in 0.12 *M* phosphate buffer, pH 6.8, is loaded into quartz cuvettes and overlaid with mineral oil. A UV spectrophotometer with a water-jacketed cuvette holder and fitted with a temperature probe can be used for thermal denaturation. For each experiment, a reference cuvette containing 0.12 $M$ phosphate buffer, pH 6.8, should be present. With all cuvettes in place, the temperature is raised to 60°C. The temperature at 60°C is allowed to equilibrate and is thereafter raised 0.3 to 0.5 degrees/min up to 100°C. The absorbance at 260 nm is monitored in each cuvette at each temperature point. The measure of hyperchromicity ($H$) is

$$\frac{A_{260}[100°] - A_{260}[60°C]}{A_{260}[100°C]} \tag{3}$$

The median dissociation temperature ($Tm$) is the temperature at which 50% hyperchromicity occurs. Using the equation

$$\%GC = 2.44(Tm - 81.5 - 16.6 \log M) \tag{4}$$

where $M$ is the concentration of monovalent cations in the solution, the %GC content of the DNA[16] can be determined.

## H. SPECTROPHOTOMETRIC DETERMINATION OF A COT CURVE

Decrease in hyperchromicity can be used to generate a Cot curve. Sonicated DNA (≅300 np) in 0.12 $M$ phosphate buffer, pH 6.8, is denatured by boiling and is then reannealed at 60°C in a water-jacketed cuvette holder of a UV spectrophotometer. Absorbance readings are taken at 260 and 320 nm during the reassociation period. To obtain a complete set of Cot points in a reasonable time, a range of initial DNA concentrations (e.g., 20 to 1500 μg/ml) should be used. With corrected absorbance values, the decrease in hyperchromicity is calculated at each Cot value and converted to the fraction of the sample that is single-stranded DNA. For example, a 10% decrease in hyperchromicity is more or less equivalent to 90% of the DNA sample being single-stranded. The results are plotted as the fraction of single-stranded DNA as a function of the log equivalent Cot. With one of the available computer programs,[7,14,15] the best-fit curves for the various kinetic components can be generated, and other kinetic information can be ascertained.

## I. DETERMINING THE EXTENT OF DUPLEX DNA IN A RENATURED DNA SAMPLE

Thermal denaturation of renatured DNA of a particular Cot value can be used to calculate the fraction of duplex DNA in the sample. The hyperchromicity of a DNA sample is proportional to the fraction of the bases that are hydrogen-bonded. Thus, the ratio of the hyperchromicity of a renatured DNA sample ($Hr$) to that of a native DNA sample ($Hn$) with DNA samples of the same fragment size gives a measure of the fraction of nucleotides of the renatured DNA that are base-paired. This ratio, however, yields an approximation of the extent of matched bases (i.e., duplex DNA). The computation is amended by taking into consideration both variation due to size[17] and the contribution of mismatched base pairs.[10] For example, short fragments (<650 nucleotides) cause an underestimation of $Tm$. The extent of this reduction is equivalent to 650/L, where L is the length of a single strand equivalent in nucleotides of the DNA sample. Mismatched base pairs in a DNA sample reduce the value of the hyperchromicity. A difference in the $Tm$ values ($\Delta Tm$) between native and renatured DNA of 1 to 1.5°C is, as a rule of thumb, considered equal to approximately 1% mismatch. To determine the fraction of a renatured DNA sample that consists of duplex DNA, the $Tm$ values for both native and renatured DNA must be determined, each $Tm$ must be corrected for length effect, if any, and the extent of mismatch must be ascertained. The form of the final equation is

$$D = \frac{Hr - 0.10}{Hn - (Hn \times M) - 0.10} \tag{5}$$

where $D$ is the fraction of the sample that comprises duplex DNA, $Hr$ and $Hn$ are the hyperchromicity values for the renatured and native DNA samples, respectively, $M$ represents the fraction of mismatch, and the value 0.10 is an arbitrary correction factor for the contribution of single-stranded DNA to the hyperchromic shift when a sample is heated from 60 to 100°C.[10]

# III. DETERMINATION OF THE COPY NUMBER OF CODING AND NONCODING GENOMIC SEQUENCES

A Cot curve of genomic DNA gives an estimate of the overall repeat values for the major repetitive DNA classes. The moderately and highly repetitive DNA fractions contain multigene families that consist of both coding and noncoding sequences. Within these families, sequences have either remained homogeneous or undergone divergence. In addition, the arrangement of the members of a multigene family can be either dispersed or tandemly arrayed. DNA blotting strategies can be invoked to determine (1) if a query sequence is a member of a multigene family, (2) how the family members are arranged in the genome, and (3) the total number of members of a multigene family.

## A. DETERMINATION OF MULTIGENE FAMILIES

### 1. Strategy

1. Use a cDNA, exon-containing, or anonymous DNA clone that has been characterized for restriction endonuclease sites as a hybridization probe. To check for nonspecific hybridization, also use the vector without the insert as a probe.
2. Treat the genomic DNA with a restriction endonuclease(s) that does not cut the cloned (insert) DNA of the probe.
3. After electrophoretic separation and blotting of the treated genomic DNA samples, hybridize under standard conditions and wash the blot at normal stringency (60°C, 0.1 × SSC where 1X SSC = 0.15 *M* NaCl, 0.015 *M* trisodium citrate, pH 6.5).
4. Examine the autoradiograph.

### 2. Analysis

If one band appears, then the target DNA sequence either exists once in the genome, or it is a highly homogeneous multigene family.

If a discrete, countable set of bands appears with either a cDNA or anonymous open reading frame (ORF) DNA probe, then a multiple gene family may be present. However, the presence of multiple bands may not be indicative of a multigene family because restriction endonuclease sites may occur within introns. A less than definitive test of a multigene family hypothesis can be carried out by stripping the original blot of the hybridized probe and carrying out a second hybridization at high stringency, e.g., 15°C below Tm. If the multiple bands are due to a diverged gene family, the relative intensity of the bands after the second hybridization should be decreased relative to the results of the first hybridization.

If a smear (i.e., no discrete bands) appears in the autoradiograph, then the target DNA is a highly repetitive, dispersed DNA family.

If a ladder of bands with sizes equivalent to multiples of the smallest fragment is evident, then the target DNA consists of a tandemly organized set of repeated elements, with the monomeric length probably being equivalent to the smallest fragment size.

Finally, if the autoradiography shows a multiple set of bands superimposed on a smear, then the probe has detected a tandemly repeated DNA family that has dispersed elements.

## B. DETERMINATION OF GENE COPY NUMBER

An estimate of the copy number of a particular genomic DNA sequence can be obtained by measuring the extent of hybridization signals of genomic bands and comparing the signal level with those produced by bands that contain known gene equivalents of the query sequence.

The rationale for calculating the quantity of DNA that corresponds to a specific copy number is the following: if $G$ is the genome size in bp and g the size of the gene (target) in bp, then $g/G$ is the proportion of the genome that carries one copy of the gene per haploid genome equivalent. Thus, in a 1 μg genomic DNA sample, the quantity of the target gene will be equal to $g/G$ μg. If $i$ is the size of the insert (query sequence) in bp in the cloning vector and $v$ is the size of the vector plus insert in bp, then solve for $X$ in the relationship

$$\frac{i}{v}X = \frac{g}{G}\mu\text{g} \tag{6}$$

The value for $X$ will give the quantity, in micrograms, of a "vector with insert" DNA sample that is equivalent to one copy per genome of the target sequence. Whenever possible, it is best to use purified insert DNA, in which case the v term in the above equation can be omitted. For quantification, run a set of samples that are equivalent to, for example, 1, 2, 5, 10, 20, and 50 copies of the target gene.

### 1. Strategy

1. Load into separate wells of an agarose gel restriction endonuclease-treated genomic DNA (1 μg) and a set of copy number standards.
2. Prepare a Southern blot and hybridize under standard conditions with purified, radiolabeled cDNA, monomeric sequence of a repetitive DNA family, or linearized plasmid with insert DNA.
3. Scan the autoradiograph with an integrating densitometer or equivalent instrument.

### 2. Analysis

On the basis of the extent of hybridization of the known copy number equivalents, determine the total copy number within the genomic DNA sample. With a family of tandem repeats, on the basis of the copy number standards, the sum of the extent of hybridization of all the bands will give an estimate of the total number of gene copies.

## C. DETERMINATION OF THE COPY NUMBER OF DISPERSED REPETITIVE DNA SEQUENCES

### 1. Strategy

1. After applying DNA to the membrane with a "slot blot" apparatus, DNA hybridization can be used to determine the copy number of dispersed, repetitive DNA sequences.[18]

2. With the membrane in place in the "slot blot" manifold and prepared according to the manufacturer's recommendations, add an aliquot of known quantity of neutral, denatured DNA to a slot. If the DNA samples were denatured with 1/40 volume of 4 *M* NaOH and then heated to 95°C, add 1/4 volume of 5 *M* ammonium acetate. The DNA samples are genomic DNA and vector with insert (query sequence).
3. Prepare a nitrocellulose membrane for hybridization by soaking it in 5X SSC (1X SSC: 0.15 *M* NaCl, 0.015 *M* trisodium citrate, pH 6.5), air drying it, and then baking it at 80°C under vacuum for 2 h. With other types of membranes, follow the manufacturer's instructions.
4. Prehybridize a nitrocellulose membrane in the standard manner. For example, immerse the membrane for 30 min at 65°C in 1% sarcosyl, 4 × SSC, 0.025 *M* phosphate buffer, 1 × Denhardt's solution, 50 µg/ml denatured salmon sperm DNA.
5. Hybridize under standard conditions with radiolabeled probe. The probe can be purified monomeric query sequence or vector carrying the query sequence in single or multiple copies.
6. Prepare an autoradiograph and scan the bands with an integrating densitometer or equivalent instrument.

## 2. Analysis

The equation[18]

$$T\% = \frac{Hg}{Hvi} \times \frac{Qvi}{Qg} \times \frac{Lv}{Li} \times 100 \tag{7}$$

gives the percent of DNA that is complementary to the probe (T%), where *Hg* and *Hvi* are the extents of hybridization, determined from the densitometric analyses of the autoradiographs for genomic and "vector plus insert" DNA samples respectively, *Qg* and *Qvi* are the quantities of genomic and "vector plus insert" DNA samples that were applied to the membrane per slot, respectively, and *Lv* and *Li* are the lengths in kb of the "vector minus insert" and insert (query) sequences, respectively. Then, the copy number can be determined from the equation

$$\text{copy number} = \frac{T\% \times \text{genome size [bp]}}{\text{monomer size [bp]}} \tag{8}$$

## REFERENCES


1. **Sorenson, J. C.,** The structure and expression of nuclear genes in higher plants, *Adv. Genet.*, 22, 109, 1984.
2. **Wimpee, C. F. and Rawson, J. R. Y.,** Characterization of the nuclear genome of pearl millet, *Biochim. Biophys. Acta*, 562, 192, 1979.
3. **Cullis, C. A.,** DNA sequence organization in the flax genome, *Biochim. Biophys. Acta*, 652, 1, 1981.
4. **SivaRaman, L., Gupta, V. S., and Ranjekar, P. K.,** DNA sequence organization in the genomes of three related millet plant species, *Plant Mol. Biol.*, 6, 375, 1986.
5. **Angerer, R. C., Davidson, E. H., and Britten, R. J.,** Single copy DNA and structural gene sequence relationships among four sea urchin species, *Chromosoma*, 56, 213, 1976.
6. **Britten, R. J. and Kohne, D. E.,** Repeated sequences in DNA, *Science*, 161, 529, 1968.
7. **Britten, R. J., Graham, D. E., and Neufeld, B. R.,** Analysis of repeating DNA sequences by reassociation, *Meth. Enzymol.*, 29, 363, 1974.

8. **Britten, R. J. and Davidson, E. H.,** Studies on nucleic acid reassociation kinetics: empirical equations describing DNA reassociation, *Proc. Natl. Acad. Sci. U.S.A.*, 73, 415, 1976.
9. **Davidson, E. H., Hough, B. R., Amenson, C. S., and Britten, R. J.,** General interspersion of repetitive with nonrepetitive sequence elements in the DNA *Xenopus*, *J. Mol. Biol.*, 77, 1, 1973.
10. **Goldberg, R. B., Crain, W. R., Ruderman, J. V., Moore, G. P., Barnett, T. R., Higgins, R. C., Gelfand, R. A., Galau, G. A., Britten, R. J., and Davidson, E. H.,** DNA sequence organization in the genomes of five marine invertebrates, *Chromosoma*, 51, 225, 1975.
11. **Graham, D. E., Neufeld, B. R., Davidson, E. H., and Britten, R. J.,** Interspersion of repetitive and non-repetitive DNA sequences in the sea urchin genome, *Cell*, 1, 127, 1974.
12. **Wetmur, J. G. and Davidson, N.,** Kinetics of renaturation of DNA, *J. Mol. Biol.*, 31, 349, 1968.
13. **Gupta, U. S., Gadre, S. R., and Ranjekar, P. K.,** Novel DNA sequence organization in rice genome, *Biochim. Biophys. Acta*, 656, 147, 1981.
14. **Pearson, W. R., Davidson, E. H., and Britten, R. J.,** A program for least squares analysis of reassociation and hybridization data, *Nucl. Acids Res.*, 4, 1727, 1977.
15. **Kells, D. I. C. and Straus, N. A.,** A rapid computer analysis for multicomponent DNA reassociation-kinetics, *Anal. Biochem.*, 80, 344, 1977.
16. **Marmur, J. and Doty, P.,** Determination of the base composition of deoxyribonucleic acid from its thermal denaturation temperature, *J. Mol. Biol.*, 5, 109, 1962.
17. **Britten, R. J., Graham, D. E., Eden, F. G., Painchaud, D. M., and Davidson, E. H.,** Evolutionary divergence and length of repetitive sequences in sea urchin DNA, *J. Mol. Evol.*, 9, 111, 1976.
18. **Cullis, C. A., Rivin, C. J., and Walbot, V.,** A rapid procedure for the determination of the copy number of repetitive sequences in eukaryotic genomes, *Plant Mol. Biol. Rep.*, 2, 24, 1984.

# 4

# Isolation and Characterization of Plant DNAs

*Brian H. Taylor, James R. Manhart, and Richard M. Amasino*

## I. INTRODUCTION

The ability to extract and analyze DNA from plant tissues is an essential aspect of plant molecular biology and may be divided into two stages. In the first stage, plant cells are disrupted to release the DNA. This is a critical stage, in that failure to disrupt all of the cells in the tissue or to protect the DNA during disruption may greatly reduce the final yield of DNA. It is at this point that a decision must be made whether to isolate total DNA or to obtain a specific DNA fraction from the nucleus, chloroplast, or mitochondria. Total DNA is acceptable for analysis of nuclear genes without closely related organellar genes, or for organellar genes without closely related nuclear homologues. Fractionation of the nuclear and organellar genomes may be useful for assigning a gene to a particular genome, for avoiding cross-hybridization of related genes, and for generating organelle-specific libraries for cloning.

In the second stage, the DNA is extracted away from the other cellular constituents such as proteins, carbohydrates, membranes, and cell walls. It is important at this stage to ensure that significant amounts of DNA are not trapped in the cell debris, and that the DNA is completely dissociated from proteins and other contaminants that might copurify and interfere with subsequent analyses.

A number of different methods for purifying plant DNA have been described.[1-7] It is beyond the scope of this chapter to consider each method and compare relative advantages and disadvantages; however, some general considerations that might be applied when evaluating a particular method are:

1. Will it work with my tissue? Some tissues have high levels of nuclease or polyphenol oxidase. DNA isolation from these tissues may require additional steps and precautions not necessary with more tractable tissues.
2. Is it relatively simple to use? A method that is reliable and easy to use is preferable to one that has a number of time-consuming or complicated steps, particularly when a large number of samples are involved.
3. Is the DNA obtained pure enough for the intended purpose? This refers to both the relative enrichment for organellar or nuclear DNA, and to the levels of contaminating protein and carbohydrate present. The presence of these contaminants can interfere with restriction endonucleases and other enzymes used in DNA analysis and cloning.

The methods described for DNA isolation in this chapter are based on the CTAB (cetyltrimethylammonium bromide) extraction procedure first described by Murray and Thompson[2] and later adapted for fresh tissue.[8-10] The CTAB precipitation step, when

0-8493-5164-2/93/$0.00+$.50
© 1993 by CRC Press, Inc.

combined with chloroform extraction, is remarkably effective for eliminating complex carbohydrates that can be problematic with other isolation protocols. Another major advantage is that, for most plants, high quality DNA and RNA can be obtained simultaneously from the same tissue sample. In this chapter, we present CTAB-based protocols for the isolation of total plant DNA (and RNA) from both large (>1 g) and miniprep sized (<0.5 g) samples of plant tissue and for chloroplast DNA isolation. An excellent miniprep procedure based on the same CTAB protocol may also be found in Rogers and Bendich.[11]

DNA isolated from plant cells typically will be used for hybridization analysis or cloning. Our protocols for restriction endonuclease digestion, gel electrophoresis, blotting, and the hybridization of labeled probes are included in the latter part of this chapter.

## II. CTAB ISOLATION PROTOCOL FOR TOTAL DNA

In this method, the plant cells are frozen in liquid nitrogen and ground with a mortar and pestle or coffee mill. The ground tissue is suspended in a buffer containing the detergent CTAB at 1% (w/v), 0.7 *M* NaCl, and 1% β-mercaptoethanol, and heated to 65°C. These conditions cause many proteins to denature and help dissociate contaminants from the DNA. β-mercaptoethanol also plays an essential role in inhibiting polyphenol oxidation, a process that causes browning in many plant tissues. After the initial incubation, the denatured protein and most of the carbohydrates are removed by two chloroform/isoamyl alcohol extractions. The aqueous phase with the DNA is then diluted two-fold with buffer containing CTAB but no salt. Under reduced salt conditions, the CTAB and nucleic acids form an insoluble complex and precipitate out of solution. This provides a two-step purification of the DNA away from carbohydrate contaminants, since most of the carbohydrates that remain in solution after the high salt chloroform extractions do not precipitate with the CTAB/nucleic acid complexes. For larger quantities of material, the final stage of purification is a cesium chloride step gradient that separates the DNA and RNA and removes any remaining contaminants.

1. The first step is to grind the tissue to release the DNA and RNA. Grind moderate amounts of tissue (1 to 10 g) with a mortar and pestle after quick freezing it in a beaker or in the mortar with liquid nitrogen. Large amounts of plant tissue (>10 g) can also be ground with a mortar and pestle; however, it is often easier to grind the frozen tissue with a coffee mill or blender. Small amounts of tissue should be ground fresh with an equal volume of 2X extraction buffer in an Eppendorf tube using a hand-held or motor-driven pestle that matches the shape of the tube (available from Kontes, Vineland, NJ).
   *Notes:* Excess liquid on the surface of the tissue should be drained or blotted away before freezing; otherwise small pieces of ice may form that will interfere with grinding. Stems, calli, and other thick tissue samples will be easier to grind if they are broken into smaller pieces before they freeze completely. Once the tissue sample has been frozen, it is imperative that it remain frozen until the extraction buffer is added (Step 2). Ensuring that the grinder or mortar and pestle are completely chilled will help extend the amount of time available before the buffer must be added. A mortar and pestle may also be placed in dry ice to provide additional time for grinding. Adding liquid nitrogen to the sample after grinding is not recommended since it tends to scatter the tissue. The degree of cell breakage after grinding can be determined by examining a sample of the ground tissue under a microscope.
2. Add an equal amount (w/v) of 65°C 2X extraction buffer to the ground tissue powder, mix well and heat to 65°C. Transfer miniprep samples directly to 65°C.
   *Notes:* Ground tissue powder in a mortar, coffee mill, or blender may be transferred directly into a beaker or tube containing 2X extraction buffer if care is taken to ensure

the powder stays frozen until it is in the buffer. Buffer added to frozen tissue in a cold mortar or blender cup will probably freeze, so it is important that it is mixed in quickly. The container and sample may then be placed in a 65°C water bath for thawing. The sample should be transferred to a centrifuge tube (28 or 50 ml Oak Ridge) as soon as it is thawed and the incubation continued until the temperature of the sample reaches 65°C. The length of incubation at 65°C may be adjusted as needed; however, extended incubation at 65°C may affect RNA yield and quality.

3. Add an equal volume of chloroform/isoamyl alcohol (24:1), and shake the samples gently for several minutes (see cautionary note below). When the samples are well mixed, centrifuge them (preferably in a swinging bucket rotor) at 10 to 16,000 × g for 20 min at room temperature. Centrifuge minipreps for 20 min at top speed in a microfuge.

   *Notes:* **CAUTION: Chloroform boils at 61°C. Allow the sample tubes to cool to 50°C or below before adding the chloroform:isoamyl alcohol and periodically vent the tubes while shaking.** Do not reduce the temperature of the samples below 15°C, since this may cause premature precipitation of the CTAB/nucleic acid complex. Be careful to avoid the interface material. If the interface is not tight or there are floating tissue particles, re-extract the supernatants from the first centrifugation with a second equal volume of chloroform:isoamyl alcohol.

4. Transfer the supernatants to clean tubes or bottles. Add 1 to 1.5 volumes of precipitation buffer. For most samples, a cloudy or stringy precipitate will be observed. The precipitate may be recovered by centrifugation immediately, or after extended room temperature incubation (see below) to enhance recovery.

   *Notes:* The amount of precipitate obtained is proportional to the amount of nucleic acid in the original tissue. Preparations from young leaves, for example, will have a much greater precipitate than preparations from calli with large, highly vacuolated cells. The nucleic acids are stable in precipitation buffer and can be left overnight at room temperature if necessary (the recovery of a very dilute precipitate is significantly enhanced by overnight precipitation).

5. Recover the precipitate by centrifugation at 4000 × g for 5 min. Remove the supernatant with a Pasteur pipette and discard.

   *Notes:* Care should be taken not to centrifuge the samples too hard, since tightly packed pellets are difficult to resuspend. Most precipitates will form good pellets at 4000 × g, but some fine precipitates may require a higher speed for complete recovery. For very small pellets, removal of the supernatant is facilitated by using glass tubes, rather than plastic, since the pellets adhere better to the side of the tube.

6a. Nongradient option: Drain the pellets well, and resuspend them in 1.0 *M* $NH_4OAc$ (200 µl (minipreps) or 0.5 ml/g of starting tissue). Adjust the $NH_4OAc$ concentration to 2.5 *M*, mix well, and precipitate the nucleic acid with 2 volumes of isopropanol. Incubate at room temperature for 10 min. For moderate to large amounts of DNA, spool the DNA out as described in Step 8 and proceed with the protocol from there. To recover DNA from miniprep samples, centrifuge them at 13,000 rpm for 5 min. Wash the pellets with 70% ethanol, dry them in a spin-vac or desiccator, and resuspend them in 20 µl TE (more or less, depending on the amount and type of tissue sampled). Confirm that the sample is completely suspended by slowly passing it through a disposable pipette tip.

   *Notes:* DNA isolated by the nongradient method should digest readily with restriction endonucleases, but will contain RNA. DNA concentrations should be determined by fluorometric quantitation using a DNA-specific dye (e.g., Hoechst 22358). For Southern blots, add 1 µl of 1 mg/ml RNAse A to each sample when performing the restriction digests, and the RNA will run below the DNA on the gel and not be a problem.

6b. Gradient Option: Resuspend the pellets in 2.5 ml of 1 *M* CsCl solution. Carefully layer the resuspended sample onto a 2 ml cushion of 5.7 M CsCl in a Beckman SW50.1 or comparable tube (*Note:* CsCl may precipitate in longer tubes, e.g., SW41 or SW27). Mark the position of the 1.0 *M*/5.7 *M* CsCl interface. Add sufficient 1 *M* CsCl to position the meniscus in each tube 2 mm from the top. Centrifuge at 36,000 rpm for 14 to 16 h at 20°C.

*Notes:* Resuspended CTAB pellets may be stored frozen at –20°C. Pellets may also be stored prior to resuspension but are more difficult to resuspend afterward. Ensure that the pellet material is completely resuspended before layering it onto the 5.7 *M* CsCl cushion. The nucleic acid is approximately 10:1 RNA to DNA, so undissolved bits of pellet will go to the bottom of the tube with the rest of the RNA. Extended centrifugation times (>17.5 h) will cause the DNA to pellet with the RNA.

7. Remove the gradient from the tube in three layers (top to midpoint between top and interface mark, midpoint to mark, and mark to bottom) changing pipettes or pipette tips between layers. Protein (RNAse) will band at the top of the gradient, so care should be taken not to allow material from the meniscus region to proceed down the tube. The DNA will band about 1/3 to 1/2 of the distance from the interface mark to the bottom of the tube. Transfer the portion of the gradient between the interface mark and the bottom to a separate tube. Care should be taken not to disturb the RNA pellet at the bottom of the tube when removing the CsCl supernatant. Invert the tube with the RNA pellet on a Kimwipe for several minutes. Wipe away any liquid that collects before proceeding.
8. Precipitate the DNA from the 5.7 *M* CsCl solution by adding 3 volumes of cold (–20°C) ethanol. Remove the DNA from the CsCl solution by spooling it onto the end of a long Pasteur pipette, the end of which as been flame sealed. Wash the DNA by dipping it into cold 70% ethanol, and allow it to dry after inverting the pipette in a test tube rack. Resuspend the DNA by inserting the pipette into an Eppendorf tube containing 400 µl TE (for Step 9) or an appropriate final volume.
9. (Optional) Extract the resuspended DNA with 400 µl of (1:1) phenol/chloroform (equilibrated with TE), and precipitate the DNA with 200 µl of 2.5 *M* $NH_4OAc$ and 1 ml isopropanol. Recover the precipitated DNA by spooling, and wash the pellet with cold 70% ethanol. Allow the pellet to air dry completely, and resuspend the DNA in an appropriate volume of TE (generally 50 to 500 µl).
10. The RNA pellet may be resuspended in 10 m*M* Tris-HCl, pH 7.6, 1 m*M* EDTA, and precipitated with 0.5 volume 7.5 *M* ammonium acetate and 2 volumes isopropanol. This RNA, when redissolved in water or TE, pH 7.6, is suitable for fractionation on oligo-dT cellulose into poly $A^+$ and poly $A^-$ RNA. Alternatively, the RNA pellet may be stored frozen in the gradient tube until needed.

SOLUTIONS:

2X Extraction Buffer
2% CTAB (Sigma H-5882)* (w/v)
100 m*M* Tris-HCl, pH 8.0
20 m*M* EDTA
1.4 *M* NaCl
2% β-mercaptoethanol+

Precipitation Buffer
1% CTAB (w/v)
50 m*M* Tris-HCl, pH 8.0
10 m*M* EDTA
1% β-mercaptoethanol+

* Cetyltrimethylammonium bromide, also called hexadecyltrimethylammonium bromide.
+ Add just prior to use.

1 *M* CsCl Solution
50 m*M* Tris-HCl, pH 8.0
5 m*M* EDTA
1.0 *M* CsCl

5.7 *M* CsCl Solution
50 m*M* Tris-HCl, pH 8.0
5 m*M* EDTA
5.7 *M* CsCl

For RNA isolation, make up CsCl solutions and TE with water treated with 0.05% diethylpyrocarbonate for 1 h. (**Note: DEPC is mutagenic. Treat the water in a fume hood and autoclave before use.**) Filter CsCl solutions through a 0.45 μm filter to remove any particulate impurities.

TE Buffer
10 m*M* Tris-HCl, pH 8.0
1 m*M* EDTA

All reagents and buffers should be made from stocks that are used for genomic DNA isolation only.

## III. CHLOROPLAST DNA ISOLATION

Three procedures for isolating chloroplast DNA and suggestions for modification of these procedures are described below. Procedures I and II are used to purify chloroplast DNA from plants that have large numbers of small chloroplasts. It is recommended that Procedure I be attempted first since it is the quickest and easiest. Procedure III separates chloroplast, mitochondrial, and nuclear DNAs from a total DNA preparation on the basis of their densities. This procedure can be used to purify chloroplast DNA from plants in which it is not possible to isolate large numbers of intact chloroplasts. It is especially useful for isolating chloroplast DNA from algae which have large and complex chloroplasts. Procedure III will not work with vascular land plants, presumably because methylation of the nuclear DNA lowers the density of the nuclear DNA to that of the chloroplast DNA.

### A. PROCEDURE I: HIGH SALT EXTRACTION PROTOCOL

This procedure is essentially that of Milligan[10] with modifications. Please refer to the original protocol for a more extensive discussion.

1. Grind a small amount of leaf tissue with the high-salt extraction buffer in a mortar and pestle. Examine using a microscope at 400X. If large numbers of starch granules are present, place plants in the dark for 1 to 4 days, or grow plants under low light conditions. If difficulties are encountered, Procedures I and II can be modified by adding 0.1 to 1% polyvinylpyrrolidone to remove tannins and 10 to 25% polyethylene glycol to maintain chloroplast structural integrity.[10,12]
2. Rinse leaves with distilled water, remove midribs, and cut leaves into pieces 1 to 3 cm in size. The size of the pieces depends on the toughness of the leaves; tougher leaves should be cut into smaller pieces.
3. Place 50 to 100 g of cut leaves in 400 ml of ice-cold isolation buffer and homogenize in a prechilled 1 liter Waring blender for three to five 5 sec bursts at high speed.
4. Filter through two layers of cheesecloth, then six layers of cheesecloth. Do not force the filtrate through by squeezing.
5. Centrifuge the filtrate at 1500 × g for 5 to 10 min at 4°C, and pour off the supernatant.
6. Resuspend the pellet in 100 ml of ice-cold isolation buffer using an artist's brush and swirling.

7. Centrifuge the filtrate at 1500 × g for 5 to 10 min at 4°C and pour off the supernatant.
8. Add a volume of 2X CTAB extraction buffer (preheated to 60°C) that is equal to the pellet volume. Loosen the pellet, and break up any clumps with an artist's brush. Incubate at 60°C for 15 to 60 min.
9. Extract the mixture two times with equal volumes of chloroform/isoamyl alcohol (24:1), centrifuging each time at 1500 × g for 5 min to separate the phases.
10. Precipitate the DNA by mixing in a 2/3 volume of ice-cold isopropanol. Place the solution at –20°C for at least 30 min.
11. Centrifuge the solution at 1500 × g for 10 min. Discard the supernatant and let the DNA air-dry for 15 min.
12. Resuspend the DNA in TE and store at 4°C short-term and –20°C long-term. The DNA is normally clean enough to cut with restriction endonucleases or amplify at this point. If further purification is desired, run the DNA on a CsCl-ethidium bromide gradient as described in Procedure III.

High Salt Isolation Buffer
1.25 *M* NaCl
50 m*M* Tris-HCl pH 8.0
5 m*M* EDTA
0.1% BSA (w/v)
0.1% β-mercaptoethanol (v/v)

## B. PROCEDURE II: SUCROSE STEP GRADIENT PROTOCOL

This procedure is taken from Palmer[12] with modifications. Please refer to the original protocol for a detailed discussion.

1-5. Perform Steps 1-5 of Procedure I using sorbitol extraction buffer in place of high salt isolation buffer.
6. Resuspend the pellet in 8 ml of ice-cold sorbitol wash buffer with an artist's brush and swirling. The chloroplasts should not be clumped.
7. Load the resuspended pellet onto a step gradient consisting of 52% sucrose (w/v), 50 m*M* Tris, pH 8.0, 25 m*M* EDTA, overlayered with 30% sucrose (w/v), 50 m*M* Tris, pH 8.0, 25 m*M* EDTA. The interface between the sucrose solutions should be somewhat diffuse since a very sharp interface will trap nuclei in the interface with the chloroplasts. The ratio of the three solutions in the gradient should be 1 part resuspended pellet : 1 part 30% sucrose solution : 3 parts 52% sucrose solution.
8. Centrifuge the step gradients at 85,000 × g for 30 to 60 min at 4°C in a swinging bucket rotor.
9. Remove the chloroplast band from the 30 to 52% interface using a wide bore pipette, dilute with 3 to 10 volumes wash buffer, and centrifuge at 1500 × g for 15 min at 4°C.
10. Pour off supernatant and go to Step 8 of Procedure I.

SOLUTIONS:

Sorbitol Extraction Buffer
0.35 *M* sorbitol
50 m*M* Tris-HCl, pH 8.0
25 m*M* EDTA
0.1% BSA (w/v)
0.1% β-mercaptoethanol (v/v)

Sorbitol Wash Buffer
0.35 *M* sorbitol
50 m*M* Tris-HCl, pH 8.0
25 m*M* EDTA

## C. PROCEDURE III. CESIUM CHLORIDE DENSITY GRADIENT PROTOCOL

This isolation procedure is designed to work with plants in which it is not possible to obtain intact chloroplasts, as is often the case for algae.[13-15] It takes advantage of the fact that most algal organelle DNAs are less dense than their corresponding nuclear DNAs and will separate on CsCl gradients, especially CsCl-bisbenzimide gradients. It should be noted that if the plant produces large amounts of complex carbohydrates, the isopropanol will precipitate these carbohydrates along with the DNA, and the DNA will not be recoverable. If this is the case, follow Steps 1-5 of the Large Scale Preparation to isolate and precipitate the DNA, and then return to Step 5 of this procedure.

1. Grind fresh material with a mortar and pestle using 2X CTAB buffer (7.5 ml buffer/ g fresh tissue). The grinding is easier if the mortar and pestle are preheated to 60°C.
2. Incubate the ground material at 60°C for 45 min to 1 h.
3. Extract once with 1/2 volume chloroform-isoamyl alcohol (25:1), and centrifuge at 1500 × g to concentrate the layers. This step removes the chlorophyll and, more importantly, many of the proteins from the aqueous layer.
4. Recover the aqueous (upper) phase, add 2/3 volume of cold isopropanol, and mix gently. This step precipitates the DNA. If there is a very large amount of precipitate at this step, it is probably mostly complex carbohydrates. Often, the only observable change is a cloudy appearance of the solution. Spin the tube in a clinical centrifuge, initially at a low speed. If nothing comes down at low speeds, try higher speeds and longer times, but be aware that the faster and longer spins will bring down more "trash", which will make it more difficult to resuspend the DNA.
5. Allow the pellet to air dry briefly and resuspend it in 2 ml of 1 *M* CsCl (0.17 g/ml).
6. Bring the CsCl concentration up to 0.76 g/ml and add 0.2 mg/ml ethidium bromide (EtBr). Place all the DNA in one tube unless there are more than 4 g of starting material per ml of CsCl solution (i.e., a 5 ml tube will handle 20 g of starting material). Centrifuge the gradients in a vertical rotor at 275,000 × g for 6 to 8 h at 20°C. Run times will have to be longer (24 h) for fixed angle rotors. The chloroplast DNA will be the heaviest upper band, and the nuclear DNA will be in one or two heavy lower bands. Mitochondrial DNA tends to run close to the chloroplast DNA, but it should be at a much lower concentration, if it is visible at all. You may want to do at least one more run to improve your separation. If the bands are well separated, go to Step 12. If there is only one band or if the bands are too close together to separate, go to Step 7.
   *Note:* **Ethidium bromide and bisbenzimide are highly mutagenic and must be handled with care.** Wear gloves and a mask when making up solutions, and dispose of waste solutions appropriately.
7. Recover the chloroplast DNA band (and other bands, until a positive identification is made) with a needle and syringe in as little volume as possible, and determine the band volume. Remove the EtBr by extracting the solution several times with isopropanol saturated with water and NaCl.
8. Bring the CsCl concentration up to 0.87 g/ml, assuming that the CsCl removed from the gradient is 0.76 g/ml.
9. Add 4 µl of 10 mg/ml bisbenzimide (Sigma B-2883) for each ml of CsCl solution. The amount of bisbenzimide added may be increased, but if too much is added, the background will be so bright that the DNA bands will not be discernible. There is usually a small amount of yellow or light green precipitate formed at this point.
10. Centrifuge the sample at 275,000 × g for 6 to 8 h at 20°C. There should be several bands present, the upper of which is usually the chloroplast DNA. There may be a faint

band near the cpDNA band, which is likely to be mitochondrial DNA. The lower and heavier band(s) will be nuclear DNA.

11. Remove the bands from the gradient with as little cross-contamination as possible and in as small a volume as possible (less than 0.5 ml). If the bands have been mixed together to any degree, it is advisable to make another run. However, do not add any more bisbenzimide to the new tubes.
12. Remove the bisbenzimide (or EtBr) from the DNA samples with several extractions of isopropanol saturated with water and NaCl.
13. Dialyze the DNA against at least three changes of TE (2 l) over a period of 1 to 2 days.
14. Store the DNA at 4°C short term, –20°C long term.

## IV. ANALYSIS OF DNA BY FILTER HYBRIDIZATION

The detection and characterization of endogenous and introduced genes is an important component of many experiments in plant molecular biology and biotechnology. Typically, this involves digesting purified DNA with a restriction endonuclease, separating the fragments by size on an agarose gel, and transferring the DNA from the gel to a filter for hybridization to a labeled DNA or RNA probe.[16] DNA digestion and gel electrophoresis are fairly straightforward and best done following the supplier's recommendations and the general protocols described in Sambrook et al.[17] We note in passing that the digestion of recalcitrant DNA is often dramatically improved by adding spermidine to 4 m*M* in the digestion reaction.[18] Unfortunately, this trick may not be used with low salt buffers (<50 m*M*) or samples that need to be refrigerated, since the spermidine may precipitate the DNA under these conditions. In this section we will focus our discussion on membrane selection, methods for transferring the DNA from the gel to the membrane, and hybridization and washing protocols.

### A. DNA TRANSFER

Our protocol for capillary transfer, a technique originally developed by Southern,[16] is presented below. One advantage of capillary blotting is that it does not require specialized equipment. Other methods for transfer include vacuum, positive pressure, or electrophoretic blotting.[17] These devices can save time and should be considered if money is no object. The best type and brand of membrane filter to use is the subject of much debate. We have experienced a great deal of lot to lot variability with certain brands of filters; thus, the optimal brand of filter may change with time, and we will not make any specific recommendations. It is clear, however, that nylon filters, particularly positively charge-modified filters, have a much greater nucleic acid binding and retention capacity than nitrocellulose and therefore exhibit much stronger hybridization signals. Moreover, nylon filters are more durable than nitrocellulose and can be subjected to multiple cycles of hybridization and washing. Other membrane materials that are claimed to have the same useful properties as nylon membranes are polyvinylidene difluoride and polysulfone. We recently tested one lot of polysulfone membranes (Biotrace HP, Gelman) in DNA blot hybridizations and found that this membrane produced a stronger hybridization signal with less background than the nylon membranes we tested.

There are many different recommended blotting solutions, and the manufacturer of a particular membrane does not always, in our experience, recommend the optimal protocol. We have found that the low-salt blotting protocol described below offers the simplicity of direct blotting after denaturation and gives the best results with nylon (and polysulfone) membranes from most sources tested; however, a direct comparison between this protocol and the manufacturer's protocol in your laboratory is advised. Some membranes (e.g., Magnagraph from MSI) appear sensitive to the base used for denaturation. An optional gel neutralization step is included in the protocol for these membranes. Our low-salt blotting protocol is modified from the method of Reed and Mann,[19] and a discussion of the possible basis for its effectiveness is presented in their paper.

The reader is referred to Sambrook et al.[17] for protocols describing gel preparation and electrophoresis. Typically, a gel will be stained with ethidium bromide and photographed under UV light prior to blotting.

1. Incubate the gel in several volumes of 0.25 *M* HCl for 10 min. (This acid treatment results in nicking of the DNA by depurination and facilitates transfer of large DNA fragments. This step may be omitted if efficient transfer of large fragments is not required.)
2a. Incubate the gel in several volumes of 0.4 *N* NaOH for 30 min.
2b. (Optional for base-sensitive membranes.) After the 0.4 *N* NaOH wash, incubate the gel in several volumes of 0.5 *N* Tris-HCl, pH 7.0, 1 m*M* EDTA for 30 min.
3. Transfer the DNA from the gel to a prewet membrane filter by capillary action using 10 m*M* Tris-HCl, pH 7.0, 1 m*M* EDTA as the blotting solution. We have found that a 4 to 5 cm thick sponge in a baking dish or plastic tray works very well as a blotting platform. (The sponge should be boiled in 50 m*M* EDTA solution and rinsed well before initial use.) Place two sheets of filter paper (Whatman 3MM) soaked in blotting solution on the sponge. Trim excess gel extending from the edges and around the wells, and place the gel upside down on the filter paper. (This prevents blotting artifacts caused by variability in the upper gel surface.) Position the membrane filter on the gel and cover it with two additional sheets of wet filter paper. Be sure to exclude all air bubbles and excess buffer between the layers. Place a weighted stack of paper towels cut to size on top as an absorbent. The blotter may be reused indefinitely by refilling the tray with fresh blotting solution. Keep covered between use.
4. After 6 h or longer of blotting (depending upon the gel concentration and thickness), remove the membrane, rinse it in TE (10 m*M* Tris-HCl, 1 m*M* EDTA, pH 7.0), and blot it dry with filter paper.
5. Bake the membrane at 65°C for 30 min. This last step "irreversibly" fixes the DNA onto the membrane. Baking in a vacuum oven is necessary only for nitrocellulose membranes. Another method to fix the DNA to the membrane is exposure to ultraviolet light. Ultraviolet fixation has been reported to result in enhanced hybridization signals;[20,21] however, the dose of irradiation must be carefully calibrated, or reduced hybridization signals will result, and the proper dose is affected by the moisture content of the membrane. In our experience, baking as recommended results in a strong hybridization signal.

## B. HYBRIDIZATION TO FILTER-BOUND NUCLEIC ACIDS

There are extensive reviews on the mechanical aspects of hybridization and probe preparation, and the effect of probe concentration, salt concentration, and temperature on the rate of formation and stability of nucleic acid duplexes.[17] We present only the hybridization and wash protocols that we have found to be most effective with nylon or polysulfone membranes. Parameters such as salt concentration and temperature can be adjusted to suit particular applications. The protocols are based on those of Church and Gilbert.[21]

Hybridization Buffer

0.25 *M* $NaHPO_4$, pH 7.2

A stock of 1 *M* $NaHPO_4$ (1 *M* in $Na^+$) is prepared by titrating 0.5 *M* $Na_2HPO_4$ to pH 7.2 with $H_3PO_4$.

2.5 m*M* EDTA

7% sodium dodecyl sulfate (SDS)

1% bovine serum albumin (BSA) or non-fat dry milk
Optional: 100 μg/ml denatured, sheared salmon sperm DNA (carrier DNA is usually not required since the SDS and protein are effective blocking agents)
Prehybridize for 1 h before adding the probe. Prehybridization and hybridization are effective at 68°C for homologous probes.

Protein blocking agents, especially nonfat milk, are likely to contain high levels of RNAase that will degrade RNA probes and RNA on blots. For hybridizations involving RNA, we dissolve the hybridization buffer ingredients at 65°C; then treat the 65°C buffer with 0.5% diethylpyrocarbonate for 1 h. DEPC breaks down rapidly in aqueous solutions at 65°C and should not be present after 1 h. Alternatively, the hybridization may be performed in the buffer described above with 50% formamide at 42°C to inactivate RNAases. We also treat the wash solutions described below with 0.05% diethylpyrocarbonate when working with RNA.

The addition of 10% polyethylene glycol (MW 6 to 8 $\times$ $10^3$) to the hybridization buffer greatly accelerates the rate of nucleic acid hybridization.[22] This is useful when the concentration of the probe is low (or the concentration of a specific probe in a mixed probe population is low), or when a short hybridization time is desired. The addition of polyethylene glycol may increase background with certain probe preparations.

If using a hybridization buffer with formamide at 42°C, polyethylene glycol can be added to the buffer without modification. When using the aqueous buffer with polyethylene glycol at 68°C, we decrease the level of phosphate buffer to 20 m*M* and add NaCl to 0.23 *M*.

Wash the membrane in Solution I at 68°C. The washing time can be adjusted to sufficiently lower the background. We generally find that three washes for 30 min each is sufficient if the wash solution is preheated before use.

Wash Buffer

| I. 0.25 *M* $NaHPO_4$, pH 7.2 | II. 0.04 *M* $NaHPO_4$, pH 7.2 |
|---|---|
| 2% SDS | 1% SDS |
| 1 m*M* EDTA | 1 m*M* EDTA |

A "stringent" wash with Solution II will eliminate mismatched hybrids ($T_m$-12°C for GC=44% and probe length=450 bases[23]). The ionic strength of the solution can be adjusted as needed; for example, a lower ionic strength may be necessary to discriminate between closely related sequences. (Note that NaSDS and EDTA (as $Na_2$EDTA) contribute to ionic strength.) We generally wash twice for 30 min each time at 68°C with preheated buffer.

### C. REMOVAL OF BOUND PROBE FOR REHYBRIDIZATION

First, rinse blots in distilled water to dilute salt. Then heat a large volume (e.g., 500 ml) of 1 m*M* Tris-HCl, pH 7.0 to 7.5, 0.1 m*M* EDTA, with 0.1% SDS to 90°C, and incubate the blot for 15 min, or incubate in 50% Formamide, 10 m*M* Tris-HCl, pH 7.0, 0.1 m*M* EDTA, 0.1% SDS at 65°C for 60 min.

## ACKNOWLEDGMENTS

This work was supported by USDA grant 91-37304-6659 to B.H.T., NSF grant BSR-8906126 to J.R.M., and NSF grant DMB-8957036 to R.M.A. R.M.A. also acknowledges the patient technical assistance of Manorama John.

## REFERENCES


1. **Bendich, A. J., Anderson, R. S., and Ward, B. L.,** Plant DNA: long, pure and simple, in *Genome Organization and Expression in Plants*, Leaver, C. J., Ed., Plenum Press, New York, 1980, 31.
2. **Murray, M. G. and Thompson, W. F.,** Rapid isolation of high molecular weight DNA, *Nucl. Acids. Res.*, 8, 4321, 1980.
3. **Rivin, C. J., Zimmer, E. A., and Walbot, V.,** Isolation of DNA and DNA recombinants from maize, in *Maize for Biological Research*, Sheridan, W. F., Ed., Plant Molecular Biology Association (Now International Society for Plant Molecular Biology), Athens, GA, 1982, 161.
4. **Dellaporta, S. L., Wood, J., and Hicks, J. B.,** A plant DNA minipreparation: version II, *Plant Mol. Biol. Rep.*, 1, 19, 1983.
5. **Lichtenstein, C. P. and Draper, J.,** Genetic engineering of plants, in *DNA Cloning, A Practical Approach*, Vol. 2, Glover, D. M., Ed., IRL Press, Oxford, U.K., 1985, 67.
6. **Watson, J. C. and Thompson, W. F.,** Purification and restriction endonuclease analysis of plant nuclear DNA, *Meth. Enzymol.*, 118, 57, 1986.
7. **Jofuku, K. D. and Goldberg, R. B.,** Analysis of plant gene structure, in *Plant Molecular Biology: A Practical Approach*, Shaw, C. H., Ed., IRL Press, Oxford, U.K., 1988, 37.
8. **Taylor, B. and Powell, A.,** Isolation of Plant DNA and RNA, *Focus (BRL)*, 4, 4, 1982.
9. **Doyle, J. J. and Dickson, E. E.,** A rapid DNA isolation procedure for small quantities of fresh leaf tissue, *Phytochem. Bull.*, 19, 11, 1987.
10. **Milligan, B. G.,** Purification of chloroplast DNA using hexadecyltrimethylammonium bromide, *Plant Mol. Biol. Rep.*, 7, 144, 1989.
11. **Rogers, S. O. and Bendich, A. J.,** Extraction of DNA from milligram amounts of fresh, herbarium and mummified plant tissues, *Plant Mol. Biol.*, 5, 69, 1985.
12. **Palmer, J. D.,** Isolation and structural analysis of chloroplast DNA, *Meth. Enzymol.*, 118, 167, 1986.
13. **Moore, L. J. and Coleman, A. W.,** The linear 20 kb mitochondrial genome of *Pandorina morum* (Volvocaceae, Chlorophyta), *Plant Mol. Biol.*, 13, 459, 1989.
14. **Manhart, J. R., Hoshaw, R. W., and Palmer, J. D.,** Unique chloroplast genome in *Spirogyra maxima* (chlorophyta) revealed by physical and gene mapping, *J. Phycol.*, 26, 490, 1990.
15. **Coleman, A. W. and Goff., L. J.,** DNA analysis of eukaryotic algal species, *J. Phycol.*, 27, 463, 1991.
16. **Southern, E. M.,** Detection of specific sequences among DNA fragments separated by gel electrophoresis, *J. Mol. Biol.*, 98, 503, 1975.
17. **Sambrook, J., Fritsch, E. F., and Maniatis, T.,** *Molecular Cloning: A Laboratory Manual*, 2nd ed., Cold Spring Harbor Press, Cold Spring Harbor, NY, 1989.
18. **Pingoud, A., Urbanke, C., Alves, J., Ehbrecht, H.-J., Zabeau, M., and Gualerzi, C.,** Effect of polyamines and basic proteins on cleavage of DNA by restriction endonucleases, *Biochemistry*, 23, 5697, 1984.
19. **Reed, K. C. and Mann, D. A.,** Rapid transfer of DNA from agarose gels to nylon membranes, *Nucl. Acids Res.*, 13, 7207, 1985.
20. **Khandjian, E. W.,** Optimized hybridization of DNA blotted and fixed to nitrocellulose and nylon membranes, *Bio/Technology*, 5, 165, 1987.
21. **Church, G. M. and Gilbert, W.,** Genomic sequencing, *Proc. Natl. Acad. Sci. U.S.A.*, 81, 1991, 1984.
22. **Amasino, R. M.,** Acceleration of nucleic acid hybridization rate by polyethylene glycol, *Anal. Biochem.*, 152, 304, 1986.
23. **Bolton, E. T. and McCarthy, B. J.,** A general method for the isolation of RNA complementary to DNA, *Proc. Natl. Acad. Sci. U.S.A.*, 48, 1390, 1962.

# 5

# Isolation and Characterization of Plant mRNA

*Judith Strommer, Robert Gregerson, and Michael Vayda*

## I. INTRODUCTION

Few studies of gene expression can proceed far without analysis of the messenger RNA intermediate between gene and protein product. In some respects, RNA is simpler to work with than either proteins or DNA. RNA is tolerant of varied solvent, salt, and temperature conditions, so inactivation by denaturation — except in the rare case of double-stranded RNAs — is not a problem. It is highly resistant to shear, and can be pipeted or vortex-mixed without detriment. The overriding concern is for the activity of ribonucleases, ubiquitous in natural environments and difficult to inactivate or eliminate. We therefore include detailed descriptions of means for controlling ribonucleases.

For any set of experiments requiring RNA, the investigator must define the appropriate tissue source and also determine the kind of isolation most appropriate for the experiments in mind. Total cellular RNA is satisfactory for most northern hybridizations, for example, and is the preferred form for most quantitative analyses. For characterization of rare messages or for cDNA synthesis, on the other hand, polyadenylated RNA may be required. For some purposes, attention is focused on actively translating, i.e., polysome-associated, mRNA. Those working with organellar gene expression may be able to use total RNA;[1] alternatively, initial purification of organelles may be required.[2,3]

In this chapter, we describe methods for controlling ribonucleases and provide protocols for isolating total, polysomal, and polyadenylated RNA from plant tissues. We also include a protocol for analyzing messenger RNA by northern blot hybridization using radiolabeled probes; alternative means of detection, together with detailed instructions for their use, are available as kits from a number of vendors. The methods we describe make use of the materials available to us, e.g., Beckman ultracentrifuge rotors SW28.1, SW41, and 70.1. Substitutions can be made as long as conditions are modified to take into account differences in maximum g forces and speeds. Similarly, we use Corex glass tubes and plastic Oak Ridge tubes for preparative centrifugation. In general, any RNAase-free tubes can be used, as long as glass tubes are made hydrophobic with a silicone coating.

There are a number of excellent cloning manuals, limited primarily by their focus on animal systems. Methods for analyzing purified mRNA are the same whatever the source, however. The methods we use for primer extension and S1 protection, for example, are precisely as described in the Molecular Cloning Manual edited by Sambrook, Fritsch, and Maniatis.[4] We have, therefore, included only the most commonly used RNA analytical tool, the northern blot.

0-8493-5164-2/93/$0.00+$.50
© 1993 by CRC Press, Inc.

# II. METHODS

## A. INACTIVATION OR ELIMINATION OF RIBONUCLEASES

There are a number of stategies employed either to inactivate or eliminate ribonucleases from solutions, glassware, and plasticware. As a start, workers should wear clean, disposable gloves to prevent fingertip contamination of vessels and solutions, and should remember where nucleases are (fingertips, benches, hair, anywhere they have not been removed specifically or chemically inactivated) and where nuclease-containing solutions have been (pipetters, sides of tubes). The aim, then, is to prevent contact between these areas and RNA. It is probably of little value to worry about ribonuclease contamination at the initial stage of RNA purification, as the plant material itself is rich in RNAase activity; subsequent stages of purification demand special thought. Except after chemical modification, inactivation should be thought of as transient, for ribonucleases readily renature when buffer conditions improve.[5]

Most commonly, diethylpyrocarbonate (DEP) is used to inactivate ribonucleases in solutions and on glass and plastic.[6] It acts by carbethoxylating amino acids, especially histidine. It also carbethoxylates single-stranded nucleic acids, which precludes its use as an inactivating agent during RNA isolation. Fortunately, DEP is unstable in water, so it can be used to irreversibly inactivate RNAases from solutions and vessels and then can be allowed to decompose before coming into contact with RNA.

Where total RNA is the desired purification product, an ionic detergent like sodium dodecyl sulfate (SDS) is commonly included in buffers and serves as an inexpensive, mild RNAase-inhibitor.[7] When the preliminary isolation of cytoplasmic RNA or polysomal RNA requires maintenance of organelles, SDS is inappropriate. In such cases, placental RNAase inhibitor (RNAsin, available from Promega Biotech through Fisher Scientific, Atlanta, GA), aurin tricarboxylic acid (ATC, Sigma Chemical Co., St. Louis, MO), or vanadyl ribonucleoside complexes,[8] can be used to inhibit ribonucleases. RNAsin and ATC have worked better than vanadyl complexes in our hands, but when possible we avoid their use altogether.

### 1. Handling of DEP

Due to the instability of DEP, a supplier's bottle should be opened once. In the hood, prepare 1-ml aliquots in microfuge tubes, and release a stream of nitrogen or argon over aliquoted samples. Tubes should then be stored at –20°C. An aliquot should be warmed to room temperature before opening; once opened, a tube should not be returned to the stockpile, but should be marked and stored separately for short-term use. As a potent alkylating agent, DEP should be used in the hood with gloved hands.

### 2. Preparation of Solutions

Add 0.1% DEP to water or buffer. Shake well, then autoclave 20 to 30 min. DEP should not be used with Tris, which has DEP-reactive groups important for buffering; Tris buffers can be prepared in baked containers with DEP-treated water and then autoclaved. Strong acids, strong bases, and nonaqueous solvents need not and should not be DEP-treated or autoclaved. Detergent solutions can be treated with DEP, but they are generally unstable under autoclaving conditions; an alternative is to add DEP, shake, and incubate the solution at 37°C overnight or at 65°C for a few hours.

### 3. Preparation of Glassware and Plasticware

Nucleic acids stick to glass unless the glass is coated with a film that renders the surface hydrophobic. Glassware should therefore be siliconized with an agent such as Sigmacote (available from Sigma Chemical Co., St. Louis, MO). Clean siliconized glassware covered with foil should be freed of RNAase contamination by baking at high temperatures, e.g.,

200°C, for several hours. Plastic caps should be treated as plasticware. Autoclave-resistant plasticware and glassware can be thoroughly rinsed in a fresh solution of DEP-water, covered or wrapped if necessary, and autoclaved. Sterile wrapped disposable plasticware can generally be assumed to be RNAase free.

### 4. Work Area

Be aware of possible sources of RNAase contamination from everything in the immediate work environment, including nearby plant material, hair, and so on. Forethought can minimize exposure of a sample to such sources.

## B. ISOLATION OF TOTAL RNA

Source material for isolation of total RNA can be freshly harvested or harvested and quick-frozen by immersion in liquid nitrogen and then stored at –70°C for subsequent isolation. Any method for total RNA extraction must allow for (1) the need to break open cells, (2) a means of inhibiting RNAase-directed degradation of RNA, and (3) a method of separating RNA from DNA, protein, and polysaccharide contaminants. If contamination with tannins is a problem (peak absorbance at 270 nm is a clue), we recommend a technique which avoids the use of chaotropic agents, for example, that of Tesniere and Vayda.[9]

To lyse cells, the material may be ground with liquid nitrogen in a mortar and pestle, prefrozen in liquid nitrogen, and then quick-ground in a coffee grinder or ground together with dry ice in a coffee grinder. The most suitable grinders we have found are those which can be inverted to deposit ground tissue in a removable lid. The Miracle Mill, marketed by Markson Scientific, Phoenix, AZ, is relatively inexpensive, reliable, efficient, and resistant to blade breakage.

After grinding, the material is thawed in a solution which inhibits ribonucleases. To prevent ribonucleolytic activity in total RNA extractions, by far the most commonly used strategies are to include either guanidinium hydrochloride[10] or guanidinium thiocyanate[11] in the initial solution. Both are strong chaotropic agents. SDS or sarkosyl, the latter more soluble at 4°C, is usually added both to disrupt membranes and to inhibit ribonucleases. Beta-mercaptoethanol is added to limit free-radical dependent crosslinking of phenolics to DNA and also to disrupt disulfide bonds present in ribonuclease molecules. To separate RNA from other cellular components, phenol extraction, differential centrifugation, multiple precipitations, or a combination of the three is routinely used. Extreme caution against ribonuclease introduction is required once chaotropic agents have been removed.

The method presented here is a combination of the single-step procedure of Chomczynkski and Sacchi,[12] and the CsCl-pelleting step introduced by Glisin et al.[13] Its virtues are the speed with which multiple samples of RNA can be prepared and the ease with which it can be scaled up or down. The protocol given is for 1 g of tissue. To scale up or down simply keep components in the same ratios.

This procedure has been employed to isolate RNA for northern blots, *in vitro* translation, and cDNA synthesis. It has worked well with as little as 50 mg of leaf tissue. Yield is highly dependent upon tissue source, varying from about 100 μg RNA/g root tissue to about one mg RNA/g leaf. The method has proven successful using roots, stems, leaves, anthers, and pollen of petunia and roots, stems, leaves, cotyledons, and root nodules of alfalfa, for example.

### 1. Solutions and Reagents

- RNA extraction buffer (Can be stored for weeks at 4°C. Heat slightly to get back into solution.):
  4 *M* guanidinium isothiocyanate

25 m*M* sodium citrate (from a 1 M stock at pH 7.0)
0.5% sarkosyl
0.1 *M* β-mercaptoethanol

- Water-saturated phenol. (Do not use buffered phenol. Molecular biology grade phenol may be used without further purification; technical grade should be redistilled and stored at –20°C. Water-saturated phenol is stable for several months at 4°C. Use phenol with great caution. By the time you notice a phenol burn it is too late to do much about it. Thoroughly clean up spills without delay. If phenol touches the skin, wash the area with soap and water immediately, and then liberally apply a skin cream such as Vaseline Intensive Care lotion.)
- Chloroform: isoamyl alcohol, 24:1
- Isopropanol
- Ethanol
- 2 *M* sodium acetate, pH 4.0, DEP-treated and autoclaved
- 3 *M* sodium acetate, pH 6.5, DEP-treated and autoclaved
- 4 *M* lithium chloride, DEP-treated and autoclaved
- 5.7 *M* cesium chloride, DEP-treated and autoclaved
- Water, DEP-treated and autoclaved

### 2. Supplies (Rinsed in Fresh DEP-Water and Autoclaved or, if Heat-tolerant, Baked)

- 30-ml Corex centrifuge tubes, siliconized, with rubber stoppers
- Ultracentrifuge tubes (open-top; if non-autoclavable, heat to drive off DEP)
- Pasteur pipettes
- Micropipet tips
- Microfuge tubes

### 3. Steps in the Procedure

1. Grind 1 g of tissue in liquid nitrogen with a mortar and pestle or coffee grinder, and transfer it to a Corex centrifuge tube. If dry ice is used instead of nitrogen, ensure that most of the ice has sublimated before extraction buffer is added; otherwise the tube may explode from built up gas pressure.
2. Add 10 ml of extraction buffer to each tube, cap with a rubber stopper, and vortex-mix for 30 sec. Release any pressure from sublimation of remaining dry ice.
3. Add 1 ml of 2 *M* sodium acetate, pH 4.0, and vortex-mix for 30 sec.
4. Add 10 ml of water-saturated phenol and vortex-mix for 30 sec.
5. Add 2 ml of chloroform:isoamyl alcohol (24:1) and vortex-mix for 30 sec. Carefully release any pressure built up from volatilized chloroform.
6. Centrifuge at room temperature for 10 min at 5000 × g. Protein and DNA will go to thick interface or nonaqueous phase.
7. Remove the aqueous phase to a new tube with a Pasteur pipette and add an equal volume of isopropanol. Mix well, and incubate at –20°C for at least 1 h.
8. Centrifuge at 4°C for 10 min at 10,000 × g.
9. For large-scale preparations, proceed to Step 11. For 1 g of starting material or less, resuspend in 500 μl of DEP water, transfer to a microfuge tube, add an equal volume of 4 *M* LiCl, and mix. Place on ice for at least 1 h.
10. Centrifuge for 5 min at full speed in a microcentrifuge. Polysaccharides should remain in the supernate. Wash the pellet twice with cold 70% ethanol, dry, and resuspend in a small volume of DEP water. Determine the concentration of RNA by reading the absorbance of a diluted aliquot at 260 nm. (One mg RNA/ml has an absorbance of 25

O.D.U. The ratio of absorbances at 260 and 280 nm should be close to 2.0; it may be slightly higher. A wavelength scan should reveal a peak absorbance at about 260 nm.) Discard the aliquot after reading.

11. For large scale preparations, resuspend the pellet in 1 to 2 ml DEP-treated water.
12. Layer the RNA solution over a 3 ml cushion of 5.7 *M* CsCl in an ultracentrifuge tube fitting a SW28.1 rotor (or other swinging bucket rotor), and fill the tube to within 3 to 5 mm of the top with DEP-treated water.
13. Centrifuge overnight in an SW28.1 rotor at 24,000 rpm and 17° C to pellet RNA. Polysaccharides, remaining DNA, and protein will not pellet through the cesium chloride cushion. To use a different swinging bucket rotor, keep the product of g-force and time equivalent to that described, roughly $112{,}000 \times g \times 16$ h $= 1.8 \times 10^6$ gh.
14. Carefully pour off the supernate or remove it with a Pasteur pipette, and rinse the pellet gently with cold 70% ethanol.
15. Allow the pellet to air-dry, and resuspend it in 200 to 400 μl of DEP-treated water. Transfer the RNA solution to a microfuge tube.
16. Add 1/10 volume 3 *M* sodium acetate, pH 6.5, and 2.5 volumes cold absolute ethanol. Leave at –20°C for at least 1 h.
17. Centrifuge for 5 min at full speed in a microfuge. Rinse the pellet with cold 70% ethanol, air-dry, and resuspend in a small volume of DEP-treated water.
18. Determine the concentration of RNA from its absorbance at 260 nm. See note in Section II.B.3.10 above regarding absorbance readings.

## C. ISOLATION OF POLYSOMES

Although the isolation of polysomes is more difficult than many other operations associated with mRNA isolation and analysis, there are reasons for choosing this route. It may be desirable, for example, to distinguish RNA molecules associated with ribosomes from those otherwise sequestered.[14] The method presented here is based on the procedures described by Mignery et al.[15] and Laroche and Hopkins.[16] It allows recovery and quantitative comparison of RNA in polysomal and nonpolysomal fractions. If the isolation of polysomes is a prelude to mRNA isolation, and neither quantitative comparisons among polysomal preparations nor analysis of the nonpolysomal fraction is of interest, steps involving addition and monitoring of M13 DNA should be omitted.

The general process involves cell lysis, pelleting of dense organelles and fragments, and pelleting of ribosomal complexes. Polysomes recovered from the latter pellet can be fractionated on a sucrose gradient to produce a profile of the polysome distribution by size and, in conjunction with northern hybridization, to determine the number of ribosomes associated with a given messenger RNA. Alternatively, total polysomal and nonpolysomal mRNA — derived from the pellet and supernate, respectively — can be purified without additional fractionation.

This method was originally developed for potato tubers; the relative amounts of tissue and buffers are reasonable for tissues moderately active in protein synthesis. EGTA is used to chelate cations implicated in RNA cleavage reactions without chelating the magnesium ions needed for ribosome stability. A gentle, nonionic detergent, which does not displace mRNA or disrupt ribosomal structure, is used in place of SDS.

For control of RNA degradation, keep solutions ice-cold and work quickly. If necessary, specific RNAse inhibitors such as those mentioned in Section II.A can be added.

### 1. Solutions and Reagents

- Water-saturated phenol (See note under Section B.1.a.)
- Chloroform

- β-mercaptoethanol
- Ethanol
- Water, DEP-treated and autoclaved
- Polysome buffer: DEP-treat and autoclave KCl, MgOAc and EGTA; make Tris, sucrose, and Triton reagents with DEP-treated, autoclaved water. Dilute reagents with DEP-treated, autoclaved water in an RNAase-free container.
  200 m*M* Tris·Cl, pH 9.0
  400 m*M* KCl
  60 m*M* MgOAc
  50 m*M* EGTA
  250 m*M* sucrose
  0.01% Triton X-100
- Sucrose cushion: Prepare as described above.
  1.5 *M* sucrose
  40 m*M* Tris.Cl, pH 9.0
  100 m*M* KCl
  30 m*M* MgOAc
  5 m*M* EGTA
  7.0 m*M* β-mercaptoethanol
- Cushion buffer: same as sucrose cushion, without sucrose

## 2. Supplies

- Weigh boats
  (**The following should be baked or DEP-treated and autoclaved.*)
- Oak Ridge centrifuge tubes
- Spatulas
- Ultracentrifuge tubes
- Pasteur pipettes

## 3. Steps in Polysome Preparation

1. Weigh out 10 g freshly harvested tissue.
2. Quick-freeze tissue in weigh boat by adding liquid nitrogen. Keep thick tissues like tubers submerged in nitrogen for about 3 min to ensure freezing.
3. Meanwhile add 13 ml ice-cold polysome buffer to an Oak Ridge centrifuge tube. To this add 14 μl β-mercaptoethanol (to a final concentration of 15 m*M*). Keep tube on ice.
4. Transfer the frozen tissue to a coffee grinder, and grind with shaking for 30 sec. Turn the grinder upside down for the last 2 sec to catch ground tissue in the lid.
5. Quickly transfer the fine powder to the centrifuge tube with a spatula. Work quickly and do not let the tissue thaw before it is in polysome buffer.
6. Cap the tube and shake it with mild force for 5 min at room temperature.
7. Centrifuge at 15,000 × g for 15 min at 4°C.
8. During the spin, dispense 3 ml cold sucrose cushion into each of two quick-seal ultracentrifuge tubes, keeping tubes on ice.
9. Remove supernatant fraction from Step 7 with a Pasteur pipette, and layer it gently atop the sucrose cushion. If necessary, gently add additional polysome buffer to fill the tube to its neck (the volume of the homogenate will be approximately 16 ml; volume of the Beckman 70.1 polyallomar quick-seal tubes is 13 ml).

10. Seal the tube, and centrifuge in a Beckman 70.1 fixed angle rotor for 4.5 h at 45,000 rpm at 4°C. Under these conditions only RNA complexes of 50S and greater should pellet; increasing the centrifugation time to 5 h results in the pelleting of free 9.5 kb RNA size standards.
11. Poke a hole in the top of the tube with a sterile 16-gauge needle. If the nonpolysomal fraction is desired, transfer the top few milliliters of supernate to a sterile Oak Ridge tube, slice off the top of the ultracentrifuge tube with a fresh razor blade, and transfer the rest of the supernate. If only the pellet is desired, transfer all the supernate carefully to a waste receptacle.
12. Invert the tube and drain it on a Kimwipe to remove the last vestiges of supernate.

At this point the polysomes may be fractionated on a sucrose gradient to produce a polysome profile or used directly as a source of RNA.

#### 4. Steps in RNA Isolation from Polysomes and Supernate

1. Resuspend the pellet in 1 ml of cushion buffer without sucrose, using a P1000 micropipette. Then add an additional 4 ml of the same buffer.
2. For normalization of polysomal and nonpolysomal RNA content, add 20 μl of a 2 ng/μl stock solution of single-stranded M13 phage DNA to both the resuspended polysomes and the supernate.
3. Dilute the supernate 2:1 with cushion buffer minus sucrose. Failure to do so will result in poor separation of phases and precipitation of sucrose in Step 7.
4. Add an equal volume of water-saturated phenol to each fraction and vortex-mix for 5 min. Add an equal volume of chloroform and vortex-mix an additional 3 min.
5. Centrifuge at 10,000 × g for 5 min at room temperature.
6. Remove the upper aqueous phases with a Pasteur pipette to fresh Oak Ridge tubes, and re-extract with an equal volume of chloroform. Centrifuge as in Step 5.
7. To the supernates, in a fresh Oak Ridge tube, add 3 *M* LiCl to a final concentration of 500 m*M*. Then add 2.5 volumes of cold ethanol and store overnight at –20°C.
8. Centrifuge at 15,000 × g for 30 min at 4°C to pellet RNA.
9. Resuspend pellets in 0.5 ml DEP-treated sterile water.
10. Determine RNA concentrations by light absorbance at 260 nm (see Section II.B.3.10).
11. Quick-freeze the samples and store them at –70°C. The amount of RNA recovered in the polysomal and nonpolysomal fractions will differ greatly. To compare, prepare a dot blot or northern blot using equal volumes of RNA from both fractions, and hybridize to a radiolabeled M13 probe. Adjusting volumes for equal intensities of M13 hybridization will standardize for relative amounts of polysomal and nonpolysomal RNA per gram of starting tissue.

### D. ISOLATION OF POLY(A)$^+$ RNA FROM TOTAL OR POLYSOMAL RNA

For many purposes, total cellular or polysomal RNA can be used directly, e.g., for most northern blot analyses. For other purposes — *in vitro* translation or cDNA cloning, for example — it may be critical to purify messenger RNA from contaminating RNA species. The polyadenylate tract at the 3′ end of nearly every eukaryotic mRNA provides a useful handle for the separation of mRNA from rRNA and tRNA. The method devised by Aviv and Leder[17] and modified by Bantle et al.[18] and Pembroke et al.[19] relies on the ability of poly(A)$^+$ tails to bind to oligo(dT)-cellulose

The basic principles to be kept in mind are the following: (1) the amount of rRNA present in total or polysomal RNA outweighs mRNA by roughly 100:1 in most cell types; (2) RNA complexes form readily; and (3) high salt concentrations favor hydrogen bond formation, while low salt conditions favor dissociation. The first point explains the need to purify mRNA for some purposes and gives an indication of the degree of purification required. Heating RNA in low salt before its application to the column is important to denature RNA complexes, which act both to block accessibility of poly(A)$^+$ regions and to permit rRNA to copurify with mRNA bound to the column. Finally, RNA is applied to the column in high salt to promote A-T hydrogen bonding between mRNA and column-bound oligo(dT). Elution is accomplished by the substitution of water, or a weakly buffered solution, which disrupts the bonds holding poly(A)$^+$ RNA on the column.

Batches of oligo(dT)-cellulose vary in the degree to which they bind RNA nonspecifically.[18] The binding capacity can be estimated from literature accompanying the product. For many commercially available preparations, 1 g of oligo(dT)-cellulose theoretically binds about 4 mg of polyadenylic acid, which translates roughly to 40 mg of mRNA. Oligo(dT)-cellulose also binds some RNA irreversibly.[18] One can assume that a gram will bind the mRNA from at least 40 mg of total RNA, and that extra oligo(dT)-cellulose will only result in decreased yields due to irreversible binding of the sample and to unnecessary dilution of the eluate. A column which works well can be reused indefinitely if it is regenerated and stored properly (see Section II.D.3.20).

The inclusion of SDS helps prevent degradation, but it presents new problems. It necessitates that columns be used at room temperature, i.e., at least 18°C. For RNA destined for cDNA cloning, inclusion of SDS in the washing and elution buffer may decrease yields of first strand synthesis. The presence of SDS also dictates the use of LiCl or NaCl in place of KCl, the buffer originally used by Aviv and Leder,[17] in which SDS is highly insoluble. We recommend the use of buffers without SDS; it may be added to a final concentration of 0.2% after autoclaving if desired.

Although some researchers feel the use of an intermediate washing buffer is unnecessary to remove nonspecifically bound RNA from the column, we occasionally wash off sizable amounts of nonpoly(A)$^+$ RNA at this step and have therefore retained it.

As noted above, rinsing in fresh DEP-water can substitute for high temperature baking to eliminate RNAase in chromatographic column supports. Buffers must be RNAase free, but DEP is not recommended for use with Tris; Section II.A.2 describes procedures recommended for preparation of Tris-containing buffers.

### 1. Solutions and Reagents

- 0.1 *N* NaOH
- Ethanol
- Water, DEP-treated and autoclaved
- Binding buffer:
  10 m*M* Tris·Cl 7.4
  1 m*M* EDTA
  0.4 *M* NaCl
- Washing buffer:
  10 m*M* Tris·Cl 7.4
  1 m*M* EDTA
  0.1 *M* NaCl
- Elution buffer:
  10 m*M* Tris·Cl 7.4
  1 m*M* EDTA
- 4 *M* NaCl

## 2. Supplies (Treated as Described for Isolation of Total RNA

- Column, commercially available or rigged from a 5-ml syringe barrel, its tip stuffed with siliconized glass wool. There should be a stopcock or tubing and clamp which can be used to control column flow.
- Glass-baked Pasteur pipettes
- Erlenmeyer flasks
- Siliconized 15 ml Corex centrifuge tubes
- Pipette tips
- Microfuge tubes
- Parafilm

## 3. Steps

1. Mix dried oligo (dT)-cellulose with an excess of elution buffer, about 20 ml buffer for 0.1 g of cellulose. Allow it to settle for several minutes; then aspirate the buffer.
2. Repeat this wash procedure two additional times.
3. Add about 10 ml 0.1 NaOH, pour the slurry into a column support, and allow the meniscus to reach the top of the column bed.
4. Wash the barrel and column matrix with 1 total-column-volume (TCV, i.e., enough to fill the chromatographic support cylinder) of 0.1 *N* NaOH, allowing all of it to pass through column before proceeding.
5. Equilibrate with 1 TCV elution buffer.
6. Wash with 2 TCV binding buffer.
7. Meanwhile, dissolve pelleted RNA in elution buffer to a final concentration of less than 0.4 mg/ml.
8. Heat RNA solution to 65°C for 5 min, quick-cool for 30 sec in an ice bath, then set at room temperature.
9. Add 1/10 volume 4 *M* NaCl to the RNA solution.
10. Apply the RNA sample to the washed, equilibrated column, and allow it to flow through the bed at a rate of 2 to 4 ml/min, collecting the flow-through in an Erlenmeyer flask. Stop chromatography when the meniscus reaches the cellulose bed.
11. Reheat the flow-through to 65°C for 5 min, quick-cool, and reapply it to the column. Collect the flow-through as "poly(A)$^-$ fraction".
12. Wash the column with 2 to 3 TCV of binding buffer and allow the last wash to reach the top of the cellulose bed.
13. Wash with 2 to 3 TCV of washing buffer, again to the top of the cellulose bed.
14. Add elution buffer heated to 45°C. Use a volume ten times greater than the starting mass of cellulose, e.g., 2 ml for 0.2 g dry cellulose. Collect the eluate as Fraction I.
15. Add the same amount again, and collect the eluate as Fraction II. Fraction I should contain about 80% of the bound RNA; Fraction II, 20%.
16. Do not combine the two fractions. To each add NaCl to 0.4 *M*, then 2 volumes of ethanol, and incubate at –20°C for several hours to precipitate the RNA.
17. Centrifuge precipitated RNA for 30 min at 15,000 × g at 4°C. Suction off the ethanol supernate, taking care to avoid the region where the pellet is expected. Drain the tube thoroughly.
18. Resuspend RNA in a small volume of DEP-treated, autoclaved water, combining the two fractions at this step. Remove an aliquot to determine the concentration and purity of the mRNA (see Section II.B.3.10 for details of spectrophotometric analysis).

19. Quick-freeze the remainder and store at –70°C indefinitely.
20. The column may be regenerated by the following means: Wash the barrel and column bed sequentially with 1 TCV each water, 0.1 *N* NaOH, water, and ethanol. Dry in vacuum oven without heat, seal, and store at –70°C. One column can be used many times; replace it when you notice a decline in yield.

## E. NORTHERN BLOT HYBRIDIZATION

Once RNA has been isolated, a number of different experimental methods can be employed to characterize the RNA. *In vitro* translation, protocols for which accompany commercially available kits, is useful to verify the overall integrity of an RNA preparation, to assess the prevalence of an mRNA species, to address problems of translational regulation, etc. Direct sequencing of RNA[20] is also possible and can provide limited sequence data without the need for additional steps. Often the production of cDNA is the desired use of isolated mRNA, either first-strand cDNA for PCR amplification [21,22] or double-stranded cDNA for library construction. Purified mRNA can also be used to define the transcriptional start site of a specific message for which cloned DNA is available, either by S1-nuclease protection[23,24] or by primer extension.[25] S1 protection can also be used to determine poly(A) addition sites in some cases, although their variability in plants has made 3′-mapping less fruitful in plant than in animal research.

The most common reasons for RNA isolation, however, are needs to assess the size, tissue specificity, and/or relative quantity of a specific transcript. All this information is available from northern blot hybridizations.[26] We present a standard protocol for northern blot analysis of RNA utilizing agarose gels, transfer to nylon membrane, and hybridization with a radiolabeled probe. A number of different membrane and probe systems are available commercially and come with detailed instructions. The following method works well in our laboratory; it is generally applicable and can be used to familiarize a novice with the basic steps . We have attempted to describe methods requiring a minimum of special equipment. Electrophoretic or vacuum transfer apparatuses, hybridization ovens, and automatic developers are useful but nonessential.

Methods are essentially unchanged whether poly(A)$^+$ or total RNA is fractionated, although much higher levels of the latter are needed to visualize a specific message. We use total RNA for quantitative estimations, since variable losses of message and ribosomal RNAs may occur during oligo (dT)-cellulose chromatography. Total RNA, however, is not always appropriate: messages expressed at very low levels will not be detected so readily; messages close in size to ribosomal RNAs will not produce hybridization bands as dense as would be expected by their concentration; and, in general, bands will be broader than those obtained from poly(A)$^+$ RNA samples. In general, if time and expense are considerations, total RNA is preferable for most purposes.

Exposure of the RNA to ribonuclease remains a danger. Materials in contact with RNA should be treated to limit ribonucleases, and clean gloves should be worn. In addition, ethidium bromide is a mutagen, and solutions or gels containing it should be handled and disposed of with care; handling of radioactivity requires training and supervision.

We normally use a horizontal gel apparatus with a tray measuring 15 × 15 cm, for which 150 ml of gel solution works well. Volumes can be scaled up or down to suit any apparatus. In our hands, using standard well-formers, 0.25 to 10 µg of poly(A)+ RNA or up to 50 µg of total RNA can be loaded in a well. Vertical gels provide better resolution but require more initial effort.

Radiolabeled probes can be prepared by nick translation[27] or by random-primer extension;[28,29] several vendors sell kits containing the reagents necessary for these procedures. Probe DNA should be purified from unincorporated radioactive nucleotides either by separation on a Sephadex G-25 column (poured in a 5 or 10 ml disposable pipet) or by two sequential ethanol precipitations.

The choice of a hybridization protocol depends on the type of membrane used. Nitrocellulose, nylon, and charge-modified nylon all are commonly used for northern hybridizations. Nylon membranes are much more tolerant of handling than nitrocellulose, which rips easily. We find that a charge-modified nylon membrane (such as Zetaprobe, Bio-Rad, Richmond, CA) produces very good signals with little or no background. The following protocol is included for use with a charge-modified nylon membrane. It employs a simple hybridization solution in which the blocking reagent, for reducing background, is SDS used at a high concentration. Every manufacturer provides recommendations for hybridization solutions and conditions, so one should check and use techniques consistent with recommended guidelines.

## 1. Solutions and Reagents

- Agarose, DNA grade
- Water, DEP-treated and autoclaved
- Formaldehyde, 37%
- 0.1 *M* NaOH
- 10 X MOPS/EDTA buffer:
  0.2 *M* MOPS (3-N-morpholinopropanesulfonic acid)
  50 m*M* sodium acetate
  10 m*M* EDTA
  Adjust to pH 7.0 and autoclave
- RNA Sample-loading buffer (1 ml):
  750 µl deionized formamide
  240 µl formaldehyde
  150 µl 10 X MOPS/EDTA
  200 µl 50% glycerol
  10 µl 10 mg/ml ethidium bromide
  0.5 mg bromphenol blue (or a small bit that sticks to a micropipette tip)
- 10 X SSC
  1.5 *M* NaCl
  0.15 *M* sodium citrate
- 20% SDS
- Hybridization solution:
  50% formamide
  0.25 *M* $NaHPO_4$, pH 7.2 (from a 1 *M* stock containing equimolar mono- and di-sodium salts, pH adjusted to 7.2)
  0.25 *M* NaCl
  7% SDS
  1 m*M* EDTA

## 2. Supplies

- RNA size markers, if information on length of specific RNA is desired; the low molecular weight ladder from BRL (Bethesda Research Labs, Gaithersburg, MD) is useful for most purposes.
- Gel electrophoresis system, vertical or horizontal; the gel tray either rinsed with DEP-water or soaked for an hour in 0.1 *M* NaOH and then rinsed with DEP-water
- Erlenmeyer flask, 500 ml, baked
- Pasteur pipettes, baked
- Microfuge tubes, rinsed with DEP-water and autoclaved
- UV transilluminator
- Glass baking dishes, baked

- Whatman 3MM filter paper (available through Fisher Scientific, Atlanta, GA)
- Parafilm
- Nylon membrane, e.g., Zetaprobe (Bio-Rad, Richmond, CA)
- Plastic wrap
- Heat-seal bags and sealer
- Plastic food storage dish with tight-fitting lid
- X-ray film cassette
- X-ray film such as Royal X-omat XAR-5 (Kodak, Rochester, NY)

## 3. Gel Electrophoresis and Transfer

1. Add 1.5 to 2.25 g of agarose to baked 500 ml flask. Typically a 1.5% agarose gel will give clear separation in the range of 0.5 to 2.5 kb RNA. Use a lower percentage of agarose for large RNA molecules, higher for small RNAs.
2. Add 110 ml DEP-water and 15 ml 10 X MOPS/EDTA buffer. Microwave on high setting until dissolved (check by swirling and looking for Schlieren patterns in the solution; use a hot pad to protect hands). If microwave oven is unavailable, set flask in a boiling water bath until agarose is thoroughly dissolved.
3. Place dissolved agarose in a water bath at 55°C.
4. When agarose is at 55°C and everything is ready, take the flask to a fume hood and add 25 ml of 37% formaldehyde. Swirl well, and pour into leveled gel tray (or between plates of a vertical gel apparatus). Note that formaldehyde is a carcinogen and should be handled carefully.
5. After the gel has hardened, set it into the electrophoresis chamber, cover it with 1X MOPS/EDTA buffer, and rinse the wells with the same buffer using a Pasteur pipette.
6. Prepare RNA samples:

   | | |
   |---|---|
   | RNA (plus DEP-treated water if necessary) | 5 µl |
   | Loading buffer | 25 µl |

7. Mix samples well, heat to 65°C for 15 min, and then place on ice.
8. Load samples into wells of the gel. Size markers should be used according to vendors' instructions.
9. Electrophorese toward the anode until the bromphenol blue dye is near the bottom of the gel. The gel can be run very slowly overnight, or up to 10V/cm (i.e., cm of gel length). If a sufficient amount of RNA is loaded to fluoresce visibly, RNA migration can be checked during the run by UV transillumination. In general, the slower the gel is run, the tighter the bands will be. The amount of RNA loaded also affects band morphology.
10. When electrophoresis is complete, remove the gel to a baked glass baking dish, and shake in DEP-water with moderate agitation for 30 min.
11. Set up the apparatus for transferring RNA from gel to membrane as follows: place a glass plate which is larger than the gel inside a baking dish, propping it up with rubber stoppers so that it is 5 to 10 cm from the bottom. (Alternatively the horizontal gel box can be used directly.)
12. Cut two strips of Whatman 3MM paper long enough to cover the glass plate, and bend the ends so that they reach the bottom of the dish.
13. Fill the dish with 10X SSC up to the bottom of the glass plate, and saturate the 3MM paper with the same buffer.
14. Place the gel on the paper, teasing out any bubbles formed between gel and filter paper, and lay full-width strips of Parafilm alongside all four edges of the gel, leaving the gel uncovered.
15. Soak the membrane, which has been cut to the same size as the gel, in distilled water

for 5 min; then lay it onto the gel. Carefully remove any bubbles from between gel and membrane.

16. Place four sheets of 3MM paper, cut to the same size as the gel, on top of the membrane, and over them set a stack of paper towels about 10 cm thick. Be sure that buffer can wick to paper towels only by passing through gel and membrane.
17. Gently cover the stack with plastic wrap, and overlay a plastic or glass plate, just heavy enough to keep the paper towels in contact with the 3MM paper and gel. Allow the transfer to proceed for 6 h or more; we usually leave it overnight.
18. Once transfer is complete, paper towels can be removed, and the gel and membrane gently inverted on 3MM paper. At this point, well locations and orientation can be marked with a sharpened, soft pencil or a membrane-marking pen, available from many vendors.
19. Carefully peel off the membrane, rinse it briefly with 2X SSC to remove any bits of agarose, and allow it to dry at room temperature. The transfer apparatus can be covered with plastic wrap and used several times.
20. Set the membrane between paper towels, and bake it in a vacuum oven at 70 to 80°C for 0.5 to 2 h. It may be stored indefinitely under vacuum at room temperature.

## 4. Hybridization

1. Place membrane inside a heat-sealable plastic bag. Add 0.15 ml of hybridization solution per square centimeter of membrane. Seal and incubate at 42°C for at least 5 min.
2. Meanwhile place the tube containing probe DNA in a boiling water bath for 2 min; then quench in an ice-bath for 1 min.
3. Pour off the solution in the plastic bag, and replace it with an equal volume of fresh solution. Add probe, at a concentration of $2 \times 10^5 - 1 \times 10^6$ cpm/ml of hybridization solution. Eliminate air bubbles as much as possible, and then seal the bag, double-sealing if possible. If hybridization will be carried out in a water bath, set this bag inside another heat-sealed bag as extra protection against leakage of radioactivity.
4. Incubate at the desired temperature overnight, or up to 24 h, in a special incubator-oven or a water bath. A good starting point is 42°C for high stringency and 30°C for low stringency.
5. After incubation, carefully cut open the bag, and pour the solution into an appropriate waste or storage container.
6. Remove the membrane and place it in a suitable dish with a snug-fitting top. Prewash the membrane with 100 ml of 2X SSC, 0.1% SDS by swirling the bag; then pour the prewash solution into a radioactive waste container.
7. The washing protocol will vary with the degree of stringency desired. For high stringency, wash twice for 15 to 30 min each in 1X SSC, 0.1% SDS at 65°C, followed with two washes for 15 to 30 min each in 0.1X SSC, 0.1% SDS. For low stringency, salt concentration of the final washes can be increased to 1 or 2X SSC, and the temperature can be lowered, for example, to 50°C. Optimal washing conditions depend on how dissimilar the cross-hybridizing sequences are and are best determined empirically.
8. Gently blot the membrane with paper toweling and wrap it in plastic wrap or seal it in a heat-seal bag. With a Geiger counter, check the membrane for the general level of radioactivity. Tape it to a solid support, such as an old piece of X-ray film, and in the dark place, it in an X-ray film cassette with an intensifying screen and unexposed X-ray film for autoradiography.
9. Set the cassette in an ultrafreezer for a length of time judged from the level of radioactivity detected with a Geiger counter. Initially, try an overnight exposure and

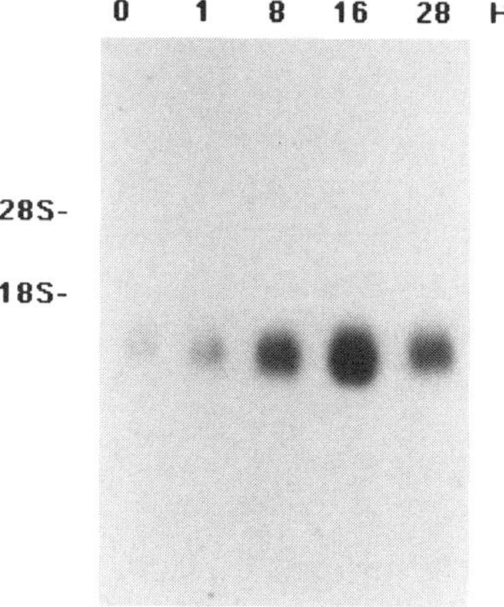


**FIGURE 1.** Autoradiogram of a northern blot illustrating the change in alcohol dehydrogenase RNA levels in leaves of petunia seedlings which have been subjected to anaerobic stress. Total RNA was prepared as described in the text from leaves maintained in argon for 0, 1, 8, 16, or 28 h. Thirty micrograms were loaded into each well. RNA was transferred to a nylon membrane and probed with a DNA fragment containing the entire petunia *Adh1* gene. Hybridization was at 42°C in 50% formamide; the blot was then washed three times in 0.2 X SSC, 0.1% SDS at 68°C.

repeat as necessary. If the membrane has been kept damp, e.g., wrapped in plastic film during exposure, it can be rewashed at higher stringency and re-exposed.

## III. DISCUSSION

The detailed procedures described here yield good quality RNA reproducibly in quantities of 1 to 2 mg/10 g of tissue. This RNA is suitable for northern blot analysis, *in vitro* translation, and cDNA synthesis. Figure 1, for example, depicts a northern blot autoradiogram of petunia alcohol dehydrogenase RNA levels in anaerobically stressed leaves. Total RNA prepared by the method described in the text was fractionated, transferred, and probed with homologous DNA. Low levels of this RNA can be detected in control leaves; in the absence of oxygen it is heavily but transiently induced. With 30 µg of RNA loaded into each well, bands are broad, but levels of Adh1-RNA can be readily compared.

For the sucrose gradient profile presented in Figure 2, polysomes of potato tubers have been fractionated to reveal the relative proportions of monosomes, disomes, and larger polysome complexes. In conjunction with northern blot hybridizations, polysome fractionation provides a means of assessing whether a specific mRNA species is associated with the translational machinery.

Figure 3 is an autoradiogram of a northern blot containing RNA associated with polysomes from nonwounded, wounded, hypoxic, and wounded-hypoxic potato tubers. The mRNA encoding patatin, the major tuber storage protein, is associated with polysomes under normal and hypoxic conditions but displaced when tubers are wounded. Messenger RNA encoding the wound-inducible enzyme phenylalanine ammonia lyase remains bound to ribosomes when wounded tubers are transferred to hypoxic conditions. Alcohol dehydrogenase mRNA is expressed and bound to polysomes under hypoxic conditions whether or not the tissue is wounded.

For the autoradiogram illustrated in Figure 4, poly(A)$^+$ RNA was isolated from developing maize kernels which were homozygous either for the nonmutant *Sh1* allele or for a mutant allele containing the transposable element *Mu-1* at the transcriptional start site. *Sh1* encodes sucrose synthase, the message for which is present at high levels at 12 days and undetectable

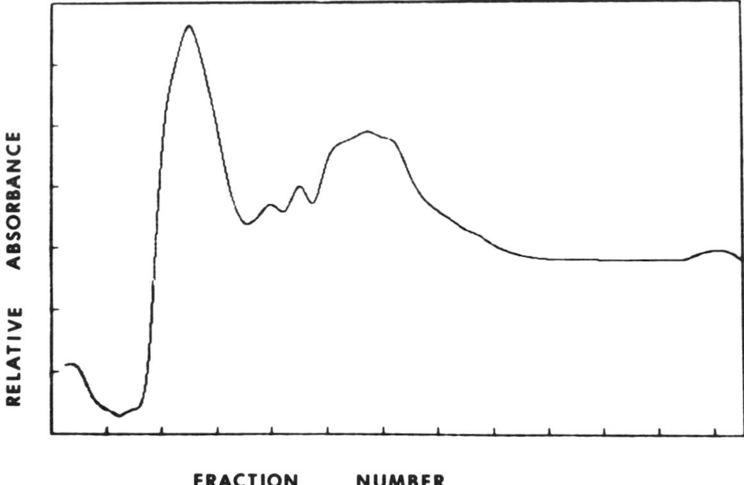


**FIGURE 2.** Profile of potato tuber polysomes prepared as described in the text. The polysomes were resuspended in cushion buffer without sucrose and fractionated at 4°C on a 5 to 20% linear sucrose gradient in an SW-41 rotor centrifuged at 39,000 rpm for 20 h. (14). RNA was collected from the top of the gradient (left), and absorbance monitored at 254 nm. The major peak represents monosomes.

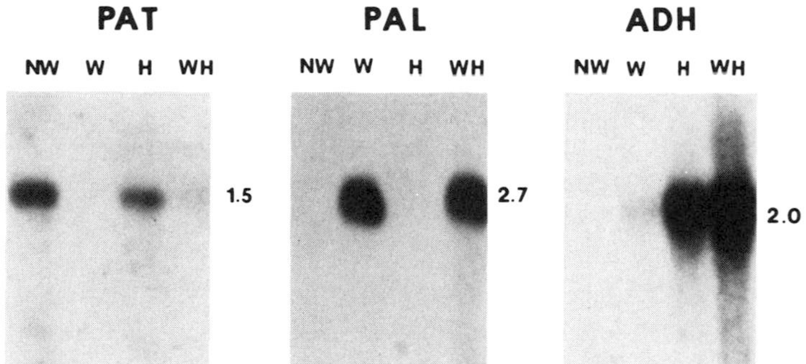


**FIGURE 3.** Northern blot autoradiogram of potato tuber RNA recovered from polysomes prepared as described in the text from control (NW), wounded (W), hypoxic (H), and wounded + hypoxic (WH) tubers. The blot was probed sequentially with homologous sequences for patatin (PAT), phenylalanine ammonia lyase (PAL), and alcohol dehydrogenase (ADH) mRNA.

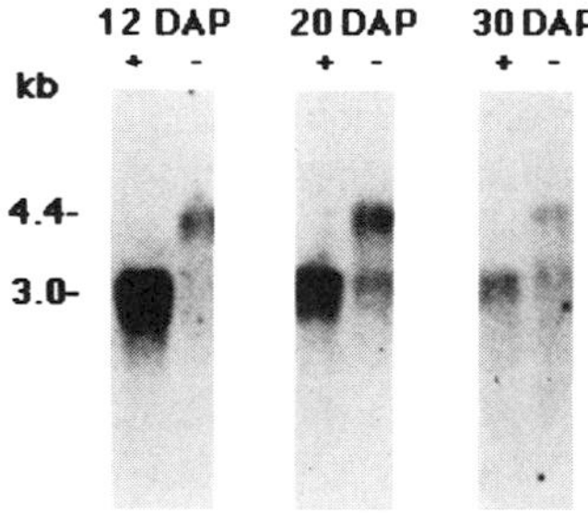


**FIGURE 4.** Autoradioagrams illustrating production of poly(A)$^+$ RNA from nonmutant and mutant alleles of the *Shrunken-1* gene of maize in developing kernels. Poly(A)$^+$ RNA was isolated from kernels of *Sh1* (+) and *sh-mu-9026* (–) plants harvested 12, 20, and 30 days after pollination. The first lane contained 1 μg RNA; other lanes of *Sh1* RNA contained 2 μg RNA; and lanes representing mutant RNA samples contained 5 μg RNA. The blot was probed with *Sh1* genomic DNA and washed as described in Figure 1.

by 40 days after pollination (data not shown). The mutant allele produces low levels of two transcripts, one of the normal size and one containing the *Mu-1* insertion. The gel lanes in this experiment contained from 1 to 5 μg of poly(A)$^+$ RNA, allowing for better resolution of bands than one would expect from total RNA (see Figure 1).

The methods we have described have also been used in the preparation of cDNA libraries and in transcript mapping. They are thus suitable for preparation of plant RNA for virtually any analytical approach currently in use. We use them because they are also the methods that have worked most quickly and most reliably in our own laboratories. While working with RNA normally requires more forethought and more practice than many other procedures used in molecular biology, and to some there is a mystique associated with its manipulation, there is really nothing more mystical in the successful handling of RNA than (1) judicious consideration of ribonucleases, (2) workable protocols, and (3) first-hand experience.

## REFERENCES


1. **Sieburth, L. E., Berry-Lowe, S., and Schmidt, G. W.,** Chloroplast RNA stability in Chlamydomonas: rapid degradation of *psbB* and *psbC* transcripts in two nuclear mutants, *Plant Cell,* 3, 175, 1991.
2. **Stern, D. B. and Newton, K. J.,** Isolation of plant mitochondrial RNA, *Meth. Enzymol.,* 118, 488, 1986.
3. **Schuster, A. M. and Sisco, P. H.,** Isolation and characterization of single-stranded and double-stranded RNAs in mitochondria, *Meth. Enzymol.,* 118, 497, 1986.
4. **Sambrook, J., Fritsch, E. F., and Maniatis, T.,** *Molecular Cloning, A Laboratory Manual,* Cold Spring Harbor Laboratory Press, Cold Spring Harbor, NY, 1989.
5. **Sela, M., Anfinsen, C. B., and Harrington, W. F.,** The correlation of ribonuclease activity with specific aspects of tertiary structure, *Biochim. Biophys. Acta,* 26, 502, 1957.
6. **Ehrenberg, L., Fedorcsak, I., and Solymosy, F.,** Diethyl pyrocarbonate in nucleic acid research, *Prog. Nucl. Acid Res. Mol. Biol.,* 16, 189, 1964.
7. **Noll, H. and Stutz, E.,** The use of sodium and lithium dodecyl sulfate in nucleic acid isolation, *Meth. Enzymol.,* 12B, 129, 1968.
8. **Berger, S. L. and Birkenmeier, C. S.,** Inhibition of intractable nucleases with ribonucleoside-vanadyl complexes: isolation of messenger ribonucleic acid from resting lymphocytes, *Biochemistry,* 18, 5143, 1979.
9. **Tesniere, C. and Vayda, M. E.,** Method for the isolation of high-quality RNA from grape berry tissues without contaminating tannins or carbohydrates, *Plant Mol. Biol. Rep.,* 9, 242, 1991.
10. **Cox, R. A.,** The use of guanidinium chloride in the isolation of nucleic acids, *Meth. Enzymol.,*12B, 120, 1968.
11. **Chirgwin, J. M., Przybyla, A. E., MacDonald, R. J., and Rutter, W. J.,** Isolation of biologically active ribonucleic acid from sources enriched in ribonuclease, *Biochemistry,* 18, 5294, 1979.
12. **Chomczynski, P. and Sacchi, N.,** Single-step method of RNA isolation by acid guanidinium thiocyanate-phenol-chloroform extraction, *Anal. Biochem.,* 162, 156, 1987.
13. **Glisin, V., Crkvenjakov, R., and Byus, C.,** Ribonucleic acid isolated by cesium chloride centrifugation, *Biochemistry,* 13, 2633, 1974.
14. **Crosby, J. S. and Vayda, M. E.,** Stress-induced translational control in potato tubers may be mediated by polysome-associated proteins, *Plant Cell,* 3, 1013–1023, 1991.
15. **Mignery, G. A., Pikaard, C. S., Hannapel, D. J., and Park, W. D.,** Isolation and sequence analysis of cDNAs for the major tuber protein patatin, *Nucl. Acids Res.,* 12, 7987–8000, 1984.
16. **Laroche, A. and Hopkins, W. G.,** Isolation and in vitro translation of polysomes from mature rye leaves, *Plant Physiol.,* 83, 371–376, 1987.
17. **Aviv, H. and Leder, L.,** Purification of biologically active globin messenger RNA by chromatography on oligothymidylic acid-cellulose, *Proc. Natl. Acad. Sci. U.S.A.,* 69, 1408, 1972.
18. **Bantle, J. A., Maxwell, I. H., and Hahn, W. E.,** Specificity of oligo (dT)-cellulose chromatography in the isolation of polyadenylated RNA, *Anal. Biochem.,* 72, 413, 1976.
19. **Pembroke, R. E., Liberti, P., and Baglioni, C.,** Isolation of messenger RNA from polysomes by chromatography on oligo(dT)-cellulose, *Anal. Biochem.,* 66, 18–28, 1975.
20. **Wiener, J. R. and Joklik, W. K.,** Comparison of the reovirus serotype 1,2 and 3 S3 genome segments encoding the nonstructural protein NS, *Virology,* 161, 332, 1987.

21. **Wang, A. M., Doyle, M. V., and Mark, D. F.,** Quantitation of mRNA by the polymerase chain reaction, *Proc. Natl. Acad. Sci. U.S.A.*, 86, 9717, 1989.
22. **Frohman, M. A., Dush, M. K., and Martin, G. R.,** Rapid production of full-length cDNAs from rare transcripts: amplification using a single gene-specific oligonucleotide primer, *Proc. Natl. Acad. Sci. U.S.A.*, 85, 8998, 1988.
23. **Weaver, R. F. and Weissman, C.,** Mapping of RNA by a modification of the Berk-Sharp procedure: the 5′ termini of 15S β-globin mRNA precursor and mature 10S β-globin mRNA have identical map coordinates, *Nucl. Acids Res.*, 7, 1175, 1979.
24. **Murray, M. G.,** Use of sodium trichloracetate and mung bean nuclease to increase sensitivity and precision during transcript mapping, *Anal. Biochem.*, 158, 165, 1986.
25. **McKnight, S. L. and Kingsbury, R.,** Transcription control signals of a eukaryotic protein-encoding gene, *Science*, 217, 315, 1982.
26. **Thomas, P. S.,** Hybridization of denatured RNA and small DNA fragments transferred to nitrocellulose, *Proc. Natl. Acad. Sci. U.S.A.*, 77, 5201, 1980.
27. **Rigby, P. W. J., Dieckmann, M., Rhodes, C., and Berg, P.,** Labeling deoxyribo-nucleic acid to high specific activity in vitro by nick translation with DNA polymerase I, *J. Mol. Biol.*, 113, 237, 1977.
28. **Feinberg, A. P. and Vogelstein, B.,** A technique for radiolabeling DNA restriction endonuclease fragments to high specific activity, *Anal. Biochem.*, 132, 6, 1983.
29. **Feinberg, A. P. and Vogelstein, B.,** Addendum: a technique for radiolabeling DNA restriction endonuclease fragments to high specific activity, *Anal. Biochem.*, 137, 266, 1984.

# 6

# Procedures for Introducing Foreign DNA into Plants

*B. L. Miki, P. F. Fobert, P. J. Charest, and V. N. Iyer*

## I. INTRODUCTION

Students of plant molecular biology vary widely in their research goals for transferring foreign DNA into plant cells. In some cases, transient expression is sufficient to meet the experimental requirements; therefore, stable integration of the transferred DNA is not a necessity. Often, however, the function of a gene product within a differentiated plant is the objective. Furthermore, transferred DNA may be used as a genetic tag, insertional mutagen, or probe for plant regulatory signals. For these kinds of studies, stable integration of the foreign DNA must occur in a predictable and efficient manner, and the recovery of transgenic plants is needed. The particular research goal determines both the chosen mode of transfer as well as the genetic and molecular structure of the DNA that has been constructed for transfer.

This chapter is concerned with concepts and procedures that enable the transfer and integration of characterized DNA molecules into plant cells. Some methods for transforming plant cells are now sufficiently well established to be exploitable. We will consider these in detail and also point out their limitations and the directions in which they need to be developed further. There are many emerging techniques of DNA transfer that are promising. We will introduce the reader to them but will not consider them in detail as they are not yet well tested.

## II. MANIPULATING THE NATURAL GENE TRANSFER SYSTEMS OF *AGROBACTERIUM*

The species of soil bacteria *A. tumefaciens* and *A. rhizogenes* are capable of transferring a discrete sequence of DNA called T-DNA from their tumor-inducing (Ti) plasmids into the genome of a broad range of gymnosperms and angiosperms. The details of the molecular biology of this natural transfer system have been reviewed previously[1-4] and are not considered in this chapter. The transfer process requires short *cis*-acting sequences called T-DNA borders and *trans*-acting functions called Vir functions, some of which are specified by the Ti plasmid and others by the bacterial chromosome. Transfer also requires wounding of plant tissues, and it has been shown that phenolic compounds in the wound exudate induce the activity of several of the bacterial genes determining the Vir functions, including those that act directly at the T-DNA borders. The sequences in the T-DNA between the T-DNA borders include genes for the plant tumorigenic symptoms (genes determining particular pathways of auxin and cytokinin biosynthesis which could interfere with plant regeneration in tissue culture) and opine synthesis. These genes are replaced by other DNA sequences, including those that

0-8493-5164-2/93/$0.00+$.50
© 1993 by CRC Press, Inc.

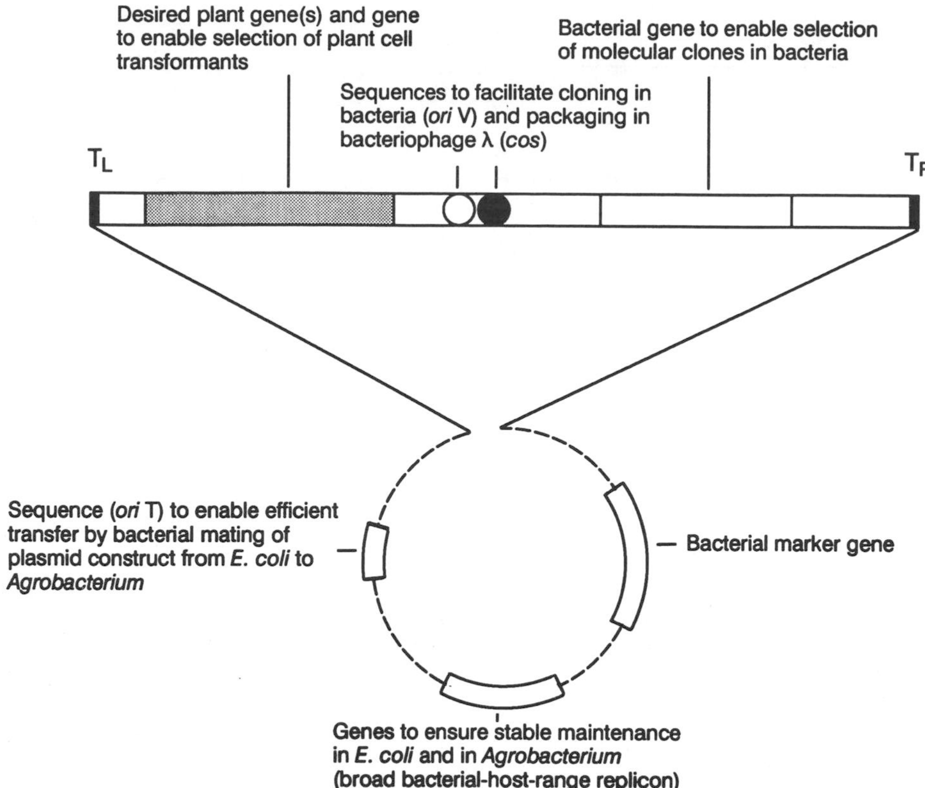


**FIGURE 1.** Desirable features of a binary plasmid vector system for use in *Agrobacterium* strains for the transformation of susceptible plants. $T_L$ and $T_R$ are the two borders defining the region that is integrated into the plant genome. The inclusion of *ori*V (origin of vegetative replication) and *cos* (cohesive sites of bacteriaphage λ) within the borders facilitates subsequent rescue of the plant sequences in *E. coli*. The bacterial gene marker within T-DNA is included for the same purpose. The genes shown on the plasmid but outside the T-DNA region are all to facilitate manipulation in the bacteria (*E. coli* or *Agrobacterium*) prior to plant transformation. This region of the plasmid is shown as a broken line as it does not survive in the plant cells.

can serve as convenient genetic markers or that may be interesting to install in a transgenic plant. Basic features of T-DNA are illustrated in Figure 1. Useful features of this transfer system include the ability of the T-DNA borders to direct insertion of the DNA sequences precisely into the recipient plant genome, the predominance of single insertions, and the accommodation of substantial amounts of DNA between the T-DNA borders, without impairing its transfer.

A feature of considerable practical significance is that the transfer process proceeds efficiently even when the T-DNA borders and the *vir* genes are on different genetic elements, such as two compatible plasmids in *Agrobacterium*. The resulting binary vector systems[5,6] permit one to engineer the T-DNA bearing vector alone and provide all other functions required for gene transfer on a separate plasmid or the bacterial chromosome, neither of which require extensive tinkering. Provided the stability of the T-DNA vector plasmid in the *Agrobacterium* can be monitored and secured, these binary systems are to be preferred to the cointegrate vector systems that one may choose to use only because of their relative stability. Cointegrate vector systems are not considered further in this chapter, and the interested reader is referred to the reviews of Rogers and Klee[7] and Klee et al.[8]

**TABLE 1**
**Disarmed Ti Helper Plasmid Vectors Useful in *Agrobacterium* Binary Gene Transfer Systems**

| Helper plasmid | Origin: Ti plasmid | Origin: Strain | Ref. |
|---|---|---|---|
| pAL4404 | pTiAch5 | Ach5 | 5 |
| pMP90RK | pTiC58 | C58 | 60 |
| pEHA101 | pTiBo542 | Bo542 | 61 |
| pTiB6S3-SE | pTiB6S3 | B6S3 | 62 |
| pTiBo542ΔT | pTiBo542 | Bo542 | 63 |

## A. STRAINS AND VECTORS

The technology of constructing chimeric DNA molecules *in vitro* has now developed to a point where there are hardly any limitations to constructing a vector. These vectors are made up of cassettes intended to fulfill different purposes. It is the design of these cassettes that will determine the potential usefulness of the vectors. In this section, we will therefore consider vectors in terms of their constituent cassettes.

If a binary *Agrobacterium* vector system is to be tested in a new plant, it would be prudent first to screen combinations of *Agrobacterium* chromosome and helper Ti-plasmids. At present, there are only a few disarmed helper Ti plasmids to choose from, and none have been derived from *A. rhizogenes*. However, the disarmed plasmids derived from *A. tumefaciens* have been shown to function in some strains of *A. rhizogenes*.[9] The short list in Table 1 suggests the need for others. These helper Ti plasmids are said to be disarmed because their own T-DNA region or the oncogenes within it have been deleted so as to prevent any possible interference with subsequent plant regeneration. They have been derived from a broad range of Ti plasmid types and are collectively suited for a wide range of plant species.

The T-DNA vectors that are part of the above binary systems typically consist of plasmids that are constructed to carry between their T-DNA borders cloning sites for the insertion of foreign DNA, markers that facilitate the selection or screening of transformed plant tissue, and the foreign DNA with appropriate sequences controlling their expression (Figure 1). An origin of replication that is functional in *Escherichia coli* and bacteriophage lambda *cos* (cohesive) sites is often included to facilitate cloning in bacteria and recovery of plant DNA near the site of insertion. Outside the T-DNA region, there will be an origin of replication that is compatible with the different origin within the T-DNA and any other element needed to ensure stable maintenance of the plasmid in *Agrobacterium*. It is also useful (but not mandatory) to include an origin of conjugative transfer (*ori*T) to facilitate conjugative transfer of the plasmid between *E. coli* and *Agrobacterium*. As a rule, the T-DNA vector is constructed, cloned, and characterized in *E. coli*, and then transferred into an appropriate *Agrobacterium* strain (carrying a suitable disarmed Ti helper plasmid), which in turn is used for transfer of the T-DNA into plants.

From the above description, it will be clear that the T-DNA borders are included in all vectors that are intended for use with *Agrobacterium*. For vectors used in direct DNA transfer systems that do not involve *Agrobacterium*, inclusion of the T-DNA borders or of the *ori* T region provides no advantage. Inclusion of both T-DNA borders in a T-DNA vector provides the major advantage that the DNA inserted into the plant is limited to regions defined by the two borders. This capability does not exist at present for non-T-DNA vectors and may be exploited for many purposes including promoter element tagging.

## B. SELECTABLE MARKER AND REPORTER GENES

All vectors must include one or more genetic markers that allow transformed plant cells to be either selected or screened. A range of dominant and selectable genetic markers for plant cells has been identified (Table 2). This enables one to select a marker that is most suitable for the plant species or tissues to be transformed. The most commonly used of these is kanamycin resistance conferred by chimeric neomycin phosphotransferase II (*npt II*) genes under the control of plant regulatory signals. The regulatory elements are often taken from viral or T-DNA genes, such as the 35S transcript of cauliflower mosaic virus and the nopaline synthase (*nos*) gene, respectively. The common selectable marker genes are frequently bacterial in origin and code for enzymes that metabolically detoxify the selective chemical agent, which may be an antibiotic or herbicide (Table 2). Notable exceptions are genes coding for mouse dihydrofolate reductase, plant 5-*enol*pyruvylshikimate-3-phosphate (EPSP) synthase, and plant acetolactate synthase (Table 2). They achieve resistance to the selective agent when overexpressed in the plant cell or when mutated to prevent binding of the chemical inhibitor.

Selectable marker genes may also be used for the genetic analysis of the inserted DNA, as described later. Different markers may be used in combination for somatic cell genetics involving protoplast fusion[10] and for studying the interactions between inserted genes.[11] Recently, a selectable marker gene lacking a promoter was used to identify insertions adjacent to gene regulatory elements.[12]

A second class of markers requires screening rather than selection of transformed plant cells. They are also shown in Table 2 and are particularly useful when one needs to quantitate or to visualize the spatial pattern of expression of a gene in specific cells, tissues, or organs. Such genes are also referred to as reporter genes because they can be fused to plant genes (or plant gene regulatory sequences) to study expression.

The bacterial gene coding for β-glucuronidase (GUS) is now widely used to analyze gene regulatory elements in both transgenic plants and in transient expression assays. A variety of detection methods exist which combine simplicity and sensitivity.[13] GUS has largely replaced chloramphenicol acetyltransferase (CAT) and neomycin phosphotransferase in such studies. The histochemical assay provides the added advantage of revealing spatial expression patterns with resolution at the level of single cells.[13] The potential applications of GUS would be greatly extended by the development of nontoxic detection procedures designed for living cells. At this time only the *Vibrio harveyi lux* gene system provides this option (Table 2). Examples of additional applications of the GUS reporter system include the study of gene interactions resulting in antisense mechanisms for control of expression[14] and the identification of expression patterns associated with the chromosomal sites of T-DNA integrations.[15]

## C. *AGROBACTERIUM*-MEDIATED LEAF DISC TRANSFORMATION

The most widespread method of transferring genes to plant cells via *Agrobacterium* infection is the leaf disc transformation-regeneration protocol developed by Horsch et al.[16] This procedure exploits the fact that wounded cells within excised plant tissues are competent for transformation by *Agrobacterium* and are also capable of developing organs *de novo* when cultured under appropriate hormonal conditions *in vitro*. The transgenic shoots eventually give rise to plants in which all cells are derived from the same initial transformation event. Transformed cells can be selected directly at the stage of shoot regeneration by incorporating appropriate selectable agents in the culture medium.

Leaf disc transformation was first described for species such as tobacco and petunia, and these remain model systems for studies of gene transfer. Recent interest in *Arabidopsis thaliana* as a model system for plant molecular genetics has prompted the development of a modified version of the leaf disc protocol to achieve transformation from root explants.[17] Transformation has also been achieved using hypocotyledons,[18] stem pieces,[19] and tubers[20] as examples of other explants. The protocol described below is for transformation of tobacco

**TABLE 2**
**Selection and Reporter Gene Systems for the Recovery and Identification of Transformed Plant Cells**

**Selectable Marker Gene[a]**

| Gene | Origin[b] | Selective Agent[c] | Ref.[d] |
|---|---|---|---|
| Neomycin phosphotransferase I | *Tn 601* | Kanamycin | 64 |
| Neomycin phosphotransferase II | *Tn 5* | Kanamycin | 64 |
| Hygromycin phosphotransferase | *E. coli* | Hygromycin | 65 |
| Gentamycin acetyl transferase | *pLG62* | Gentamycin | 66 |
| Streptomycin phosphotransferase | *Tn 5* | Streptomycin | 67 |
| Aminoglycoside-3′-adenyl transferase | *E. coli aad A* | Spectinomycin | 68 |
| Bleomycin resistance determinant | *Tn 5* | Bleomycin | 69 |
| Dihydrofolate reductase | Mouse | Methotrexate | 70 |
| Phosphinothricin acetyltransferase | *S. hygroscopicus* | Phosphinothricin | 71 |
| Acetolactate synthase | Plants | Sulfonylureas | 72 |
| Bromoxynil nitrilase | *K. ozaenae bxn* | Bromoxynil | 73 |
| EPSP synthase[e] | Petunia | Glyphosate | 74 |
| Dihydropteroate synthase | pR46 | Sulfonamide | 75 |

**Reporter Genes[a]**

| Gene | Origin[b] | Ref.[d] |
|---|---|---|
| β-Glucuronidase | *E. coli uid A* | 13 |
| β-Galactosidase | *E. coli lacZ* | 76 |
| Luciferase | Firefly | 77 |
| Luciferase | *V. harveyi luxA,B* | 78 |
| Chloramphenicol acetyltransferase | *Tn 9* | 79 |
| Neomycin phosphotransferaseI I | *Tn 5* | 80 |

[a] The list of selectable markers and reporters is not comprehensive.
[b] In many cases the genes have been obtained from different sources.
[c] The selectable marker genes often convey resistance to more than one selective agent. Only one selective agent is presented for each.
[d] Only single references are provided for each gene; however, several may exist for each.
[e] 5-enolpyruvylshikimate-3-phosphate synthase.

(*Nicotiana tabacum*) leaf discs using kanamycin selection. With this protocol, we routinely obtain several transgenic plants from each leaf disc infected. The specific conditions used for explant preparation, as well as shoot regeneration and selection, may not be optimal for all species. Potential factors affecting plant transformation and regeneration which should be considered when developing protocols for new species are considered in more detail in Section IV and in Draper et al.[21] When establishing procedures for transformation of other species, it is advisable first to gain some experience with these model systems.

### 1. General Procedures, Media Preparation, and Growth Conditions

All steps in the procedure must be performed using sterile conditions, preferably in a laminar flow hood. Instruments are sterilized by flaming in 70% ethanol and permitted to cool prior to handling the plant material. Sterilization should be performed frequently during the procedure. All media are sterilized by autoclaving for 17 min at 121°C and are permitted to cool to 50 to 55°C prior to dispensing. Although growth regulators may be added prior to sterilization, antibiotics are filter-sterilized (0.2 μm cellulose acetate filter) and added following autoclaving, once media have cooled to 50 to 55°C. The antibiotics can be dissolved in distilled water. We keep frozen stocks of filter-sterilized kanamycin sulfate (Sigma, St Louis, MO) and carbenicillin disodium (Pyopen, Ayerst Laboratories, Montréal) at 25 mg/ml in 2 ml aliquots which are allowed to thaw just prior to use. Cefotaxime (Claforan, Roussel Canada, Inc., Montréal) stocks of 50 mg/ml are kept shielded from light at 4°C for up to 1 month.

The basic culture medium consists of MS salts[22] with B5 vitamins.[23] Details of media composition and preparation are given in Tables 3 and 4. We prepare 5× frozen stocks of MSB5 (100 ml) in Whirl-Pack bags and add appropriate amounts of water, growth regulators, agar, and sucrose to the stocks to make up the desired culture media. A growth-regulator-free medium (MSO) containing 1× MSB5, 2% sucrose, and 0.8% Difco Bacto agar, pH 5.7, is used to germinate seeds, to culture *in vitro*-grown plants, and for rooting kanamycin-resistant shoots. Leaf discs are cultured, and shoots regenerated on 1× MSB5 supplemented with 1 μg/ml 6-benzylaminopurine (BAA), 0.1 μg/ml α-naphthalene acetic acid (NAA), 3% sucrose, and 0.8% Bacto agar, pH 5.7 (MS104). Tobacco cell suspensions used as nurse cultures are grown in 25 to 35 ml of MSB5 medium supplemented with 2 mg/ml 2,4-dichlorophenoxy acetic acid (2,4-D) and 3% sucrose (MS2D) in 125 ml Erlenmeyer flasks covered with aluminum foil.

All plant material is maintained in a controlled environment at 25°C with low to moderate illumination (200 to 400 $\mu E/m^2/s$) supplied by fluorescent lights. We routinely grow our cultures with continuous illumination, although a 16 h photoperiod is also sufficient. Leaf discs are cultured in 100 × 25 mm Petri plates containing 25 ml MS104 medium and sealed with Parafilm, while plants and shoot cultures are grown in Magenta GA7 boxes (Magenta Corp., Chicago, Il) on 50 ml of MSO medium. As an alternative to Magenta boxes, autoclavable glass jars (ca. 16 oz, 100 × 125 mm) can be used. Suspension cultures are grown with continuous agitation (125 to 150 rpm) on a New Brunswick Scientific (Edison, NJ) gyrotory shaker. These are transferred at weekly intervals by permitting the cells to settle, decanting the supernatant, and pouring approximately 15 ml of the remaining suspension into an Erlenmeyer flask containing 25 ml of MS2D.

*Agrobacterium tumefaciens* strains used to infect leaf discs are grown at 28°C with moderate agitation in 125 ml Erlenmeyer flasks containing 25 ml of Luria broth (LB; 10 g/l Bactotryptone, 5 g/l yeast extract, 5 g/l NaCl) supplemented with appropriate antibiotics to maintain transformation vectors.

## 2. Explant Source and Sterilization

For most applications, we prepare leaf discs from plants aseptically grown *in vitro*. These provide a convenient and uniform source of starting material and minimize the chances of contamination. It is advisable to maintain a continuous supply of starting material. For most applications, a sufficient number of explants can be prepared from five to ten Magenta boxes. Leaf discs may also be prepared from young plants grown in the greenhouse. These must be sterilized (10% commercial bleach for 20 min, followed by extensive rinsing in sterile, distilled water) before being cultured *in vitro*, which may damage the tissue and reduce the efficiency of transformation. Insufficient sterilization will result in contamination. However, leaf discs prepared from greenhouse-grown plants are more convenient if very large numbers of explants are required.

### *a. Seed Sterilization*

1. Place seeds in a filter paper bag made from Whatman No. 2 filter paper and stapled at the ends.
2. Immerse the bags in 70% ethanol for 1 min and transfer to a large beaker containing 2 liters of bleach (6% sodium hypochlorite, Javex, Bristol-Myers) and a few drops of Tween 20. Sterilize for 25 min with gentle mixing on a stir-plate placed in a laminar

**TABLE 3**
**Preparation of Media Components for Plant Tissue Culture and Transformation**

**MS Micronutrients (Prepare 100× stocks of 100 ml in Whirl-Pack bags at –20°C)**

| | 1× (l) (in mg) | For 1 liter of 100× (g) |
|---|---|---|
| $H_3BO_3$ | 6.2 | 0.62 |
| $MnSO_4 \cdot 4H_2O$ | 22.3 | 2.23 |
| $ZnSO_4 \cdot 7H_2O$ | 8.6 | 0.86 |
| $Na_2MoO_4 \cdot 2H_2O$ | 0.25 | 0.025 |
| $CuSO_4 \cdot 5H_2O$ | .025 | 0.0025 |
| $CoCl_2 \cdot 6H_2O$ | .025 | 0.0025 |

**B5 Vitamins (Prepare 100× stocks of 100 ml in Whirl-Pack bags at –20°C)**

| | 1× (l)(in mg) | For 1 liter of 100× stock (g) |
|---|---|---|
| Myo-inositol | 100.0 | 10.0 |
| Nicotinic acid | 1.0 | 0.1 |
| Pyridoxine·HCl | 1.0 | 0.1 |
| Thiamine·HCl | 10.0 | 1.0 |

**MS Macronutrients (Prepare each time 5× MSB5 stock is made)**

| | 1× (l) (g) | For 2 liter of 5× stock (g) |
|---|---|---|
| $NH_4NO_3$ | 1.65 | 16.5 |
| $KNO_3$ | 1.90 | 19.0 |
| $MgSO_4 \cdot 7H_2O$ | 0.37 | 3.7 |
| $KH_2PO_4$ | 0.17 | 1.7 |
| Fe-330 | 0.04 | 0.4 |

**$CaCl_2 \cdot 2H_2O$ (Store in bottle in refrigerator)**

| 1× (l) (g) | For 100 ml of stock (g) |
|---|---|
| 0.4 | 15.0 |

**KI (Store in bottle in refrigerator)**

| 1× (l) (in mg) | For 100 ml of stock (g) |
|---|---|
| 0.83 | 0.075 |

flow hood. To ensure that the filter paper bags remain submersed, a smaller beaker may be placed on top of the bleach solution.

3. Rinse the bags several times (four to five times) in 2 liters of sterile distilled water with gentle mixing.
4. Remove the bags from the water using sterile forceps, place in a sterile Petri plate, and allow to dry under laminar flow (usually overnight with the lid of the Petri plate left slightly open).

### *b. Culture of Plants In Vitro*

1. Spread several seeds on top of a Petri plate containing 25 ml of MSO medium, seal with Parafilm, and incubate 3 to 4 weeks.
2. Remove individual seedlings from the Petri plates, and place upright in Magenta boxes

**TABLE 4**
**Preparation of Media for Plant Tissue Culture and Transformation**

**Preparation of 5× MSB5**

1. Prepare 5× stock of macronutrient in 1.5 l water
2. Add the following to macronutrient solution:100 ml of 100× micronutrient stock (1 bag), 100 ml of 100× vitamin stock (1 bag), 10 ml of KI stock 29.5 ml $CaCl_2 \cdot 2H_2O$ stock
3. Adjust pH to 5.7 using KOH
4. Adjust volume to 2 l, and dispense in Whirl-Pack bags (100 ml)

**Preparation of MSO Medium (500 ml)**

1. Thaw 100 ml of 5× MSB5 stock in ca. 300 ml $H_2O$
2. Add 10 g sucrose (2%)
3. Once the sucrose has dissolved, pH to 5.7 using KOH
4. Adjust volume to 500 ml
5. Add 4 g Bacto agar or 3 g agarose, autoclave

**Preparation of MS104 Medium (500 ml)**

1. Thaw 100 ml of 5× MSB5 stock in approximately 300 ml $H_2O$
2. Add 500 µl of BAA (1.0 mg/ml stock)
3. Add 250 µl of NAA (0.2 mg/ml stock)
4. Add 15 g sucrose (3%)
5. Once the sucrose has dissolved, pH to 5.7 using KOH
6. Adjust volume to 500 ml
7. Add 4 g Bacto agar or 3 g agarose, autoclave

**Preparation of MSD2 (500 ml)**

1. Thaw 100 ml of 5× MSB5 stock in approximately 300 ml $H_2O$
2. Add 2 ml of 2,4-D (0.5 mg/ml stock water)
3. Add 15 g sucrose (3%)
4. Once the sucrose has dissolved, pH to 5.7 using KOH
5. Adjust volume to 500 ml, and dispense 25 ml per 125 ml Erlenmeyer flask

containing 50 ml MSO medium. Whole seedlings may be transferred, or stems of seedlings may be cut using a sharp scalpel and placed in Magenta boxes. Culture one to two seedlings per box.

3. After 3 to 4 weeks, the shoot apex of the plants should be near the top of the Magenta box. Cut the stem beneath the second leaf and transfer the apical bud to a fresh Magenta box containing MSO.

3a. For rapid establishment of numerous sterile plants, axillary buds may also be excised by cutting the stem a few mm above and below the petiole and transferring to fresh medium.

### 3. Preparation of Leaf Discs and *Agrobacterium* Infection

We find it convenient to cut the leaf into strips using a sterile scalpel. If desired, discs may be prepared using a paper punch or cork borer. Although listed as optional by other groups,[16] we choose to preculture the leaf strips prior to *Agrobacterium* infection, and use a feeder-layer for the first 4 to 5 days of culture. Although the leaf discs may be cocultured with *Agrobacterium* for up to 1 week,[24] we find that incubation for more than 2 to 3 days increases the rates of contamination. Finally, selection of transformed shoots is performed using kanamycin at 100 μg/ml. We find that control discs which are not infected with *Agrobacterium*, or which are infected with a strain containing a helper plasmid alone, will not form shoots at this concentration of kanamycin. It is recommended that these controls be performed in parallel to the transformation-selection experiment, as well as controls where the discs are cultured in the absence of selection pressure.

1. Remove leaves from Magenta boxes and place on a sterile surface (e.g., Petri dish). As leaves will wilt rapidly under laminar flow, remove only one leaf at a time and work rapidly. It is also advisable to keep the Magenta boxes covered throughout the procedure to prevent wilting of the plants.
2. Remove the midvein and cut strips of approximately 1 $cm^2$ from the leaf blade. As *Agrobacterium* interacts with wounded cells, cutting a thin strip along the outer edge of the leaf will increase the number of cells competent for transformation.
3. Place the strips (or discs), epidermis side down, on nurse culture dishes. These are prepared by spreading 3 ml of a 4 to 5 day-old suspension culture over 25 ml MS104 medium, which is then covered with two layers of sterile Whatman No. 2 filter paper. If needed, trim the filter paper to fit into the Petri dishes prior to sterilization.
4. Seal dishes with Parafilm and incubate for 2 days.
5. The same day as the discs are prepared, start one or a few cultures of the appropriate *Agrobacterium* strain(s) in LB medium supplemented with appropriate antibiotics to ensure that the vector is maintained.
6. Pour a small amount (10 ml) of the bacterial culture into a 25 × 60 mm Petri dish. If desired, the cells may be washed and resuspended in liquid MSO medium and diluted appropriately; however, we routinely infect leaf discs directly in the undiluted LB cultures containing antibiotics.
7. Transfer a few leaf discs (three to ten) to the Petri dish, and quickly immerse in the *Agrobacterium* culture. Ensure that the entire cut surface is brought in contact with the bacterial culture, but do not submerge the discs for more than a few seconds, as this could damage the tissue and does not increase transformation frequencies.
8. Transfer the discs onto sterile Whatman No. 2 filter paper, and gently blot to remove excess bacterial culture. Leaf strips often curl after a few days in culture, and it is important to ensure that liquid is blotted from areas that are hard to access.
9. Return explants to the nurse culture dishes, epidermis side down, seal with Parafilm, and incubate for 2 to 3 days.
10. Transfer discs to MS104 medium, supplemented with 500 μg/ml cefotaxime or carbenicillin to inhibit bacterial growth and 100 μg/ml kanamycin to select for transformed shoots.
11. Transfer leaf discs to fresh medium every 2 to 3 weeks until callus and shoots arise.

## 4. Excision and Rooting of Kanamycin Resistant Shoots

Callus and shoots first start to appear after 3 to 5 weeks in culture. We usually separate independent areas of growth at this stage by cutting the leaf discs into sectors. To ensure that every shoot represents an independent transformation event, only one shoot is excised from each of these sectors. Although it is tempting to excise shoots as soon as they appear, it is advisable to permit them to grow until a well-defined stem, which can be cut, develops. Explants containing small shoots can, however, be transferred from regeneration medium to Magenta boxes containing MSO supplemented with 100 μg/ml kanamycin and 500 μg/ml cefotaxime or carbenicillin to accelerate shoot elongation. This transfer also minimizes the amount of time to which tissues are exposed to growth regulators, thereby reducing the chances of inducing somaclonal variations. Once transgenic shoots have rooted, they may be transferred to the greenhouse to flower and set seeds. Generally, we vegetatively propagate original transgenic plants by apical bud cutting prior to transferring the plant to the greenhouse and maintain one clone *in vitro* for further analysis or in the event that the clone transferred to the greenhouse does not survive.

1. Cut the stem beneath the second leaf, and transfer to a Magenta box containing MSO supplemented with 100 μg/ml kanamycin and 500 μg/ml cefotaxime or carbenicillin. Care should be taken to remove all traces of callus from the shoot. Do not excise bleached, variegated, vitrified, or otherwise morphologically abnormal shoots.
2. Observe the shoots periodically for root formation. Should a callus form at the cut edge, or if there are no signs of root formation after 3 weeks in culture, cut the stem a few millimeters above the cut edge and return to a fresh Magenta box. Discard shoots which fail to form roots after this point. We routinely observe that more than 1/3 of kanamycin-resistant shoots fail to produce roots in the presence of this antibiotic.
3. Once a root system appears, remove the plant from the Magenta box. A gentle touch is required so as to not snap the stem.
4. Thoroughly remove agar from roots under running water tap or in a water basin.
5. Transfer plants to soil in pots (approximately 5 in. diameter) and water thoroughly.
6. Place a plastic bag over the pot to retain humidity, and allow plants to acclimate to conditions of lower humidity. Remove the plastic bag after 5 to 7 days.

### 5. Confirmation of Transformation and Analysis of T-DNA Copy Number and Structure

Before any detailed analysis of transgene expression is performed on transformed plants, it is highly recommended that the copy number and the extent of structural rearrangements within the T-DNA be assessed. It is desirable to analyze the expression of transgenes within plants which contain only single, intact copies of the T-DNA. However, several instances of multiple or abberant T-DNA transformation events have been reported.[2,6,25,26] Factors that can influence T-DNA integration patterns are too numerous to list. They include most steps of explant preparation, culture, transformation, and selection, as well as the type of transformation vector.

The number of loci segregating for the presence of functional T-DNA insertions can be readily determined by analysis of the kanamycin-resistance marker in the selfed progeny of the original transformants. To ensure self-fertilization, we place wax paper bags over the flower heads prior to pollen anthesis. Capsules are collected from the transformants and surface-sterilized as described in Section II.C.2.A. Seeds are germinated on MSO medium containing 100 μg/ml kanamycin. After 5 to 6 weeks, the number of kanamycin resistant (green) to kanamycin sensitive (bleached) seedlings is recorded and analyzed by the chi-squared test.

Southern blot analysis using T-DNA sequences as hybridization probes represents a more direct approach in determining T-DNA copy number. This method also enables one to assess structural rearragements which may have occurred within the T-DNA during transfer, such as deletions[25] and the formation of tandem repeats.[26] T-DNA copy number is best estimated by analyzing the number of "border" fragments generated by hybridizing DNA from transgenic plants with probes specific to the ends of the T-DNA (for an example of this, see Fobert et al.[15]). The size of such fragments depends on the location of appropriate restriction sites within the flanking chromosomal DNA, and varies between different integration events. As the formation of tandem repeats is a common occurrence with certain transformation vectors,[26] it is advisable to calculate beforehand the expected sizes of border fragments which would be generated by direct or indirect repeats. The use of multiple hybridization probes, recognizing both ends of the T-DNA, is also highly recommended.

## III. DIRECT GENE TRANSFER

Despite the fact that the plant host-range of *Agrobacterium* species is known to be broad, a present limitation to its use is that some major cereal crop species are recalcitrant to this mode of gene transfer for reasons that are not sufficiently understood. It is necessary, therefore, to consider alternative modes. A number of such methods, dubbed collectively as "direct transfer" methods, exist. Among these, our preference is to advocate those that allow the use of intact cells or tissues rather than plant protoplasts because they enable one to bypass long-standing difficulties in regenerating plants from the protoplasts of the particular species that have been recalcitrant to *Agrobacterium.* Table 5 provides an introduction to several attempted or emerging techniques for direct gene transfer into whole cells, tissues, or protoplasts. Recently, methods based on the use of DNA-coated microprojectiles (Figure 2) have advanced significantly. This will be the second procedure which we will describe and assess in detail.

### A. MICROPROJECTILE-MEDIATED TRANSFORMATION

This method was invented by J. C. Sanford, T. M. Klein, E. D. Wolf, and N. Allen in 1984 at Cornell University (U.S.A.). It is based on the acceleration of microprojectiles (1 to 4 μm) to speeds (300 to 600m/s) sufficient for the penetration of plant cell walls and membranes.[27-29] The methodology was first used to deliver DNA and RNA into the large epidermal onion (*Allium cepa*) cells with expression of a chimaric CAT gene and transcription of the complete tobacco mosaic virus (TMV) RNA genome.[30]

The novelty of the microprojectile-mediated delivery method is its potential universality. It has been used with tissues of plants (Table 6) and animals,[31-33] green algae,[34-37] and yeast.[38-40] Monocot and conifer species that are recalcitrant to *Agrobacterium*-mediated DNA transfer have been transformed with this method. In some cases, transgenic plants were recovered (Table 6). Furthermore, successful delivery of genes into chloroplasts[35,37,41-44] and mitochondria[38,39,41] has been demonstrated.

The mechanics of the method can be divided into two components: the propelling mechanism and the type of particles to be accelerated (Table 7). Two processes for particle acceleration have been used for plant species: (1) acceleration by a macroprojectile and (2) acceleration by transferred impulse (Table 7). Both processes generate sufficient speed for the microprojectiles to penetrate cells. The most commonly used microprojectiles are gold or tungsten microparticles of 1 to 4 μm diameter. The DNA or RNA is coated to the surface of these particles by precipitation with $CaCl_2$, spermidine, or PEG. Variation in the propelling force and the design of the acceleration device have resulted in six different apparati (Table 7) used for gene delivery into plant cells. The Biolistic™ device is the most widely used design and employs either gun powder or helium gas to generate the propelling force. Both gold and tungsten microparticles can be used with the gun powder device; however, only gold can be used with the helium device. The other types of propelling forces are nitrogen gas, compressed air, and electric discharge.

Compared with other methods, microprojectile-mediated delivery is simple and rapid to employ. It has been adapted for studying gene expression in a variety of tissues, such as pollen, anthers, and petals of both tomato and tobacco[45] or the aleurone and embryos of *Zea mays.*[46,47] The major drawback of the method is the inherent variability caused by factors such as variation in the size of the microprojectiles, inconsistency in the strength of the vacuum and propelling force, and the uneven surfaces and morphology of the target tissue. All of these factors are very difficult to control within and among experiments, and therefore the relative level of gene expression is difficult to compare among tissues with certainty. For detailed comparisons with other DNA delivery methods, the reader should consult Howe,[48] Potrykus,[49-51] and Sanford.[27-29]

**TABLE 5**
**Procedures for the Direct Transfer of DNA Molecules into Plant Cells Without the Intervention of *Agrobacterium***

| Host cell type | Procedure | Comments |
|---|---|---|
| Intact cells or tissues | Microperforation of cells with silicon carbide fibers coated with DNA.[81] The perforation is accomplished by vigorous mixing | Transient gene expression is reported for corn and tobacco<br>Transgenic plants were not obtained<br>Evaluation is incomplete |
| | Use of laser microbeams on cells prior to the addition of DNA[82] | Transient gene expression is reported for embryos of rapeseed. Transgenic plants have not been obtained. Evaluation is incomplete. |
| | Deposition or injection of DNA into floral parts. In one procedure involving rice, DNA is deposited on a style at a position where the stigma was cut off.[83,84] In another procedure involving rye, a syringe is used for the injection of DNA into floral tillers[85] | Transgenic rice has been reported from the first procedure. The reported frequency of transformation is lower using the latter procedure. Both procedures need further evaluation |
| | Microinjection of DNA into proembryos in culture[86] | Transgenic rapeseed plants have been reported but the evaluation is incomplete |
| | Soaking seeds in aqueous solutions of DNA[87] | Seeds of several plant species were tested but only transient gene expression has been reported. |
| Protoplasts | Protoplasts treated with polyethylene glycol (PEG) or polyvinyl alcohol (PVA) are subjected to a high voltage electrical pulse in the presence of DNA. The procedure is referred to as electroporation.[88,89]<br>PEG or PVA have also been used without electroporation in which case transformation frequencies tend to be lower | Transgenic plants have been obtained from corn, rice, lettuce, rapeseed, and tobacco. The procedure is suitable for plant cells from which protoplasts can be prepared and regenerated. Transient expression has been observed in many other cases |
| | Protoplasts fused to liposomes in which DNA is encapsulated[90] | Regenerant transgenic plants have been obtained from tobacco |
| | Microinjection of DNA into protoplasts[91,92] | Transformed tobacco and alfalfa calli have been obtained. Transformation frequencies are high. Procedures are technically demanding |
| | Fusion of plant protoplasts with bacterial spheroplasts[93,94] | *Vinca rosea* and tobacco have been transformed. Transgenic plants were not regenerated |

Although all reporter genes may be employed, genes such as GUS and *lux* that allow the detection of single-cell transformation events by histochemical means are particularly useful. One measurement used for evaluating differences in gene expression levels has been the number of islands of cells (expression units[52]) that express the introduced reporter gene (GUS). If the level of gene expression exceeds a certain threshold, differences in the level

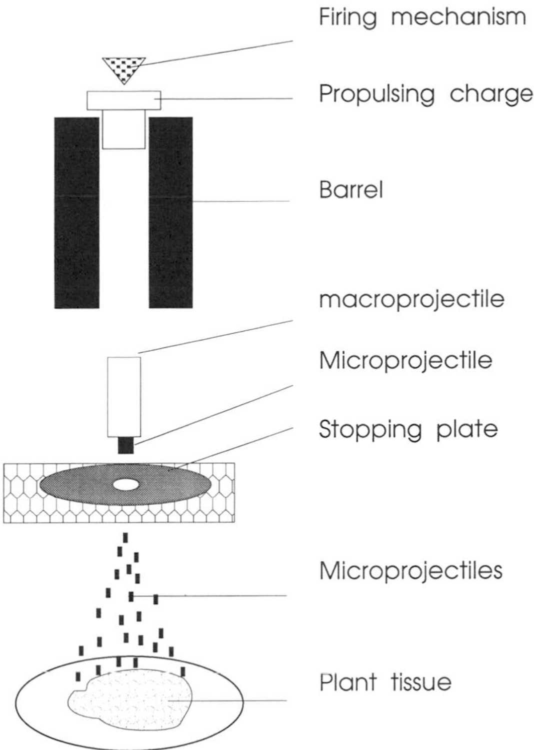


**FIGURE 2.** Schematic of the Biolistic™ microprojectile gun. The whole set up is enclosed in a vacuum chamber. Generally, the strength of vacuum is 800 mm of Hg. The microprojectiles are tungsten or gold, coated with DNA.

**TABLE 6**
**Microprojectile-Mediated Transformation**

| Recipient tissues[a] | Plants | Tissues recovered | Ref. |
|---|---|---|---|
| Cell suspension culture | Cotton | Plants | 95 |
| | Tobacco | Plants | 96 |
| | Corn | Plants | 97 |
| | Sorghum | Calli | 98 |
| Callus cultures | Tobacco | Calli | 99 |
| | Hybrid poplar | Plants | 100 |
| Embryogenic calli | Norway spruce | Calli | 53 |
| | Corn | Plants | 46, 97 |
| | Papaya | Plants | 101 |
| Immature embryos | Soybean | Plants | 102, 103, 104 |
| | Papaya | Plants | 101 |
| Hypocotyl segments | Papaya | Plants | 101 |
| Leaves | Tobacco | Plants | 96 |

[a] A wide variety of cells, tissues and organs have been used to examine transient expression in many species; however, they are not included in this table.

of gene expression cannot be measured.[53] In such cases, the fluorescent assay is appropriate (Charest, unpublished).

The following protocol has been used successfully in the author's laboratory for stable transformation of tobacco and hybrid poplar or for transient expression studies in somatic

**TABLE 7**
**Microprojectile Delivery Systems**

| Propelling force | Propelling Mechanism: Macroprojectiles | Propelling Mechanism: Transferred impulse | Microprojectile | Ref. |
|---|---|---|---|---|
| Gun powder | Teflon | — | Tungsten or gold | 27, 30 |
| Helium | — | Kapton disc | Gold | 105 |
| Nitrogen | Teflon | — | Gold | 106 |
| Compressed air | Wool pads | — | Tungsten | 107 |
| Compressed air | Plastic | — | Gold | 108 |
| Electric discharge | — | Retaining screen | Gold | 109 |
| Compressed air | — | — | Liquid microdroplets Gold | 99 |

embryos of white spruce, black spruce, red spruce, hybrid larch, European larch, Japanese larch, and cell suspensions and explants of several hybrid poplars, willow, and aspen. The apparatus is the Biolistic™ PD-1000 from Dupont that uses gunpowder as the propelling force (Figure 2). The particles that gave the best results are the 1.6 μm gold particles. The procedure can be separated into two parts: the preparation of the microprojectiles and the bombardment of the tissue.

## 1. Preparation of Sterile Particles and Precipitation of DNA

1. Add 1 ml of absolute ethanol to 60 mg of gold (or tungsten) particles in an Eppendorf tube. Store at –20°C for several hours or overnight.
2. Once the particles are settled at the bottom of the tube, remove the ethanol and resuspend the particles in 1 ml of sterile distilled water.
3. Centrifuge for 2 min at 10,000 g. Remove the water, and repeat the washing procedure twice.
4. Resuspend the particles in 500 μl of sterile distilled water, and put on ice.
5. After vigorously vortexing the particles, remove 25 μl, and to this aliquot add 10 μl of DNA (1 μg/ml). Vortex the mixture.
6. Add 50 μl of $CaCl_2$ (2.5 m*M*), and vortex the mixture again. To this, add 20 μl of ice cold spermidine (0.1 m*M*), and vortex the solution.
7. Allow settling at room temperature for 5 to 10 min. Remove 80 μl of the supernatant without disturbing the particles at the bottom of the tube. The particles are ready for use and should be kept on ice. They should not be used after 4 h.

## 2. Bombardment of the Plant Tissues

1. Clean the chamber of the apparatus with 95% ethanol before use. The stopping plates and the macroprojectiles are sterilized by washing overnight with 95% ethanol.
2. Spread the plant tissue over an area of 4 cm (diameter) in a 100 × 15mm Petri dish. The spread can be done over filter paper for easy transfer or directly on tissue culture medium. A screen can be used on top of the tissue (e.g., leaf explants) to immobilize them. Record the exact weight of the tissue for expression studies. The age and type of tissue will greatly affect the results and must be optimized for each species and type of tissue.
3. Place the Petri dish in the chamber, 13 cm from the stopping plate (Figure 2).

4. Place the stopping plate in the holder, and tighten the retaining screws.
5. Vortex the particle-DNA mixture and deposit 2.5 µl onto the macroprojectile. This is loaded into the barrel of the unit later.
6. Place a 0.22 caliber blank charge (730 to 735 mg) in the barrel behind the macroprojectile.
7. Engage the firing pin mechanism on top of the barrel, and apply a partial vacuum of 800 mm of Hg.
8. Fire and vent the apparatus. The machine must be cleaned every five bombardments to remove residual gunpowder and tissue material.

The variables that must be optimized for this method are of two kinds. First, the physical variables include the size and type of microprojectiles, the distance of the stopping plate, the distance of the sample from the stopping plate, the strength of the vacuum, and the disposition of the sample on the Petri dish. Second, the biological variables comprise the vector construction, the quantity of DNA, the cell cycle, and the subculture cycle.

## IV. TRANSFORMED SPECIES

The host ranges of a limited number of *A. tumefaciens* strains indicate a wide variety of species that may be potentially transformed by *Agrobacterium*-mediated systems. Since the surveys providing this information depend on the appearance of tumors,[54] they no doubt underestimated the range of species that may be amenable to *Agrobacterium*-mediated systems. For instance, cereals were believed to be recalcitrant to infection by *Agrobacterium*; however, viral cDNA inserted into a T-DNA vector was delivered by the process of agroinfection,[55] which depends on the Ti plasmid encoded delivery system. Furthermore, the host-ranges of naturally occurring strains and Ti plasmids are known to differ,[56-59] yet vectors and helper Ti plasmids have only been produced from a few of them (Table 1). As a first step towards improving the delivery systems, the genes governing the transfer process must be better understood so they may be combined eventually in more favorable combinations and aided by other genes. For now, species such as the cereals and gymnosperms are not transformed routinely by *Agrobacterium*-mediated systems, and therefore, alternatives such as microprojectile-mediated systems are needed.

After transformation events have occurred, a major obstacle to the recovery of transgenic plants is the effectiveness of culture systems for growth and regeneration of the cells. Table 8 provides examples of species for which transgenic plants have been generated. The prominent role played by *Agrobacterium*-mediated systems is evident in this table. A list of species for which transformation has been obtained without transgenic plant regeneration is too long to present in this chapter and includes many species considered to be difficult to work with. For many research applications, plant regeneration is not necessary.

In the future, transgenic crops will play a major role in agriculture. The influence of the inserted genes and of the processes for regenerating transgenic plants on the agronomic performance of a crop can only be assessed under field conditions. In Canada and the United States, the list of transgenic field trials (Table 9) is relatively modest compared with the list of transformed species; however, the use of transgenic plants in agriculture is increasing rapidly.

## ACKNOWLEDGMENTS

The authors wish to thank T. White for critical review of the manuscript. Their research is supported by Agriculture Canada, the Natural Sciences and Engineering Research Council of Canada (NSERC), and Forestry Canada. P. Fobert was the recipient of an NSERC postgraduate scholarship, Plant Research Center Contribution Number 1382.

**TABLE 8**
**Examples of Transgenic Plants Which Have Been Generated**

| Plants[a] | Methods[b] | Ref.[c] |
|---|---|---|
| | **Dicots** | |
| Cruciferous | | |
| Arabidopsis | *Agrobacterium* | 17 |
| Mustard | *Agrobacterium* | 110 |
| Rapeseed | *Agrobacterium* | 19 |
| Flax | *Agrobacterium* | 111 |
| Leguminous | | |
| Alfalfa | *Agrobacterium* | 112 |
| Pea | *Agrobacterium* | 113 |
| Soybean | *Agrobacterium* | 114 |
| Soybean | Microprojectiles | 104 |
| Trefoil | *Agrobacterium* | 115 |
| White clover | *Agrobacterium* | 116 |
| Solanaceous | | |
| Eggplant | *Agrobacterium* | 117 |
| Petunia | *Agrobacterium* | 16 |
| Potato | *Agrobacterium* | 118 |
| Tobacco | *Agrobacterium* | 16 |
| Tomato | *Agrobacterium* | 119 |
| Others | | |
| Apple | *Agrobacterium* | 120 |
| Aspen | *Agrobacterium* | 121 |
| Belladona | *Agrobacterium* | 122 |
| Black currant | *Agrobacterium* | 123 |
| Carrot | *Agrobacterium* | 124 |
| Celery | *Agrobacterium* | 125 |
| Cotton | *Agrobacterium* | 126 |
| Cucumber | *Agrobacterium* | 127 |
| Grape | *Agrobacterium* | 128 |
| Horseradish | *Agrobacterium* | 129 |
| Lemon | DNA uptake | 130 |
| Lettuce | *Agrobacterium* | 131 |
| Morning glory | *Agrobacterium* | 132 |
| Muskmelon | *Agrobacterium* | 133 |
| Neem | *Agrobacterium* | 134 |
| Poplar | *Agrobacterium* | 135 |
| Strawberry | *Agrobacterium* | 136 |
| Sugar beet | *Agrobacterium* | 137 |
| Sunflower | *Agrobacterium* | 138 |
| Walnut | *Agrobacterium* | 139 |
| | **Monocots** | |
| Asparagus | *Agrobacterium* | 140 |
| Corn | DNA uptake | 141 |
| Corn | Microprojectiles | 142 |
| Orchard grass | DNA uptake | 143 |
| Rice | DNA uptake | 144 |
| Rye | Injection | 85 |

[a] The plants cited do not represent a comprehensive listing of transgenic plants reported, but rather examples.

[b] The methods cited represent those most commonly used. For plants such as tobacco or rapeseed, transgenic plants have been generated by a variety of methods. The references cited are examples only.

**TABLE 9**
**Transgenic Field Trials Approved in Canada and the United States Prior to 1990**

| Crop | Number of applications approved[a] |
|---|---|
| Alfalfa | 8 |
| Canola (rapeseed) | 26 |
| Cantaloupe/squash | 7 |
| Corn | 5 |
| Cotton | 13 |
| Cucumber | 2 |
| Flax | 10 |
| Poplar | 1 |
| Potato | 21 |
| Rice | 3 |
| Soybean | 8 |
| Tomato | 30 |
| Walnut | 1 |

[a] Data obtained from Agriculture Canada and the U.S. Department of Agriculture.

# REFERENCES


1. **Nester, E. W. and Kosuge, T.,** Plasmids specifying plant hyperplasias, *Annu. Rev. Microbiol.*, 35, 531, 1981.
2. **Zambryski, P.,** Basic processes underlying *Agrobacterium*-mediated DNA transfer to plant cells, *Annu. Rev. Genet.*, 22, 1, 1988.
3. **Miki, B. and Iyer, V. N.,** Fundamentals of gene transfer in plants, in *Plant Physiology, Biochemistry and Molecular Biology*, Dennis, D. T. and Turpin, D. H.,.Eds., Longman, U.K., 1990 p. 481.
4. **Charest, P. J. and Michel, M.-F.,** Basics of plant genetic engineering and its potential application to tree species, Information Report PI-X-104, Petawawa Natl. Forestry Institute, Forestry Canada, 1991.
5. **Hoekema, A., Hirsch, P. R., Hooykaas, P. J. J., and Schilperoort, R. A.,** A binary plant vector strategy based on separation of VIR- and T-region of the *Agrobacterium tumefaciens* Ti-plasmid, *Nature*, 303, 179, 1983.
6. **Bevan, M.,** Binary *Agrobacterium* vectors for plant transformation, *Nucl. Acids Res.*, 12, 8711, 1984.
7. **Rogers, S. G. and Klee, H.,** Pathways to plant genetic manipulation employing *Agrobacterium*, in *Plant DNA Infectious Agents*, Hohn, T. and Schell, J., Eds., Springer-Verlag, New York, 1987, p. 179.
8. **Klee, H. J., Horsch, R., and Rogers, S.,** *Agrobacterium*-mediated plant transformation and its further applications to plant molecular biology, *Annu. Rev. Genet.*, 20, 467, 1987.
9. **Simpson, R. B., Spielmann, A., Margossian, L., and McKnight, T. D.,** A disarmed binary vector from *Agrobacterium tumefaciens* functions in *Agrobacterium rhizogenes*, *Plant Mol. Biol.*, 6, 403, 1986.
10. **Sproule, A., Donaldson, P., Dijak, M., Bevis, E., Pandeya, R., Keller, W. A., and Gleddie, S.,** Fertile somatic hybrids between transgenic *Nicotiana tabacum* and transgenic *N. debneyi* selected by dual antibiotic resistance, *Theor. Appl. Genet.*, 82, 450, 1991.
11. **Matzke, M. A. and Matzke, A. J. M.,** Differential inactivation and methylation of a transgene in plants by two suppressor loci containing homologous sequences, *Plant Mol. Biol.*, 16, 821, 1991.
12. **Koncz, C., Martini, N., Mayerhofer, R., Koncz-Kalman, Z., Körber, H., Redei, G. P., and Schell, J.,** High frequency T-DNA-mediated gene tagging in plants, *Proc. Natl. Acad. Sci. U.S.A.*, 86, 8467, 1989.
13. **Jefferson, R. A.,** Assaying chimeric genes in plants: the GUS gene fusion system, *Plant Mol. Biol. Rep.*, 5, 387, 1987.
14. **Robert, L. S., Donaldson, P. A., Ladaigue, C., Altosaar, I., Arnison, P. G., and Fabijanski, S. F.,** Antisense RNA inhibition of β-glucuronidase gene expression in transgenic tobacco can be transiently overcome using a heat-inducible β-glucuronidase gene construct, *Bio/Technology*, 8, 459, 1990.
15. **Fobert, P. R., Miki, B. L., and Iyer, V. N.,** Detection of gene regulatory signals in plants revealed by T-DNA-mediated fusions, *Plant Mol. Biol.*, 17, 837, 1991.

16. **Horsch, R. B., Fry, J. E., Hoffman, N. L., Eichholtz, D., Rogers, S. G., and Fraley, R. T.,** A simple and general method for transferring genes into plants, *Science,* 227, 1229, 1985.
17. **Valvekens, D., Van Montagu, M., and Vijsebettens, M.,** *Agrobacterium tumefaciens*-mediated transformation of *Arabidopsis thaliana* root explants by using kanamycin selection, *Proc. Natl. Acad. Sci. U.S.A.*, 85, 5536, 1988.
18. **De Block, M., De Brouwer, D., and Tenning, P.,** Transformation of *Brassica napus* and *Brassica oleracea* using *Agrobacterium tumefaciens* and the expression of the *bar* and *neo* genes in the transgenic plants, *Plant Physiol.*, 91, 694, 1989.
19. **Fry, J., Barnason, A., and Horsch, R. B.,** Transformation of *Brassica napus* with *Agrobacterium tumefaciens* based vectors, *Plant Cell Rep.*, 6, 321, 1987.
20. **Sheerman, S. and Bevan, M. W.,** A rapid transformation method for *Solanum tuberosum* using binary *Agrobacterium tumefaciens* vectors, *Plant Cell Rep.*, 7, 13, 1988.
21. **Draper, J., Scott, R., Armitage, P., and Walden, R.,** *Plant Genetic Transformation and Gene Expression, A Laboratory Manual*, Oxford: Blackwell Scientific Publications, 1988.
22. **Murashige, T. and Skoog, F.,** A revised medium for rapid growth and bioassays with tobacco tissue cultures, *Physiol. Plant,* 15, 473, 1962.
23. **Gamborg, O. L., Miller, R. A., and Ojima, K.,** Nutrient requirements of suspension cultures of soybean root cells, *Exp. Cell Res.*, 50, 151, 1978.
24. **Burow, M. D., Chlan, C. A., Sen, P., Lisca, A., and Murai, N.,** High-frequency generation of transgenic tobacco plants after modified leaf disk cocultivation with *Agrobacterium tumefaciens, Plant Mol. Biol. Rep.*, 8, 124, 1990.
25. **Deroles, S. C. and Gardner, R. C.,** Analysis of the T-DNA structure in a large number of transgenic petunias generated by *Agrobacterium*-mediated transformation, *Plant Mol. Biol.*, 11, 365, 1988.
26. **Jorgensen, R., Snyder, C., and Jones, J. D. G.,** T-DNA is organized predominantly in inverted repeat structures in plants transformed with *Agrobacterium tumefaciens* C58 derivatives, *Mol. Gen. Genet.*, 207, 471, 1987.
27. **Sanford, J. C., Klein, T. M. Wolf, E. D., and Allen, N.,** Delivery of substances into cells and tissues using a particle bombardment process, *Part. Sci. Technol.*, 5, 27, 1987.
28. **Sanford, J. C.,** The biolistic process, *Trends Biotech.*, 6, 299, 1988.
29. **Sanford, J. C.,** Biolistic plant transformation, *Physiol. Plant*, 79, 206, 1990.
30. **Klein, T. M., Wolf, E. D., Wu, R., and Sanford, J. C.,** High-velocity microprojectiles for delivering nucleic acids into living cells, *Nature*, 327, 70, 1987.
31. **Titomirov, A. V. and Zelenin, A. V.,** New methods of tranfection of mammalian cells: a minireview, *Mol. Biol.*, 359, 1155, 1989.
32. **Zelenin, A. V., Titomirov, A. V., and Kolesnikov, V. A.,** Genetic transformation of mouse cultured cells with the help of high-velocity mechanical DNA injection, *FEBS Lett.*, 244, 65, 1989.
33. **Yang, N. S., Burkholder, J., Roberts, B., Martinell, B., and McCabe, D.,** *In vitro* and *in vivo* gene transfer to mammalian somatic cells by particle bombardment, *Proc. Natl. Acad. Sci. U.S.A.*, 87, 9568, 1990.
34. **Debuchy, R., Purton S., and Rochaix, J.,** The arginosuccinate lyase gene of *Chlamydomonas reinhardtii*: an important tool for nuclear transformation and for correlating the genetic molecular maps of the ARG7 locus, *EMBO J.*, 8, 2803, 1989.
35. **Blowers, A. D., Bogorad, L., Shark K., and Sanford, J. C.,** Studies on *Chlamydomonas* chloroplast transformation: foreign DNA can be stably maintained in the chromosome, *Plant Cell,* 1, 123, 1989.
36. **Kindle, K. L., Schnell, R. A., Fernandez, R., and Lefebvre, P. A.,** Stable nuclear transformation of *Chlamydomonas* using the *Chlamydomonas* gene for nitrate reductase, *J. Cell. Biol.*, 109, 2589, 1989.
37. **Boynton, J. E., Gillham, N. W., Harris, E. H., Hosier, J. P., Johnson, A. M., Jones, A. R., Randolph-Anderson, B. L., Robertson, D., Klein, T. M., Shark, K. B., and Sanford, J. C.,** Chloroplast transformation in *Chlamydomonas* with high velocity microprojectiles, *Science,* 240, 1534, 1988.
38. **Fox, T. D., Sanford, J. C., and McMullin, T. W.,** Plasmids can stably transform yeast mitochondria lacking endogenous mt DNA, *Proc. Natl. Acad. Sci. U.S.A.*, 85, 7288, 1988.
39. **Johnston, S. A., Anzsiano, P. Q., Shark, K., Sanford, J. C., and Butow, R. A.,** Mitochondrial tranformation in yeast by bombardment with microprojectiles, *Science,* 240, 1538, 1988.
40. **Armaleo, D., Ye, G. N., Klein, T. M., Shark, K. B., Sanford, J. C., and Johnston, S. A.,** Biolistic nuclear transformation of *Saccharomyces cerevisiae* and other fungi, *Curr. Genet.*, 17, 97, 1990.
41. **Butow, R. A. and Fox, T. D.,** Organelle transformation: shoot first, ask questions later, *Trends Biotech.*, 15, 465, 1990.
42. **Daniell, H., Vivekananda, J., Nielsen, B. L., Ye, G. N., Tervari, K. K., and Sanford, J. C.,** Transient foreign gene expression in chloroplasts in cultured tobacco cells after biolistic delivery of chloroplast vectors, *Proc. Natl. Acad. Sci. U.S.A.*, 87, 88, 1990.
43. **Goldschmidt-Clermont, M.,** Transgenic expression of aminoglycoside adenine transferase in the chloroplast: a selectable marker for site directed transformation of *Chlamydomonas, Nucl. Acids Res.*, 15, 4083, 1991.

44. **Ye, G. N., Daniell, H., and Sanford, J. C.,** Optimization of delivery of foreign DNA into higher-plant chloroplasts, *Plant Mol. Biol.,* 809, 1990.
45. **Twell, D., Klein, T. M., Fromm, M. E., and McCormick, S.,** Transient expression of chimeric genes delivered into pollen by microprojectile bombardment, *Plant Physiol.,* 91, 1270, 1989.
46. **Goff, S. A., Klein, T. M., Roth, B. A., Fromm, M. E., Cone, K. C., Radicella, J. P., and Chandler, V. L.,** Transactivation of anthocyanin biosynthetic genes following transfer of B regulatory genes into maize tissues, *EMBO J.,* 2517, 1991.
47. **Ludwig, S. E., Bowen, B., Beach, L., and Wessler, S. R.,** A regulatory gene as a novel visible marker for maize transformation, *Science,* 247, 449, 1990.
48. **Howe, C. J.,** Organelle transformation, *Trends Genet.,* 4, 150, 1988.
49. **Potrykus, I.,** Gene transfer to cereals: an assessment, *Trends Biotech.,* 7, 269, 1989.
50. **Potrykus, I.,** Gene transfer to cereals: an assessment, *BioTech.,* 8, 535, 1990.
51. **Potrykus, I.,** Gene transfer to plants: assessment of published approaches and results, *Annu. Rev. Plant Physiol.,* 42, 205, 1991.
52. **Klein, T. M., Gradziel, T., Fromm, M. E., and Sanford, J. C.,** Factors influencing gene delivery into *Zea mays* cells by high-velocity microprojectiles, *Bio/Technology,* 6, 559, 1988.
53. **Robertson, D., Weissinger, A. K., Ackley, R., Glover, S., and Sederoff, R. R.,** Transient expression and stable transformation of Norway spruce (*Picea abies* L. Karst) somatic embryos using microprojectile bombardment, *Plant Mol. Biol.,* 19, 925, 1992.
54. **Decleene, M. and DeLey, J.,** The host range of crown gall, *Bot. Rev.,* 42, 389, 1976.
55. **Grimsley, N. and Bisaro, D.,** Agroinfection, in *Plant DNA Infectious Agents,* Hohn, T. and Schell, J., Eds., Springer-Verlag, New York, 1987.
56. **Yanofsky, M. F. and Nester, E. W.,** Molecular characterization of a host-range determining locus from *Agrobacterium tumefaciens, J. Bacteriol.,* 68, 244, 1986.
57. **Knauf, V., Yanofsky, M., Montoya, A., and Nester, E. W.,** Physical and functional map of an *Agrobacterium tumefaciens, J. Bacteriol.,* 160, 564, 1986.
58. **Close, T. J., Tait, R. C., and Kado, C. I.,** Regulation of Ti plasmid virulence genes by a chromosomal locus of *Agrobacterium tumefaciens, J. Bacteriol.,* 164, 774, 1985.
59. **Huang, M. L. W., Cangelosi, G. A., Halperin, W., and Nester, E. W.,** A chromosomal *Agrobacterium tumefaciens* gene required for effective plant signal transduction, *J. Bacteriol.,* 173, 1814, 1990.
60. **Koncz, C. and Schell, J.,** The promoter of TL-DNA gene 5 controls the tissue-specific expression of chimaeric genes carried by a novel type of *Agrobacterium* binary vector, *Mol. Gen. Genet.,* 204, 383, 1986.
61. **Hood, E. E., Helmer, G. L., Fraley, R. T., and Chilton, M. D.,** The hypervirulence of *Agrobacterium tumefaciens* A281 is encoded in a region of pTiB0542 outside of T-DNA, *J. Bacteriol.,* 178, 1291, 1986.
62. **Fraley, R. T., Rogers, S. G., Horsch, R. B., Eichholtz, D. A., Flick, J. S., Fink, C. L., Hoffman, N. L., and Sanders, P. R.,** The SEV system: a new disarmed Ti plasmid vector system for plant transformation, *Bio/Technology,* 3, 629, 1985.
63. **Lazo, G. R., Stein, P. A., and Ludwig, R. A.,** A DNA transformation-competent *Arabidopsis* genomic library in *Agrobacterium, Bio/Technology,* 9, 963, 1991.
64. **Fraley, R. T., Rogers, S. G., Horsch, R. B., Sanders, P. R., Flick, J. S., Adams, S. P., Bittner, M. L., Brand, L. A., Fink, G. L., Fry, J. S., Gallupi, G. R., Goldberg, S. B., Hoffman, N. L., and Woo, S. C.,** Expression of bacterial genes in plant cells, *Proc. Natl. Acad. Sci. U.S.A.,* 80, 4803, 1983.
65. **Vanden Elzen, P. J. M., Townsend, J., Lee, K. Y., and Bedbrook, J. R.,** A chimaeric hygromycin resistance gene as a selectable marker in plant cells, *Plant Mol. Biol.,* 5, 299, 1985.
66. **Hayford, M. B., Medford, J. I., Hoffman, N. L., Rogers, S. G., and Klee, J. H.,** Development of a plant transformation selection system based on expression of genes encoding gentamycin acetyltransferase, *Plant Physiol.,* 86, 1216, 1988.
67. **Jones, J. D. G., Svab, Z., Harper, E. C., Hurwitz, C. D., and Maliga, P.,** A dominant nuclear streptomycin resistance marker for plant cell transformation, *Mol. Gen. Genet.,* 210, 86, 1987.
68. **Svab, Z., Harper, E. C., Jones, J. D. G., and Maliga, P.,** Aminoglycoside-3″-adenyl transferase confers resistance to spectinomycin and streptomycin in *Nicotiana tabacum, Plant Mol. Biol.,* 14, 197, 1990.
69. **Hille, J., Verheggen, F., Roelvink, P., Franssen, H., Van Kammen, A., and Zabel, P.,** Bleomycin resistance: a new dominant selectable marker for plant cell transformation, *Plant Mol. Biol.,* 7, 171, 1986.
70. **Eichholtz, D. A., Rogers, S. G., Horsch, R. B., Klee, H. J., Hayford, M., Hoffman, N. L., Braford, S. B., Fink, C., Flick, J., O'Connell, K. M., and Fraley, R. T.,** Expression of mouse dihydrofolate reductase gene confers methotrexate resistance in transgene petunia plants, *Somatic Cell Mol. Genet.,* 13, 67, 1987.
71. **De Block, M., De Brouwer, D., and Tenning, P.,** Transformation of *Brassica napus* and *Brassica oleracea* using *Agrobacterium tumefaciens* and the expression of the *bar* and *neo* genes in the transgenic plants, *Plant Physiol.,* 91, 694, 1989.
72. **Charest, P. J., Hattori, J., De Moor, J., Iyer, V. N., and Miki, B. L.,** *In vitro* study of transgenic tobacco expressing *Arabidopsis* wild type and mutant acetohydroxyacid synthase genes, *Plant Cell Rep.,* 8, 643, 1990.

73. **Stalker, D. M., McBride, K. E., and Malyj, L. D.,** Herbicide resistance in transgenic plants expressing a bacterial detoxification gene, *Science,* 242, 419, 1988.
74. **Shah, D. M., Horsch, R. B., Klee, H. J., Kishore, G. M., Winter, J. A., Tumer, N. E., Hironaka, C. M., Sanders, P. R., Gasser, C. S., Aykent, S., Siegel, N. R., Rogers, S. G., and Fraley, R. T.,** Engineering herbicide tolerance in transgenic plants, *Science,* 233, 478, 1986.
75. **Guerineau, F., Brooks, L., Meadows, J., Lucy, A., Robinson, C., and Mullineaux, P.,** Sulfonamide resistance gene for plant transformation, *Plant Mol. Biol.,* 15, 127, 1990.
76. **Teeri, T. H., Lehväslaiho, H., Franck, M., Uotila, J., Heino, P., Palva, E. T., Van Montagu, M., and Herrera-Estrella, L.,** Gene fusions to *lacZ* reveal new expression patterns of chimeric genes in transgenic plants, *EMBO J.,* 8, 343, 1989.
77. **Ow, D. W., Wood, K. V., De Luca, M., De Wet, J. R., Helinski, D. R., and Howell, S. H.,** Transient and stable expression of the firefly luciferase gene in plant cells and transgenic plants, *Science,* 234, 856, 1986.
78. **Koncz, C., Olsson, O., Langridge, W. H. R., Schell, J., and Szalay, A. A.,** Expression and assembly of functional bacterial luciferase in plants, *Proc. Natl. Acad. Sci. U.S.A.,* 84, 131, 1987.
79. **De Block, M., Herrera-Estrella, L., Van Montagu, M., Schell, J., and Zambryski, P.,** Expression of foreign genes in regenerated plants and in their progeny, *EMBO J.,* 3, 1681, 1984.
80. **Sanders, P. R., Winter, J. A., Barnason, A. R., Rogers, S. G., and Fraley, R. T.,** Comparison of cauliflower mosaic virus 35S and nopaline synthase promoters in transgenic plants, *Nucl. Acids Res.,* 15, 1543, 1987.
81. **Kaeppler, H. F., Gu, W., Somer, D. A., Rines, H. W., and Cockburn, A. F.,** Silicon carbide fiber-mediated DNA delivery into plant cells, *Plant Cell Rep.,* 9, 415, 1990.
82. **Weber, G., Monajembashi, S., Geulich, K. O., and Wolfrum, J.,** Uptake of DNA in chloroplasts of *Brassicas napus* (L.) facilitated by a UV-laser microbeam, *Eur. J. Cell Biol.,* 49, 73, 1989.
83. **Duan, X. and Chen, S.,** Variation of the characters in rice (*Oryza sativa*) induced by foreign DNA uptake, *China Agr. Sci.,* 3, 6, 1985.
84. **Luo, Z. and Wu, R.,** A simple method for the transformation of rice via the pollen tube pathway, *Plant Mol. Biol. Rep.,* 7, 69, 1989.
85. **De Le Pena, J., Lorz, H., and Schell, J.,** Transgenic rye plants obtained by injecting DNA into young floral tillers, *Nature,* 325, 274, 1987.
86. **Neuhaus, G., Spangenberg, G., Mittelsten-Scheid, O., and Schweiger, H.-G.,** Transgenic rapeseed plants obtained by the microinjection of DNA into microspore-derived embryoids, *Theor. Appl. Genet.,* 75, 30, 1987.
87. **Topfer, F., Gronenborn, B., Schell, J., and Steinbiss, H. H.,** Uptake and transient expression of chimeric genes in seed-derived embryos, *Plant Cell,* 1, 133, 1989.
88. **Shillito, R. D., Saul, M. W., Paszkowski, J., Muller, M., and Potrykus, I.,** High efficiency direct gene transfer to plants, *Bio/Technology,* 3, 1099, 1985.
89. **Fromm, M. E., Taylor, L. P., and Walbot, V.,** Stable transformation of maize after gene transfer by electroporation, *Nature,* 319, 791, 1986.
90. **Deshayes, A., Herrera-Estrella, L., and Caboche, M.,** Liposome-mediated transformation of tobacco mesophyll protoplasts by an *Escherichia coli* plasmid, *EMBO J.,* 4, 2731, 1985.
91. **Reich, T. J., Iyer, V. N., and Miki, B. L.,** Efficient transformation of alfalfa protoplasts by the intranuclear microinjection of alfalfa protoplasts, *Bio/Technology,* 4, 1001, 1986.
92. **Crossway, A., Oakes, J. V., Irvine, J. M., Ward, B., Knauf, V. C., and Shewmaker, C. K.,** Integration of foreign DNA following microinjection of tobacco mesophyll protoplasts, *Mol. Gen. Genet.,* 202, 179, 1986.
93. **Hasezawa, S., Nagatat, and Syono, K.,** Transformation of *Vinca* protoplasts mediated by *Agrobacterium* spheroplasts, *Mol. Gen. Genet.,* 182, 206, 1981.
94. **Hain, R., Steinbiss, H. H., and Schell, J.,** Fusion of *Agrobacterium* and *E. coli* spheroplasts with *Nicotiana tabacum* protoplasts - direct gene transfer from microorganism to higher plants, *Plant Cell Rep.,* 3, 60, 1984.
95. **Finer, J. J. and McMullen, M. D.,** Transformation of cotton (*Gossypium hirsutum* L.) via particle bombardment, *Plant Cell Rep.,* 8, 586, 1990.
96. **Klein, T. M. Harper, E. C., Svab, Z., Sanford, J. C., Fromm, M. E., and Maliga, P.,** Stable genetic transformation of intact *Nicotiana* cells by the particle bombardment process, *Proc. Natl. Acad. Sci. U.S.A.,* 85, 8502, 1988.
97. **Fromm, M. E. Morrish, F., Armstrong, C., Williams, R., Thomas, J., and Klein, T. M.,** Inheritance and expression of chimeric genes in the progeny of transgenic maize plants, *Bio/Technology,* 8, 833, 1990.
98. **Hagio, T., Blavers, A. D., and Earle, E. D.,** Stable transformation of sorghum cell cultures after bombardment with DNA-coated microprojectiles, *Plant Cell Rep.,* 10, 260, 1991.
99. **Sautter, C., Waldner, H., Neuhaus-Url, C., Galli, A., Neuhaus, G., and Potrykus, I.,** Microtargeting: high efficiency gene transfer using a novel approach for the acceleration of micro-projectiles. *Bio/Technology,* 9, 1080, 1991.
100. **McCown, B. H., McCabe, D. E., Russell, D. R., Robinson, D. J., Barton, K. A., and Raffa, K. F.,** Stable transformation of *Populus* and incorporation of pest resistance by electric discharge particle acceleration, *Plant Cell Rep.,* 9, 590, 1991.

101. **Fitch, M. M. M., Manshardt, R. M., Gonsalves, D., Slighton J. L., and Sanford, J. C.,** Stable transformation of papaya via microprojectile bombardment, *Plant Cell Rep.*, 9, 189, 1990.
102. **Christou, P., McCabe, D. E., and Swain, W. F.,** Stable transformation of soybean callus by DNA-coated gold particles, *Plant Physiol.*, 87, 671, 1988.
103. **Christou, P., Swain, W. F., Yang, N. S., and McCabe, D. E.,** Inheritance and expression of foreign genes in transgenic soybean plants, *Proc. Natl. Acad. Sci. U.S.A.*, 86, 7500, 1989.
104. **McCabe, D. E., Swain, W. F., Martinell, B. J., and Christou, P.,** Stable transformation of soybean (*Glycine max*) by particle acceleration, *Bio/Technology*, 6, 923, 1988.
105. **Johnston, S. A.,** Biolistic transformation: microbes to mice, *Nature*, 346, 776, 1990.
106. **Morikawa, H., Iida, A., and Yamada, Y.,** Transient expression of foreign genes in plant cells and tissues obtained by a simple biolistic device (particle-gun), *Appl. Microbiol. Biotech.*, 31, 320, 1989.
107. **Oard, J. H., Paige, D. F., Simmonds, J. A., and Gradziel, T. M.,** Transient gene expression in maize, rice, and wheat cells using an airgun apparatus, *Plant Physiol.*, 92, 334, 1990.
108. **Iida, A., Morikawa, H., and Yamada, Y.,** Stable transformation of cultured tobacco cells by DNA-coated gold particles accelerated by gas-pressure-driven particle gun, *Appl. Microbiol. Biotech.*, 33, 560, 1990.
109. **Christou, P.,** Soybean transformation by electric discharge particle acceleration, *Physiol. Plant.*, 79, 210, 1990.
110. **Mathews, H., Bharathan, N., Litz, R. E., Narayanan, K. R., Rao, P. S., and Bhatia, C. R.,** Transgenic plants of mustard *Brassica jaucea* (L.) Czern and Coss, *Plant Sci.*, 72, 245, 1990.
111. **Jordan, M. C. and McHughen, A.,** Glyphosate tolerant flax plants from *Agrobacterium*-mediated gene transfer, *Plant Cell Rep.*, 7, 281, 1988.
112. **Shahin, E. A., Spielmann, A., Sukhapinda, K., Simpson, R. B., and Yashar, M.,** Transformation of cultivated alfalfa using disarmed *Agrobacterium tumefaciens*, *Crop Sci.*, 26, 1235, 1986.
113. **Puonti-Kaerlas, J., Eriksson, J., and Engström, P.,** Production of transgenic pea (*Pisum sativum* L.) plants by *Agrobacterium tumefaciens* - mediated gene transfer, *Theor. Appl. Genet.*, 80, 246, 1990.
114. **Hinchee, M. A., Connor-Ward, D. V., Newell, C. A., McDonnell, R. E., Sato, S. J., Gasser, C. S., Fischhoff, D. A., Re, O. B., Fraley, R. T., and Horsch, R. B.,** Production of transgenic soybean plants using *Agrobacterium*-mediated DNA transfer, *Bio/Technology*, 6, 915, 1988.
115. **Stougaard-Jensen, J., Marcker, K. A., Otten, L., and Schell, J.,** Nodule specific expression of a chimeric soybean leghaemoglobin gene in transgenic *Lotus corniculatus*, *Nature*, 321, 669, 1986.
116. **White, D. W. R. and Greenwood, D.,** Transformation of the forage legume *Trifolium repens* L., using binary *Agrobacterium* vectors, *Plant Mol. Biol.*, 8, 461, 1987.
117. **Filippone, E. and Lurquin, P. F.,** Stable transformation of eggplant (*Solanum melongena* L.) by cocultivation of tissues with *Agrobacterium tumefaciens* carrying a binary plasmid vector, *Plant Cell Rep.*, 8, 370, 1989.
118. **Ooms, G., Burrell, M. M., Karp, A., Bevan, M., and Hille, J.,** Genetic transformation in two potato cultivars with T-DNA from disarmed *Agrobacterium*, *Theor. Appl. Genet.*, 73, 744, 1987.
119. **McCormick, S., Niedermeyer, J., Fry, J., Barnason, A., Horsch, R., and Fraley, R.,** Leaf disc transformation of cultivated tomato (*L. esculentum*) using *Agrobacterium tumefaciens*, *Plant Cell Rep.*, 5, 81, 1986.
120. **James, D. J., Passey, A. J., Barbara, D. J., and Bevan, M.,** Genetic transformation of apple (*Malmas pumila* Mill.) using a disarmed Ti binary vector, *Plant Cell Rep.*, 7 658, 1989.
121. **De Block, M.,** Factors influencing the tissue culture and the *Agrobacterium tumefaciens*-mediated transformation of hybrid aspen and poplar clones, *Plant Physiol.*, 93, 1110, 1990.
122. **Mathews, H., Bharathan, N., Litz, R. E., Narayanan, K. R., Rao, P. S., and Bhatia, C. R.,** The promotion of *Agrobacterium* mediated transformation in *Atropa belladona* L. by acetosyringone, *J. Plant Physiol.*, 136, 404, 1990.
123. **Graham, J. and McNicol, R. J.,** Regeneration and transformation of Ribes, *Plant Cell Tissue Organ Cult.*, 24, 91, 1991.
124. **Scott, R. J. and Draper, J.,** Transformation of carrot tissues derived from proembryogenic suspension cells: a useful model system for gene expression studies in plants, *Plant Mol. Biol.*, 8, 265, 1987.
125. **Catlin, D., Ochoa, O., McCormick, S., and Quiros, C. F.,** Celery transformation by *Agrobacterium tumefaciens*: cytological and genetic analysis of transgenic plants, *Plant Cell Rep.*, 7, 100, 1988.
126. **Firoozabady, E., Deboer, D. L., Merlo, D. J., Halk, E. L., Amerson, L. N., Rashka, K. E., and Murray, E. E.,** Transformation of cotton (*Gossypium hirsutum* L.) by *Agrobacterium tumefaciens*: cytological and genetic analysis of transgenic plants, *Plant Mol. Biol.*, 10, 105, 1987.
127. **Chee, P.,** Transformation of *Cucumis sativus* tissue by *Agrobacterium tumefaciens* and the regeration of transformed plants, *Plant Cell Rep.*, 9, 245, 1990.
128. **Mullins, M. G., Tang, F. C. A., and Facciotti, D.,** *Agrobacterium*-mediated genetic transformation of grapevines: transgenic plants of *Vitis rapestris* schule and buds of *Vitis vinifera* L., *Bio/Technology*, 8, 1041, 1990.
129. **Noda, T., Tanaka, N., Mano, Y., Nabeshima, S., Ohkawa, H., and Matsui, C.,** Regeneration of horseradish hairy root incited by *Agrobacterium rhizogenes* infection, *Plant Cell Rep.*, 6, 283, 1987.

130. **Vardi, A., Bleichman, S., and Aviv, D.,** Genetic transformation of *citrus* protoplasts and regeneration of transgenic plants, *Plant Sci.,* 69, 199, 1990.
131. **Michelmore, R., Marsh, E., Seely, S., and Landry, B.,** Transformation of lettuce (*Lactuca sativa*) mediated by *Agrobacterium tumefaciens, Plant Cell Rep.,* 6, 439, 1987.
132. **Slightom, J. L., Jouanin, L., Leach, F., Drong, R. F., and Tepfer, D.,** Isolation and identification of TL-DNA/plant junctions in *Convolvulus arvensis* transformed by *Agrobacterium rhizogenes* strain A4, *EMBO J.,* 4, 3069, 1985.
133. **Fang, G. and Grumet, R.,** *Agrobacterium tumefaciens* mediated transformation and regeneration of transgenic muskmelon plants, *Plant Cell Rep.,* 9, 160, 1990.
134. **Naina, N. S., Gupta, P. K., and Mascarenhas, A. F.,** Genetic transformation and regeneration of transgenic neem (*Azadirachta indica*) plants using *Agrobacterium tumefaciens, Curr. Sci.,* 58, 184, 1989.
135. **Fillati, J. A., Sellmer, J., McCown, B., Haissig, B., and Comai, L.,** *Agrobacterium*-mediated transformation and regeneration of *Populus, Mol. Gen. Genet.,* 206, 192, 1987.
136. **Nehra, N. S., Chibbar, R. N., Kartha, K. K., Datla, R. S. S., Crosby, W. L., and Stushnoff, C.,** *Agrobacterium* mediated transformation of strawberry calli and recovery of transgenic plants, *Plant Cell Rep.,* 9, 10, 1990.
137. **Lindsay, K. and Gallois, P.,** Transformation of sugar beet (*Beta vulgaris*) by *Agrobacterium tumefaciens, J. Exp. Bot.,* 41, 529, 1990.
138. **Everett, N. P., Robinson, K. E. P., and Mascarenhas, D.,** Genetic engineering of sunflower (*Helianthus annuus L.*), *Bio/Technology,* 5, 1201, 1987.
139. **McGranahan, G. H., Leslie, C. A., Unatsu, S. L., Martin, L. A., and Dandekar, A. M.,** *Agrobacterium*-mediated transformation of walnut somatic embryos and regeneration of transgenic plants, *Bio/Technology,* 6, 800, 1988.
140. **Bytebier, B., Deboeck, F., deGreve, H., VanMontagu, M., and Hernalsteens, J. P.,** T-DNA organization in tumor cultures and transgenic plants of the monocotyledon *Asparagus officinalis, Proc. Natl. Acad. Sci. U.S.A.,* 84, 5345, 1987.
141. **Rhodes, C. A., Pierce, D. A., Mettler, I. J., Mascarenhas, D., and Detmer, J. J.,** Genetically transformed maize plants from protoplasts, *Science,* 240, 204, 1988.
142. **Gordon-Kamm, W. J., Spencer, T. M., Mangano, M. L., Adams, T. R., Daines, R. J., Start, W. G., O'Brien, J. V., Chambers, S. A., Adams, W. R., Willetts, M. G., Rice, T. B., Mackey, C. J., Kraeger, R. W., Kausch, A. P., and Lemaux, P. G.,** Transformation of maize cells and regeneration of fertile transgenic plants, *Plant Cell,* 2, 603, 1990.
143. **Horn, M. E., Shillito, R. D., Conger, B. V., and Harms, C. T.,** Transgenic plants of orchard grass (*Dactylis glomerata* L.) from protoplasts, *Plant Cell Rep.,* 7, 469, 1988.
144. **Shimamoto, K., Terada, R., Izarva, T., and Fujimoto, H.,** Fertile transgenic rice plants regenerated from transformed protoplasts, *Nature,* 338, 274, 1989.

# 7

# Vectors for Plant Transformation

*Margaret Y. Gruber and William L. Crosby*

## I. INTRODUCTION

The past decade has witnessed the rapid development of methods for the transformation of plants and plant cells. One can now choose from a wide array of existing vectors, or alternatively, one can select an existing framework important to the transformation mechanism and incorporate readily available gene sequences in a vector designed specifically to one's needs. The choice of a transformation system and its accompanying vector(s) is really dictated by the purpose of the transformation experiment, and the reader is advised to be clear in this regard.

Here, we review the development of plant transformation vectors and outline some of the functional aspects of vector selection or design which a novice to plant transformation might consider before undertaking gene transfer experiments with a particular plant species. The scope, by necessity, is not exhaustive, either from the point of view of methods or examples of available transformation vectors. In the assembly of this review, we offer regrets to those respondents who provided material which could not be included. The reader is directed to any of several recent reviews which have addressed the subject of plant transformation technology.[1-12]

### A. GENERAL CONSIDERATIONS FOR PLANT TRANSFORMATION

The ability to stably introduce defined genetic constructs into plant cells is an essential component of molecular approaches to plant biology. The ultimate objectives for such approaches are as varied as the biologists undertaking them, but most include the desirability of regenerating and recovering normal, fertile plants which stably transmit the introduced gene of interest to their progeny.

Transformation, then, may be defined operationally as the introduction of genetic material into plant cells, resulting in chromosomal (nuclear or organellar) integration and the stable heritability of that material through meiosis. This definition distinguishes transformation from transient expression, where genetic material is introduced (generally) through physical uptake approaches and may exhibit a short-lived mitotic stability in cells. We have largely avoided the subject of plant viral vectors as a special case involving epichromosomal replication and gene expression which exhibits variable properties with respect to meiotic stability. However, we do include the use of viral replicating sequences and expression systems where they have affected the subject of *Agrobacterium*-mediated gene delivery to monocots.

The development of transformation methods for plants has closely paralleled the corresponding development of efficient cell propagation, selection, and regeneration protocols for individual plant species. Almost invariably, the application of existing transformation methods to a recalcitrant species will involve one or more problems with aspects of *in vitro* culture.

0-8493-5164-2/93/$0.00+$.50
© 1993 by CRC Press, Inc.

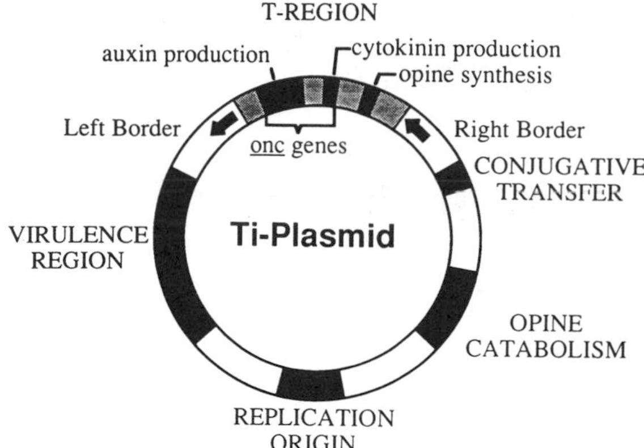


**FIGURE 1.** Genetic map of an octopine Ti plasmid. The position of the left and right border sequences are indicated by arrows, various other elements are indicated by boxes. (From *Oxford Surveys of Plant Molecular and Cell Biology*, B. J. Miflin, Ed. Reprinted by permission of Oxford University Press.)

As well as developing effective protocols in this area, the investigator must address other prerequisites for transformation. These include selection in favor of transformed cells in the presence of a suitable antimetabolite and under tissue culture conditions allowing regeneration of transformed cells. The combined selection pressure and tissue culture techniques should permit normal development of the plant to seed and stability of the newly inserted gene through meiosis to progeny.

## II. VECTORS FOR BIOLOGICAL TRANSFORMATION

### A. *AGROBACTERIUM*-MEDIATED TRANSFER

*Agrobacterium tumefaciens* and *A. rhizogenes* are pathogenic soil bacteria which, in the wild, genetically transform cells in the crown or roots of plants. The agents responsible are 200 to 500 Kbp circular plasmids, the Ti plasmid from *A. tumefaciens* and the Ri plasmid from *A. rhizogenes* (Figure 1). The mechanism underlying plant transformation by these species has been extensively reviewed.[1,6,7,13-16] *Agrobacterium* is attracted to plants by chemotactic compounds arising from wounded plant cells. The net result of this attraction is the movement of a segment of the plasmid called the transfer DNA (T-DNA) and its insertion into the plant chromosome.

The T-DNA consists of two 25 bp repeated sequences encompassing auxin, cytokinin, and opine biosynthetic genes. While the physical nature of the transferred DNA and the transfer mechanism are not completely understood, the main body of evidence supports the 5′ to 3′ synthesis of a single-stranded copy of the T-DNA, called the T-strand, beginning at the right border. Virulence genes (*vir* and *chv*) encoded on the plasmid and bacterial chromosome determine the *Agrobacterium* host range and direct production of the T-strand and its transfer out of the bacterium into the plant nucleus. Specifically, it is thought that virE2, a single-stranded DNA-binding protein, coats and protects the T-strand[17-19] after T-DNA nicking by virD enzymes and T-strand synthesis.[20-22] virD2, which covalently binds to the 5′ end of the T-strand, appears to function as a nuclear targeting protein.[23-25] The *vir*B gene products are thought to form a membrane-associated transport mechanism.[26,27] The production of the T-strand (and, consequently, DNA transfer) is strongly stimulated to occur leftward through the T-DNA by the presence of a "transfer enhancing element" called *overdrive*, which is situated 3′ to the right repeat and binds the *vir*C1 gene product.[28-30]

In the transformed dicot plant cell, auxin, cytokinin, and opine biosynthetic genes encoded within the integrated T-DNA region reprogram the plant biosynthetic apparatus for uncontrolled cell division and the biosynthesis and excretion of opines. The opines are then transported into the bacteria and catabolized by opine utilization gene products encoded on the nontransferred region of the plasmid, providing a major source of metabolites. The Ti phenotype is a plant with a tumor-like mass or gall. Since the T-DNA from an Ri plasmid has only auxin or no hormone biosynthetic genes, the Ri phenotype is, at most, a plant with a "hairy" root mass and a somewhat altered morphology.

Exposure of a monocot plant to *Agrobacterium* does not result in tumor formation.[8] This finding led to the widely held belief that monocots do not fall within the host range of *Agrobacterium* — a belief that was recently discarded with the discovery that infected monocot plants can produce opines[31,32] and that *Agrobacterium* can be exploited to deliver cloned cDNA copies of viral chromosomes to plants, resulting in a productive infection.[33,34] Agroinfection is now being used as a sensitive marker for T-DNA transfer to characterize the physiological parameters which distinguish monocot and dicot plant responses to *Agrobacterium*.[35]

## B. *AGROBACTERIUM* VECTOR SYSTEMS

*Agrobacterium* plasmids have been exploited as vectors for biological delivery of foreign DNA to plants; this is the most widespread transformation strategy in use today. Two types of vector systems have been developed, cointegrating and binary.[9,36-38] The early vectors arose from plasmids mutated in the T-DNA hormone-biosynthetic region, the discovery that homologous recombination occurred in *Agrobacterium* between resident and engineered Ti-plasmids, and by using triparental mating to mobilize cloned genes from *Escherichia coli* to *Agrobacterium*.[39] When the hormone biosynthetic genes within the T-DNA borders were fully deleted (disarmed plasmid), it became possible to regenerate transgenic plants in the laboratory rather than only create transgenic tumorous tissue.

Binary shuttle vectors are usually smaller than cointegrating vectors and, consequently, are easier to maintain and manipulate in *E. coli*. These plasmids have a 10,000-fold greater frequency of transfer from *E. coli* to *Agrobacterium* than cointegrating plasmids, since there is no T-DNA integration step. The presence of genes encoded in the T-DNA of a binary plasmid in *Agrobacterium* is confirmed easily by plasmid restriction digests, rather than by Southern hybridization or PCR, which is required to detect large cointegrated plasmids.[40] As a result of these features, a greater number of binary vectors have been developed.

Biological transformation results in one to three copies of the T-DNA, which are stably integrated in the plant chromosome with relatively few insert rearrangements.[30,41,42] Characterization of plant chromosomal integration sites has resulted in a model of random integration compatible with illegitimate recombination,[43] although in at least one instance the DNA integrated into a region of the plant genome which had homology with vector DNA.[44] This simple integration pattern may reflect the possibility that *Agrobacterium* presents a specific DNA-protein structure to the plant chromosome, ensuring more faithful integration. Alternatively, the border sequences may act as a buffer for the inserts and incur most of the rearrangements coincidental to illegitimate recombination.[30]

Although *Agrobacterium* has been used widely to produce transgenic plants, the procedure and its associated tissue culture practices are constantly under revision to reflect the specific details inherent with each prospective host plant. Currently, a number of approaches have been shown to improve transformation frequencies and to widen the host range of susceptibility to *Agrobacterium*. These include the use of highly virulent strains,[45-47] strains with high copy number Ti plasmids,[47] engineered strains expressing extra copies of *vir*G,[45] and over-expression of the *vir*D1 and *vir*D2 genes.[48] For example, derivatives of pTiBo542 have proven useful with legume species which cannot be transformed with octopine strains (M. P. Gordon, personal communication). A specific virulence background may also be the reason why the pCGN (Calgene) vectors are particularly effective for *Brassica* species (M. Moloney, personal communication).

Ri plasmid vectors are becoming increasingly popular because of their wide host range and the regenerative capacity of wild-type *A. rhizogenes* transformants.[6,49-51] Coinoculation with *A. tumefaciens* wild strains is also being exploited for this purpose.[52] In addition, wounding with microprojectiles strongly increases the frequency of *Agrobacterium*-mediated transfer.[42] Recently, a method based on opine production near the wound site for testing the susceptibility of plant species to *Agrobacterium* was published.[53]

### 1. Cointegrating Vectors

The cointegrating system features two independent plasmids: a Ti plasmid in *Agrobacterium* and an intermediate vector in *E. coli*. Both plasmids have a region of homology which undergoes recombination to form a large, cointegrated plasmid after conjugation between *Agrobacterium* and *E. coli*. Genes which are to be introduced into plants are cloned and manipulated in *E. coli* and, after recombination with the Ti plasmid in *Agrobacterium*, are situated between two T-DNA border repeats.[54] The *E. coli* plasmid has no origin of replication for maintenance in *Agrobacterium* and is not retained without the recombination step.

The split-end vector (SEV) system is a more efficient cointegrative vector system in which the left and right border sequences reside each on one of the independent plasmids[55,56] (Figure 2a,b). These plasmids form the cointegrate following a single recombination event (Figure 2c), whereas other cointegrating systems may involve one or two. With both systems, the *vir* gene products are able to transfer the T-DNA to wounded plants, whereupon it becomes integrated into the plant genome (Figure 2d).

A common SEV intermediate plasmid is represented by pMON200 (Figure 2b).[56] Its framework encodes the right border from a nopaline plasmid, a nopaline synthase gene which can be detected biochemically, spectinomycin/streptomycin resistance ($Str^R/Spc^R$)genes for bacterial selection, and an LIH region of homology with the Ti plasmid. The kanamycin selectable plant marker, neomycin phosphotransferase II (*npt*II), uses nopaline synthase (*nos*) expression signals and contains a multiple cloning site for the insertion of foreign DNA. The corresponding *Agrobacterium* plasmid pTiB6S3SE contains the left border, the LIH sequence, *vir* genes, and a kanamycin resistance ($Kan^R$) gene for bacterial selection.

A variety of intermediate Ti and Ri vectors have been developed to include alternative selectable markers and basic expression cassettes, as well as improvements to the plasmid skeleton.[40,51,57] A contemporary derivative of pMON200 is the plasmid pCIT30, illustrated in Figure 3.[310] With this vector, a hygromycin resistance ($Hyg^R$) gene has replaced the *npt*II gene, and a *cos* site and polylinker have been added. The *cos* site permits the cloning and insertion of 25 to 40 Kbp of plant DNA in a phage λ packaging system, without the need for extra subcloning steps. The polylinker contains restriction sites necessary to clone and release insert DNA. This vector also contains T7 and SP6 bacteriophage promoters essential for the synthesis of end-specific RNA probes used with chromosome walking.

Stringent size selection of potential inserts between 25 to 40 Kbp is important when using pCIT30. Smaller fragments encourage the insertion of concatemers; as well, the potential exists for transcription of vector-containing RNA probes. An alternative strategy to size selection involves the ligation of half-site restriction sites between *Bam*HI-cohesive and *Xho*I or *Sal*I sites residing on either the vector or plant DNA.[58] This strategy prevents problematic self-ligation, since the half filled restriction sites are no longer complementary. Several cosmid libraries have been constructed in this vector. The libraries have since been used both to isolate genes by chromosome walking and to complement a plant mutant following transformation.

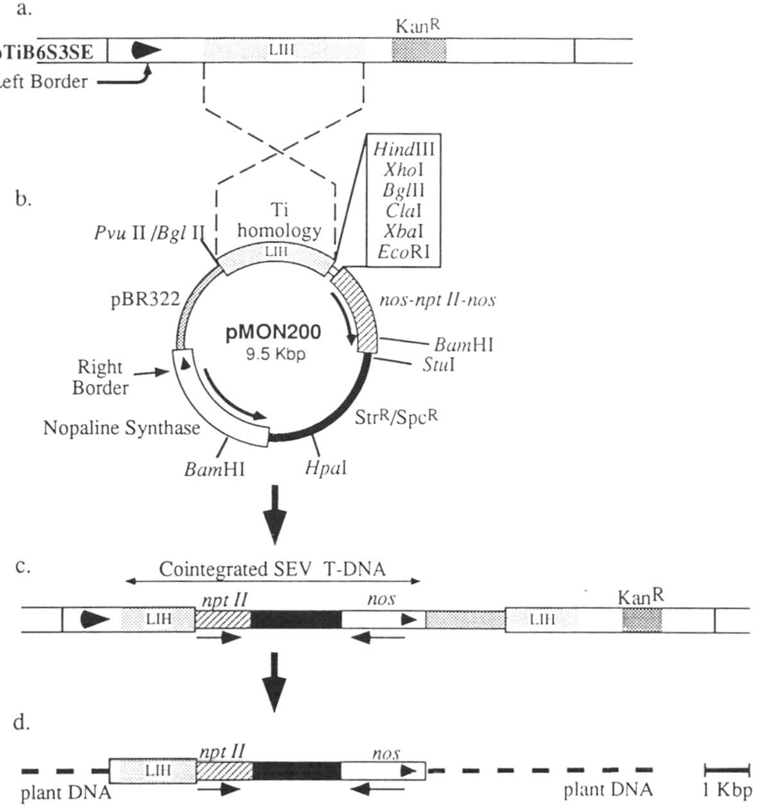


**FIGURE 2.** Cointegrate formation and plant integration of plasmid pMON200 using the SEV system: (a) pTiB6S3SE, an engineered Ti plasmid residing in *Agrobacterium*; (b) restriction map of intermediate vector pMON200 residing in *E. coli*; (c) cointegrated plasmid residing in *Agrobacterium* after conjugation; (d) T-DNA integrated into plant chromosome after transformation; other elements are represented by boxes. (From Rogers et al., Gene transfer in plants: production of transformed plants using Ti plasmid vectors, in *Methods for Plant Molecular Biology*, Weissbach A. and Weissbach H., Eds., Academic Press, 1988, pp. 423. With permission.)

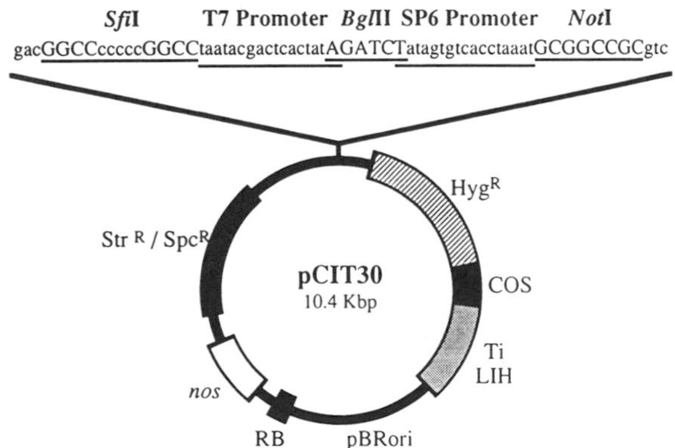


**FIGURE 3.** Map of cointegrating vector pCIT30. The direction of transcription is indicated by arrows; various other elements are indicated by boxes. (From Ma, H., Yanofsky, M. F., Klee, H. J., Bowman, J. L., and Meyerowitz, E. M. Vectors for plant transformation and cosmid libraries, *Gene,* 117, 161, 1992. With permission.)

The stability of the cointegrated plasmid may be a factor when designing a vector system, particularly if plant libraries are to be maintained in *Agrobacterium*. The cointegrated Ri plasmid is considered unstable in *Agrobacterium* without selective pressure.[51] This is not so for the cointegrated Ti plasmid. It has been suggested that this instability could be exploited by testing potentially hazardous genes (e.g., a viral sequence) using the cointegrated *A. rhizogenes* system.[51] The transformants would be recognized easily by the hairy root phenotype. Later, the gene could be reconstructed in a binary vector, if so desired.

## 2. Binary Vector Systems

The binary system features two plasmids which coexist autonomously in *Agrobacterium* after conjugation, a shuttle plasmid and a Ti plasmid.[7,9] A typical shuttle plasmid contains one or both T-DNA border repeats, as well as broad-host-range replication and mobilization functions (*ori*V, *ori*T, and *trf*A from plasmid pRK2 or pSA), and a bacterial selectable marker such as ampicillin, enabling the plasmid to function in either *E. coli* or *Agrobacterium*. In order to reduce the size of the shuttle, *trf*A can be positioned on either the bacterial chromosome or on the Ti plasmid. Recently constructed shuttles contain a *col*E1 origin of replication (pUC or pBR based), so that large amounts of plasmid can be purified.[47,59] These determinants collectively form the skeleton of the plasmid.

Shuttle plasmids encode plant selectable markers, expression signals, and polylinkers for the subcloning of foreign genes within the two T-DNA borders. Only the right border is mandatory for T-strand synthesis and for T-DNA transfer. It is likely that transfer from a small plasmid with only a right border will be successful regardless of the direction in which T-strand synthesis occurs. However, both borders in correct orientation and complete with the *overdrive* sequence[47] should be included in the case of large T-DNA inserts to ensure the accurate synthesis and transfer of a full-length T-strand.[30,60]

The *Agrobacterium* plasmid in a binary system is an engineered Ti plasmid with plasmid *vir* genes, an *Agrobacterium* origin of replication (*oriA*), but no T-DNA borders. After conjugation, the two plasmid partners coexist autonomously with selective pressure in *Agrobacterium*. When *Agrobacterium* infects a wounded plant, the *vir* genes on the Ti plasmid interact with the right border on the shuttle plasmid in *trans* to transfer the T-DNA into the plant genome. These latter two features define the binary vector strategy.[61] Wild type plasmids containing T-DNA can be used to supply *vir* functions, but some of the resulting transformants will be tumorous from the transfer of wild type hormone biosynthetic genes. *Vir* genes of *A. tumefaciens* or *A. rhizogenes* can function in *trans* with T-DNA from either plasmid type.

As with cointegrating vectors, a core of basic binary vectors has been developed over time to contain T-DNA expression cassettes, encoding a variety of plant expression signals, polylinkers for transcriptional and translational fusions, and plant selectable markers. A useful strategy among these vectors has been to include a polylinker within the *lac*Z gene to facilitate the direct selection of cloned expression cassettes.[59] Another practical scheme is to include a *cos* sequence to clone large inserts or plant libraries, an approach which was adopted early in the history of binary vectors.[47,62,63] One of these latter constructs encoded an Ri origin, a measure intended to stabilize large cosmids in *Agrobacterium* without selective pressure, but employed with little success.[62] The Ri origin of replication, as well as the pVS1 origin, may only stabilize smaller binary plasmids in *Agrobacterium*.[40,51,59]

The gene cassettes within the T-DNA region can also be arranged for optimum transfer and plant expression. Genes of interest should be situated directly adjacent to the right border where integration is faithful, leaving the plant selectable marker closer to the left border where incorrect integration has been known to occur (M. De Block, personal communication). Dual promoters can also be subcloned in opposite orientation in the center of the T-DNA, so that there is less influence on the expression of introduced genes from flanking plant genomic DNA.[9,64]

Two examples of contemporary binary vectors are pCIT104[310] and pLZ03[65] (Figure 4). pLZ03 is a 30.1 Kbp cosmid and has an insert capacity of 13 to 22 Kb. Plasmid pCIT104 (7 Kb), as a small cosmid, can be used either to clone up to 30 to 46 Kbp genomic inserts,

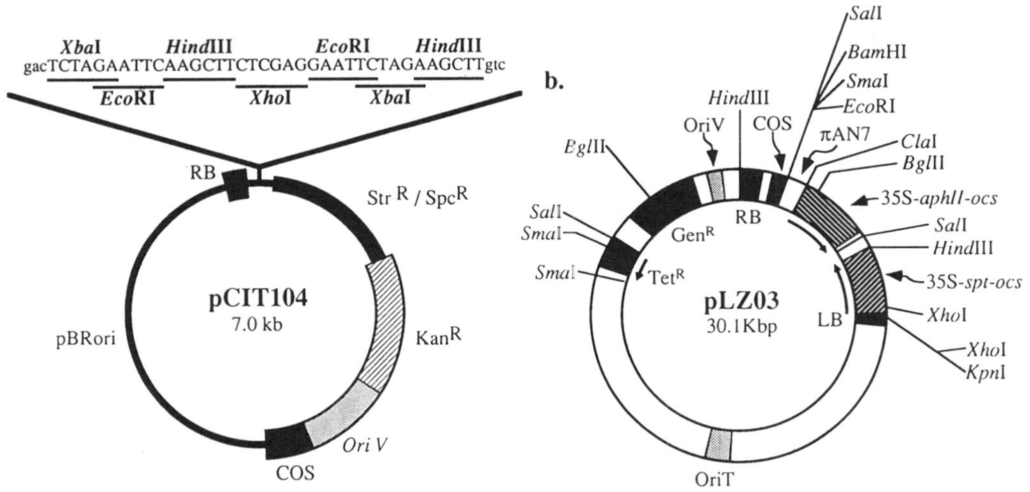


**FIGURE 4.** Restriction maps of binary vectors pCIT104[310] and pLZ03. ([a] from *Gene*, 117, 161, 1992. By permission of Elsevier. [b] from *Bio/Technology*, 9, 963, 1991. By permission of Bio/Technology Magazine/Nature America.)

or alternatively, it can be used to manipulate smaller genes. Plasmid pCIT104 can be maintained as a cointegrate if the *Agrobacterium* host does not have a *trf*A gene. Plasmid pCIT104 encodes a kanamycin resistance plant selectable marker, while pLZ03 contains the plant selectable *aph*II and *spt* genes. Several other binary vectors have been developed using the same skeleton as pCIT104, including one which contains the polylinker encoded in the integrative vector pCIT30 (Figure 3).[310]

Each vector uses a different strategy to reduce the possibility of cloning vector or insert concatemers in phage λ. With pLZ03, vector monomers are too small and dimers too large to be packaged efficiently. With pCIT104, the *Xho*I half-site can be ligated with *Bam*HI, *Bgl*II, or *Sau*3A half-site genomic fragments.

Plasmid pCIT104 only encodes the right border, whereas pLZ03 has both borders and the *overdrive* sequence; hence, pCIT104 is unlikely to be as good a vector as pLZ03 for transforming plants with the large inserts originating from genomic libraries. A transformation-competent *Arabidopsis* library has been cloned in pLZ03 and stored in *Agrobacterium* and is now being used to transform *Arabidopsis* phenotypic mutant plants in a gene rescue strategy. Plasmid pCIT104 has not been tested functionally yet.

## III. VECTORS FOR PHYSICAL TRANSFORMATION

### A. CELL TRANSFORMATION

Methods for the physical delivery of DNA to plants have developed in parallel to *Agrobacterium* systems.[5,10,66-69] *Agrobacterium* plasmids were used as vectors in early physical transformation experiments, since they contain hormone and opine biosynthetic genes which function as selectable and reporter-gene markers.[70-74] These have been largely replaced by vectors without the biological determinants for plant transfer.

By far the greatest number of experiments have involved direct uptake of DNA into protoplasts using polyethylene glycol, electroporation, or microprojectile-mediated delivery to tissues using a biolistic device. Biolistic delivery of DNA into meristematic tissue, in particular, bypasses the long period of regeneration from protoplasts required after PEG or electroporation, and thus has been particularly effective for those species lacking useful tissue culture protocols. A drawback of the biolistic method involving tissue explants is the production of chimeric tissue. All of these methods have been used to stably transform recalcitrant species such as barley,[75] rice,[76,77] maize,[78,79] sugar cane,[80] orchard grass,[81] wheat,[82] soybean,[83,84] and cotton.[85] Microinjection,[86,87] sonication of the target cells,[88] liposome or spheroplast fusion,[89,90] and direct uptake into protoplasts using $CaCl_2$ precipitation,[91] polyvinyl alcohol, or poly-L-ornithine[73] have also been reported. Some of these methods have also been used to study transient gene expression.

Physical vectors are usually 4 to 7 Kbp in size. Plasmids as large as 9 and 11 Kbp have been used as vectors without compromising transgene stability.[82,92] Larger vectors may be somewhat unstable, and in one case the selectable markers encoded on an 18.7 Kbp plasmid were not all expressed within each transformant.[93]

DNA conformation plays a role in the frequency of physical transformation. Both linear and supercoiled plasmids are used as vectors, and there is evidence that the linear form transforms plants more efficiently.[94,95] Single-stranded plasmid has also been shown to increase transformation frequencies compared with double-stranded DNA.[96]

Physical vectors are distinguished by features unrelated to their method of delivery. The simplest vectors consist of rudimentary plant expression cassettes and plant selectable markers subcloned into pBR322, which provides an *E. coli* replicon and a bacterial selectable marker.[83,97] Expression cassettes have also been subcloned into the *lac*Z polylinker of pUC or pGEM plasmids,[76,77,92,98] a strategy which permits direct selection of the insert and a higher plasmid copy number in *E. coli* compared with pBR322-based vectors.

Vectors can be constructed using a phage backbone (M13) in order to produce single- or double-stranded DNA.[96] Expression cassettes can also be subcloned into the multiple cloning site of a phagemid.[78] This latter design offers distinct advantages, including direct selection of the insert, *in vitro* synthesis of RNA probes, production of single- or double-stranded DNA for transformation experiments, and facilitation of DNA sequencing.

Plant expression cassettes in vectors used for physical delivery include multiple selectable markers,[99,100] tandem promoters,[91] and transcription enhancing elements.[76,78,98] Transformation events, particularly in cereals, are more likely to be detected in the future as optimized promoter/expression cassettes become more available.

Transgenic calli or plants have been recovered from physical delivery experiments. In some cases their patterns of DNA integration have been analyzed by Southern hybridization and marker enzyme activity,[78,84,99,101-103] and the stability and segregation of the integrated DNA were followed through self-fertilization or backcross fertilization. From these data it can be inferred that the DNA from all physical transfer methods display a wide variation in integration pattern consistent with illegitimate recombination, including single and multiple integration sites and a wide range of copy number.

As the number of species that can be successfully transformed by physical means continues to increase, the optimization of transformation efficiency and frequency will play a strong role in transformation experiments. These parameters are highly dependent on the tissue culture system and the developmental competence of the target tissue. They are also dependent, but to a lesser extent, on the vector design. Following this theme, it has been shown that the transformation frequency can be raised by up to 100-fold using synchronized protoplast cultures.[104,105] In addition, the transformation frequency of *Petunia* protoplasts is raised 20-fold when a genomic transformation booster sequence (TBS) is included in the plant vector.[106] Integration patterns were not altered as long as the TBS construct was used to transform protoplasts during mitosis.[107]

## B. ORGANELLE TRANSFORMATION

Until recently, genetic engineering of plant organelles was limited to the expression of cytoplasmically synthesized proteins targeted to subcellular organelles; however, the advent of the biolistic device has raised the real possibility of consistent higher plant organelle transformation.[3,108,109] Genes providing resistance to antibiotics which only affect the chloroplast ribosome, such as streptomycin, spectinomycin, chloramphenicol, and erythromycin, can now be used to select chloroplast transformants[3] (Table 1). The activities of the chloroplast *psb*A, *rbc*L, and *atp*B promoters have been tested using chloramphenicol acetyltransferase (CAT) and β-glucuronidase (GUS) in transient expression systems.[110,111] Similar approaches are being used to develop a transformation system for mitochondria using a mitochondrial *atp*9 promoter coupled with the CAT gene,[111] although to date there have been no reports of plant mitochondrial transformation.

One design for a chloroplast transformation vector is to incorporate chloroplast-specific replication sequences and chloroplast-driven selectable markers in an *E. coli* plasmid framework. It has been shown recently that a spinach *psb*A promoter-CAT construct coupled to a chloroplast replicon continued to express maximum CAT activity in tobacco suspension cells after the CAT activity stimulated by a replicon-less plasmid declined.[111] Neither plasmid expressed CAT after introduction into tobacco by electroporation, indicating the importance of the biolistic method in penetrating the chloroplast membrane. Regarding this approach, it is our opinion that cells transformed without DNA integration into the chloroplast genome will likely require ongoing selective pressure and therefore have value only as an experimental system.

Stable transformation of the tobacco plastid genome has been achieved by combining biolistic transfer with plasmids which take advantage of the strong natural recombination

**TABLE 1**
**Genetic Selection Markers for Plant Transformation**

| Marker | Antimetabolite | Ref. |
|---|---|---|
| *npt*II | Aminoglycoside Antibiotics (kan, neo, G418) | 115,179 |
| *hyg* | Hygromycin | 291 |
| *gent* | Gentamycin | 292 |
| *bleo* | Bleomycin | 293 |
| *blas* | Blasticidin | 294 |
| *aat* | Streptomycin, Spectinomycin (Aminoglycoside adenyl-transferase expression) | 295 |
| *str/spc* | Streptomycin, Spectinomycin (mutant rRNA) | 296 |
| *dHFR* | Mutant Methotrexate | 297 |
| *cat* | Chloramphenicol | 3, 148 |
| *bar* | Phosphinothricin | 125 |
| *bxn* | Bromoxynil | 298 |
| *sul* | Sulfonamide | 299 |

ability of the chloroplast. An example of a homologous recombination vector is the plasmid pJS75 (Figure 5a), which encodes 6.2 Kbp of a tobacco chloroplast repeat sequence, engineered to contain six RFLP markers and two chloroplast-specific selectable markers.[112] The markers are scattered throughout the length of the sequence in locations not expected to interfere with chloroplast function. Photosynthetically competent transformed cell lines containing the complete recombinant sequence were recovered by growth on selective medium after transformation with pJS75. Stable transformants were detected by monitoring the frequency of the nonselected RFLP markers. All of the markers in one of the transformed lines were maintained upon regeneration, upon selfing, and when the maternal transformed plant was used for crossing. A GUS gene has been expressed in chloroplasts using this vector. The authors present a model to illustrate the formation of transgenic plastid genomes using this vector (Figure 5b).

The above experiment showed that the plant chloroplast is capable of being transformed stably without disturbing functional regions, and that the transformed chloroplasts are able to replicate and provide a selective advantage to the plant. However, extensive selective pressure and monitoring of the markers may be required for transformed chloroplasts to compete with the large number of untransformed chloroplasts, as well as to guard against the strong potential that exists within the chloroplast for further recombination and loss of selectable markers. It also is not clear how large a foreign gene can be integrated stably using this strategy. These considerations are relevant additionally if mitochondrial vectors are constructed from homologous sequences derived from the plant mitochondrial genome.[113]

The *Agrobacterium* binary vector system has been exploited as a vehicle for delivery of homologous DNA into the chloroplast genome.[114] We believe that biolistic delivery is likely to give more consistent rates of chloroplast transformation than the *Agrobacterium* system, since it appears that the T-DNA is transferred bound to virD2, a nuclear targeting protein.[23,24] However, it may be useful to test the activity of virD2 with chloroplast and mitochondrial membranes, in the event that it has a more general membrane translocation activity useful in organelle transformation.

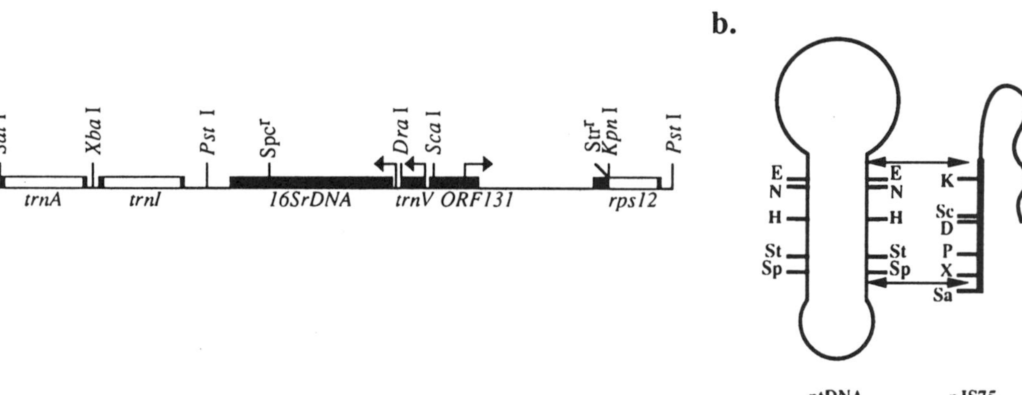


**FIGURE 5.** Restriction map of a chloroplast homologous recombination vector system: (a) restriction map of engineered chloroplast DNA region of plasmid pJS75 (cloned in pUC19); (b) region of recombination between pJS75 and a plastid genome. (From Staub, J. M. and Maliga, P., *Plant Cell,* 4, 39, 1992. By permission of ASPP.)

# IV. SELECTABLE MARKERS AND REPORTER GENES

## A. MARKERS FOR GENETIC SELECTION

A prominent consideration in the identification of a useful transformation vector is the choice of a marker gene suitable for positive genetic selection of transformants and effective in combination with existing tissue culture protocols. The choice of a marker is dictated by the antimetabolite chosen as a counterselection agent, which should be useful in preventing growth of nontransformed cells. Other factors must be considered in choosing a marker gene, including that (1) expression of the marker gene does not metabolically disrupt the transformed host cell, (2) expression of the marker gene effectively protects the host cell from the growth-inhibitory properties of the selection agent and provides a clear phenotypic growth distinction between transformed and nontransformed cells, and (3) exposure of transformed cells to the antimetabolite should have minimal effects on subsequent growth and development of an intact, fertile plant.

Table 1 lists several genetic selection markers which have been employed successfully in plant transformation experiments. The gene *npt*II, encoding neomycin phosphotransferase (NPT), *npt*II is the most commonly used, and has proven particularly effective in transformation experiments involving Solanaceous species including tobacco,[115] potato,[116,117] and tomato.[116,118] However, *npt*II has proven less useful with legumes[57] and monocot species.[119,120]

The *npt*II gene product acts to enzymatically phosphorylate and inactivate aminoglycoside antibiotics such as kanamycin, neomycin, and G418 (geniticin). Some early constructs, involving *npt*II as the selectable marker, were found to harbor a mutation which reduced the enzymic efficiency of the NPT gene product.[121] *npt*II has also been observed to function as a marker for chloroplast transformation,[122] although markers affecting 70S ribosome structure, including *blas*, *str*/*spc*, and *cat*, are currently favored for this purpose (Table 1).

Several selectable marker gene products function to detoxify antimetabolites including hygromycin and gentamycin, while other markers encode an altered target which is insensitive to the inhibitor (e.g., *str*/*spc*, involving a mutant organellar rRNA; *sul*, encoding an mutant sulfonamide-resistant dihydropteroate enzyme and others; see Table 1). Marker genes encoding tolerance to herbicides have become generally available and have proven particularly effective in recalcitrant species, including monocots.[78,120,123-126] One approach to the identification of relatively rare transformation events has been devised, in which the marker gene encodes a dominant constitutive regulator of the *Zea mays* anthocyanin pigmentation pathway active at the cellular level.[127] In addition, nitrate reductase (chlorate sensitivity) has been developed recently as a suicide or counter-selection marker, useful in the production of clonal lines.[128]

## B. BIOCHEMICAL MARKERS

As a secondary indication of transformation, it is often comforting to incorporate and be able to assess biochemically the expression of an independent reporter gene under the regulation of a plant-specific promoter. Biochemical marker genes are also used extensively in a host of other applications, including the assessment of promoter expression characteristics and the analysis of subcellular compartmentation studies. The ideal plant reporter gene should have a number of inherent characteristics, including (1) its product should be unique and nontoxic to the host plant cell, (2) the marker enzyme should exhibit a high degree of post-translational stability, (3) a convenient, inexpensive, sensitive, and specific enzyme assay for the reporter gene product should be available, and (4) it should be amenable to translational fusions with extraneous polypeptides, while retaining enzymatic activity.[100,129-131] In all respects, the *uid*A gene product of *E. coli*, β-glucuronidase (Table 2), fulfills these requirements for a reporter gene and is arguably the most extensively used gene for plant transformation and gene expression studies.

**TABLE 2**
**Biochemical Markers as Reporter Genes in Plant Transformation**

| Biochemical marker | Gene product | Ref. | Assay |
|---|---|---|---|
| *gus* (*E. coli uid*A) | β-Glucuronidase | 100, 129–131 | 129, 300, 301 |
| *modified gus* | β-Glucuronidase; GUS::NPT | 132–135 | |
| *lac* (*E. coli lac*Z) | β-Galactosidase | 23, 168 | 302 |
| *cat* | Chloramphenicol Acetyltransferase | 136, 139, 141, 142 | 145, 146, 303, 304 |
| opine | Opine biosythesis (e.g., octopine, nopaline) | 160,165 | |
| *npt*II | Neomycin Phosphotransferase | 171 | 169, 170, 305, 306 |
| *blas* | Blasticidin S deaminase | 294 | |
| *lux* | Luciferase (*P. pyralis*) | 149, 151 | 149, 307 |
| | luciferase (*Vibrio* spp.) | 154 | 308 |

The GUS enzyme can be conveniently assayed using any of several commercially available substrates, including indole-β-glucuronide substrates for quantitative spectrophotometric as well as cytochemical applications. Use of a commercially available methyl-umbelliferone β-glucuronide substrate permits fluorometric detection of picomole quantities of reaction product over extended reaction times in many species.[129] Several new substrates and assay protocols have become available for this enzyme (Table 2). In order to facilitate the plant-specific use of this reporter gene, plant intron-containing *uid*A derivatives have been developed.[132,133] Sequences surrounding the translational initiation codon have been engineered to reflect an optimal plant context.[134] A bifunctional *gus::npt* fusion marker gene containing enzymatically active domains from both GUS and NPTII has also been described.[135]

A second reporter gene extensively used in plant studies is *cat* or chloramphenicol acetyltransferase.[136-144] The activity of this enzyme is generally assayed by combined thin-layer chromatography and autoradiography following incorporation of radiolabeled acetyl moieties into chloramphenicol, although more quantitative ELISA and fluorescent assays are also available.[145,146] In general, *cat* is less post-translationally stable than *gus*, and some reports of endogenous activity or uncharacterized inhibitors have appeared.[147,148]

In addition to *gus* and *cat*, genes encoding luciferase (*lux*) enzymes from insect[149-153] and prokaryote[154-158] sources have found favor as plant reporter genes, principally for their high degree of specificity and sensitivity of the assay. The pros and cons of using *lux* reporter genes have been reviewed recently.[159] *E. coli* β-galactosidase (*lac*Z), as well as opine biosynthesis[160-167] has also been used as a biochemical marker gene in some plant species[23,168] where, like *gus*, either may be used additionally as a cytochemical stain. As indicated in Table 2, the *npt*II genetic selection marker is amenable to radiometric assay. This assay ensures the transfer of labeled phosphorus to the antibiotic substrate kanamycin.[169-171] However, it is not generally used for quantitation and serves principally to confirm the expression of the genetic selectable marker in putative transformed tissue.

## V. SEQUENCES INFLUENCING GENE EXPRESSION

The expression of a transgene is dependent on the presence of transcriptionally-efficient 5′-regulatory regions, a 5′-untranslated leader sequence, translational start sequences (AUG) with plant favorable context, and a transcription termination/polyadenylation sequence. These and other determinants contribute to mRNA and protein production, stability, and turnover and, as such, they are all important components to consider in a strategy to develop a plant vector.

## A. 5′ REGULATORY SEQUENCES

In addition to a selectable marker, most vectors destined for use in plant transformation will contain a plant-active 5′-regulatory sequence flanking a polylinker site for the convenient cloning and expression of a chimeric foreign gene. The diversity of available promoters is both wide and expanding rapidly. Where transformation of a recalcitrant species or cultivar is difficult, and where effective expression of the selectable marker is suspect, it is also important to consider the expression potential of the selectable marker in combination with different promoter sequences.

### 1. Constitutive Promoters

Both Cauliflower Mosaic Virus (CaMV) and the closely related Figwort Mosaic Virus[172] are circular duplex DNA viruses which replicate via transcription of a full-length (35S) genomic RNA intermediate. For direct use as a biological vector, these viruses suffer from several technical problems, including a restricted host range, packaging constraints limiting the size of a foreign DNA insert, and the need to remove all extraneous 5′ and 3′ DNA from the cloned gene. However, the strong promoter responsible for the genomic replication of the CaMV virus (the 35S promoter) has been extensively exploited for the expression of heterologous genes in plants.

The first such experiments involved cloning of a foreign gene into the virus molecule.[12,173] Subsequent experiments used the excised promoter in chimeric constructs.[32,174,175] Other promoters derived from genes originating in the T-DNA segment of *A. tumefaciens* (e.g., the nopaline synthase gene[176-178]) have also been exploited for the expression of foreign genes in plants.[115,168,179]

Both the *nos* and CaMV 35S promoters are constitutive in their expression, although a certain degree of cell- and tissue-specificity has been observed for both sequences.[180-183] These two promoters have been compared for their relative expression potential in different hosts and tissues, with the 35S promoter generally exhibiting higher expression potential across a broader range of host plant species.[184,185]

The CaMV 35S promoter has been studied intensively with respect to functional domains which define this sequence as a promoter,[152,181,182,186,187] leading to the development of derivatives in which the intrinsic enhancer element has been duplicated.[188] A construct containing such a tandem 35S promoter exhibits a four-fold increase in steady-state expression of various marker genes relative to the unmodified 35S promoter, in electroporation experiments involving plant species as diverse as maize, *Picea glauca* (White Spruce), and *Brassica napus* (Figure 6; J. Sanford, personal communication). Because of its early availability to plant molecular biologists, broad host-range applicability, and biological characterization, the 35S promoter has found wide use in a variety of vector constructs,[12,174,189-192] including chimeric constructs which combine regulatory elements from different promoters.[193] It has also been reported to initiate transcription and expression of marker gene constructs in *E. coli* and the fission yeast *Schizosaccharomyces pombe*,[194-196] and it may prove useful in constructs used for rapid 'shuttling' between plants and these alternate hosts.

Unfortunately, both the 35S and the *nos* promoters do not usually function as well in monocots as in dicot species,[197,198] although the 35S seems to be at least as, if not more, efficient in rice than in tobacco.[199,200] As well, chimeric genes expressed from 35S display a lower level of expression in *Medicago* protoplasts compared with tobacco.[198] Hence, other promoter elements which are efficiently expressed in monocots have been incorporated into vector constructs targeted to these species.[201-203] In that regard, plasmid pEmu, which contains multiple copies of an anaerobic response element from the maize *Adh*1 gene and elements from the octopine synthase gene from *A. tumefaciens*, has been shown to exhibit ten- to fifty-fold greater expression in five monocot species, as well as in *Nicotiana plumbaginifolia*.[201]

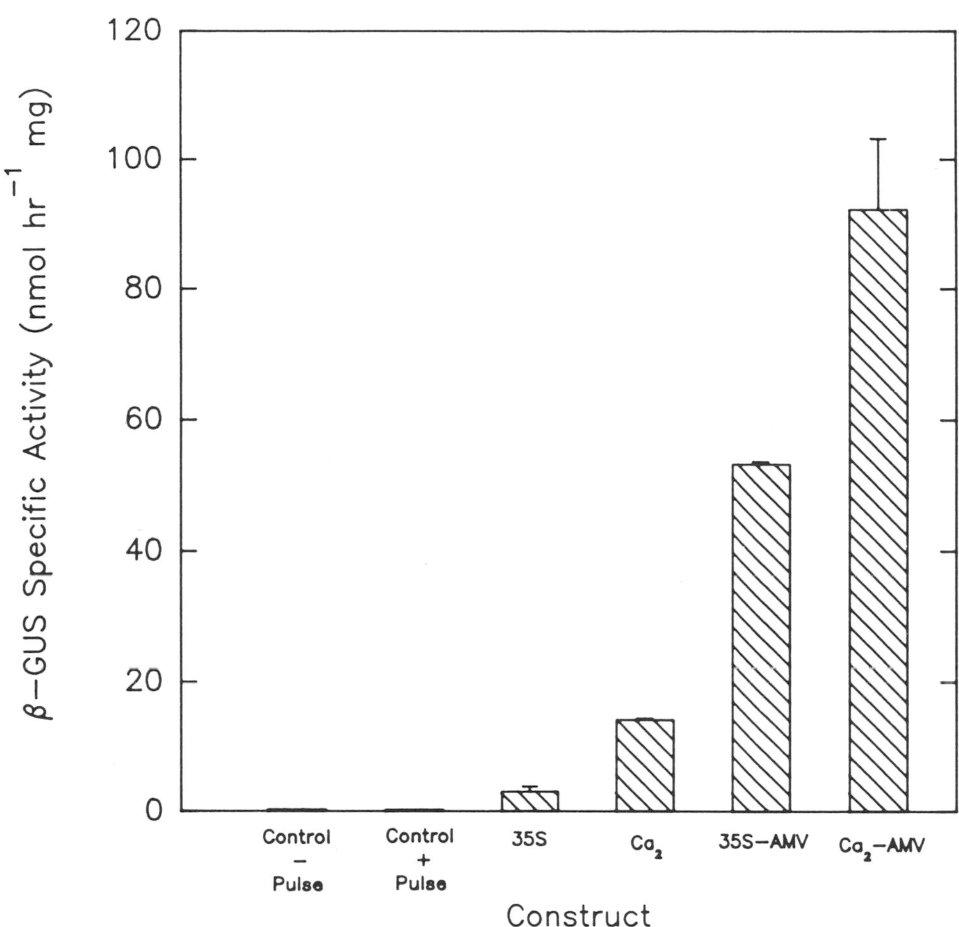


**FIGURE 6.** Transient *gus* activity of AMV and control constructs in white spruce protoplasts. Protoplasts were prepared from 6 to 7 day old suspension cultures of White Spruce (*Picea glauca* [Moench] Voss) in VAE medium as described. Electroporation was carried out at 300 V/cm using plasmids described in Table 3 at a DNA concentration of 0.625 mg/ml. The mean of duplicate values for various promoter/leader combinations is shown. Error bars indicate the standard error to 95% confidence limits. (From Bekkaoui, F., Pilon, M., Laine, E., Raju, D. S. S., Crosby, W. L., and Dunstan, D. I., *Plant Cell Rep.*, 7, 481, 1988. With permission.)

## 2. Inducible Promoters

There are few conditional promoters suitable for heterologous gene expression in plants, if the precise *in situ* physiological significance of the conditionally expressed gene must be determined. The interpretation of experiments using most conditional promoters can be problematic, since many of these are induced in ways which can profoundly alter the physiology of the host cell. However, despite this problem, it may still be advantageous to express some foreign genes under inducible conditions, for example antioxidation genes to protect plants under conditions of aerobic stress, or a heat-inducible antibiotic resistance gene to select mutants in the heat-inducible signal transduction pathway.[204]

Some inducible promoters which have been characterized recently in transgenic plants include heat-shock promoters,[205,206] a nitrate-inducible promoter derived from the spinach

**TABLE 3**
**Plasmid Constructs**

| Construct | Promoter element | 5′ Leader | Ref. |
|---|---|---|---|
| pBI-508 | 35S | Native | 137 |
| pBI-538 | 35S | AMV | — |
| pBI-364 | Tandem 35S | None | 188 |
| pBI-505 | Tandem 35S | AMV | — |

*Note*: All constructs are assembled in a pUC9 or pUC19 backbone (Pharmacia-LKB). The 44 bp synthetic AMV-derived leader sequence is comprised of the sequence: 5′-AGATCTTTTTATTTTTAATTTTCTTTCAAATACTTCCACCATGG-3′ and includes restriction sites for cloning (5′ *Bgl*II and 3′ *Hind*III).

nitrite reductase gene,[207] hormone-inducible sequences,[208,209] and light-inducible promoters associated with the small subunit of RuBP carboxylase and LHCP gene families.[210-215] A recent development that might circumvent the complications associated with such promoters involves the introduction into plants of a regulatory sequence containing a human glucocorticoid response element. Since plants do not naturally respond to steroid hormones, gene induction by a glucocorticosteroid hormone may provide a truly benign induction system with which to test gene expression.[216]

### 3. Tissue-Specific/Developmentally Regulated Promoters

Along with increased interest in plants as model systems for developmental molecular biology, numerous genes and associated promoters have been described which exhibit a wide range of tissue and/or developmental expression patterns. Some examples include genes which are specifically expressed in pollen,[217-219] flower,[220,221] phloem,[222,223] root,[224-226] and seed.[227-230]

Strategies have been designed recently for the tagging and recovery of tissue/developmental specific promoters. In these experiments, marker genes lacking a functional promoter element, and therefore genetically quiescent, are mobilized and inserted into plant chromosomes using *Agrobacterium*-mediated approaches.[231,232] Transgenic plants are regenerated and analyzed for genetic activation of the marker gene, with particular interest directed to developmental or tissue-specific expression patterns. It should be mentioned that the *gus::npt* bifunctional fusion gene (see Section IV.B) is well suited to experiments of this type, since transgenic plants can be directly selected and subsequently analyzed through development as a consequence of genetic activation of the same marker gene (W. Crosby, unpublished).

## B. OTHER 5′-REGULATORY SEQUENCES

It is now known that 5′ nontranslated sequences from transcripts of Tobacco Mosaic Virus and Alfalfa Mosaic Virus (AMV) confer a selective advantage upon mRNAs that contain them during *in vitro* translation.[233-237] The preferential translation of these chimeric mRNAs is mediated at the level of the cytoplasmic ribosome.[72] Figure 6 compares the expression potential of 35S promoter-based constructs (Table 3), with or without a 44 bp synthetic translational enhancer sequence derived from AMV. In each case, constructs containing the AMV leader sequence exhibited an approximate five-fold increase in steady-state expression. Thus, a combination of the duplicated-enhancer 35S promoter[188] with the AMV leader results in an approximate 20-fold increased expression relative to an unmodified 35S construct. The tandem CaMV 35S/AMV promoter is available as a cassette in the plasmid vector pBI-524, which also contains a polylinker

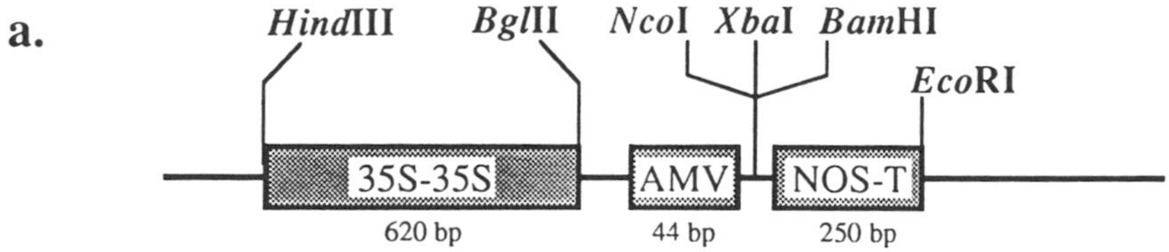


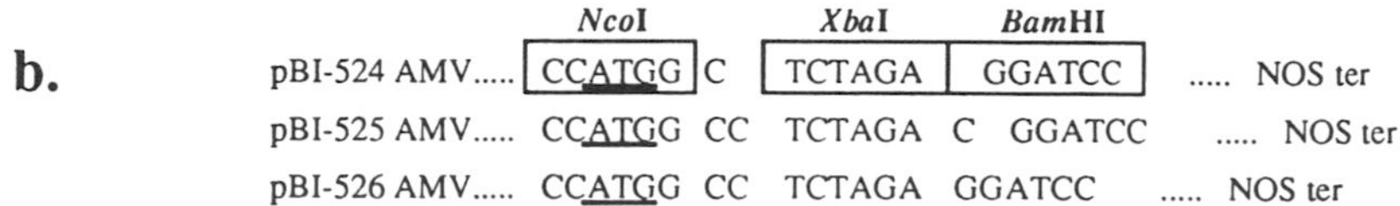


**FIGURE 7.** High expression promoter constructs for transgenic plants. All plasmids were assembled in pUC9 (colE1, Amp$^R$) and total plasmid size is 2.9 Kb (sizes are approximate). Boxed restriction sites are unique to the construct and *Bam*HI, *Hin*dIII, *Eco*RI originate from pUC9: (a) restriction map containing tandem 35S-AMV promoter cassette and *NOS* terminator; (b) boxed restriction sites in the polylinker of plasmid pBI-524 are in two alternate reading frames in pBI-525 and pBI-526; the AMV sequence can be found in the legend to Table 3.

and *nos* 3′ regulatory sequences for cloning and expression of a foreign gene (Figure 7a). Plasmids pBI-525 and pBI-526 (Figure 7b) also contain the tandem 35S/AMV promoter cassette, but the polylinker in these plasmids is constructed to provide restriction sites in alternate reading frames. These vectors facilitate translational fusions with inserts that lack a translation start codon.

## C. 3′-REGULATORY SEQUENCES

### 1. Transcription Termination Sequences

Although most of the research pertaining to gene expression and vector development has focused on 5′ regulatory regions, the 3′ end of a gene is also important in the optimization of gene expression. Sequence comparison studies have shown that the structure of the sequence that mediates termination of transcription and polyadenylation of the resulting mRNA varies widely in plants, and may diverge significantly from the eukaryotic consensus sequence, 5′-AAUAAA-3′.[238,239] Examples have been described where plant genes are naturally associated with multiple cryptic polyA signals, presumably as a means to improve efficient 3′ end formation.[240] While all 3′ ends are functional, the level of stable mRNA accumulated differs when transgenes are expressed using 3′ ends purified from different plant genes.[241] In addition, spatial effects have been described where the extension of the 3′ nontranslated region of a chalcone synthase gene resulted in a 20-fold increase in steady-state expression.[241,242]

## D. INTRON/EXON INFLUENCES

The expression of intron-containing genes can be strongly influenced by the sequence content (e.g., translation stop codons) and the presence of functional gene regulatory sequences in an intron. Several examples have been described where a plant intron may contain a position-dependent enhancer sequence, for example in the maize *Adh*1 and rice *Act*1 actin structural genes.[98,243-247] In chimeric constructs combining the 35S promoter with the rice *Act*1 enhancer, expression was enhanced up to 40-fold in transformed rice cells.[202] This strategy has been particularly successful using the alcohol dehydrogenase and sucrose synthase (*Shrunken*1) introns when engineering genes for monocot expression.[243,246,248,249] The source, size, and spatial positioning of the intron have been shown to strongly influence gene expression,[243,250] as well as the combination of intron enhancer with a surrounding exon region.[243,247,251]

The spatial position and origin of exon sequences can strongly influence gene expression in chimeric constructs, for example during elicitor- and light-activation of a parsley defense gene encoding 4-coumarate:CoA ligase.[252] An additional example is found in the maize *Shrunken*1 gene, where *cis*-active regulatory elements have been localized to the first exon.[243] Chimeric genes containing combinations of the *Shrunken*1 exon1 and intron1 sequences,[243] the pEmu promoter,[201] and a variety of reporter genes are currently being tested for their expression characteristics with barley (O. Olsen and D. von Wettstein, personal communication). Vectors designed with such multiple regulatory elements, together with physical delivery into meristematic tissue and improved culture practices, will contribute to the development of stable transformation protocols for barley, a particularly recalcitrant species.

Aside from containing transcriptional regulatory elements, exonic sequences can encode in-frame termination codons, which have been shown to have a deleterious effect on steady state mRNA accumulation.[251] Small in-frame deletions within a coding region have been observed to affect protein but not mRNA accumulation.[251]

## E. OTHER INFLUENCES ON THE OPTIMIZATION OF GENE EXPRESSION

Numerous examples in the literature point to a role for methylation in the regulation of gene expression in plants, including transgenes.[253-259] The structural features, sequence, or chromosomal context of a transgene which favor hypermethylation and the corresponding reduced gene expression are unknown. The chromosomal context of independent insertions, or "position effect" is known to exert a strong influence — either positive or negative — over the expression potential of the integrated construct.[167,254,260-264] These phenomena are probably distinct from *cis*-active effects involving homologous convergent transcriptional units.[265] Rather, they may involve unknown interactions of the inserted expression construct with higher-order chromosomal structures.[266] Repressive effects on gene expression can also be derived by the subsequent introduction of multiple-copies of a given construct, either through tandem transformation or sexual crossing of independent lines.[167,254,262-264]

As well as factors which directly determine the steady-state accumulation of a chimeric transcript, the codon selection bias of the associated coding region involved has, in some instances, been shown to have a dramatic effect on steady-state accumulation of a polypeptide gene product.[267] An example can be found in experiments in which a modified *Bacillus thuringiensis* cryIA(b) insecticidal protein gene was synthesized *in vitro* to reflect a plant codon bias and to remove internal AT rich sequences, resulting in an approximately 100-fold increase in protein expression and corresponding insect resistance.

## F. SUBCELLULAR TARGETING SEQUENCES

In addition to considerations involving transcription and effective translation of foreign genes, the investigator may wish to direct the gene product to a particular subcellular compartment. There exists a vast and growing literature base describing deductive experiments which have defined structural peptide domains required for the targeting of a given polypeptide to a particular subcellular compartment, including chloroplasts,[268-273] mitochondria,[273-275] peroxisomes,[275,276] vacuole,[277-281] and nucleus.[282] In some instances, a transit peptide defined in this way has proven useful in directing the specific compartmentation of a foreign gene product.[139,282,283] However, it is important to appreciate the complexity of interaction between individual polypeptides and specific cellular transport processes, such as the possibility that fundamentally different mechanisms are involved in transport to the chloroplast outer membrane compared with the thylakoid membrane.[284]

For the most part, transit peptide sequences are localized to the amino-terminus of propeptides destined for chloroplast and mitochondrial transport, where such sequences exist for mitochondrial proteins. However, targeting sequences are largely restricted to carboxy-terminal domains for peptides destined for transport to glyoxysomes and peroxisomes. The requirement for vacuolar targeting has been found to involve a complex series of domains in barley proaleurain.[278] In the case of nuclear-targeted proteins, the putative nuclear translocation sequence may share strong similarity to analogous domains from other similarly targeted peptides.[23] On the other hand, independent gene products which are targeted to chloroplasts nevertheless can display highly divergent amino-terminal transit peptide sequences, suggesting complex structural requirements for recognition and transport.[126]

## VI. NEW APPROACHES

### A. BIOLOGICAL TRANSFORMATION

The biology of *Agrobacterium*-mediated transfer of T-DNA to plant cells is not yet fully understood. However, as details of the process are revealed, new opportunities for manipulating the host range with respect to stable chromosomal transformation may yet be realized. In the meantime, host/vector systems which optimize the expression of essential factors required for T-DNA transfer, stabilization, compartmentation to the nucleus, and eventual integration to plant chromosomes may contribute to the facility of this natural system.

Further insight into the molecular biology and chromosome mechanics of plant viruses, both RNA and DNA in origin, will doubtless open new vistas for their use in the transfer and stable expression of chimeric genes. Of particular interest in this regard is the biology of viral packaging, replication, and gene expression mechanisms which, in combination, will likely yield new opportunities for epichromosomal approaches to chimeric gene expression in plants.

### B. IMPROVED DIRECT-UPTAKE STRATEGIES FOR TRANSFORMATION

Improvements in transformation mediated by direct-uptake mechanisms are being achieved by the refinement of the DNA-introduction mechanism itself. For example, biolistic devices are being developed which introduce DNA-coated micro-particles to a more consistent tissue depth or to a more defined area, with a view to maximizing DNA introduction to the most regenerative tissue of an explant.[83,103,285]

With respect to vector design, few physical vectors have been specifically designed to favor the high frequency integration of foreign DNA to the chromosome. In principle, vectors based on *Tam*, *Ac*, *Ds*, *En*/*Spm* transposons,[286-289] including the *Ac* transposase,[290] could be introduced physically and exploited for the random, stable integration of foreign constructs. As the molecular biology of T-DNA transit and integration becomes better understood, there may emerge direct-uptake strategies for the introduction of "artificial T-DNA" ribonucleoprotein complexes (e.g., involving physical vectors in conjunction with *vir*D and/or *vir*E gene products) which exhibit improved uptake to the nucleus and chromosomal integration.

Future work will likely include the manipulation of plant genomic sequences as tools for chromosomal integration at a specific location, rather than the random integration currently associated with plant transformation. Developments in this direction might serve to improve substantially direct-uptake transformation frequencies, as well as to mimic expression profiles of the chromosomal position in question. Such vectors would also minimize the occurrence of random rearrangements which derive from the illegitimate recombination inherent with present day direct-uptake vectors. A more in depth understanding of plant recombination mechanisms may help to address a major limitation in plant molecular biology at the present time: the lack of an efficient, targeted gene-displacement protocol suitable for the insertional inactivation or conversion of specific genes *in situ*.

## ACKNOWLEDGMENTS

The authors would like to acknowledge the graphics expertise of Christian Gruber and Ralph Underwood.

## REFERENCES


1. **Kado, C. I.,** Molecular mechanisms of crown gall tumorigenesis, *Crit. Rev. Plant. Sci.,* 10, 1, 1991.
2. **Potrykus, I.,** Gene transfer to plants - assessment of published approaches and results, *Annu. Rev. Plant Physiol.,* 42, 205, 1991.
3. **Haring, M. A. and De Block, M.,** New roads towards chloroplast transformation in higher plants, *Physiol Plant.,* 79, 218, 1990.
4. **Potrykus, I.,** Gene transfer to plants: assessment and perspectives, *Physiol. Plant.,* 79, 125, 1990.
5. **Sanford, J. C.,** Biolistic plant transformation, *Physiol. Plant.,* 79, 206, 1990.
6. **Tepfer, D.,** Genetic transformation using *Agrobacterium rhizogenes*, *Physiol. Plant.,* 79, 140, 1990.
7. **Walden, R., Koncz, C., and Schell, J.,** The use of gene vectors in plant molecular biology, *Methods Mol. Cell Biol.,* 1, 175, 1990.
8. **Hooykaas, P. J. J.,** Transformation of plant cells via *Agrobacterium*, *Plant Mol. Biol.,* 13, 327, 1989.
9. **Klee, H. J. and Rogers, S. G.,** Plant gene vectors and genetic transformation: plant transformation systems based on the use of *Agrobacterium tumefaciens*, in *Cell Culture and Somatic Cell Genetics of Plants*, Schell, J. and Vasil, I. K., Eds., Academic Press, Toronto, 1989, 1.
10. **Paszkowski, J., Saul, M. W., and Potrykus, I.,** Plant gene vectors and genetic transformation: DNA-mediated direct gene transfer to plants, in *Cell Culture and Somatic Cell Genetics of Plants*, Schell, J. and Vasil, I. K., Eds., Academic Press, Toronto, 1989, 51.
11. **Tempe, J. and Casse-Delbart, F.,** Plant gene vectors and genetic transformation: *Agrobacterium* Ri plasmids, in *Cell Culture and Somatic Cell Genetics of Plants*, Schell, J. and Vasil, I. K., Eds., Academic Press, Toronto, 1989, 25.
12. **Brisson, N. and Hohn, T.,** Plant virus vectors: cauliflower mosaic virus, in *Methods for Plant Molecular Biology*, Weissbach, A. and Weissbach, H., Eds., Academic Press, Toronto, 1988, 437.
13. **Howard, E., Citovsky, V., and Zambryski, P.,** The T-complex of *Agrobacterium tumefaciens*, in *Plant Gene Transfer*, Lamb, C. J. and Beachy, R. N., Eds., Alan R. Liss, Toronto, 1990, 1.
14. **Zambryski, P., Tempe, J., and Schell, J.,** Transfer and function of T-DNA genes from *Agrobacterium* Ti and Ri plasmids in plants, *Cell*, 56, 193, 1989.
15. **Binns, A. N.,** Cell biology of *Agrobacterium* infection and transformation of plants, *Annu. Rev. Microbiol.,* 42, 575, 1988.
16. **Howard, E. and Citovsky, V.,** The emerging structure of the *Agrobacterium* T-DNA transfer complex, *Bioessays*, 12, 103, 1990.
17. **Citovsky, V., Wong, M. L., and Zambryski, P.,** Cooperative interaction of *Agrobacterium Vir*E2 protein with single-stranded DNA: implications for the T-DNA transfer process, *Proc. Natl. Acad. Sci. U. S. A.,* 86, 1193, 1989.
18. **Christie, P. J., Ward, J. E., Jr., Gordon, M. P., and Nester, E. W.,** A gene required for transfer of T-DNA to plants encodes an ATPase with autophosphorylating activity, *Proc. Natl. Acad. Sci. U. S. A.,* 86, 9677, 1989.
19. **Christie, P. J., Ward, J. E., Winans, S. C., and Nester, E. W.,** The *Agrobacterium tumefaciens vir*E2 gene product is a single-stranded DNA-binding protein that associates with T-DNA, *J. Bacteriol.,* 170, 2659, 1988.
20. **Wang, K., Stachel, S. E., Timmerman, B., van Montagu, M., and Zambryski, P. C.,** Site-specific nick in the T-DNA border sequence as a result of *Agrobacterium Vir* gene expression, *Science*, 235, 587, 1987.
21. **Jayaswal, R. K., Veluthambi, K., Gelvin, S. B., and Slightom, J. L.,** Double-stranded cleavage of T-DNA and generation of single-stranded T-DNA molecules in *Escherichia coli* by a *vir*D-encoded border-specific endonuclease from *Agrobacterium tumefaciens*, *J. Bacteriol.,* 169, 5035, 1987.
22. **Stachel, S. E., Timmerman, B., and Zambryski, P.,** Activation of *Agrobacterium-tumefaciens Vir* gene expression generates multiple single-stranded T-strand molecules from the pTiA6 T-region. Requirement for 5′ *vir*-D gene products, *EMBO J.,* 6, 857, 1987.
23. **Herrera-Estrella, A., van Montagu, M., and Wang, K.,** A bacterial peptide acting as a plant nuclear targeting signal: the amino-terminal portion of *Agrobacterium Vir*D2 protein directs a beta-galactosidase fusion protein into tobacco nuclei, *Proc. Natl. Acad. Sci. U.S.A.,* 87, 9534, 1990.
24. **Herrera-Estrella, A., Chen, Z. M., van Montagu, M., and Wang, K.,** *Vir*D proteins of *Agrobacterium tumefaciens* are required for the formation of a covalent DNA-protein complex at the 5′ terminus of T-strand molecules, *EMBO J.,* 7, 4055, 1988.
25. **Howard, E. A., Zupan, J. R., Citovsky, V., and Zambryski, P. C.,** The *Vir*D2 protein of *A.-tumefaciens* contains a C-terminal bipartite nuclear localization signal — implications for nuclear uptake of DNA in plant cells, *Cell*, 68, 109, 1992.
26. **Ward, J. E., Jr., Dale, E. M., Nester, E. W., and Binns, A. N.,** Identification of a *vir*B10 protein aggregate in the inner membrane of *Agrobacterium tumefaciens*, *J. Bacteriol.,* 172, 5200, 1990.
27. **Ward, J. E., Dale, E. M., and Binns, A. N.,** Activity of the *Agrobacterium* T-DNA transfer machinery is affected by *vir*B gene products, *Proc. Natl. Acad. Sci. U.S.A.,* 88, 9350, 1991.

28. **Toro, N., Datta, A., Carmi, O. A., Young, C., Prusti, R. K., and Nester, E. W.,** The *Agrobacterium tumefaciens vir*C1 gene product binds to overdrive, a T-DNA transfer enhancer, *J. Bacteriol.*, 171, 6845, 1989.
29. **Peralta, E. G., Hellmiss, R., Ji, J. M., Berger, W. H., and Ream, W.,** Overdrive, a T-DNA transmission enhancer on the *A. tumefaciens* tumor-inducing plasmid, in *Molecular Genetics of Plant Microbe Interactions*, Verma, D. P. S. and Brisson, N., Eds., Kluwer, Boston, 1987, 20.
30. **Miranda, A., Janssen, G., Hodges, L., Peralta, E. G., and Ream, W.,** *Agrobacterium tumefaciens* transfers extremely long T-DNAs by a unidirectional mechanism, *J. Bacteriol.*, 174, 2288, 1992.
31. **Graves, A. C. and Goldman, S. L.,** *Agrobacterium tumefaciens*-mediated transformation of the monocot genus Gladiolus: detection of expression of T-DNA-encoded genes, *J. Bacteriol.*, 169, 1745, 1987.
32. **Koziel, M. G., Adams, T. L., Hazlet, M. A., Damm, D., Miller, J., Dahlbeck, D., Jayne, S., and Staskawicz, B. J.,** A cauliflower mosaic virus promoter directs expression of kanamycin resistance in morphogenic transformed plant cells, *J. Mol. Appl. Genet.*, 2, 549, 1984.
33. **Grimsley, N., Hohn, B., Hohn, T., and Walden, R.,** "Agroinfection," an alternative route for viral infection of plants by using the Ti plasmid, *Proc. Natl. Acad. Sci. U.S.A.*, 83, 3282, 1986.
34. **Grimsley, N., Hohn, T., Davies, J. W., and Hohn, B.,** *Agrobacterium*-mediated delivery of infectious maize streak virus into maize plants, *Nature*, 325, 177, 1987.
35. **Grimsley, N., Jarchow, E., Oetiker, J., Schlaeppi, M., and Hohn, B.,** Agroinfection as a tool for the investigation of plant-pathogen interactions, in *Plant Molecular Biology*, Herrmann, R. G. and Larkins, B., Eds., Plenum Press, New York, 1991, 225.
36. **Lurquin, P. F.,** Foreign gene expression in plant cells, *Prog. Nucl. Acid Res. Mol. Biol.*, 34, 143, 1987.
37. **Schell, J.,** Transgenic plants as tools to study the molecular organization of plant genes, *Science*, 237, 1176, 1987.
38. **Walden, R. and Schell, J.,** Techniques in plant molecular biology — progress and problems, *Eur. J. Biochem.*, 192, 563, 1990.
39. **Van Haute, E., Joos, H., Maes, M., Warren, G., van Montagu, M., and Schell, J.,** Intergeneric transfer and exchange recombination of restriction fragments cloned in pBR322: a novel strategy for the reversed genetics of the Ti plasmids of *Agrobacterium tumefaciens*, *EMBO J.*, 2, 411, 1983.
40. **Deblaere, R., Reynaerts, A., Hofte, H., Hernalsteens, J.-P., Leemans, J., and van Montagu, M.,** Vectors for cloning in plant cells, in *Methods in Enzymology*, Wu, R. and Grossman, L., Eds., Academic Press, New York, 1987, 277.
41. **Wallroth, M., Gerats, A. G. M., Rogers, S. G., Fraley, R. T., and Horsch, R. B.,** Chromosomal localization of foreign genes in Petunia hybrida, *Mol. Gen. Genet.*, 202, 6, 1986.
42. **Bidney, D., Scelonge, C., Martich, J., Burrus, M., Sims, L., and Huffman, G.,** Microprojectile bombardment of plant tissues increases transformation frequency by *Agrobacterium tumefaciens*, *Plant Mol. Biol.*, 18, 301, 1992.
43. **Gheysen, G., Villarroel, R., and van Montagu, M.,** Illegitimate recombination in plants: a model for T-DNA integration, *Genes Dev.*, 5, 287, 1991.
44. **Matsumoto, S., Ito, Y., Hosoi, T., Takahashi, Y., and Machida, Y.,** Integration of *Agrobacterium* T-DNA into a tobacco chromosome: possible involvement of DNA homology between T-DNA and plant DNA, *Mol. Gen. Genet.*, 224, 309, 1990.
45. **Jin, S. G., Komari, T., Gordon, M. P., and Nester, E. W.,** Genes responsible for the supervirulence phenotype of *Agrobacterium tumefaciens* A281, *J. Bacteriol.*, 169, 4417, 1987.
46. **Charest, P. J., Iyer, V. N., and Miki, B. L.,** Virulence of *Agrobacterium tumefaciens* strains with *Brassica napus* and *Brassica juncea*, *Plant Cell Rep.*, 8, 303, 1989.
47. **Zyprian, E. and Kado, C. I.,** *Agrobacterium*-mediated plant transformation by novel mini-T vectors in conjunction with a high-copy vir region helper plasmid, *Plant Mol. Biol.*, 15, 245, 1990.
48. **Wang, K., Herrera-Estrella, A., and van Montagu, M.,** Overexpression of *vir*D1 and *vir*D2 genes in *Agrobacterium tumefaciens* enhances T-complex formation and plant transformation, *J. Bacteriol.*, 172, 4432, 1990.
49. **Stougaard, J., Abildsten, D., and Marcker, K. A.,** The *Agrobacterium rhizogenes* pRi TL DNA segment as a gene vector system for transformation of plants, *Mol. Gen. Genet.*, 207, 251, 1987.
50. **Casse-Delbart, F., Jouanin, L., Pautot, V., Robaglia, C., Tepfer, M., Tourneur, J., and Vilaine, F.,** Transformation of plants by agropine-type *Agrobacterium rhizogenes*. Organization of the transferred DNA T-DNA and its use to introduce new genes into plants, *Symbiosis*, 2, 319, 1986.
51. **Robaglia, C., Vilaine, F., Pautot, V., Raimond, F., Amselem, J., Jouanin, L., Casse-Delbart, F., and Tepfer, M.,** Expression vectors based on the *Agrobacterium rhizogenes* Ri plasmid transformation system, *Biochimie*, 69, 231, 1987.
52. **Brasileiro, A. C., Leple, J. C., Muzzin, J., Ounnoughi, D., Michel, M. F., and Jouanin, L.,** An alternative approach for gene transfer in trees using wild-type *Agrobacterium* strains, *Plant Mol. Biol.*, 17, 441, 1991.

53. **Hooykaas, P. J. J. and Schilperoort, R. A.,** Detection of monocot transformation via *Agrobacterium tumefaciens*, in *Methods in Enzymology*, Wu, R. and Grossman, L., Eds., Academic Press, New York, 1987, 305.
54. **Shaw, C. H., Leemans, J., van Montagu, M., and Schell, J.,** A general method for the transfer of cloned genes to plant cells, *Gene*, 23, 315, 1983.
55. **Rogers, S. G., Horsch, R. B., and Fraley, R. T.,** Gene transfer in plants: production of transformed plants using Ti plasmid vectors, in *Methods in Plant Molecular Biology*, Weissbach, A. and Weissbach, H., Eds., Academic Press, Toronto, 1988, 423.
56. **Fraley, R. T., Rogers, S. G., Horsch, R. B., Eichholtz, D. A., Flick, J. S., Fink, C. L., Hoffmann, N. L., and Sanders, P. R.,** The SEV system: a new disarmed Ti plasmid vector system for plant transformation, *Bio/Technology*, 3, 629, 1985.
57. **Rogers, S. G., Klee, H. J., Horsch, R. B., and Fraley, R. T.,** Improved vectors for plant transformation: expression cassette vectors and new selectable markers, in *Methods in Enzymology*, Wu, R. and Grossman, L., Eds., Academic Press, New York, 1987, 253.
58. **Zabarovsky, E. R. and Allikmets, R. L.,** An improved technique for the efficient construction of gene libraries by partial filling-in of cohesive ends, *Gene*, 42, 119, 1986.
59. **Mcbride, K. E. and Summerfelt, K. R.,** Improved binary vectors for *Agrobacterium*-mediated plant transformation, *Plant Mol. Biol.*, 14, 269, 1990.
60. **Peralta, E. G. and Ream, L. W.,** Sequences signaling T-DNA ends on the *Agrobacterium tumefaciens* tumor-inducing plasmid, in *Advances in the Molecular Genetics of the Bacteria-Plant Interaction*, Szalay, A. A. and Legocki, R. P., Eds., Cornell University Press, Ithaca, 1985, 124.
61. **Hoekema, A., Hirsch, P. R., Hooykaas, P. J. J., and Schilperoort, R. A.,** A binary plant vector strategy based on separation of *vir*- and T-region of the *Agrobacterium tumefaciens* Ti-plasmid, *Nature*, 303, 179, 1983.
62. **Simoens, C., Alliotte, T., Mendel, R., Muller, A., Schiemann, J., Van Lijsebettens, M., Schell, J., van Montagu, M., and Inze, D.,** A binary vector for transferring genomic libraries to plants, *Nucl. Acids Res.*, 14, 8073, 1986.
63. **An, G.,** Binary Ti vectors for plant transformation and promoter analysis, in *Methods in Enzymology*, Wu, R. and Grossman, L., Eds., Academic Press, New York, 1987, 292.
64. **Velten, J. and Schell, J.,** Selection-expression plasmid vectors for use in genetic transformation of higher plants, *Nucl. Acids Res.*, 13, 6981, 1985.
65. **Lazo, G. R., Stein, P. A., and Ludwig, R. A.,** A DNA transformation-competent *Arabidopsis* genomic library in *Agrobacterium*, *Bio/Technology*, 9, 963, 1991.
66. **Perani, L., Radke, S., Wilke-Douglas, M., and Bossert, M.,** Gene transfer methods for crop improvement. Introduction of foreign DNA into plants, *Physiol. Plant.*, 68, 566, 1986.
67. **Birch, R. G. and Franks, T.,** Development and optimisation of microprojectile systems for plant genetic transformation, *Aust. J. Plant Physiol.*, 18, 453, 1991.
68. **Paszkowski, J. and Saul, M. W.,** Direct gene transfer to plants, in *Methods for Plant Molecular Biology*, Academic Press, New York, 1988, 447.
69. **Shillito, R. D. and Potrykus, I.,** Direct gene transfer to protoplasts of dicotyledonous and monocotyledonous plants by a number of methods, including electroporation, in *Methods in Enzymology*, Wu, R. and Grossman, L., Eds., Academic Press, New York, 1987, 313.
70. **Krens, F. A., Mans, R. M. W., Van Slogteren, T. M. S., Hoge, J. H. C., Wullems, G. J., and Schilperoort, R. A.,** Structure and expression of DNA transferred to tobacco via transformation of protoplasts with Ti-plasmid DNA: co-transfer of T-DNA and non T-DNA sequences, *Plant Mol. Biol.*, 5, 223, 1985.
71. **Krens, F. A., Molendijk, L., Wullems, G. J., and Schilperoort, R. A.,** *In vitro* transformation of plant protoplasts with Ti-plasmid DNA, *Nature*, 296, 72, 1982.
72. **Gallie, D. R., Walbot, V., and Hershey, J. W. B.,** The ribosomal fraction mediates the translational enhancement associated with the 5′-leader of tobacco mosaic virus, *Nucl. Acids Res.*, 16, 8675, 1988.
73. **Draper, J., Davey, M. R., Freeman, J. P., Cocking, E. C., and Cox, B. J.,** Ti plasmid homologous sequences present in tissues from *Agrobacterium* plasmid-transformed Petunia protoplasts, *Plant Cell Physiol.*, 23, 451, 1982.
74. **Langridge, W. H. R., Li, B. J., and Szalay, A. A.,** Electric field mediated stable transformation of carrot protoplasts with naked DNA, *Plant Cell Rep.*, 4, 355, 1985.
75. **Lazzeri, P. A., Brettschneider, R., Luhrs, R., and Lörz, H.,** Stable transformation of barley via PEG-induced direct DNA uptake into protoplasts, *Theor. Appl. Genet.*, 81, 437, 1991.
76. **Cao, J., Wang, Y.-C., Klein, T. M., Sanford, J. C., and Wu, R.,** Transformation of rice and maize using the biolistic process, in *Plant Gene Transfer*, Lamb, C. J. and Beachy, R. N., Eds., Alan R. Liss, Toronto, 1990, 21.
77. **Peng, J. Y., Kononowicz, H., and Hodges, T. K.,** Transgenic Indica rice plants, *Theor. Appl. Genet.*, 83, 855, 1992.

78. **Walters, D. A., Vetsch, C. S., Potts, D. E., and Lundquist, R. C.,** Transformation and inheritance of a hygromycin phosphotransferase gene in maize plants, *Plant Mol. Biol.,* 18, 189, 1992.
79. **Lyznik, L. A., Ryan, R. D., Ritchie, S. W., and Hodges, T. K.,** Stable co-transformation of maize protoplasts with *gusA* and *neo* genes, *Plant Mol. Biol.,* 13, 151, 1989.
80. **Chen, W. H., Garland, K. M. A., Davey, M. R., Sotak, R., Garland, J. S., Mulligan, B. J., Power, J. B., and Cocking, E. C.,** Transformation of sugarcane protoplasts by direct uptake of a selectable chimeric gene, *Plant Cell Rep.,* 6, 297, 1987.
81. **Horn, M. E., Shillito, R. D., Conger, B. V., and Harms, C. T.,** Transgenic plants of orchardgrass *Dactylis glomerata* L. from protoplasts, *Plant Cell Rep.,* 7, 469, 1988.
82. **Vasil, V., Brown, S. M., Re, D., Fromm, M. E., and Vasil, I. K.,** Stably transformed callus lines from microprojectile bombardment of cell suspension cultures of wheat, *Bio/Technology,* 9, 743, 1991.
83. **McCabe, D. E., Swain, W. F., Martinell, B. J., and Christou, P.,** Stable transformation of soybean (*Glycine max*) by particle acceleration, *Bio/Technology,* 6, 923, 1988.
84. **Christou, P., Swain, W. F., Yang, N.-S., and McCabe, D. E.,** Inheritance and expression of foreign genes in transgenic soybean plants, *Proc. Natl. Acad. Sci. U.S.A.,* 86, 7500, 1989.
85. **Finer, J. J. and Mcmullen, M. D.,** Transformation of cotton (*Gossypium hirsutum* L.) via particle bombardment, *Plant Cell Rep.,* 8, 586, 1990.
86. **De la Pena, A., Lorz, H., and Schell, J.,** Transgenic rye plants obtained by injecting DNA into young floral tillers, *Nature,* 325, 274, 1987.
87. **Reich, T. J., Iyer, V. N., and Miki, B. L.,** Efficient transformation of alfalfa protoplasts by the intranuclear microinjection of Ti plasmids, *Bio/Technology,* 4, 1001, 1986.
88. **Zhang, L.-J., Chen, L.-M., Xu, N., Zhao, N.-M., Li, C.-G., Yuan, J., and Jia, S.-R.,** Efficient transformation of tobacco by ultrasonication, *Bio/Technology,* 9, 1991.
89. **Deshayes, A., Herrera-Estrella, L., and Caboche, M.,** Liposome-mediated transformation of tobacco mesophyll protoplasts by an *Escherichia coli* plasmid, *EMBO J.,* 4, 2731, 1985.
90. **Christou, P., Murphy, J. E., and Swain, W. F.,** Stable transformation of soybean by electroporation and root formation from transformed callus, *Proc. Natl. Acad. Sci. U.S.A.,* 84, 3962, 1987.
91. **Hain, R., Stabel, P., Czernilofsky, A. P., Steinbi, H.-H., Herrera-Estrella, L., and Schell, J.,** Uptake, integration, expression and genetic transmission of a selectable chimaeric gene by plant protoplasts, *Mol. Gen. Genet.,* 199, 161, 1985.
92. **Potrykus, I., Saul, M. W., Petruska, J., Paszkowski, J., and Shillito, R. D.,** Direct gene transfer to cells of a graminaceous monocot, *Mol. Gen. Genet.,* 199, 183, 1985.
93. **Mccown, B. H., McCabe, D. E., Russell, D. R., Robison, D. J., Barton, K. A., and Raffa, K. F.,** Stable transformation of *Populus* and incorporation of pest resistance by electric discharge particle acceleration, *Plant Cell Rep.,* 9, 590, 1991.
94. **Shillito, R. D., Saul, M. W., Paszkowski, J., Muller, M., and Potrykus, I.,** High efficiency direct gene transfer to plants, *Bio/Technology,* 3, 1099, 1985.
95. **West, J. A., Mettler, I. J., and Dietrich, P. S.,** Vector linearization affects transient gene expression and transformation frequency in BMS cells, in *Molecular Biology of Plant Growth and Development; ISPMB Abstracts,* Hallick, R. B., Ed., University of Arizona, Tucson, 1991, 427.
96. **Rodenburg, K. W., De Groot, M. J. A., Schilperoort, R. A., and Hooykaas, P. J. J.,** Single-stranded DNA used as an efficient new vehicle for transformation of plant protoplasts, *Plant Mol. Biol.,* 13, 711, 1989.
97. **Lörz, H., Baker, B., and Schell, J.,** Gene transfer to cereal cells mediated by protoplast transformation, *Mol. Gen. Genet.,* 199, 178, 1985.
98. **Callis, J., Fromm, M., and Walbot, V.,** Introns increase gene expression in cultured maize cells, *Genes Dev.,* 1, 1183, 1987.
99. **Tomes, D. T., Weissinger, A. K., Ross, M., Higgins, R., Drummond, B. J., Schaaf, S., Maloneschoneberg, J., Staebell, M., Flynn, P., Anderson, J., and Howard, J.,** Transgenic tobacco plants and their progeny derived by microprojectile bombardment of tobacco leaves, *Plant Mol. Biol.,* 14, 261, 1990.
100. **Jefferson, R. A., Kavanagh, T. A., and Bevan, M. W.,** GUS fusions: β-glucuronidase as a sensitive and versatile gene fusion marker in higher plants, *EMBO J.,* 6, 3901, 1987.
101. **Shimamoto, K., Terada, R., Izawa, T., and Fujimoto, H.,** Fertile transgenic rice plants regenerated from transformed protoplasts, *Nature,* 338, 274, 1989.
102. **Seki, M., Shigemoto, N., Komeda, Y., Imamura, J., Yamada, Y., and Morikawa, H.,** Transgenic *Arabidopsis thaliana* plants obtained by particle-bombardment-mediated transformation, *Appl. Microbiol. Biotechnol.,* 36, 228, 1991.
103. **Christou, P., McCabe, D. E., and Swain, W. F.,** Stable transformation of soybean callus by DNA-coated gold particles, *Plant Physiol.,* 87, 671, 1988.
104. **Okada, K., Takebe, I., and Nagata, T.,** Expression and integration of genes introduced into highly synchronized plant protoplasts, *Mol. Gen. Genet.,* 205, 398, 1986.

105. **Meyer, P., Walgenbach, E., Bussmann, K., Hombrecher, G., and Saedler, H.,** Synchronized tobacco protoplasts are efficiently transformed by DNA, *Mol. Gen. Genet.*, 201, 513, 1985.
106. **Meyer, P., Kartzke, S., Niedenhof, I., Heidmann, I., Bussmann, K., and Saedler, H.,** A genomic DNA segment from *Petunia hybrida* leads to increased transformation frequencies and simple integration patterns, *Proc. Natl. Acad. Sci. U.S.A.*, 85, 8568, 1988.
107. **Kartzke, S., Saedler, H., and Meyer, P.,** Molecular analysis of transgenic plants derived from transformations of protoplasts at various stages of the cell cycle, *Plant Sci.*, 67, 63, 1990.
108. **Chasan, R.,** Taming the plastid genome, *Plant Cell*, 4, 1, 1992.
109. **Butow, R. A. and Fox, T. D.,** Organelle transformation: shoot first, ask questions later, *Trends Biol. Sci.*, 15, 465, 1990.
110. **Ye, G. N., Daniell, H., and Sanford, J. C.,** Optimization of delivery of foreign DNA into higher-plant chloroplasts, *Plant Mol. Biol.*, 15, 809, 1990.
111. **Daniell, H., Vivekananda, J., Nielsen, B. L., Ye, G. N., Tewari, K. K., and Sanford, J. C.,** Transient foreign gene expression in chloroplasts of cultured tobacco cells after biolistic delivery of chloroplast vectors, *Proc. Natl. Acad. Sci. U.S.A.*, 87, 88, 1990.
112. **Staub, J. M. and Maliga, P.,** Long regions of homologous DNA are incorporated into the tobacco plastid genome by transformation, *Plant Cell*, 4, 39, 1992.
113. **Wissinger, B., Hiesel, R., Schobel, W., Unseld, M., Brennicke, A., and Schuster, W.,** Duplicated sequence elements and their function in plant mitochondria, *Z. Naturforsch Sect. C.*, 46, 709, 1991.
114. **Venkateswarlu, K. and Nazar, R. N.,** Evidence for T-DNA mediated gene targeting to tobacco chloroplasts, *Bio/Technology*, 9, 1103, 1991.
115. **Bevan, M. W., Flavell, R. B., and Chilton, M.-D.,** A chimaeric antibiotic resistance gene as a selectable marker for plant cell transformation, *Nature*, 304, 184, 1983.
116. **An, G., Watson, B. D., and Chiang, C. C.,** Transformation of tobacco, tomato, potato, and *Arabidopsis thaliana* using a binary Ti vector system, *Plant Physiol.*, 81, 301, 1986.
117. **De Block, M.,** Genotype-independent leaf disc transformation of potato (*Solanum tuberosum*) using *Agrobacterium tumefaciens*, *Theor. Appl. Genet.*, 76, 767, 1988.
118. **McCormick, S., Niedermeyer, J., Fry, J., Barnason, A., Horsch, R., and Fraley, R.,** Leaf disc transformation of cultivated tomato *Lycopersicon esculentum* using *Agrobacterium tumefaciens*, *Plant Cell Rep.*, 5, 81, 1986.
119. **Huang, Y.-W. and Dennis, E. S.,** Factors influencing stable transformation of maize protoplasts by electroporation, *Plant Cell Tissue Organ Cult.*, 18, 281, 1989.
120. **Hauptmann, R. M., Vasil, V., Ozias-Akins, P., Tabaeizadeh, Z., Rogers, S. G., Fraley, R. T., Horsch, R. B., and Vasil, I. K.,** Evaluation of selectable markers for obtaining stable transformants in the Graminae, *Plant Physiol.*, 86, 602, 1988.
121. **Yenofsky, R. L., Fine, M., and Pellow, J. W.,** A mutant neomycin phosphotransferase II gene reduced the resistance of transformants to antibiotic selection pressure, *Proc. Natl. Acad. Sci. U.S.A.*, 87, 3435, 1990.
122. **De Block, M., Schell, J., and van Montagu, M.,** Chloroplast transformation by *Agrobacterium tumefaciens*, *EMBO J.*, 4, 1367, 1985.
123. **Gordon-Kamm, W. J., Spencer, T. M., Mangano, M. L., Adams, T. R., Daines, R. J., Start, W. G., O'Brien, J. V., Chambers, S. A., Adams, W. R. J., Willetts, N. G., Rice, T. B., Mackey, C. J., Krueger, R. W., Kausch, A. P., and Lemaux, P. G.,** Transformation of maize cells and regeneration of fertile transgenic plants, *Plant Cell*, 2, 603, 1990.
124. **Spencer, T. M., Obrien, J. V., Start, W. G., Adams, T. R., Gordonkamm, W. J., and Lemaux, P. G.,** Segregation of transgenes in maize, *Plant Mol. Biol.*, 18, 201, 1992.
125. **D'Halluin, K., De Block, M., Denecke, J., and Janssens, J.,** The *bar* gene as selectable and screenable marker in plant engineering, *Methods Enzymol.*, in press, 1992.
126. **Haughn, G. W., Smith, J., Mazur, B., and Somerville, C.,** Transformation with a mutant *Arabidopsis* acetolactate synthase gene renders tobacco resistant to sulfonylurea herbicides, *Mol. Gen. Genet.*, 211, 266, 1988.
127. **Ludwig, S. R., Bowen, B., Beach, L., and Wessler, S. R.,** A regulatory gene as a novel visible marker for maize transformation, *Science*, 247, 449, 1990.
128. **Nussaume, L., Vincentz, M., and Caboche, M.,** Constitutive nitrate reductase: a dominant conditional marker for plant genetics, *Plant J.*, 1, 267, 1991.
129. **Jefferson, R. A.,** Assaying chimeric genes in plants: the GUS gene fusion system, *Plant Mol. Biol. Rep.*, 5, 387, 1987.
130. **Jefferson, R. A., Burgess, S. M., and Hirsh, D.,** β-Glucuronidase from *Escherichia coli* as a gene-fusion marker, *Proc. Natl. Acad. Sci. U.S.A.*, 83, 8447, 1986.
131. **Jefferson, R. A., Bevan, M., and Kavanagh, T. A.,** The use of the *Escherichia coli* β-glucuronidase gene as a gene fusion marker for studies of gene expression in higher plants, *Biochem. Soc. Trans.*, 15, 17, 1986.

132. **Vancanneyt, G., Schmidt, R., O'Connor-Sanchez, A., Willmitzer, L., and Rocha-Sosa, M.,** Construction of an intron-containing marker gene. Splicing of the intron in transgenic plants and its use in monitoring early events in *Agrobacterium* mediated plant transformation, *Mol. Gen. Genet.*, 220, 245, 1990.
133. **Ohta, S., Mita, S., Hattori, T., and Nakamura, K.,** Construction and expression in tobacco of beta-glucuronidase (Gus) reporter gene containing an intron within the coding sequence, *Plant Cell Physiol.*, 31, 805, 1990.
134. **Kato, T., Shirano, Y., Kawazu, T., Tada, Y., Itoh, E., and Shibata, D.,** A modified β-glucuronidase gene: sensitive detection of plant promoter activities in suspension-cultured cells of tobacco and rice, *Plant Mol. Biol.*, 9, 333, 1991.
135. **Datla, R. S. S., Hammerlindl, J. K., Pelcher, L. E., Crosby, W. L., and Selvaraj, G.,** A bifunctional fusion between beta-glucuronidase and neomycin phosphotransferase — a broad-spectrum marker enzyme for plants, *Gene*, 101, 239, 1991.
136. **Herrera-Estrella, L., Depicker, A., van Montagu, M., and Schell, J.,** Expression of chimaeric genes transferred into plant cells using a Ti-plasmid-derived vector, *Nature*, 303, 209, 1983.
137. **Fromm, M. E., Taylor, L. P., and Walbot, V.,** Expression of genes transferred into monocot and dicot plant cells by electroporation, *Proc. Natl. Acad. Sci. U.S.A.*, 82, 5824, 1985.
138. **Werr, W. and Lörz, H.,** Transient gene expression in a Graminaeae cell line. A rapid procedure for studying plant promoters, *Mol. Gen. Genet.*, 202, 471, 1986.
139. **Boutry, M., Nagy, F., Poulsen, C., Aoyagi, K., and Chua, N.-H.,** Targeting of bacterial chloramphenicol acetyltransferase to mitochondria in transgenic plants, *Nature*, 328, 340, 1987.
140. **Takamatsu, N., Ishikawa, M., Meshi, T., and Okada, Y.,** Expression of bacterial chloramphenicol acetyltransferase gene in tobacco plants mediated by TMV-RNA, *EMBO J.*, 6, 307, 1987.
141. **Bekkaoui, F., Pilon, M., Laine, E., Raju, D. S. S., Crosby, W. L., and Dunstan, D. I.,** Transient gene expression in electroporated *Picea glauca* protoplasts, *Plant Cell Rep.*, 7, 481, 1988.
142. **Klein, T. M., Fromm, M., Wessinger, A., Tomes, D., Schaaf, S., Sletten, M., and Sanford, J. C.,** Transfer of foreign genes into intact maize cells with high-velocity microprojectiles, *Proc. Natl. Acad. Sci. U.S.A.*, 85, 4305, 1988.
143. **Prols, M., Topfer, R., Schell, J., and Steinbiß, H.-H.,** Transient gene expression in tobacco protoplasts. I. Time course of CAT appearance, *Plant Cell Rep.*, 7, 221, 1988.
144. **Topfer, R., Prols, M., Schell, J., and Steinbiß, H.-H.,** Transient gene expression in tobacco protoplasts: II. Comparison of the reporter gene systems for *CAT*, *NPT II*, and *GUS*, *Plant Cell Rep.*, 7, 225, 1988.
145. **Young, S. L., Barbera, L., Kaynard, A. H., Haugland, R. P., Kang, H. C., Brinkley, M., and Melner, M. H.,** A nonradioactive assay for transfected chloramphenicol acetyltransferase activity using fluorescent substrates, *Anal. Biochem.*, 197, 401, 1991.
146. **Gendloff, E. H., Bowen, B., and Buchholz, W. G.,** Quantitation of chloramphenicol acetyl transferase in transgenic tobacco plants by ELISA and correlation with gene copy number, *Plant Mol. Biol.*, 14, 575, 1990.
147. **Balazs, E. and Bonneville, J. M.,** Chloramphenicol acetyltransferase activity in *Brassica* spp., *Plant Sci.*, 50, 65, 1987.
148. **Charest, P. J., Iyer, V. N., and Miki, B. L.,** Factors affecting the use of chloramphenicol acetyltransferase as a marker for *Brassica* genetic transformation, *Plant Cell Rep.*, 7, 628, 1989.
149. **Ow, D. W., Wood, K. V., DeLuca, M., DeWet, J. R., Helinski, D. R., and Howell, S. H.,** Transient and stable expression of the firefly luciferase gene in plant cells and transgenic plants, *Science*, 234, 856, 1986.
150. **Ow, D. W., Wood, K. V., DeWet, J. R., Jacobs, J., DeLuca, M., and Helinski, D. R.,** Monitoring gene activity with luciferase gene fusions, *J. Cell Biochem. Suppl.*, (11C), 130, 1987.
151. **Riggs, C. D. and Chrispeels, M. J.,** Luciferase reporter gene cassettes for plant gene expression studies, *Nucl. Acids Res.*, 15, 8115, 1987.
152. **Ow, D. W., Jacobs, J. D., and Howell, S. H.,** Functional regions of the cauliflower mosaic virus 35S RNA promoter determined by use of the firefly luciferase gene as a reporter of promoter activity, *Proc. Natl. Acad. Sci. U.S.A.*, 84, 4870, 1987.
153. **Quandt, H. J., Broer, I., and Puhler, A.,** Tissue-specific activity and light-dependent regulation of a soybean rbcS promoter in transgenic tobacco plants monitored with the firefly luciferase gene, *Plant Sci.*, 82, 59, 1992.
154. **Boylan, M., Pelletier, J., and Meighen, E. A.,** Fused bacterial luciferase subunits catalyze light emission in eukaryotes and prokaryotes, *J. Biol. Chem.*, 264, 1915, 1989.
155. **Kirchner, G., Roberts, J. L., Gustafson, G. D., and Ingolia, T. D.,** Active bacterial luciferase from a fused gene: expression of a *Vibrio harveyi luxAB* translational fusion in bacteria, yeast and plant cells, *Gene*, 81, 349, 1989.
156. **Logocki, R. P., Legocki, M., Baldwin, T. O., and Szalay, A. A.,** Bioluminescence in soybean root nodules: demonstration of a general approach to assay gene expression *in vivo* by using bacterial luciferase, *Proc. Natl. Acad. Sci. U.S.A.*, 83, 9080, 1986.

157. **Olsson, O., Nilsson, O., and Koncz, C.,** Novel monomeric luciferase enzymes as tools to study plant gene regulation *in vivo, J. Biolumin. Chemilumin.*, 5, 79, 1990.
158. **Koncz, C., Martini, N., Koncz-Kalman, Z., Olsson, O., Radermacher, A., Szalay, A., and Schell, J.,** Genetic tools for the analysis of gene expression in plants, in *Tailoring Genes for Crop Improvement: an Agricultural Perspective*, Bruening, G., Harada, J., Kosuge, T., Hollaender, A., Kuny, G., and Wilson, C. M., Eds., Plenum Press, New York, 1985, 197.
159. **Koncz, C., Langridge, W. H. R., Olsson, O., Schell, J., and Szalay, A. A.,** Bacterial and firefly luciferase genes in transgenic plants: advantages and disadvantages of a reporter gene, *Dev. Genet.*, 11, 224, 1990.
160. **van Slogteren, G. M. S., Hooykaas, P. J. J., Planoue, K., and De Groot, B.,** The lysopine dehydrogenase gene used as a marker for the selection of octopine crown gall cells, *Plant Mol. Biol.*, 1, 133, 1982.
161. **Buchmann, I., Marner, F.-J., Schroder, G., Waffenschmidt, S., and Schroder, J.,** Tumour genes in plants: T-DNA encoded cytokinin biosynthesis, *EMBO J.*, 4, 853, 1985.
162. **Christou, P., Platt, S. G., and Ackerman, M. C.,** Opine synthesis in wild-type plant tissue, *Plant Physiol.*, 82, 218, 1986.
163. **Holbrook, L. A. and Miki, B. L.,** *Brassica* crown gall tumorigenesis and in-vitro of transformed tissue, *Plant Cell Rep.*, 4, 329, 1986.
164. **Hood, E. E., Chilton, W. S., Chilton, M. D., and Fraley, R. T.,** T-DNA and opine synthetic loci in tumors incited by *Agrobacterium tumefaciens* A281 on soybean and alfalfa plants, *J. Bacteriol.*, 168, 1283, 1986.
165. **Swanson, E. B. and Erickson, L. R.,** Haploid transformation in *Brassica napus* using an octopine-producing strain of *Agrobacterium tumefaciens*, *Theor. Appl. Genet.*, 78, 831, 1989.
166. **Hood, E. E., Clapham, D. H., Ekberg, I., and Johannson, T.,** T-DNA presence and opine production in tumors of *Picea abies* (L.) Karst induced by *Agrobacterium tumefaciens* A281, *Plant Mol. Biol.*, 14, 111, 1990.
167. **Goring, D. R., Thomson, L., and Rothstein, S. J.,** Transformation of a partial nopaline synthase gene into tobacco suppresses the expression of a resident wild-type gene, *Proc. Natl. Acad. Sci. U.S.A.*, 88, 1770, 1991.
168. **Helmer, G., Casadaban, M., Bevan, M., Kayes, L., and Chilton, M.-D.,** A new chimeric gene as a marker for plant transformation: the expression of *Escherichia coli* β-galactosidase in sunflower and tobacco cells, *Bio/Technology*, 2, 520, 1984.
169. **Roy, P. and Sahasradbudhe, N.,** A sensitive and simple paper chromatographic procedure for detecting neomycin phosphotransferase II NPTII gene expression, *Plant Mol. Biol.*, 14, 873, 1990.
170. **Platt, S. G. and Yang, N.-S.,** Dot assay for neomycin phosphotransferase activity in crude cell extracts, *Anal. Biochem.*, 162, 529, 1987.
171. **Radke, S. E., Andrews, B. M., Moloney, M. M., Crouch, M. L., Kridl, J. C., and Knauf, V. C.,** Transformation of *Brassica napus* L. using *Agrobacterium tumefaciens*. Developmentally regulated expression of a reintroduced napin gene, *Theor. Appl. Genet.*, 75, 685, 1988.
172. **Sanger, M., Daubert, S., and Goodman, R. M.,** Characteristics of a strong promoter from Figwort Mosaic Virus — comparison with the analogous 35S promoter from Cauliflower Mosaic Virus and the regulated mannopine synthase promoter, *Plant Mol. Biol.*, 14, 433, 1990.
173. **Brisson, N., Paszkowski, J., Penswick, J. R., Gronenborn, B., Potrykus, I., and Hohn, T.,** Expression of a bacterial gene in plants by using a viral vector, *Nature*, 310, 511, 1984.
174. **Balazs, E., Bouzoubaa, S., Guilley, H., Jonard, G., Paszkowski, J., and Richards, K.,** Chimeric vector construction for higher-plant transformation, *Gene*, 40, 343, 1985.
175. **Hamill, J. D., Robins, R. J., Parr, A. J., Evans, D. M., Furze, J. M., and Rhodes, M. J.,** Over-expressing a yeast ornithine decarboxylase gene in transgenic roots of *Nicotiana rustica* can lead to enhanced nicotine accumulation, *Plant Mol. Biol.*, 15, 27, 1990.
176. **An, G.,** Development of plant promoter expression vectors and their use for analysis of differential activity of nopaline synthase promoter in transformed tobacco cells, *Plant Physiol.*, 81, 86, 1986.
177. **Bruce, W. B. and Gurley, W. B.,** Functional domains of a T-DNA promoter active in crown gall tumors, *Mol. Cell Biol.*, 7, 59, 1987.
178. **Kononowicz, H., Wang, Y. E., Habeck, L. L., and Gelvin, S. B.,** Subdomains of the octopine synthase upstream activating element direct cell-specific expression in transgenic tobacco plants, *Plant Cell*, 4, 17, 1992.
179. **Bevan, M.,** Binary *Agrobacterium* vectors for plant transformation, *Nucl. Acids Res.*, 12, 8711, 1984.
180. **An, G., Costa, M. A., Mitra, A., Ha, S. B., and Marton, L.,** Organ-specific and developmental regulation of the nopaline synthase promoter in transgenic tobacco plants, *Plant Physiol.*, 88, 547, 1988.
181. **Benfey, P. N., Ren, L., and Chua, N. H.,** Tissue-specific expression from CaMV 35S enhancer subdomains in early stages of plant development, *EMBO J.*, 9, 1677, 1990.
182. **Benfey, P. N., Ren, L., and Chua, N. H.,** The CaMV 35S enhancer contains at least two domains which can confer different developmental and tissue-specific expression patterns, *EMBO J.*, 8, 2195, 1989.
183. **Yang, N.-S. and Christou, P.,** Cell type specific expression of a CaMV 35S-Gus gene in transgenic soybean plants, *Dev. Genet.*, 11, 289, 1990.

184. **Harpster, M. H., Townsend, J. A., Jones, J. D., Bedbrook, J., and Dunsmuir, P.,** Relative strengths of the 35S cauliflower mosaic virus, 1′, 2′, and nopaline synthase promoters in transformed tobacco sugarbeet and oilseed rape callus tissue, *Mol. Gen. Genet.,* 212, 182, 1988.
185. **Sanders, P. R., Winter, J. A., Barnason, A. R., Rogers, S. G., and Fraley, R. T.,** Comparison of cauliflower mosaic virus 35S and nopaline synthase promoters in transgenic plants, *Nucl. Acids Res.,* 15, 1543, 1987.
186. **Odell, J. T., Nagy, F., and Chua, N. H.,** Identification of DNA sequences required for activity of the cauliflower mosaic virus 35S promoter, *Nature,* 313, 810, 1985.
187. **Benfey, P. N. and Chua, N.-H.,** The cauliflower mosaic virus 35S promoter: combinatorial regulation of transcription in plants, *Science,* 250, 959, 1990.
188. **Kay, R., Chan, A., Daly, M., and McPherson, J.,** Duplication of CaMV 35S promoter sequences creates a strong enhancer for plant genes, *Science,* 236, 1299, 1987.
189. **Hirochika, H. and Hayashi, K.,** A new strategy to improve a cauliflower mosaic virus vector, *Gene,* 105, 239, 1991.
190. **Timmermans, M. C. P., Maliga, P., Vieira, J., and Messing, J.,** The pFF plasmids: cassettes utilising CaMV sequences for expression of foreign genes in plants, *J. Biotechnol.,* 14, 333, 1990.
191. **Tautorus, T. E., Bekkaoui, F., Pilon, M., Datla, R. S. S., Crosby, W. L., Fowke, L. C., and Dunstan, D. I.,** Factors affecting transient gene expression in electroporated Black Spruce (*Picea mariana*) and Jack Pine (*Pinus banksiana*) protoplasts, *Theor. Appl. Genet.,* 78, 531, 1989.
192. **Futterer, J., Bonneville, J. M., and Hohn, T.,** Cauliflower mosaic virus as a gene expression vector for plants, *Physiol. Plant.,* 79, 154, 1990.
193. **Comai, L., Moran, P., and Maslyar, D.,** Novel and useful properties of a chimeric plant promoter combining CaMV 35S and MAS elements, *Plant Mol. Biol.,* 15, 373, 1990.
194. **Assaad, F. F. and Signer, E. R.,** Cauliflower mosaic virus P35S promoter activity in *Escherichia coli, Mol. Gen. Genet.,* 223, 517, 1990.
195. **Pobjecky, N., Rosenberg, G. H., Dinter-Gottlieb, G., and Kaufer, N. F.,** Expression of the beta glucuronidase gene under the control of the CaMV 35S promoter in *Schizosaccharomyces pombe, Mol. Gen. Genet.,* 220, 314, 1990.
196. **Gmunder, H. and Kohli, J.,** Cauliflower mosaic virus promoters direct efficient expression of a bacterial G418 resistance gene in *Schizosaccharomyces pombe, Mol. Gen. Genet.,* 220, 95, 1989.
197. **Hauptmann, R. M., Ozias-Akins, P., Vasil, V., Tabaeizadeh, Z., Rogers, S. G., Horsch, R. B., Vasil, I. K., and Fraley, R. T.,** Transient expression of electroporated DNA in monocotyledonous and dicotyledonous species, *Plant Cell Rep.,* 6, 265, 1987.
198. **Larkin, P. J., Taylor, B. H., Gersmann, M., and Bretell, R. I. S.,** Direct gene transfer to protoplasts, *Aust. J. Plant Physiol.,* 17, 291, 1990.
199. **Battraw, M. J. and Hall, T. C.,** Histochemical analysis of CaMV 35S promoter-β-glucuronidase gene expression in transgenic rice plants, *Plant Mol. Biol.,* 15, 527, 1990.
200. **Terada, R. and Shimamoto, K.,** Expression of CaMV35S-GUS gene in transgenic rice plants, *Mol. Gen. Genet.,* 220, 389, 1990.
201. **Last, D. I., Brettell, R. I. S., Chamberlain, D. A., Chaudhury, A. M., Larkin, P. J., Marsh, E. L., Peacock, W. J., and Dennis, E. S.,** pEmu — an improved promoter for gene expression in cereal cells, *Theor. Appl. Genet.,* 81, 581, 1991.
202. **Mcelroy, D., Blowers, A. D., Jenes, B., and Wu, R.,** Construction of expression vectors based on the rice Actin-1 (*Act*1) 5′ region for use in monocot transformation, *Mol. Gen. Genet.,* 231, 150, 1991.
203. **Mcelroy, D., Zhang, W., Cao, J., and Wu, R.,** Isolation of an efficient actin promoter for use in rice transformation, *Plant Cell,* 2, 163, 1990.
204. **Severin, K. and Schoffl, F.,** Heat-inducible hygromycin resistance in transgenic tobacco, *Plant Mol. Biol.,* 15, 827, 1990.
205. **Ou-Lee, T.-M., Turgeon, R., and Wu, R.,** Expression of a foreign gene linked to either a plant virus or a *Drosophila* promoter, after electroporation of protoplasts of rice, wheat, and sorghum, *Proc. Natl. Acad. Sci. U.S.A.,* 83, 6815, 1986.
206. **Ainley, W. M. and Key, J. L.,** Development of a heat shock inducible expression cassette for plants: characterization of parameters for its use in transient expression assays, *Plant Mol. Biol.,* 14, 949, 1990.
207. **Back, E., Dunne, W., Schneiderbauer, A., de Framond, A., Rastogi, R., and Rothstein, S. J.,** Isolation of the spinach nitrite reductase gene promoter which confers nitrate inducibility on *GUS* gene expression in transgenic tobacco, *Plant Mol. Biol.,* 17, 9, 1991.
208. **Yamaguchi-Shinozaki, K., Mino, M., Mundy, J., and Chua, N. H.,** Analysis of an ABA-responsive rice gene promoter in transgenic tobacco, *Plant Mol. Biol.,* 15, 905, 1990.
209. **Kares, C., Prinsen, E., Van Onckelen, H., and Otten, L.,** IAA synthesis and root induction with *iaa* genes under heat shock promoter control, *Plant Mol. Biol.,* 15, 225, 1990.
210. **Kuhlemeier, C., Strittmatter, G., Ward, K., and Chua, N. H.,** The pea *rbc*-S-3A promoter mediates light responsiveness but not organ specificity, *Plant Cell,* 1, 471, 1989.

211. **Feinbaum, R. L., Storz, G., and Ausubel, F. M.,** High intensity and blue light regulated expression of chimeric chalcone synthase genes in transgenic *Arabidopsis thaliana* plants, *Mol. Gen. Genet.*, 226, 449, 1991.
212. **Weisshaar, B., Armstrong, G. A., Block, A., da Costa, S., and Hahlbrock, K.,** Light-inducible and constitutively expressed DNA-binding proteins recognizing a plant promoter element with functional relevance in light responsiveness, *EMBO J.*, 10, 1777, 1991.
213. **Lam, E. and Chua, N.-H.,** GT-1 binding site confers light responsive expression in transgenic tobacco, *Science*, 248, 471, 1990.
214. **Castresana, C., Garcia-Luque, I., Alonso, E., Malik, V. S., and Cashmore, A. R.,** Both positive and negative regulatory elements mediate expression of a photoregulated CAB gene from *Nicotiana plumbaginifolia*, *EMBO J.*, 7, 1929, 1988.
215. **Schulze-Lefert, P., Dangl, J. L., Becker-Andre, M., Hahlbrock, K., and Schulz, W.,** Inducible *in vivo* DNA footprints define sequences necessary for UV light activation of the parsley chalcone synthase gene, *EMBO J.*, 8, 651, 1989.
216. **Schena, M., Lloyd, A. M., and Davis, R. W.,** A steroid-inducible gene expression system for plant cells, *Proc. Natl. Acad. Sci. U.S.A.*, 88, 10421, 1991.
217. **Albani, D., Altosaar, I., Arnison, P. G., and Fabijanski, S. F.,** A gene showing sequence similarity to pectin esterase is specifically expressed in developing pollen of *Brassica napus*. Sequences in its 5′ flanking region are conserved in other pollen-specific promoters, *Plant Mol. Biol.*, 16, 501, 1991.
218. **Twell, D., Yamaguchi, J., Wing, R. A., Ushiba, J., and McCormick, S.,** Promoter analysis of genes that are coordinately expressed during pollen development reveals pollen-specific enhancer sequences and shared regulatory elements, *Genes Dev.*, 5, 496, 1991.
219. **Hamilton, D. A., Roy, M., Rueda, J., Sindhu, R. K., Sanford, J., and Mascarenhas, J. P.,** Dissection of a pollen-specific promoter from maize by transient transformation assays, *Plant Mol. Biol.*, 18, 211, 1992.
220. **van der Meer, I. M., Spelt, C. E., Mol, J. N., and Stuitje, A. R.,** Promoter analysis of the chalcone synthase (*chsA*) gene of *Petunia hybrida*: a 67 bp promoter region directs flower-specific expression, *Plant Mol. Biol.*, 15, 95, 1990.
221. **Goodrich, J., Carpenter, R., and Coen, E. S.,** A common gene regulates pigmentation pattern in diverse plant species, *Cell*, 68, 955, 1992.
222. **DeWitt, N. D., Harper, J., and Sussman, M. R.,** The promoter for a gene encoding the *Arabidopsis thaliana* plasma membrane proton pump ($H^+$-ATPase) directs reporter gene expression predominantly to phloem cells in transgenic plants, *J. Cell Biochem. Suppl.*, 15A, 69, 1991.
223. **Yang, N.-S. and Russell, D.,** Maize sucrose synthase-1 promoter directs phloem cell-specific expression of *GUS* gene in transgenic tobacco plants, *Proc. Natl. Acad. Sci. U.S.A.*, 87, 4144, 1990.
224. **Depater, B. S. and Schilperoort, R. A.,** Structure and expression of a root-specific rice gene, *Plant Mol. Biol.*, 18, 161, 1992.
225. **Vanderzaal, E. J., Droog, F. N. J., Boot, C. J. M., Hensgens, L. A. M., Hoge, J. H. C., Schilperoort, R. A., and Libbenga, K. R.,** Promoters of auxin-induced genes from tobacco can lead to auxin-inducible and root tip-specific expression, *Plant Mol. Biol.*, 16, 983, 1991.
226. **Oppenheimer, D. G., Haas, N., Silflow, C. D., and Snustad, D. P.,** The beta-tubulin gene family of *Arabidopsis thaliana*: preferential accumulation of the beta 1 transcript in roots, *Gene*, 63, 87, 1988.
227. **Knutzon, D. S., Thompson, G. A., Radke, S. E., Johnson, W. B., Knauf, V. C., and Kridl, J. C.,** Modification of *Brassica* seed oil by antisense expression of a stearoyl-acyl carrier protein desaturase gene, *Proc. Natl. Acad. Sci. U.S.A.*, 89, 2624, 1992.
228. **Bustos, M. M., Begum, D., Kalkan, F. A., Battraw, M. J., and Hall, T. C.,** Positive and negative *cis*-acting DNA domains are required for spatial and temporal regulation of gene expression by a seed storage protein promoter, *EMBO J.*, 10, 1469, 1991.
229. **Lam, E. and Chua, N. H.,** Tetramer of a 21-base pair synthetic element confers seed expression and transcriptional enhancement in response to water stress and abscisic acid, *J. Biol. Chem.*, 266, 17131, 1991.
230. **Stayton, M.., Harpster, M., Brosio, P., and Dunsmuir, P.,** High-level, seed-specific expression of foreign coding sequences in *Brassica napus*, *Aust. J. Plant Physiol.*, 18, 507, 1991.
231. **Kertbundit, S., De Greve, H., Deboeck, F., van Montagu, M., and Hernalsteens, J. P.,** *In vivo* random beta-glucuronidase gene fusions in *Arabidopsis thaliana*, *Proc. Natl. Acad. Sci. U.S.A.*, 88, 5212, 1991.
232. **Fobert, P. R., Miki, B. L., and Iyer, V. N.,** Detection of gene regulatory signals in plants revealed by T-DNA-mediated fusions, *Plant Mol. Biol.*, 17, 837, 1991.
233. **Gallie, D. R. and Kado, C. I.,** A translational enhancer derived from tobacco mosaic virus is functionally equivalent to a Shine-Dalgarno sequence, *Proc. Natl. Acad. Sci. U.S.A.*, 86, 129, 1989.
234. **Browning, K. S., Lax, S. R., Humphreys, J., Ravel, J. M., Jobling, S. A., and Gehrke, L.,** Evidence that the 5′-untranslated leader of mRNA affects the requirement for wheat germ initiation factors 4A, 4F, and 4G, *J. Biol. Chem.*, 263, 9630, 1988.

235. **Jobling, S. A. and Gehrke, L.,** Enhanced translation of chimaeric messenger RNAs containing a plant viral untranslated leader sequence, *Nature,* 325, 622, 1987.
236. **Gallie, D. R., Sleat, D. E., Watts, J. W., Turner, P. C., and Wilson, T. M. A.,** The 5′-leader sequence of tobacco mosaic virus RNA enhances the expression of foreign gene transcripts in vitro and in vivo, *Nucl. Acids Res.,* 15, 3257, 1987.
237. **Gallie, D. R., Sleat, D. E., Watts, J. W., Turner, P. C., and Wilson, T. M. A.,** A comparison of eukaryotic viral 5′-leader sequences as enhancer of mRNA expression in vivo, *Nucl. Acids Res.,* 15, 8693, 1987.
238. **Hunt, A. G., Chu, N. M., Odell, J. T., Nagy, F., and Chua, N. H.,** Plant cells do not properly recognize animal gene polyadeylation signals, *Plant Mol. Biol.,* 8, 23, 1987.
239. **Joshi, C. P.,** Putative polyadenylation signals in nuclear genes of higher plants: a compilation and analysis, *Nucl. Acids Res.,* 15, 9627, 1987.
240. **Hunt, A. G.,** Identification and characterization of cryptic polyadenylation sites in the 3′ region of a pea ribulose-1,5-bisphosphate carboxylase small subunit gene, *DNA,* 7, 329, 1988.
241. **Ingelbrecht, I. L. W., Herman, L. M. F., Dekeyser, R. A., Van Montagu, M. C., and Depicker, A. G.,** Different 3′ end regions strongly influence the level of gene expression in plant cells, *Plant Cell,* 1, 671, 1989.
242. **Ingelbrecht, I. L. W., van Montagu, M., and Depicker, A. G.,** Posttranscriptional processes influence the level of gene expression in transgenic plants, *J. Cell Biochem.,* 15, 260, 1991.
243. **Maas, C., Laufs, J., Grant, S., Korfhage, C., and Werr, W.,** The combination of a novel stimulatory element in the first exon of the maize *Shunken*-1 gene with the following intron 1 enhances reporter gene expression up to 1000-fold, *Plant Mol. Biol.,* 16, 199, 1991.
244. **Ellis, J. G., Llewellyn, D. J., Dennis, E. S., and Peacock, W. J.,** Maize *Adh*-1 promoter sequences control anaerobic regulation. Addition of upstream promoter elements from constitutive genes is necessary for expression in tobacco, *EMBO J.,* 6, 11, 1987.
245. **Lee, L., Fenoll, C., and Bennetzen, J. L.,** Construction and homologous expression of a maize *adh*-1 based NcoI cassette vector, *Plant Physiol.,* 85, 327, 1987.
246. **Bennetzen, J. L., Swanson, J., Taylor, W. C., and Freeling, M.,** DNA insertion in the first intron of maize *Adh*1 affects message levels: cloning of progenitor and mutant *Adh*1 alleles, *Proc. Natl. Acad. Sci. U.S.A.,* 81, 4125, 1984.
247. **McCullough, A. J., Lou, H., and Schuler, M. A.,** In vivo analysis of plant pre-mRNA splicing using an autonomously replicating vector, *Nucl. Acids Res.,* 19, 3001, 1991.
248. **Mascarenhas, D., Mettler, I. J., Pierce, D. A., and Lowe, H. W.,** Intron-mediated enhancement of heterologous gene expression in maize, *Plant Mol. Biol.,* 15, 913, 1990.
249. **Luehrsen, K. R. and Walbot, V.,** Intron enhancement of gene expression and the splicing efficiency of introns in maize cells, *Mol. Gen. Genet.,* 225, 81, 1991.
250. **Goodall, G. J. and Filipowicz, W.,** The minimum functional length of pre-mRNA introns in monocots and dicots, *Plant Mol. Biol.,* 14, 727, 1990.
251. **Vancanneyt, G., Rosahl, S., and Willmitzer, L.,** Translatability of a plant-mRNA strongly influences its accumulation in transgenic plants, *Nucl. Acids Res.,* 18, 2917, 1990.
252. **Douglas, C. J., Hauffe, K. D., Ites-Morales, M. E., Ellard, M., Paszkowski, U., Hahlbrock, K., and Dangl, J. L.,** Exonic sequences are required for elicitor and light activation of a plant defense gene, but promoter sequences are sufficient for tissue specific expression, *EMBO J.,* 10, 1767, 1991.
253. **Klaas, M. and Amasino, R. M.,** DNA methylation is reduced in DNaseI-sensitive regions of plant chromatin, *Plant Physiol.,* 91, 451, 1989.
254. **Hobbs, S. L. A., Kpodar, P., and Delong, C. M. O.,** The effect of T DNA copy number position and methylation on reporter gene expression in tobacco transformants, *Plant Mol. Biol.,* 15, 851, 1991.
255. **Doerfler, W.,** DNA methylation — eukaryotic defense against the transcription of foreign genes, *Microb. Pathol.,* 12, 1, 1992.
256. **Hershkovitz, M., Gruenbaum, Y., Renbaum, P., Razin, A., and Loyter, A.,** Effect of CPG methylation on gene expression in transfected plant protoplasts, *Gene,* 94, 189, 1990.
257. **Weber, H., Ziechmann, C., and Graessmann, A.,** In vitro DNA methylation inhibits gene expression in transgenic tobacco, *EMBO J.,* 9, 4409, 1990.
258. **John, M. C. and Amasino, R. M.,** Extensive changes in DNA methylation patterns accompany activation of a silent T-DNA *ipt* gene in *Agrobacterium tumefaciens*-transformed plant cells, *Mol. Cell Biol.,* 9, 4298, 1989.
259. **Vanyushin, B. F. and Kirnos, M. D.,** DNA methylation in plants, *Gene,* 74, 117, 1988.
260. **Dean, C., Jones, J., Favreau, M., Dunsmuir, P., and Bedbrook, J.,** Influence of flanking sequences on variability in expression levels of an introduced gene in transgenic tobacco plants, *Nucl. Acids Res.,* 16, 9267, 1988.
261. **Peach, C. and Velten, J.,** Transgene expression variability (position effect) of CAT and GUS reporter genes driven by linked divergent T-DNA promoters, *Plant Mol. Biol.,* 17, 49, 1991.

262. **Jorgensen, R.,** Altered gene expression in plants due to *trans* interactions between homologous genes, *Trends Biotechnol.,* 8, 340, 1990.
263. **Napoli, C., Lemieux, C., and Jorgensen, R.,** Introduction of a chimeric chalcone synthase gene into *Petunia* results in a reversible co-suppression of homologous genes in *trans, Plant Cell,* 2, 279, 1990.
264. **van der Krol, A. R., Mur, L. A., Beld, M., Mol, J. N. M., and Stuitje, A. R.,** Flavonoid genes in *Petunia.* Addition of a limited number of gene copies may lead to a suppression of gene expression, *Plant Cell,* 2, 291, 1990.
265. **Paszty, C. J. R. and Lurquin, P. F.,** Inhibition of transgene expression in plant protoplasts by the presence in *cis* of an opposing 3′-promoter, *Plant Sci.,* 72, 69, 1990.
266. **Wilson, C., Bellen, H. J., and Gehring, W. J.,** Position effects on eukaryotic gene expression, in *Annual Review of Cell Biology, Vol. 6,* Palade, G. E., Ed., Annual Reviews, Inc., Palo Alto, CA, 1990, 679.
267. **Perlak, F. J., Fuchs, R. L., Dean, D. A., McPherson, S. L., and Fischhoff, D. A.,** Modification of the coding sequence enhances plant expression of insect control protein genes, *Proc. Natl. Acad. Sci. U.S.A.,* 88, 3324, 1991.
268. **Schreier, P. H. and Schell, J.,** Use of chimeric genes harboring small subunit transit peptide sequences to study transport in chloroplasts, *Phil. Trans. R. Soc. Lond. B. Biol. Sci.,* 313, 429, 1986.
269. **Kavanagh, T. A., Jefferson, R. A., and Bevan, M. W.,** Targeting a foreign protein to chloroplasts using fusions to the transit peptide of a chlorophyll a/b protein, *Mol. Gen. Genet.,* 215, 38, 1988.
270. **Klosgen, R. B. and Weil, J.-H.,** Subcellular location and expression level of a chimeric protein consisting of the maize *waxy* transit peptide and the beta glucuronidase of *Escherichia coli* in transgenic potato plants, *Mol. Gen. Genet.,* 225, 297, 1991.
271. **VandenBroeck, G., Timko, M. P., Kausch, A. P., Cashmore, A. R., van Montagu, M., and Herrera-Estrella, L.,** Targeting of a foreign protein to chloroplasts by fusion to the transit peptide from small subunit of ribulose-1,5-bisphosphate carboxylase, *Nature,* 313, 358, 1985.
272. **Reiss, B., Wasmann, C. C., and Bohnert, H. J.,** Regions in the transit peptide of SSU essential for transport into chloroplasts, *Mol. Gen. Genet.,* 209, 116, 1987.
273. **Whelan, J., Knorpp, C., and Glaser, E.,** Sorting of precursor proteins between isolated spinach leaf mitochondria and chloroplasts, *Plant Mol. Biol.,* 14, 977, 1990.
274. **White, J. A. and Scandalios, J. G.,** Deletion analysis of the maize mitochondrial superoxide dismutase transit peptide, *Proc. Natl. Acad. Sci. U.S.A.,* 86, 3534, 1989.
275. **Scandalios, J. G.,** Targeting, import, and processing of nuclear gene-encoded proteins into mitochondria and peroxisomes, *Prog. Clin. Biol. Res.,* 344, 515, 1990.
276. **Gould, S. J., Keller, G. A., Schneider, M., Howell, S. H., Garrard, L. J., Goodman, J. M., Distel, B., Tabak, H., and Subramani, S.,** Peroxisomal protein import is conserved between yeast, plants, insects and mammals, *EMBO J.,* 9, 85, 1990.
277. **Hunt, D. C. and Chrispeels, M. J.,** The signal peptide of a vacuolar protein is necessary and sufficient for the efficient secretion of a cytosolic protein, *Plant Physiol.,* 96, 18, 1991.
278. **Holwerda, B. C., Padgett, H. S., and Rogers, J. C.,** Proaleurain vacuolar targeting is mediated by short contiguous peptide interactions, *Plant Cell,* 4, 307, 1992.
279. **Chrispeels, M. J. and Raikhel, N. V.,** Short peptide domains target proteins to plant vacuoles, *Cell,* 68, 613, 1992.
280. **Vitale, A. and Chrispeels, M. J.,** Sorting of proteins to the vacuoles of plant cells, *Bioessays,* 14, 151, 1992.
281. **Hofte, H., Faye, L., Dickinson, C., Herman, E. M., and Chrispeels, M. J.,** The protein-body proteins phytohemagglutinin and tonoplast intrinsic protein are targeted to vacuoles in leaves of transgenic tobacco, *Planta,* 184, 431, 1991.
282. **Van der Krol, A. and Chua, N.-H.,** The basic domain of plant B-ZIP proteins facilitates import of a reporter protein into plant nuclei, *Plant Cell,* 3, 667, 1991.
283. **Klosgen, R. B., Saedler, H., and Weil, J.-H.,** The amyloplast-targeting transit peptide of the waxy protein of maize also mediates protein transport *in vitro* into chloroplasts, *Mol. Gen. Genet.,* 217, 155, 1989.
284. **Li, H.-M., More, T., and Keegstra, K.,** Targeting of proteins to the outer envelope membrane uses a different pathway than transport into chloroplasts, *Plant Cell,* 3, 709, 1991.
285. **Potrykus, I.,** Micro-targeting of microprojectiles to target areas in the micrometre range, *Nature,* 355, 568, 1992.
286. **Koncz, C., Koncz-Kalman, Z., and Schell, J.,** Transposon Tn-5 mediated gene transfer into plants, *Mol. Gen. Genet.,* 207, 99, 1987.
287. **Schwarz-Sommer, Z. and Saedler, H.,** Can plant transposable elements generate novel regulatory systems?, *Mol. Gen. Genet.,* 209, 207, 1987.
288. **Peterson, P. A. and Laughnan, J. R.,** Mobile elements in plants, *Crit. Rev. Plant Sci.,* 6, 105, 1987.
289. **Spena, A.,** Unstable liaisons: the use of transposons in plant genetic engineering, *Trends Genet.,* 6, 76, 1990.

290. **Rommens, C. M. T., Van Haaren, M. J. J., Buchel, A. S., Mol, J. N. M., van Tunen, A. J., Nijkamp, H. J. J., and Hille, J.,** Transactivation of *Ds* by *Ac*-transposase gene fusions in tobacco, *Mol. Gen. Genet.*, 231, 433, 1992.
291. **Waldron, C., Murphy, E. B., Roberts, J. L., Gustafson, G. D., Armour, S. L., and Malcolm, S. K.,** Resistance to hygromycin B — a new marker for plant transformation studies, *Plant Mol. Biol.*, 5, 103, 1985.
292. **Hayford, M. B., Medford, J. I., Hoffman, N. L., Rogers, S. G., and Klee, H. J.,** Development of a plant transformation selection system based on expression of genes encoding gentamicin acetyltransferases, *Plant Physiol.*, 86, 1216, 1988.
293. **Hille, J., Verheggen, F., Roelvink, P., van Kammen, H. F. A., and Zabel, P.,** Bleomycin resistance — a new dominant selectable marker for plant cell transformation, *Plant Mol. Biol.*, 7, 171, 1986.
294. **Kamakura, T., Yoneyama, K., and Yamaguchi, I.,** Expression of the blasticidin S deaminase gene (*bsr*) in tobacco: fungicide tolerance and a new selective marker for transgenic plants, *Mol. Gen. Genet.*, 223, 332, 1990.
295. **Svab, Z., Harper, E. C., Jones, J. D. G., and Maliga, P.,** Aminoglycoside-3″-adenyltransferase confers resistance to spectinomycin and streptomycin in *Nicotiana tabacum*, *Plant Mol. Biol.*, 14, 197, 1990.
296. **Etzold, T., Fritz, C. C., Schell, J., and Schreier, P. H.,** A point mutation in the chloroplast 16s ribosomal RNA gene of a streptomycin resistant *Nicotiana tabacum*, *FEBS Lett.*, 219, 343, 1987.
297. **Eichholtz, D. A., Roger, S. G., Horsch, R. B., Klee, H. J., Hayford, M., Hoffmann, N. L., Bradford, S. B., Fink, C., Flick, J., and O'Connel, K. M.,** Expression of mouse dihydrofolate reductase gene confers methotrexate resistance in transgenic petunia plants, *Somatic Cell Mol. Genet.*, 13, 67, 1987.
298. **Stalker, D. M., Mcbride, K. E., and Malyj, L. D.,** Herbicide resistance in transgenic plants expressing a bacterial detoxification gene, *Science*, 242, 419, 1988.
299. **Guerineau, F., Brooks, L., Meadows, J., Lucy, A., Robinson, C., and Mullineaux, P.,** Sulfonamide resistance gene for plant transformation, *Plant Mol. Biol.*, 15, 127, 1990.
300. **Naleway, J. J., Zhang, Y. Z., Bonnett, H., Galbraith, D. W., and Haugland, R. P.,** Detection of *GUS* gene expression in transformed plant cells with new lipophilic, fluorogenic β-glucuronidase substrate, *J. Cell Biol.*, 115, 151a, 1991.
301. **Martin, T., Schmidt, R., Altmann, T., and Frommer, W. B.,** Non-destructive assay systems for detection of β-glucuronidase activity in higher plants, *Plant Mol. Biol.*, 10, 37, 1992.
302. **Casadaban, M. J., Chou, J., and Cohen, S. N.,** *In vitro* gene fusions that join an enzymatically active β-galactosidase segment to amino-terminal fragments of exogenous proteins: *Escherichia coli* plasmid vectors for the detection and cloning of translational initiation signals, *J. Bacteriol.*, 143, 971, 1980.
303. **Levitz, R., Klar, A., Sar, N., and Yagil, E.,** A new locus in the phosphate specific transport (PST) region of *Escherichia coli*, *Mol. Gen. Genet.*, 197, 98, 1984.
304. **Gorman, C. M., Moffat, L. F., and Howard, B. H.,** Recombinant genomes which express chloramphenicol acetyltransferase in mammalian cells, *Mol. Cell Biol.*, 2, 1044, 1982.
305. **McDonnell, R., Clark, R., Smith, W., and Hinchee, M.,** A simplified method for the detection of neomycin phosphotransferase activity in transformed plant tissues, *Plant Mol. Biol. Rep.*, 5, 380, 1987.
306. **Weide, R., Koornneef, M., and Zabel, P.,** A simple, nondestructive spraying assay for the detection of an active kanamycin resistance gene in transgenic tomato plants, *Theor. Appl. Genet.*, 78, 169, 1989.
307. **Craig, F. F., Simmonds, A. C., Watmore, D., Mccapra, F., and White, M. R. H.,** Membrane-permeable luciferin esters for assay of firefly luciferase in live intact cells, *Biochem. J.*, 276, 637, 1991.
308. **Soly, R. R., Mancini, J. A., Ferri, S. R., Boylan, M., and Meighen, E. A.,** A new *lux* gene in bioluminescent bacteria codes for a protein homologous to the bacterial luciferase subunits, *Biochem. Biophys. Res.*, 155, 351, 1988.
309. **Melchers, L. S. and Hooykaas, P. J. J.,** Virulence of *Agrobacterium*, *Oxford Surv. Plant Mol. Cell Biol.*, 4, 167, 1987.

# 8

# Construction of λ Clone Blanks

*Jerry L. Slightom, Roger F. Drong, and Paula P. Chee*

## I. INTRODUCTION

The combination of recombinant DNA cloning techniques for the isolation of specific coding or noncoding DNA region(s) of an organism's genome and ever improving nucleotide sequencing methods has been extremely productive in providing information regarding gene organization and structure. The ability to clone and purify specific coding and noncoding DNA regions became feasible on a large scale with the development of *E. coli* phage λ vectors capable of growth after the cloning of foreign DNA. λ phage vectors were first used to clone partially purified target cDNAs or genomic DNAs.[1,2] The development of improved λ vectors for the construction cDNA or genomic clone banks along with rapid λ screening procedures has made possible the construction of more complete cDNA and genomic recombinant λ phage banks.[3-6] It is now a routine process to isolate clones of mRNAs and single copy DNA regions from the respective λ clone banks that contain several million recombinant phage clones.

With the recent development of direct genome sequencing using the polymerase chain reaction (PCR) technique,[7] some have questioned the need to construct recombinant DNA clone banks. Direct sequencing of PCR-amplified cDNAs and genomic DNAs has already added greatly to our ability to obtain nucleotide sequence information rapidly.[8-10] With the existence of two different procedures that can be used to analyze coding and noncoding DNAs, researchers must decide which method is better suited for their goals. For example, if the cDNA and corresponding gene region is small, 1 to 2 kb, the PCR approach may well be the best approach; on the other hand if the cDNA or gene region is large, above 10 kb, then the cloning approach may be better suited. In the case of cloning cDNAs that represent very rare mRNA species (about one copy/cell) the combined use of both techniques can be very useful.[11] Which method to use also depends on the experience and expertise of the researcher.

The major goal in constructing a recombinant clone bank that contains either cDNA or genomic-derived DNA inserts is to obtain a sufficient number of recombinant clones to ensure representation of the total mRNA population or the entire genome of a species (see Section II). The clone bank vector should allow long term storage, easy distribution, and minimal rearrangement of cloned inserts (mostly deletions) during clone propagation, and it should aid the isolation of overlapping clones (referred to as cDNA or chromosome walking). In addition, the construction and screening of the bank should be relatively easy and straightforward to allow first-time users to succeed without spending large amounts of time troubleshooting technical steps. For genomic DNA cloning, two major types of cloning vector systems have been developed that have retained their viability because they satisfy many of these requirements; the above-mentioned bacteriophage λ and the related cosmid vector systems.[12,13] The major advantage that the cosmid system has over the λ system is a larger

0-8493-5164-2/93/$0.00+$.50
© 1993 by CRC Press, Inc.

insert cloning capacity (35 to 45 kb vs. a maximum of about 20 kb for λ). Thus fewer cosmid clones would be needed to walk across a larger region of genomic DNA. However, the cosmid system requires a higher degree of skill for screening (which is the difference between screening phage plaques versus bacterial colonies), and their large insert size increases their potential of containing genomic DNA elements that are prone to being deleted during growth in *E. coli*. Because of these disadvantages, it is recommended that experience first be gained using the λ vector system.

## II. λ VECTORS

### A. λ REPLACEMENT VECTORS AND THE CLONING OF GENOMIC DNA

Bacteriophage λ is a useful cloning vector system because the center third of the viral genome, the regions between the *J* and *N* genes (Figure 1), is not essential for lytic growth. Foreign DNAs can be inserted within this region, or this region can be replaced (referred to as λ replacement vectors) with foreign DNA. Such recombinant λ vectors retain their ability for growth in select *E. coli* hosts, provided the size of the recombinant λ genome satisfies specific size restrictions, being larger than about 40 kb but not larger than about 52 kb.[14] As a result of many genetic and recombinant DNA manipulations, a large number of λ replacement-type vectors have been developed that contain single or multiple restriction enzyme sites surrounding different dispensable center fragments (referred to as the stuffer fragment); many of these λ vectors are listed in Table 1. Several different stuffer fragments have been used to facilitate their later removal (by either physical or genetic means) from the essential λ genes that remain within the flanking DNA regions; these flanking regions are linked through the λ *cohesive end* (*cos*) site, and they are referred to as the λ right and left arms (Figure 1). Construction of a recombinant λ clone bank requires either physical or genetic removal of the stuffer fragment so that the original λ vector will not represent a large fraction of the clone bank. Physical removal of the stuffer fragment is achieved by digestion at restriction enzyme sites that flank and are located within the stuffer fragment, followed by separation of the arms from the stuffer fragments by a salt or sucrose velocity sedimentation gradient.[18-20] This method is used for the preparation of Charon 4A or 32 (*Eco*RI digest) and Charon 35 λ arms (*Eco*RI or *Bam*HI digest).[3,15] In the case of Charon 40, the stuffer fragment contains multiple copies (80 repeats) of a 235 bp fragment that are flanked by the *Nae*I restriction enzyme site (Figure 1), so that upon *Nae*I digestion the stuffer fragment is essentially destroyed.[16]

For the genetic approach, the stuffer fragments are not removed prior to cloning the target DNA fragments, because the stuffer fragment contains genetic loci that inhibit the phage from growing in selected *E. coli* hosts and thus from being represented within the clone bank. This is done by using stuffer fragments that include functional *gam* and *red* loci, such as the λ EMBL vector series (Figure 1). λ clones that contain the stuffer fragment will produce $gam^+$ phages, the growth of which can be suppressed by propagation in an *E. coli* (P2) host such as Q359.[14] Sensitivity to P2 interference (the $Spi^+$ phenotype) appears to be due in part to inactivation of the *E. coli rec*BC nuclease by the *gam* gene product. Infection of a P2 lysogen by $gam^+$ λ produces a $recBC^-$ phenotype that leads to inhibition of both protein synthesis and DNA replication by the P2 prophage. However, infection by $gam^-$ phage permits replication, provided that the phages are $Chi^+$, which is the case for the EMBL λ vectors.

The EMBL λ vector systems have been subjected to many modifications to enhance their usefulness. The most useful modifications involve the addition of more restriction enzyme recognition sites flanking the stuffer fragment and the addition of SP6, T3, or T7 bacteriophage RNA polymerase promoters. These bacteriophage RNA polymerase promoters are used to synthesize $^{32}$P-labeled RNAs that can be used as probes for the identification of overlapping λ clones. Many of the λ vectors described above are available from or have been developed by commercial vendors; some of these are listed in Table 1.

The Charon 32 to 40 λ vector systems have not been improved since their development; however, they still offer several advantages, including use of the *gam* locus and a 16 restriction enzyme site polylinker for Charon 40 (Figure 1).[15,16] Recombinant *gam*$^+$ Charon 32 to 40 phage vectors are capable of growth in *rec*A$^-$ *E. coli* hosts, such as ED8767 (*sup*E, *sup*F, *hsd*S$^-$, *met*$^-$, *rec*A56), and thus they are useful for the construction of λ clone banks that contain duplicated genetic regions which may be unstable in λ vectors grown in a *rec*A$^+$ *E. coli* host.[16] For example, the DNA region that encodes the duplicated fetal globin genes (a 5 kb direct repeat) of higher primates has been very difficult to clone in λ vectors grown in a *rec*A$^+$ host,[21] but it has been readily cloned into λ Charon 32 and 40 vectors.[22,23]

The availability of many different λ vector systems from commercial sources has essentially removed the need to prepare λ vector arms. However, λ arms from Charon 4A and 40 are not commercially available presently; those selecting to prepare λ vector arms (such as Charon 4 or 40) should consult one of the following references.[18-20,24] We have extensive experience in the use of the Charon series of λ vectors for the construction of many recombinant plant-λ phage clone banks which have been used to isolate many plant gene regions. Table 2 lists the recombinant plant-λ clone banks that we have constructed.

The construction of a recombinant λ phage genomic clone bank requires the isolation of genomic DNA in the size range of about 50 kb, followed by a series of partial restriction enzyme digestions to obtain a complete or pseudorandom (depending on the restriction enzyme used) fragmentation of the target genome. Randomization of the partial digest can be increased by using different time points or restriction enzyme concentrations in a series of digestions ranging from a minimum of about 10% up to a maximum of about 70% of completion (see Section III.B). A series of digests is useful because not all restriction enzyme recognition sites (for a specific restriction enzyme) are equally susceptible to cleavage. After obtaining the digestion series, the individual digests are pooled and size-fractionated on a velocity sedimentation gradient. Target DNA fragments between 15 to 20 kb in length are selected for λ cloning, thereby reducing the chances of cloning multiple fragments, which are generally noncontiguous in the genome of the target species. The cloning of two fragments between 15 to 20 kb in size will produce a recombinant λ molecule that is too large (>52 kb) to be encapsulated[14] and thus will not be represented in the clone bank.

The selection of which restriction enzyme is used to prepare the target DNA is important because of its impact on whether or not the clone bank will contain overlapping fragments that theoretically represent the complete target species genome. For example, λ clone banks constructed with *Eco*RI-digested target DNAs are incomplete because the 6 bp recognition site of *Eco*RI (or any other enzyme that recognizes a 6 bp site) is not represented at a frequency that ensures complete randomness of the digested target genomic DNA. On average, an *Eco*RI site is expected every $4^6$ or 4096 bp; however, in generating a partial *Eco*RI digest, not all sites are cut, and skipping just a few consecutive sites would result in *Eco*RI fragments that exceed the cloning capacity of a λ vector. In addition, the distribution of *Eco*RI sites is not truly random; thus there are some regions of a target species genome for which even a complete *Eco*RI digest will yield fragments that exceed the capacity of a λ vector.

Construction of the first random λ clone bank was achieved using the procedure described by Lawn et al.,[25] in which the target DNA was digested with two different restriction enzymes, each possessing a 4 bp recognition site. However, this clone bank construction required the use of many additional enzymatic steps (*Eco*RI site methylation and the ligation of linkers). These additional enzymatic steps can be avoided by using restriction enzymes with 4 bp and 6 bp recognition sites that share identical single-stranded core sequences. The most frequently

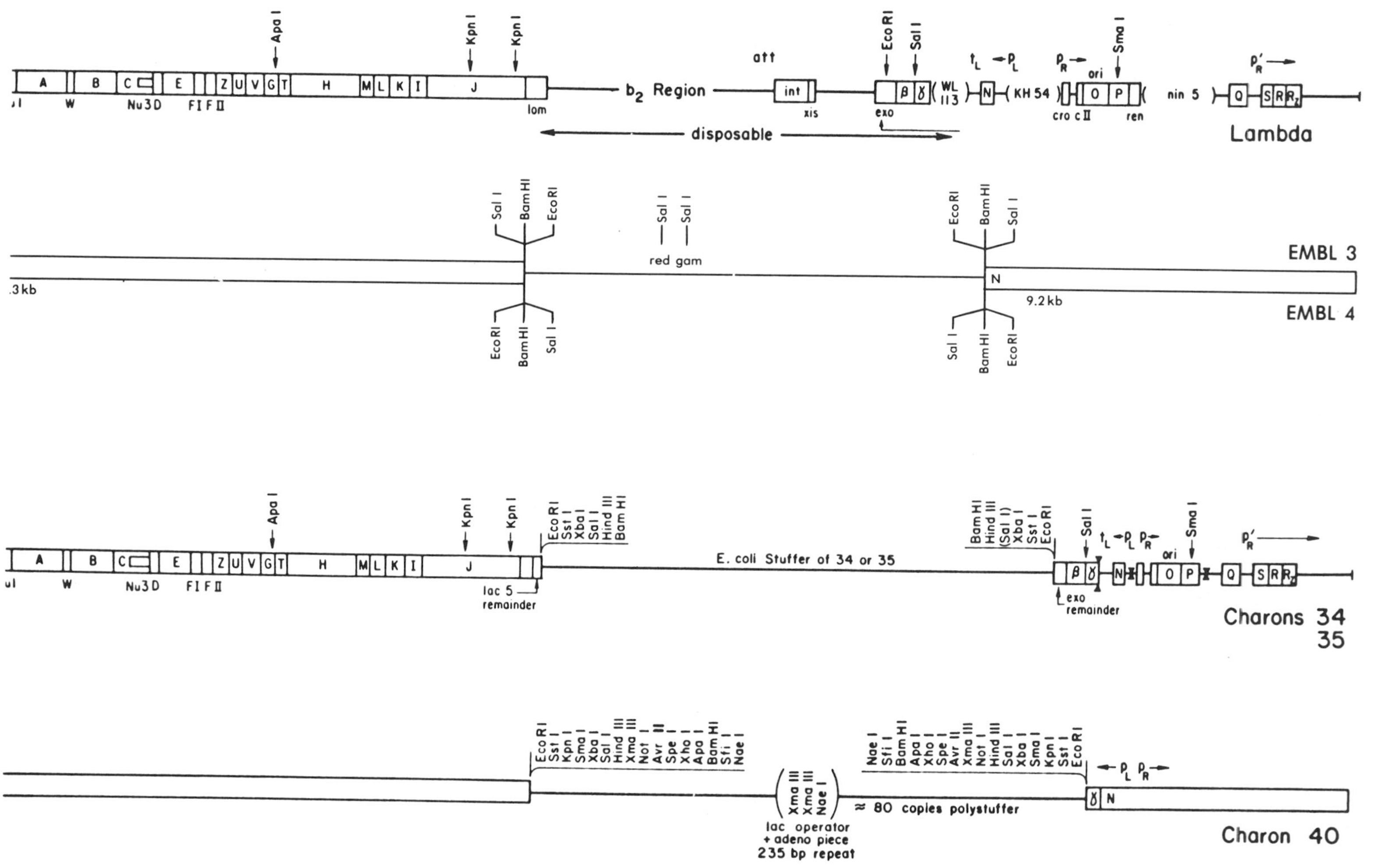

Lambda
b2 Region
att
int
xis
exo
disposable
Apa I
Kpn I
Eco RI
Sal I
Sma I
lom
nin 5
cro cII
ori
ren
EMBL 3
EMBL 4
red gam
Bam HI
9.2 kb
Charons 34
35
E. coli Stuffer of 34 or 35
lac 5 remainder
exo remainder
Charon 40
≈ 80 copies polystuffer
lac operator
+ adeno piece
235 bp repeat

**FIGURE 1.** Genetic and restriction enzyme site maps of wild-type λ and λ vectors EMBL3 and 4, and of Charon 34, 35, and 40. Maps of wild-type λ and Charon vectors are from Dunn and Blattner[15] and maps of the EMBL vectors are from Frischauf et al.[17] The DNA region between the indicated polylinkers in the λ cloning vectors is the disposable stuffer that can be replaced by the target DNA for library construction. The stuffer region of the EMBL vector includes the loci for *gam* and *red* gene function that are used for $Spi^+$ phenotype selection by growth on *E. coli* (P2) host. The stuffer region of Charon 40 consists of 80 copies of a specific 235 bp fragment; it is referred to as a polystuffer. Removal of the polystuffer and preparation of Charon 40 arms first require digestion with *Nae*I, followed by a second digest with any of the 15 other restriction enzymes that have recognition sites within the polylinker. Charon 40 clones are $gam^+$; thus they can be grown on *E. coli* hosts that are *rec*$A^-$ such as ED8767. (Reprinted by permission of Kluwer Academic Publishers.)

**TABLE 1**
**λ Replacement Vectors**

| Vector | Features | Cloning sites | Ref./Vendor |
|---|---|---|---|
| Charon 4A | *A*am32 | *Eco*RI | Blattner et al.[3] not commercially available |
| Charon 32 | *gam*, growth on *rec*A⁻ *E. coli* host | *Eco*RI | Loenen and Blattner[15] not commercially available |
| Charon 35 | Same as Charon 32 | *Eco*RI, *Sst*I *Xba*I, *Sal*I, *Hin*dIII, *Bam*HI | Loenen and Blattner[15] not commercially available |
| Charon 40 | Same as Charon 32 | 16 sites, including *Eco*RI, *Kpn*I, *Xba*I, *Sal*I, *Hin*dIII, *Not*I *Spe*I, *Bam*HI, & others | Dunn and Blattner[16] not commercially available |
| EMBL3 | Stuffer (*red, gam*) Spi⁺/P2 selection | *Sal*I, *Bam*HI, *Eco*RI | Frischauf et al.[17] Clontech, Promega, Stratagene |
| EMBL4 | Same as EMBL3 | *Eco*RI, *Bam*HI, *Sal*I | Frischauf et al.[17] Clontech, Promega, Stratagene |
| EMBL12 | Same as EMBL3 | Additional sites for *Xba*I and *Sac*I | Boehringer-Mannheim |
| λ Fix™II | Stuffer (*red, gam*) Spi⁺/P2 selection Flanking T3 and T7 Bacteriophage promoters *Xho*I partial fill-in | *Xho*I, *Sal*I, *Eco*RI | Stratagene |
| λ Dash™II | Stuffer (*red, gam*) Spi⁺/P2 selection | *Eco*RI, *Bam*HI, *Hin*dIII *Sal*I, *Xho*I | Stratagene |
| λGem™-11 | Stuffer (*red, gam*) Spi⁺/P2 selection Flanking T7 and SP6 Bacteriophage promoters *Xho*I partial fill-in | *Sal*I, *Xho*I, *Bam*HI *Avr*II, *Eco*RI | Promega |
| λGem™-12 | Same as λGem™-11 | Same as λGem™-11 except *Not*I site replaces *Avr*II site | Promega |

used restriction enzyme combinations utilize λ vectors that contain a *Bam*HI cloning site, which has the 6 bp recognition site,

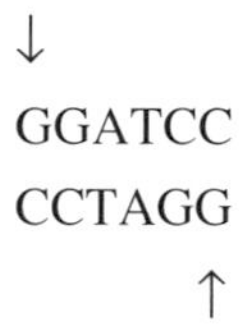


to clone target genomic DNA digested with either *Sau*3AI or *Mbo*I, which has the 4 bp recognition site,

**TABLE 2**
**Genomic Plant DNA Clone Banks Constructed in Our Laboratory**

| Plant species cultivar | Vector | # of original phage clones | Ref. |
|---|---|---|---|
| *Phaseolus vulgaris* | | | |
| Tendergreen | Charon 24 | $2 \times 10^6$ | Sun et al.[27] |
| *Glycine max* | | | |
| Corsoy | λ1059 | $1 \times 10^6$ | Tierney et al.[28] |
| Corsoy | Charon 34 | $2 \times 10^6$ | |
| Wayne | Charon 35 | $2 \times 10^6$ | Hong et al.[29] |
| Wayne | Charon 40 | $2 \times 10^6$ | Hong et al.[29] |
| Asgrow A3373 | Charon 40 | $3 \times 10^6$ | H. Quemada, J. L. Slightom, unpublished data |
| Asgrow A3205 | Charon 40 | $2 \times 10^6$ | H. Quemada, J. L. Slightom, unpublished data |
| *Zea mays* | | | |
| W64A | Charon 32 | $3 \times 10^6$ | Kriz et al.[30] Thompson et al.[31] |
| *Triticum aestivum* | | | |
| Yamhill | Charon 32 | $1.5 \times 10^6$ | Murray et al.[32] |
| *Convolvulus arvensis* | Charon 4A | $1 \times 10^6$ | Slightom et al.[33] |
| *Gossypium hirsutum* | | | |
| Coker 201 | Charon 35 | $2 \times 10^6$ | Chlan et al.[34] |

Both recognition sites share the identical single-stranded sequence GATC. A recent modification of this cloning strategy uses the partial fill-in of the *Xho*I site on the λ arms, followed by the cloning of *Sau*3AI or *Mbo*I-cut target DNAs (Table 1). Although these strategies do not allow recutting at the cloning site, analysis of the cloned insert is not difficult because many of the newer λ vectors contain additional restriction enzyme recognition sites flanking the *Bam*HI or *Xho*I cloning sites (Figure 1 and Table 1).

Assuming that the complete target species genome is clonable, the number of λ clones needed to contain the complete genome can be calculated using the following equation described by Clark and Carbon:[26]

$$N = \ln(1 - P)/\ln(1 - x/y) \tag{1}$$

where the number of clones needed ($N$) can be calculated for a set probability ($P$) knowing the size of the cloned fragments ($x$) and the total size of the target genome ($y$). In making this calculation, two assumptions must be made: first, that the fragmentation of the target genomic DNA is completely random, and second, that the exact size of the cloned inserts is known. Our analyses of many cloned inserts isolated from λ clone banks (Table 2) indicate that the average insert size is about 17 kb. If we assume a genome size of $3 \times 10^9$ bp and an insert size of 17 kb, the number of clones needed to obtain a complete λ clone bank at a probability of $P = 0.99$ is nearly $1 \times 10^6$. The calculated number of recombinant λ clones needed to obtain a complete clone bank at a probability of 0.99 for some selected plant species is presented in Table 3.

### B. cDNA CLONING AND λ VECTORS

The mRNA population in the cells of most plant organs is very complex. For example, Kamalay and Goldberg[37] estimated that the cells of tobacco express about 25,000 different mRNAs and that about 6000 of these are uniquely expressed in the cells of specific plant organs (leaf, root, stem, petal, anther, ovary, etc.). Thus, the construction of a double-stranded

**TABLE 3**
**Plant Genome Sizes and Number of Phage for a Clone Bank**

| Plant species | Genome size[a] | # Recombinant phage[b] |
|---|---|---|
| *Arabidopsis thaliana* | $7.7 \times 10^7$ bp | $2.1 \times 10^4$ |
| *Glycine max* | $8.7 \times 10^8$ bp | $2.4 \times 10^5$ |
| *Phaseolus vulgaris* | $1.8 \times 10^9$ bp | $4.9 \times 10^5$ |
| *Petunia hybrida* | $1.9 \times 10^9$ bp | $5.1 \times 10^5$ |
| *Nicotiana tabacum* | $3.8 \times 10^9$ bp | $1.0 \times 10^6$ |
| *Zea mays* | $3.9 \times 10^9$ bp | $1.1 \times 10^6$ |
| *Triticum aestivum* | $1.5 \times 10^{10}$ bp | $4.1 \times 10^6$ |

[a] Genome sizes are from Rogers and Bendich,[35] pg were converted to bp using the conversion of 1 pg = $0.965 \times 10^9$ bp.[36]
[b] Insert size is assumed to be 17 kb and probability = .99.

(ds)-cDNA λ clone bank that contains a representative of each mRNA species is similar to the construction of a complete genomic λ clone bank. The number of ds-cDNA clones needed to ensure that low-abundant mRNA species (≤ 10 copies/cell) are included within a cDNA λ clone bank can be calculated by the following probability equation:

$$N = \ln(1 - P)/\ln(1 - 1/n) \tag{2}$$

where $N$ = the number of λ clones needed, $P$ = the set probability, and $n$ = the fraction of the total mRNA population that the low-abundance mRNA species represents. Because the population of low-abundance mRNAs is not known for most plant cells, we will assume that the mRNA population contains about 10,000 unique, low-abundance mRNA species. Thus, the low-abundance mRNA represents about 40% of the total members in the mRNA population; and thus, $n = 0.4/25{,}000$. Therefore, a ds-cDNA λ clone bank that has a 99% probability of containing a ds-cDNA clone for each low-abundance mRNA species will require the construction of nearly 300,000 independent ds-cDNA clones.

However, there is a possibility that some mRNA species are represented at levels of one copy or less per cell. Isolating a ds-cDNA λ clone that represents such a rare mRNA species is not likely. The cloning of very low-abundance mRNA species should not be attempted without using an enrichment technique, such as reassociation kinetics, to normalize the mRNA population (by removing abundant mRNAs),[38] or genomic hybridization selection and PCR to amplify the selected mRNA species.[11] The cloning of very rare mRNAs is beyond the scope of the present chapter, and those interested in this technology should consult more appropriate reference material.[8,9,11,38]

Improvements in the construction of cDNA clone banks have come from both the synthesis of ds-cDNAs (see Section IV) and from the use of improved λ vectors and λ *in vitro* packaging extracts (see Section V). The most widely used λ vectors for the construction of ds-cDNA banks are λgt10 and λgt11.[6,39,40] However, the number and capabilities of related λ vectors have grown considerably in the last few years. Many of these vectors are capable of expression in *E. coli*, which permits functional screening using antibodies.[39,40] Recently several newer λ vectors have been constructed that contain a multifunction λ-M13 capability for automatic excision to a phagemid vector. These vectors also have multicloning sites that allow for directional cloning of ds-cDNA inserts. Table 4 lists λ vectors that can be used for the construction of ds-cDNA λ clone banks.

**TABLE 4**
**λ Vectors for the Construction of cDNA Clone Banks**

| Vector | Features | Cloning sites | Ref./Vendor |
|---|---|---|---|
| λgt10 | *lacZ* | *Eco*RI | Young and Davis[39,40]<br>Huynh et al.[6]<br>Bethesda Research Laboratories, Clontech, InVitrogen, Promega, Stratagene |
| λgt11 | *lacZ*<br>Product screening (antibody) | *Eco*RI | same as λgt10 |
| λgt11*Sfi-Not* | *lacZ*<br>Directional cloning<br>Product screening (antibody) | *Eco*RI, *Not*I, *Sfi*I | Promega |
| λgt22 | *lacZ*<br>Product screening (antibody) | *Eco*RI, *Not*I, *Sal*I<br>*Spe*I, *Xba*I | Han and Rutter[41]<br>Bethesda Research Laboratories |
| λgtWES.λB | *lacZ*<br>Large capacity (2 to 17 kb) | *Eco*RI, *Sal*I, *Sst*I<br>*Xho*I | Bethesda Research Laboratories |
| λMaxI™ | Directional cloning<br>Conversion to yeast plasmid, pYEUra3<br>Insert expression in yeast | *Bam*HI, *Eco*RI,<br>*Sal*I, *Xba*I, *Xho*I | Clontech |
| λBlueMid™ | Directional cloning<br>Conversion to M13 phagemid<br>T3 and T7 promoters | *Eco*RI, *Hin*dIII<br>*Not*I, *Sal*I, *Xho*I | Clontech |
| λGEM™-2 | Directional cloning<br>SP6 and T7 promoters<br>Subtractive clone<br>Bank construction | *Eco*RI, *Sac*I, *Xba*I | Promega |
| λGEM™-4 | Same as λGEM™-2<br>Ligation conversion to plasmid pGEM™-1 | Same as λGEM™-2 | Promega |
| λZAP™-II | *lacZ*<br>Product screening (antibody)<br>Conversion to pBluescript™ phagemid vector | *Eco*RI, *Not*I, *Sac*I<br>*Spe*I, *Xba*I, *Xho*I | Stratagene |
| UNI-ZAP™ XR | Same as λZAP™II<br>Directional cloning with ZAP-cDNA™ kit (see Table 5) | *Eco*RI, *Xho*I,<br>For directional cloning | Stratagene |

# III. CONSTRUCTION OF GENOMIC DNA-λ CLONE BANKS

## A. PURIFICATION OF PLANT GENOMIC DNA FOR CLONING

The following plant DNA isolation procedure is based on the method described by Murray and Thompson,[42] in which a solution of 1% hexadecyltrimethylammonium bromide (CTAB) is used to obtain plant DNAs that are relatively free of their major copurifying contaminants, carbohydrates (starches, pectins, charged polysaccharides, etc.). This original procedure was modified by Saghai-Maroof et al.[43] to allow the rapid isolation of DNAs while retaining a relatively high degree of purity without the need for any CsCl-ethidium bromide banding steps. Although plant DNAs obtained by this procedure are relatively pure, we strongly recommend that they be further purified by at least one CsCl-ethidium bromide gradient banding. We have used the procedure described below to isolate DNAs from the leaf tissues of many different plant species, including soybean,[44] cucumber,[45,46] tobacco,[47] and *Phaseolus vulgaris*.[48] Thus, this DNA isolation procedure appears to be well suited to many plant species.

1. Collect leaves and wrap in cheese cloth, mark with appropriate identification numbers, and then freeze in liquid nitrogen.
2. Transfer frozen tissue bundles into a freeze dryer container, and then connect to the freeze-dryer before the leaves begin to thaw. Freeze dry until the container reaches ambient room temperature. (If cool to the touch, $H_2O$ is still present in the tissue.) This freeze drying step can usually be done overnight depending on the vacuum.
   *Note:* Care must be taken in connecting the containers to the freeze-dryer because of the expanding nitrogen gas. We have obtained excellent freeze-drying results using 0.5 l Virtis containers (thick wall) and a Virtis Freeze Dryer Model 12.
3. Use a Thomas-Wiley mill to grind tissues to a fine powder; obtain between 0.5 to 1 g of finely ground plant tissue.
   *Note:* If a Thomas-Wiley mill is not available, the plant tissues can be ground using a pestle and mortar and ground glass.
4. Weigh out about 400 mg of the finely ground plant tissue, and transfer to a 15 ml polypropylene centrifuge tube. Distribute the tissue powder along the sides of the tube to avoid a clump of dry tissue at the bottom, and then add 9.0 ml CTAB extraction buffer.

   **CTAB Extraction Buffer (Stock 1 liter)**

   | | |
   |---|---|
   | $H_2O$ | 730 ml (sterile, double-distilled) |
   | 1 *M* Tris-HCl, pH 7.5 | 100 ml |
   | 5 *M* NaCl | 140 ml |
   | 0.5 *M* EDTA, pH 8.0 | 20 ml |
   | CTAB | 10 g |
   | β-mercaptoethanol | 10 ml |

   *Note:* CTAB = hexadecylcetyltrimethylammonium bromide (Sigma #H-5882); add to solution and stir at warm temperature until completely dissolved. Add β-mercaptoethanol to buffer just prior to use; 140 m*M* = 1% by volume.
5. Mix several times by inversion and vortex briefly. Incubate for 90 min in a 65°C water bath; mix samples every 30 min by inverting several times.
6. Remove tubes from bath, wait 4 to 5 min to allow the tubes to cool, and then to each tube add 4.5 ml chloroform/octanol (24:1). Mix gently by inverting several times.
7. Spin at room temperature in a table-top centrifuge for 10 min at 3000 to 5000 rpm.
8. Pour off the top, aqueous layer into new 15 ml polypropylene centrifuge tubes, add 4.5 ml chloroform/octanol (24:1), mix gently, and repeat Step 7.
9. Pipette the top aqueous layer into 15 ml polypropylene tubes containing 6.0 ml cold isopropanol (2-propanol). Mix **very** gently by inverting.
10. After 10 to 15 min, remove precipitated DNA with a glass hook, and transfer to 5 ml polypropylene tubes containing 3 ml of 76% EtOH and 0.2 *M* sodium acetate (NaOAc) pH 6.3. Leave DNA on hooks in tubes for about 20 min.
11. Rinse DNA on hook briefly in 1 to 2 ml of 76% EtOH and 10 m*M* $NH_4OAc$, and immediately transfer DNA to 5 ml polypropylene tubes containing 1.0 ml TEN buffer (pH 8.0).

**TEN–pH 8.0 Buffer (Stock 100 ml)**

| | |
|---|---|
| 1 *M* Tris-HCl, pH 8.0 | 1.0 ml |
| 0.5 *M* EDTA, pH 8.0 | 0.2 ml |
| 1 *M* NaCl | 1.0 ml |
| $H_2O$ | 97.8 ml (sterile, double-distilled) |

12. After the DNA is resuspended, determine the DNA concentration by the use of a TKO 100 Mini-Fluorometer (Hoefer Scientific Instruments).
    *Note:* This DNA preparation generally yields between 100 to 200 μg of DNA. A genomic clone bank can be constructed with as little as 100 μg of starting DNA, but it is recommended that at least 200 μg be obtained prior to starting. If the amount of DNA obtained is insufficient, repeat the DNA isolation procedure.
13. Band the DNA in a CsCl-ethidium bromide gradient. Measure the volume of the DNA solution and add 1 g of CsCl/ml. After dissolving the CsCl, transfer the solution into a 12 ml Oak Ridge tube, and fill with a CsCl solution (in TEN buffer) of the same specific density ($\rho = 1.50$). Add ethidium bromide to a final concentration of 250 mg/ml (use a stock solution of 30 mg/ml of DMSO). Establish a gradient by spinning at 38,000 rpm for 40 h. Carefully remove the DNA band from the gradient; if Oak Ridge tubes are used, this can be done from the top using a large bore pipette. Extract the ethidium bromide using a solution of NaCl-saturated isopropyl alcohol, and then dialyze the DNA sample against TEN buffer.

## B. PARTIAL DIGESTION AND ISOLATION OF SIZE-FRACTIONATED DNAs

The purity of the target genomic DNA is very important to any cloning experiment because impurities can greatly affect restriction enzyme digestion and, more importantly, the λ *in vitro* packaging system. General methods for the isolation of plant DNAs, such as that outlined above, yield DNAs of sufficient purity for restriction enzyme digestion and cloning; however, if contaminants remain, the DNA can be purified further by subjecting it to a neutral CsCl gradient centrifugation.

1. Check the *Sau*3AI digestibility of the target DNA sample by titering it with the enzyme. Prepare a DNA mix sample by first adding 10 μg of target DNA, 1/10 volume of enzyme buffer, and sufficient $H_2O$ to obtain a final volume of 150 μl in a 1.5 ml microfuge tube.
2. Dispense 30 μl of this DNA mix into a sterile 1.5 ml microfuge tube (tube No. 1) and set on ice, then add 15 μl of this DNA mix into each of six other tubes (Nos. 2 to 7). Chill all tubes on ice, and then add 4 U of *Sau*3AI into tube No. 1, mix, and briefly spin down the solution. Remove 15 μl and add to tube No. 2, mix well, and continue the twofold serial dilution process through tube No. 6. Tube No. 7 is the nondigested control. Incubate tubes Nos. 1 through 6 at 37°C for 30 min, and stop the reaction quickly by adding diethylpyrocarbonate (DEPC) and EDTA to a final concentration of 0.1% and 10 m*M*, respectively, and incubating the samples at 65°C for 5 min. Analyze the digestion results by electrophoresis through a 0.7% agarose gel.
    *Note:* Evaluate the digestion results, and estimate enzyme concentration and incubation times needed to generate a series of partial digest samples in the range of 10, 20, 30, 40, and up to 60% of completion. The dilution series will yield the following ratio of *Sau*3AI enzyme (1 h U) per microgram of DNA: tube 1 = 1 U/μg; tube 2 = 0.5 U/μg; tube 3 = 0.25 U/μg; tube 4 = 0.125 U/μg; tube 5 = 0.06 U/μg; and tube 6 = 0.03 U/μg. The amount of enzyme required for digestion will also be an indication of the purity of the target DNA. If no digestion is observed except at the higher enzyme concentrations, then the target DNA is not pure and could be difficult to clone.

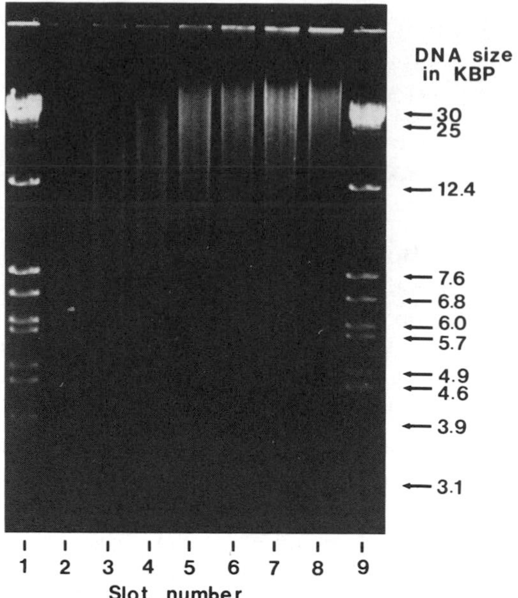


**FIGURE 2.** Pseudorandom partial *Sau*3AI digestion of total soybean DNA. Five aliquots containing 50 μg of soybean DNA each were digested as described in Section III.B, Step 4. After termination of the digestions, one μg of DNA from each aliquot was electrophoresed through a 0.7% agarose gel. After electrophoresis, the gel was stained with ethidium bromide and the DNA staining pattern photographed. The lanes contain the following: lanes 1 and 9, DNA size standard[3]; lane 2, blank; lane 3, estimated 50 to 60% of completion; lane 4, estimated 40 to 50% of completion; lane 5, estimated 30 to 40% completion; lane 6, estimated 20 to 30% of completion; lane 7, estimated 10 to 20% of completion; and lane 8, undigested total soybean DNA. (Reprinted by permission of Kluwer Academic Publishers.)

3. Using this digestion data, prepare a series of large-scale digests to obtain digested samples in the range of 10, 20, 30, 40, and 60% of completion. Depending on the availability of the target DNA, digest between 20 to 50 μg per reaction sample. Incubate the reaction at 37°C for the prescribed times, then rapidly inactivate the enzyme by adding 0.1% final volume of DEPC and 10 m*M* EDTA and incubating at 65°C for 5 min.
4. Analyze the digestion results by removing 0.5 to 1.0 μg of DNA and electrophoresing through a 0.7% agarose gel. After staining with ethidium-bromide and photographing, estimate the extent of each digestion, as shown in Figure 2, for the *Sau*3AI digestion of total soybean DNA. If the expected digestion profiles are obtained, proceed to the next step; however, if the samples are underdigested, add fresh 10X enzyme buffer and *Sau*3AI enzyme and repeat the digest. [The DEPC will not inhibit the freshly added enzyme because it is rapidly converted at 65°C to $CO_2$ and ethanol.] If the samples are overdigested, keep the samples nearest 40 to 60% of completion, and set up three new digestions using lower concentrations of *Sau*3AI enzyme or reduced time.
5. If the samples are within the expected digestion ranges, phenol-extract and ethanol-precipitate the DNA. Resuspend each sample in 50 μl of TEN buffer (Section III.A, step 11) and then pool; the volume of the pooled samples should not exceed 300 μl and should not contain more than 300 μg of DNA.
6. Using a linear gradient maker, form a 5 to 20% NaCl gradient in Beckman polyallomer tubes (1.43 cm × 8.9 cm; used in a SW41 rotor). Before loading the gradient, heat the pooled samples to 37°C for 10 min to ensure that the enzyme sites are free and not annealed. Centrifuge at 35,000 rpm for 6 to 7 h.

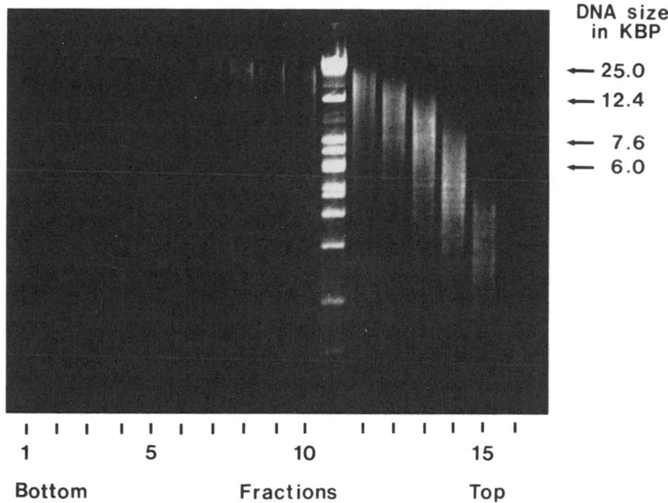


**FIGURE 3.** Analysis of 5 to 20% NaCl sedimentation gradient of pseudorandom partial *Sau*3AI-digested total soybean DNA. Partial *Sau*3AI digests of the soybean DNA were analyzed as shown in Figure 2; these samples were pooled and loaded onto a 5 to 20% NaCl gradient. After centrifugation, fractions (0.4 ml each) were collected, and 5 μl from each fraction was loaded directly onto a 0.7% agarose gel and electrophoresed. After electrophoresis, the gel was stained with ethidium bromide, and the resulting DNA pattern was photographed. Fractions containing *Sau*3AI-digested soybean DNA fragments in the size range of 15 to 25 kb were identified in fractions 7 to 10 (lanes 7 to 10). These fractions were pooled and concentrated by ethanol precipitation as described in Section III.B, Steps 8 to 10. DNA sizes were estimated using the DNA size standard[3] located in the unlabeled lane that follows lane 10. (Reprinted by permission of Kluwer Academic Publishers.)

7. After centrifugation, fractionate the gradient from the bottom using tubing connected to a peristaltic pump. Fit the pump tubing with a glass capillary, and carefully place it into the centrifuge tube, from the top, until it reaches the bottom; then pump out 0.3 to 0.4 ml fractions into sterile 1.5 ml microfuge tubes. Remove between 5 to 10 μl from each fraction, and analyze by electrophoresis through a 0.7% agarose gel. An example of the results obtained is shown in Figure 3 for the fractionation of *Sau*3AI-digested soybean DNA.
8. Pool the fractions that contain *Sau*3AI-cut target DNA of 15 to 25 kb, measure the volume, and then add an equal volume of sterile $H_2O$ and mix. Then add 2.5 volumes of cold 100% ethanol, transfer each sample to a siliconized Corex tube, and incubate at –70°C for at least 2 h and preferably overnight.
9. Pellet the DNA by centrifuging in a SS-34 rotor at 10,000 rpm and –10°C for 2 h. After spinning, carefully remove the supernatant solution and save. Resuspend the DNA pellet in 0.4 ml of 0.3 *M* NaOAc, and transfer the sample to a sterile 1.5 ml microfuge tube, and add 1 ml of 100% ethanol. Chill the sample at –70°C for at least 10 min; then microfuge at 4°C for 10 min. Carefully remove the supernatant solution, and dry the DNA sample under vacuum. Resuspend the dried DNA pellet in TEN buffer (Section III.A, Step 11) to obtain a final concentration of 0.5 to 1.0 mg/ml.
10. Remove about 1/20 of the size-fractionated target DNA, electrophorese through a 0.7% agarose gel alongside a DNA size standard, and estimate the average size of the fragments and the DNA concentration of the sample. The *Sau*3AI-cut target DNA is ready to be cloned into *Bam*HI-cut λ arms (see below).

## C. LIGATION OF SIZE-FRACTIONATED GENOMIC DNAS WITH λ ARMS

Ligation of the λ arms and size-selected *Sau*3AI-cut target genomic DNA is achieved by mixing them together at a concentration that favors the formation of concatenated λ DNA.

Concatenated λ DNAs are the most efficient substrate for the *in vitro* packaging reaction; monomeric λ DNA molecules are not efficiently packaged. Two important parameters to consider when setting up a λ ligation reaction are (1) the ratio of λ arm cloning sites to insert DNA cloning sites, and (2) the concentration of each DNA species. Optimal values for these parameters can be theoretically estimated, assuming that all the DNA molecules will participate in the ligation reaction. To obtain the best substrate for the *in vitro* packaging reaction (λ left arm-insert-λ right arm)$_N$, the type of available cloning sites (*cs*) must be considered. Each λ arm (right and left) contains two different types of cohesive termini, a *cos* site that is only compatible with its complement on the opposite arm, and one *cs* that is compatible with both termini of the size-fractionated target DNA (also referred to as *cs*) and the *cs* of the opposite λ arm. The ligation reaction should contain equal-molar concentrations of the *cs* ends for each of the three different types of DNA molecules present, and because each λ arm contains only one *cs* end, the ratio of molecules in a ligation reaction should be 2:1:2 (λ left arm : insert : λ right arm) or a molar ratio of 2:1 for annealed λ arms to size-selected target DNA. At a first approximation, ligation reactions should contain about 4 μg of annealed λ arms (31 kb for most replacement λ vectors) for every 1 μg of size-selected target DNA, assuming an average size-selected value of 17 kb. However, if *in vitro* packaging efficiencies are not found to be as expected (which could be due to damaged *cos* or *cs* sites, errors in estimating DNA concentrations, etc.) a series of test ligation and *in vitro* packaging reactions should be carried out to determine the best ratio of λ arms to size-selected target DNA for each particular batch of λ arms and target DNAs.

The second parameter, the total DNA concentration in the ligation mixture, is important because the DNA concentration must be high enough to ensure that intermolecular ligations, which lead to concatamers, are favored over self-ligation that reduces the yield of viable recombinant λ DNA molecules. A theoretical discussion concerning the effects of DNA concentration and size on the ligation reaction has been presented previously by Manaitis et al.[19] The results of these calculations are that the concentration of annealed arms should be about 135 μg/ml and the size-selected target DNA concentration should be about 43 μg/ml.

Prior to ligating λ arms with size-selected target DNA, the λ arms should be tested by self-ligation, followed by *in vitro* packaging to determine the amount of background phage that can be expected per μg of arms. Purified λ arms should not yield more than $5 \times 10^4$ background plaque forming units (PFU)/μg.

1. λ vector arms and size-selected target DNAs should be resuspended in TEN buffer (Section III.A, Step 11) at a concentration of about 500 μg/ml. Add 8 μl (4 μg) of λ vector arms and 2 μl (1 μg) of size-selected target DNA to a sterile microfuge tube. Incubate at 37°C for 5 min; then cool on ice. Add 2.5 μl of 10X ligase salts, and adjust the total volume to 25 μl by the addition of sterile double-distilled $H_2O$. Remove a 2 μl aliquot prior to adding ligase and store at 4°C for later agarose gel analysis. Add ligase, generally about 1 μl, if its concentration is about 10 U/μl.

**10X Ligase Buffer**

0.8 *M* Tris-HCl, pH 8.0
0.2 *M* $MgCl_2$
150 m*M* DTT
10 m*M* ATP

2. Incubate at 37°C for 1 h, then at 4°C overnight. After ligating, heat-inactivate the ligase by incubating at 65°C for 5 min; then remove another 2 μl aliquot and analyze (along with the aliquot removed in step 1) by agarose gel electrophoresis. The agarose gel

**TABLE 5**
**cDNA Synthesis Kits for λ Vectors**

| Name | Features | Vendor |
|---|---|---|
| λgt10 and λgt11 cloning systems | Adaptor cloning strategy, dephosphorylated λgt10 arms, avoids methylation, λ *in vitro* packaging mix provided (see Table 6) | Amersham |
| SuperScript™ | Primer-adaptor strategy for directional cloning onto λgt22 | Bethesda Research Laboratories |
| λLibrarian™X | Generation of ds-cDNA for cloning into the *Eco*RI site of λgt10 | InVitrogen |
| λLibrarian™XI | Generation of ds-cDNA for for cloning into the *Eco*RI site of λgt11 | InVitrogen |

analysis should show few DNA fragments in the size range of the original size-selected target DNA for the ligated sample; most of the DNAs in this sample should be larger than wild-type λ (50 kb). Store ligation mix at 4°C while waiting for gel analysis results.

3. If the gel analysis shows that most of the λ arms and insert target DNAs are ligated, proceed to Section V for *in vitro* packaging of the recombinant λ molecules.

## IV. PREPARATION OF PLANT cDNAs

Methods for the synthesis of ds-cDNA for the construction of cDNA clone banks have been improved considerably in the last few years, which also improves the chances of cloning ds-cDNAs that are representative of low-abundance mRNA species. A major improvement has been the use of second-strand synthesis techniques that employ the simultaneous use of RNase H and reverse transcriptases (RT).[49,50] This method has increased the efficiency of cloning full-length ds-cDNAs because it eliminates the need to use S1 nuclease to clip the hairpin loop used to prime second-strand synthesis. Improved efficiency has also resulted from the availability of purer enzymes and the use of RNAsin (a potent RNase inhibitor) in RNA isolation and first-strand synthesis procedures. There are presently available from commercial sources many different types of cDNA synthesis kits that are designed for cloning into specific λ cDNA cloning vectors. Table 5 lists many of these kits, the specific λ vector they are designed for, and their commercial sources. However, in the following subsections we have included a description of a ds-cDNA synthesis procedure that is commonly referred to as the single-tube synthesis method; this procedure is relatively simple and yields high quality ds-cDNAs without the need of any elaborate steps. This procedure should be adequate for a first try; however, if it is not successful we recommend that the reader consult Table 5 and select a commercially available ds-cDNA synthesis kit.

### A. cDNA SYNTHESIS

For this procedure, we are assuming the use of between 1 to 5 μg of purified poly(A)$^+$ mRNA in a total volume of 50 μl. Before first-strand cDNA synthesis is initiated, the mRNA sample is denatured in a denaturation-annealing buffer to ensure that full-length, single-strand (ss)-cDNA copies can be synthesized. After removal or dilution of the denaturation-annealing buffer, ss-cDNA synthesis is initiated by adding the four nucleotide triphosphates, RT, and the reaction buffer. The molecules of ss-cDNA are immediately used as a template for ds-cDNA synthesis, especially if the goal is to obtain a large amount of ds-cDNA. This is done by adding the components for the RNase H-RT reaction directly into the ss-cDNA reaction tube (without inactivation of the previously added enzyme). Thus, the complete synthesis of ds-cDNAs can be done in a single reaction tube.

1. Denature mRNA and anneal with oligo-dT primer by adding mRNA (1 to 5 μg) and oligo-$(dT)_{12-18}$ (100 to 500 ng) into a sterile 1.5 ml microfuge tube, and dry in a Savant SpeedVac.
   *Note:* If a directional ds-cDNA synthesis approach is being used, add the specific oligo-(dT) primer. (This primer usually contains the recognition site of a rare cutting restriction enzyme.)
2. Resuspend in 50 μl of denaturation-annealing buffer.

**Denaturation-Annealing Buffer**
4.0 m*M* sodium phosphate (pH 7.2)
2.0 m*M* EDTA
15% DMSO (ultrapure, from Aldrich)

3. Denature mRNA by incubating for 5 min at 65°C.
4. Add 5 μl of 1 *M* NaCl that has been preheated to 65°C.
5. Slowly cool to 42°C over a time span of about 30 min.
6. Remove the annealed mRNA-oligo-$(dT)_{12-18}$ complex from the denaturation-annealing buffer by phenol-chloroform-isoamyl alcohol (1:1:0.04) extraction, followed by ethanol precipitation and vacuum drying.
7. Initiate ss-cDNA synthesis by resuspending the mRNA-oligo$(dT)_{12-18}$ complex pellet in the following:

   2.5 μl, 1 *M* Tris-HCl (pH 8.3)
   5.0 μl, 60 m*M* $MgCl_2$
   2.5 μl, 1 *M* NaCl
   5.0 μl, 5.0 m*M* dNTP mix (5.0 m*M* each dGTP, dATP, dTTP)
   5.0 μl, 2.5 m*M* dCTP
   5.0 μl, 70 m*M* DTT
   5 to 10 μl, [α-$^{32}$P]dCTP (400 Ci/mmol)

   Before adding enzyme, remove two 1 μl aliquots for analysis of $^{32}$P incorporation (total counts and TCA-precipitable counts).
8. Add the following: 250 U RNasin and 200 U RT (AMV or M-MLV).
9. Incubate at 42°C for 45 min.
10. Add 5 more U of RT, and incubate at 50°C for 15 min.
11. Stop the reaction by placing on ice for at least 5 min; then remove two 1 μl aliquots for analysis of incorporated $^{32}$P label by TCA precipitation.
    *Note:* Estimate the amount of ss-cDNA synthesized from the amount of $^{32}$P label incorporated by spotting 1 μl of the reaction on a 3MM filter disk, dry, and then submerge the filter in a solution of cold 10% TCA. Wash twice in $H_2O$, and do a final rinse in ethanol. Count using a scintillation counter, and estimate the amount of the original counts incorporated into the TCA insoluble ss-cDNA.
12. Dry the ss-cDNA synthesis reaction sample using a Savant SpeedVac.
13. Initiate RNase H-RT synthesis of ds-cDNA by resuspending the dried ss-cDNA-mRNA duplex pellet in 10 μl of 0.1X TE buffer, followed by the addition of the following:

10 μl 1 *M* Tris-HCl (pH 8.3)
10 μl 60 m*M* $MgCl_2$
10 μl 70 m*M* DTT
5 μl 1 *M* NaCl
10 μl dNTP mix (same as in step 7)
5 μl 2.5 m*M* dCTP
5 to 10 μl [α-$^{32}$P]dCTP (400 Ci/mmol)

14. Add RNase H, 2 U/μg of ss-cDNA-RNA duplex, then add AMV (or M-MLV)-RT, 0.4 U/μl. The volume should be adjusted to 100 μl by the addition of sterile $H_2O$.
15. Incubate at 12°C for 1 h, then at 42°C for an additional hour. For the final 15 min, add an additional 2.5 U of RT.
16. Terminate second-strand synthesis by adding EDTA and SDS to a final concentration of 20 m*M* and 0.1%, respectively. Extract with phenol-chloroform-isoamyl alcohol (1:1:0.04), with chloroform-isoamyl alcohol (1:0.04), and then with ethanol precipitate. Dry the pellet under vacuum.
17. If linkers (*Eco*RI or others) are to be added, the ds-cDNA will need to be methylated to protect indigenous sites from digestion. However, if linkers contain the recognition site of a rare cutting restriction enzyme (e.g., *Not*I or *Sfi*I), the methylation step can be avoided.

## B. LINKER LIGATION AND LIGATION OF cDNAs WITH λ VECTOR ARMS

Preparing the newly synthesized ds-cDNA molecules for cloning into the λ vector arms generally involves the ligation of oligonucleotide linkers that include the sequence of a restriction enzyme recognition site. The most commonly used linkers contain the recognition site for *Eco*RI. However, if the λ vector system allows directional cloning, the 3′ recognition site is generally included in the oligomer primer used to initiate first strand DNA synthesis (see Tables 4 and 5), followed by the addition of a linker containing the second restriction enzyme recognition site. In this latter method, rare restriction enzyme sites are generally used so that the restriction enzyme site methylation step can be avoided. In the following, we present the steps needed for the addition of *Eco*RI linkers. However, if a direction cDNA clone bank is being constructed, the vendor's instructions should be consulted at this time to ensure that the steps are similar and done in the correct order.

1. Before *Eco*RI linkers are added, the ds-cDNA molecules must be blunt-ended. This can be done by a number of general methods, which include nuclease (S1 or mung bean) digestion and either chew-back by bacteriophage T4 polymerase or fill-in by *E. coli* polymerase. These blunt-end methods have been described previously.[19,20]
2. Prior to starting, the blunt-ended ds-cDNA should be freed of contaminating enzymes and dried under vacuum using a Savant SpeedVac.
3. Resuspend the dried ds-cDNA (1 to 2 μg) in 20 μl of *Eco*RI methylase buffer.

**EcoRI Methylase Buffer**
50 m*M* Tris-HCl (pH 7.5)
1 m*M* EDTA
5 m*M* DTT

4. Add 2 μl of 100 m*M* S-adenosyl-L-methionine and 10 U of *Eco*RI methylase per microgram of ds-cDNA. Incubate at 37°C for 15 min.
5. Heat-inactivate the *Eco*RI methylase by incubating at 70°C for 10 min. Remove proteins

by extracting once with phenol-chloroform-isoamyl alcohol (1:1:0.04), once by chloroform-isoamyl alcohol (1:0.04), followed by ethanol precipitation. Dry the pellet under vacuum.

6. Resuspend *Eco*RI-methylated blunt-ended ds-cDNA pellet (0.1 to 2 μg) in 5 μl of sterile, double-distilled $H_2O$, and add the following: 0.1 to 2 μl *Eco*RI linkers (use a mass ratio equal to the estimated mass of ds-cDNA), 0.5 μl of 10X ligase buffer (see Section III.C, Step 1), and 0.5 to 1.0 μl of T4 ligase (10 U/μl, Collaborative Research, Inc.). Mix, spin down solution, and incubate at 10 to 14°C for 24 to 48 h.
7. Heat inactivate the T4 ligase by incubating at 70°C for 10 min, then add *Eco*RI digestion buffer, 2 to 4 μl of *Eco*RI (20 U/μl), and incubate at 37°C for 3 h.
8. Stop the *Eco*RI digestion reaction by heating to 70°C for 10 min, or by extracting once with phenol-chloroform-isoamyl alcohol (1:1:0.04) and once with chloroform-isoamyl alcohol (1:0.04), followed by ethanol precipitation. Dry the pellet under vacuum, then resuspend in 20 μl of sterile TEN buffer (see Section III.A, Step 11).
9. Separate *Eco*RI-linkered ds-cDNA molecules from free *Eco*RI linkers using Bio-Rad A-50 M, or equivalent, chromatography.
   *Note:* The size of the *Eco*RI-linkered ds-cDNA molecules can be analyzed by loading 10,000 to 20,000 cpm from each collected fraction on a 7% polyacrylamide-20%-glycerol gel, or a 1% agarose gel. After electrophoresis, autoradiograph the gel (dry the agarose gel prior to exposing to film) to locate the fractions containing the largest sized ds-cDNA molecules.
10. Pool the fractions containing the largest sized ds-cDNA molecules, extract once with phenol-chloroform-isoamyl alcohol (1:1:0.04), once with chloroform-isoamyl alcohol (1:0.04), ethanol precipitate, and dry the pellet under vacuum. Resuspend the linkered ds-cDNA pellet in TEN buffer to obtain a final concentration of about 10 ng/μl.
11. Initiate the ligation of the *Eco*RI-linkered ds-cDNA with λ vector arms by adding 1 μg (1 to 2 μl) of phosphatase-treated λ vector arms (Table 4) to a microfuge tube, followed by the addition of 10 to 50 ng (1 to 5 μl) of *Eco*RI-linkered ds-cDNA.
12. Adjust the reaction volume to between 5 and 10 μl by the addition of sterile, double-distilled $H_2O$, mix, spin-down, and then heat the sample to 37°C for 5 min. Add 0.5 to 1 μl of DNA ligase buffer (Section III.C, Step 1) and 1 ml of T4 ligase (10 U/μl); the final volume should be adjusted to 10 μl by the addition of sterile, double-distilled $H_2O$. Mix, spin-down, and incubate at 14 to 16°C for 8 to 21 h.
13. Heat-inactivate the ligase by incubating at 70°C for 5 min. Proceed to Section V for *in vitro* packaging of recombinant λ-ds-cDNA molecules into λ phage particles. If sample is not to be used immediately, store at –20°C until needed.

## V. *IN VITRO* PACKAGING OF RECOMBINANT λ MOLECULES

The high efficiency of the λ vector cloning systems is in part due to the λ infection mechanism, which is used to transfer the recombinant λ DNA molecules into the *E. coli* host cells. However, to take advantage of this efficient transformation mechanism, the recombinant λ molecules must first be packaged into λ phage particles, which is achieved by reassembling the necessary phage components *in vitro*. Several procedures for the isolation of λ packaging components have been described,[19,51] and the λ *in vitro* packaging components isolated are capable of yielding packaging efficiencies in the range of $10^7$ to $10^8$ PFU per μg using wild-type λ DNA. The preparation of these λ packaging components is time-consuming, and each component must be subjected to quality control testing prior to use. However, this can now be avoided because many different commercial vendors supply these λ *in vitro* packaging components, and they have been able to increase packaging efficiencies up to $10^9$ PFU per

**TABLE 6**
**Commercially Available 1 *In Vitro* Packing Mixes**

| Name | Efficiency | Vendor |
|---|---|---|
| Amersham λ packaging kit | 1 to 2 × $10^9$ PFU/μg vector | Amersham |
| DNA packaging kit | 2 × $10^8$ PFU/μg vector | Boehringer-Mannheim |
| SuperScript™ | 5 × $10^8$ PFU/μg vector | Bethesda Research Labs |
| Packagene™ | 2 × $10^8$ PFU/μg vector | Promega |
| Gigapack™ II XL | 1 × $10^9$ PFU/μg vector | Stratagene |
| Gigapack™ Gold | 2 × $10^9$ PFU/μg vector | Stratagene |

μg of wild-type λ DNA. Table 6 presents a list of vendors and the efficiencies of their λ *in vitro* packaging systems. Such high efficiencies are desirable since they allow one to construct a complete recombinant λ clone bank (1 × $10^6$ clones) from as little as 1 μg of size-selected DNA. Because these superior λ *in vitro* packaging components can be obtained from commercial sources, their isolation is not described in this chapter.

1. Obtain λ *in vitro* packaging extracts from commercial source, and *in vitro* package 1/2 of the ligation mix as instructed by the vendor. This generally requires a 2 to 3 h incubation step for packaging of the recombinant λ molecules into the phage particles. After packaging, add 1 ml of λ dilution buffer and 20 μl of chloroform. Mix gently and titer the supernatant solution at dilutions of $10^{-3}$ and $10^{-4}$ using the appropriate *E. coli* host. If the ligation and *in vitro* packaging reactions are successful, the titer should be between $10^5$ and $10^6$ PFU/ml. Titers as high as $10^7$ can be expected.

**λ Dilution Buffer**
10 m*M* Tris-HCl, pH 7.4
10 m*M* $MgSO_4$
0.01% gelatin

2. The next step depends on the titer obtained and the goals of the experiment. If the titer is barely sufficient to contain the complete genome at a probability of 0.99, it should be amplified directly before screening (see Section VII). However, if the library contains five to ten genome equivalents, one to two genome equivalent(s) can be used for direct screening (see Section VI), and the remaining part can be stored for later use or amplification (Section VII).

## VI. SCREENING OF λ CLONE BANKS

The λ vector system is a highly efficient system for obtaining a large number of recombinant clones; however, its power lies in the efficiency with which it can be screened for clones that contain DNA regions (coding and noncoding) represented at low copy number. Benton and Davis[5] were the first to describe a highly effective and efficient phage plaque *in situ* screening procedure, which involves contact between the phage plaques and nitrocellulose or nylon-based membrane. Treatment of these filters with 0.5 *M* NaOH lyses the phage, and the exposed recombinant λ DNAs are denatured and fixed *in situ* on the filters. Contact between the filter and the λ plaques on the agar surface does not destroy or greatly distort the location of each λ plaque; thus the λ phage remaining on the surface of the agar is viable and can be recovered. After fixing the recombinant λ DNAs onto the filter, the filter is hybridized against $^{32}$P-labeled DNA or RNA probes, and the resulting film exposures can

**TABLE 7**
**Agar Plate Sizes and Number of λ Phage Plaque Conveniently Plated[a]**

| Size of Plate | Total area (cm$^2$) | Volume of bottom agar (ml) | Volume of indicator bacteria (ml) | Volume of top agar (ml) | Maximum number plaques/plate |
|---|---|---|---|---|---|
| 9 cm Round plate | 64 | 25 to 30 | 0.1 | 3.0 | 15,000 |
| 9 × 9 cm Square plate | 81 | 35 to 40 | 0.1 | 4.0 | 20,000 |
| 15 cm Round plate | 177 | 70 to 80 | 0.3 | 8.0 | 50,000 |
| 23 × 23 cm Square plate | 552 | 300 | 4.0 | 40 | 250,000 |

[a] Reprinted by permission of Kluwer Academic Publishers.

be used to locate the recombinant λ plaque responsible for a specific hybridization signal. After a series of plaque purification steps (repeated plating and plaque isolation), an individual λ recombinant clone can be purified to homogeneity.

The number of recombinant λ phage plaques that must be screened to obtain a specific DNA region depends on the size of the target genome, on whether the DNA region is single copy or repetitive, and on the background of nonrecombinant λ phage contained within the clone bank. For example, to isolate a single-copy DNA region from a recombinant λ-*Phaseolus vulgaris* clone bank requires screening at least $10^6$ phage plaques (Table 3); however, to compensate for background nonrecombinant λ phage and other discrepancies during construction of a clone bank, two to three times the minimum number of phage plaques should be screened. This raises the question of how to screen $2 \times 10^6$ recombinant λ phage plaques without using an unmanageable number of agar plates. The numbers of phage plaques that can be effectively plated and screened on different size culture plates are listed in Table 7. The smaller plates should not be used for the first round of screening for a single-copy DNA region; they are more suited for the isolation of clones containing highly repetitive DNA or for later purification steps (see below). The largest plates that are convenient for screening are the 23 × 23 cm plates supplied by Nunc (Vanguard International, Inc.), which are referred to as kiloplates (KPs) in the procedure presented below. These plates are extremely convenient because they are square (easy to fit with the filter membrane) and because, after exposure of the hybridized filter membrane to film, hybridizing plaques can be identified by placing the agar plate over the exposed film on a light box.

The following procedure is for screening of recombinant λ clone banks that have been constructed using a λ replacement vector such as Charon 40, which would be plate-propagated by infecting the *rec*A$^-$ *E. coli* host ED8767:

1. Prior to starting, do the following:
   a. Obtain an overnight growth (50 ml NZC medium) of the appropriate *E. coli* host; in this case for plating Charon 40, use ED8767 cells.
   b. Prepare agar, both top and bottom, for pouring the necessary number of KPs; $2 \times 10^6$ phage can be screened on 8 KPs, which requires about 2.5 liters of bottom agar (1.5%), and phage spreading requires about 300 ml of top agar (1%). Pour plates at least 1 day before plating the λ clone bank.
   c. Determine the titer of the recombinant λ library if it is not already known.

| **NZC Bottom Agar** | **NZ Top Agar** |
|---|---|
| 10 g NZ amine | 10 g NZ amine |
| 5 g NaCl | 5 g NaCl |
| 5 g casamino acids | 10 g agar |
| 15 g agar | $H_2O$ up to 1 liter |
| $H_2O$ up to 1 liter | |

2. Pour the required number of KPs prior to starting the *E. coli* overnight growth. Allow the plates to set overnight at room temperature to ensure that the KPs are free of excess moisture. If necessary place in a biological hood, remove cover, and wipe away excess moisture off of the cover.
3. Set-up preinfections of the *E. coli* host and λ clone bank; for each KP add the following to a sterile 50 ml polystyrene tube:
   a. 4 ml of overnight growth of *E. coli* host (ED8767 for Charon 40);
   b. 4 ml of λ infection salts; 0.01 *M* $MgCl_2$ and 0.01 *M* $CaCl_2$ solution;
   c. The appropriate volume of the recombinant λ clone bank expected to produce 250,000 PFU. Incubate the preinfection mixture at 37°C for 15 min.
4. Add 40 ml of top agar (stored in a 50°C water bath) to the preinfection tubes (one at a time), gently invert two to three times, pour evenly over the surface of a KP, and gently spread by rocking the plate before the agar sets. Avoid moving the plate after the agar starts to set. Repeat this step until all of the preincubation tubes have been plated. Allow the top agar to harden before moving (15 to 20 min should be adequate).
5. Place the KPs in a 37°C incubator (do not invert), and do not stack; this avoids the accumulation of condensation on the lids. Use as large an incubator as possible; if available, use a warm room. Allow the *E. coli* and phage to grow overnight to obtain full plaque development.
6. Carefully remove the KPs from the incubator, and if condensation is present on the lids, remove the lids and either shake the condensation off or wipe it off with a Kimwipe. Allow the KPs to cool to room temperature, and store in a cold room. Prior to placing filters over the plaques, place the KPs on ice; use a large smooth-bottom photo tray filled with ice, and cool plates for about 30 min.
7. While cooling the KPs, set up two smooth bottom photo trays (available from VWR, #TR71707-40), one with denaturation (0.5 *M* NaOH and 1.5 *M* NaCl) and the other with neutralization (0.5 *M* Tris-HCl pH 7.2, 3.0 *M* NaCl and 1 m*M* EDTA) solution. Place two sheets of 3MM paper in the bottom of each tray, and saturate them with the appropriate solution.
8. Cut filter membranes to 22 × 22 cm and treat according to the vendor's instruction. Generally, filters are labeled in permanent ink with the KP number, λ vector used, and name of target species. First soak the filters in distilled $H_2O$, followed by soaking in a solution of 1 *M* NaCl. Lightly blot the filters dry on 3MM paper to remove excess moisture just prior to using.
9. Carefully place the filter membrane on top of the appropriate KP, and mark the position of the filter by using a blunt needle dipped in water-resistant India ink and stabbing it through the membrane into the agar below. Traces of ink should remain in the agar and on the filter. The marks should be placed about every 4 to 5 cm to provide easy alignments for later picking of the phage plaque responsible for a particular hybridization signal.
10. Allow the filter to remain in contact with the λ phage plaques for about 4 min; then carefully remove (avoid lifting the top agar) and place the filter, plaque side up, onto

3MM paper previously saturated with denaturation solution. Remove trapped air bubbles; after incubating for 4 min, transfer the filter onto 3MM paper previously saturated with neutralization solution and incubate for an additional 4 min. After neutralization, place the filter on dry 3MM paper, and allow it to air-dry. Repeat these steps for each KP used and for each replica filter obtained from the same KP.

11. Treatment of the filter membrane to ensure fixing of the recombinant λ DNAs depends on the type of membrane used. Nitrocellulose requires only heating (68°C overnight), while nylon requires UV light treatment (see vendor's instructions).
12. After fixing DNA onto the filters, prehybridize them in a smooth bottom photo tray (25 × 25 cm). For eight filters, prehybridize in about 600 ml of hybridization solution, and incubate at 65 to 68°C, with shaking (50 rpm), for at least 2 h. Cover the photo tray with plastic wrap and a glass plate to prevent evaporation.

**Hybridization Solution**

6 × SSC (1 × SSC = 0.15 *M* NaCl, 0.015 *M* sodium citrate, pH 7.2)
0.02% bovine serum albumin
0.02% Ficoll ($M_r$ = 400,000)
0.02% polyvinylpyrolidone (PVP, $M_r$ = 36,000)
1% SDS

13. Denature the radioactive-labeled probe by adding 1/20 volume of 5 *M* NaOH, incubate at 68°C for 15 min followed by the addition of 1/20 volume of 5 *M* HCl and 1/10 volume of 1 *M* Tris-HCl (pH 7.4). Mix, then quickly add to the hybridization solution that contains 50 μg/ml poly rA (Sigma P9403), which is used to reduce nonspecific binding of the probe to the filter. Hybridize the eight filters in about 200 ml of hybridization solution that contains at least $10^5$ cpm of the denatured $^{32}$P-labeled probe (specific activity = $1 \times 10^8$ cpm/μg) per ml. Pour off the prehybridization solution, and add the probe-containing hybridization solution, incubate at 68°C with slow shaking (50 rpm) for at least 12 h or overnight. The tray should be covered with plastic wrap and a glass plate to prevent evaporation.
*Note:* Denatured sonicated calf-thymus or salmon sperm DNA (20 μg/ml) can be used in place of poly rA.
14. Decant the hybridization solution and rinse briefly with 200 ml of 3X SSC plus 0.5% SDS. Replace with 300 ml of fresh 3X SSC solution, and incubate at 68°C with shaking for 30 min to 1 h.
15. Remove the filters from the wash solution, and place on 3MM paper. Allow the filter to dry, and then scan the filter with a survey meter (Geiger counter), and if the counts are below 200 cpm, obtain film exposures of the filters. Expose filters to film using enhancer screens (Du Pont Quanta III) at low temperatures (–70°C). If the initial scan or later film exposures indicate high background signals, rewash the filters in a solution of lower ionic strength, 1X SSC, and if necessary, down to 0.1X SSC (see hybridization solution for formula for 1X SSC).
16. After obtaining film exposures, align the corresponding films and filters, and transfer the reference ink marks onto the film. By placing the film on a light box and the corresponding KP on top of the film, align the reference marks on the film with those in the agar. Align the reference marks closest to a particular hybridization signal (λ clone candidate), and pick the plaque region using the large end of a sterile Pasteur pipette. The agar plug can be excised from the plate by placing a finger over the small end of the pipette, loosing the plug with rotary movement, and lifting the plug free using "built-up" vacuum. Phage plugs are placed in 1 ml of phage storage buffer and

numbered with the KP number followed by the pick number from a particular KP. Thus, the first plaque pick from KP 1 is referred to as KP 1.1.

**Phage Storage Buffer**

0.1 *M* NaCl
10 m*M* Tris-HCl pH 7.4
0.05% gelatin
10 m*M* $MgCl_2$
Saturate with chloroform after autoclaving

*Note:* After removing agar plugs that correspond to hybridizing phage, the remaining λ clone bank can be recovered (see Section VII) for later use, long term storage, and distribution.

17. The recombinant λ phage responsible for the hybridization signal must be purified from the other phage contained within the agar plug. Purification of hybridizing phage clone is achieved by using a series of successive plating and hybridization reactions similar to those just described for the KPs, but on smaller agar plates (see Table 7). An agar plug can contain well over 100 independent phage clones; thus the next purification plating should involve the plating of about 500 PFU. This can usually be achieved by plating a $10^{-4}$ dilution from the original pick (KP1.1, etc.). Plating is done using the same technique used to obtain a phage titer; 100 μl of *E. coli* host cells, 100 μl of λ infection salts, and the appropriate dilution volume of the phage sample. The sample is preincubated at 37°C for 10 min, followed by the addition of top agar and plating (use 3 ml of top agar for round 9 cm plates and 4 ml for square 9 cm plates).
18. Plaque purification plates are screened using procedures similar to that described above for KPs, but instead of using the large end of the Pasteur pipette for picking plaques, the small end is used to reduce the number of contaminating phage carried along with the hybridizing λ phage plaque. These small phage plugs are placed into 1 ml of phage storage buffer and numbered. The secondary purified phage pick for KP 1.1 is referred to as KP 1.1.1, and so on for the other KPs and multiple picks from each KP. The procedure of successive plating, screening, and picking is continued until a plating is obtained where every phage plaque yields a hybridization signal with the $^{32}$P-labeled probe. For each successive plating step, add an additional number (.1) to distinguish it from any of the previous purification phage plaque picks.

Once the recombinant λ clone(s) that hybridizes to the $^{32}$P-labeled probe is plaque-pure, a primary (1°) phage growth should be started from a single hybridizing plaque. From this 1° growth, a large scale λ growth can be initiated, and the recombinant λ DNA can be purified.[18,19,24]

## VII. AMPLIFICATION AND STORAGE OF λ CLONE BANKS

Recovery of a screened λ clone bank or amplification of a minimal λ clone bank (a clone bank that contains only one genome equivalent at a probability of 0.99) is important so that it can be used for the isolation of other cDNAs or genomic DNAs. Also, amplified λ clone banks can be stored (we have stored such clone banks for more than 5 years), and it is a convenient form (liquid extract) for their distribution. Generally we prefer to construct cDNA or genomic clone banks that contain several genome equivalents (at a probability of 0.99), as one λ clone bank equivalent can be screened directly for the coding or noncoding DNA region of interest while another λ clone bank equivalent can be amplified without screening.

Screening of a clone bank can introduce bacterial and/or fungal contamination, which can shorten its storage life. However, if necessary, the phage remaining after a λ clone bank is screened can be recovered and rescreened for other cloned cDNAs or genomic DNAs.

Amplification of any λ clone bank should be done using plate lysis rather than liquid lysis to prevent individual phage clones that have better growth characteristics from being over-represented in the resulting amplified λ clone bank. The dynamics of liquid lysis is subject to such over-representation; however, preferential growth of an individual phage clone on an agar plate lysis is limited to the size of the plaque, which usually varies by only several fold, while in liquid culture preferential growth could vary by several orders of magnitudes. Plate lyses are extracted using phage storage buffer (see Section VI, Step 16), and the extract can be stored at 4°C for about 5 years. Longer term storage can be achieved using liquid nitrogen.[52]

Typically, for a genome the size of soybean (Table 3), we would plate 1 to $2 \times 10^6$ PFU, and the resulting plate amplification could yield a total of $10^{10}$ PFU, a 10,000-fold amplification.

1. The initial steps in plate-amplifying a recombinant λ clone bank are identical to those used for the screening of a clone bank (see Section VI, Steps 1 to 5). Four to eight KPs should be used for amplification, and not more than 250,000 PFU should be plated on each plate.
2. After plate lysis, remove the plates from the incubator, and place them on a bench or in a biological hood. Obtain a large spatula or a flat edge instrument about 8 to 10 cm wide (a putty knife will do). Place the instrument in ethanol and flame-sterilize.
3. Using the flat edge instrument, scrape the top agar off the KPs and place it into a 250 or 500-ml large-mouth, sterile, centrifuge bottle. For each KP scraped, add 25 ml of phage storage buffer (Section VI, Step 16). Fit centrifuge bottles into a shaker, and shake at about 100 rpm for 2 to 3 h at room temperature. (Temperatures as high as 37°C can also be used.)
4. Pellet agar by centrifugation at 4000 rpm for 10 min at 4°C. Decant the supernatant solution into a sterile glass or polypropylene bottle. The volume of liquid collected should be about 100 to 200 ml. Add 0.5 ml of chloroform to ensure that microbial growth (from the host *E. coli*, other bacterial, or fungal contaminants) is inhibited.
5. Titer the amplified λ clone bank using the appropriate dilutions. Dilutions in the range of $10^{-7}$ to $10^{-8}$ should be necessary.
6. Label the bottle appropriately with the name of the target species, λ vector used, titer results, and date of amplification. Store at 4°C.
   *Note:* λ clone banks stored in this manner are stable for several years. Generally one can expect a factor of 10 will be lost from the original titer for each year of storage. Thus a clone bank with a titer of $10^{10}$ PFU/ml could be stored for 4 years and still have a titer of $10^6$ PFU/ml. Assuming that the titer loss is random, only 1 ml of this 4 year old clone bank is needed to represent one genome equivalent of the soybean genome. Once the titer of a stored clone bank degenerates to the level of about $10^6$ PFU/ml, it should be reamplified.
7. Alternatively, for longer term storage, a λ clone bank can be stored in sterile dimethylsulfoxide (DMSO) at a final DMSO concentration of 7% (v/v). Add the λ clone bank-DMSO mixture to cold-storage vials or bottles, and quick freeze using liquid nitrogen. Vials can be stored in liquid nitrogen tanks or in a –70°C freezer. Phage titers for samples stored at such low temperatures are very stable.[52]

## REFERENCES


1. **Leder, P., Tilghman, S. M., Tiemeier, D. C., Polisby, F. I., Seidman, J. G., Edgell, M. L., Enquist, L. W., Leder, A., and Norman, B.,** The cloning of a mouse globin and surrounding gene sequences in bacteriophage λ,*Cold Spring Harbor Symp. Quant. Biol.,* 42, 915, 1977.
2. **Tonegawa, S., Brock, C., Hozumi, N., and Schuller, R.,** Cloning of an immunoglobin variable region gene from mouse embryo, *Proc. Natl. Acad. Sci. U.S.A.,* 74, 3518, 1977.
3. **Blattner, F. R., Williams, B. G., Blechl, A. E., Denniston-Thompson, K., Faber, H. E., Furlong, L.-A. Grunwald, D. J., Kiefer, D. O., Moore, D. D., Schumm, J. W., Sheldon, E. L., and Smithies, O.,** Charon phages: safer derivatives of bacteriophage λ for DNA cloning, *Science,* 196, 161, 1977.
4. **Blattner, F. R., Blechl, A. E., Denniston-Thompson, K., Faber, H. E., Richards, J. E., Slightom, J. L., Tucker, P. W., and Smithies, O.,** Cloning human fetal γ globin and mouse α-type globin DNA: preparation and screening of shotgun collections, *Science,* 202, 1279, 1978.
5. **Benton, W. D. and Davis, R. W.,** Screening λgt recombinant clones by hybridization to single plaques in situ, *Science,* 196, 180, 1977.
6. **Huynh, T. V., Young, R. A., and Davis, R.,** Constructing and screening cDNA libraries in λgt10 and λgt11, in *DNA Cloning, Vol.I, A Practical Approach,* Glover, D. M., Ed., IPL Press, Washington, D.C., 1985, 49.
7. **Saiki, R. K., Gelfand, D. H., Stoffel, S., Scharf, S., Higuchi, R., Horn, G. T., Mullis, K. B., and Erlich, H. A.,** Primer-directed enzymatic amplification of DNA with a thermostable DNA polymerase, *Science,* 239, 487, 1988.
8. **Erlich, H. A.,** *PCR Technology: Principles and Applications for DNA Amplification,* Stockton Press, New York, 1989.
9. **Innis, M. A., Gelfand, D. H., Sninsky, J. J., and White, T. J.,** *PCR Protocols: a Guide to Methods and Applications,* Academic Press, New York, 1989.
10. **Kocher, T. D., Thomas, W. K., Meyer, A., Edwards, S. V., Paabo, S., Villablanca, F. X., and Wilson, A. C.,** Dynamics of mitochondrial DNA evolution in animals: amplification and sequencing with conserved primers, *Proc. Natl. Acad. Sci. U.S.A.,* 86, 6196, 1989.
11. **Parimoo, S., Patanjali, S. R., Shukla, H., Chaplin, D. D., and Weissman, S. M.,** cDNA selection: efficient PCR approach for the selection of cDNAs encoded in large chromosomal DNA fragments, *Proc. Natl. Acad. Sci. U.S.A.,* 88, 9623, 1991.
12. **Collins, J.,** *Escherichia coli* plasmids packagable *in vitro* in λ bacteriophage particles, *Methods in Enzymology,* 68th ed., Wu, R., Ed., Academic Press, New York, 1979, 309.
13. **Little, P. F. R. and Cross, S. H.,** A cosmid vector that facilitates restriction enzyme mapping, *Proc. Natl. Acad. Sci. U.S.A.,* 82, 3159, 1985.
14. **Murray, N. E.,** Phage λ and molecular cloning. in *Lambda II,* Hendrix, R. W., Roberts, J. W., Stahl, F. W., and Weisberg, R. A., Eds., Cold Spring Harbor Laboratory Press, Cold Spring Harbor, NY, 1983, 395.
15. **Loenen, W. A. M. and Blattner, F. R.,** λ Charon vectors (Ch32, 33, 34, and 35) adapted for DNA cloning in recombination-deficient hosts, *Gene,* 26, 171, 1983.
16. **Dunn, I. S. and Blattner, F. R.,** Charons 36 to 40: multienzyme, high capacity, recombination deficient replacement vectors with polylinkers and polystuffers, *Nucl. Acids Res.,* 15, 2677, 1987.
17. **Frischauf, A.-M., Lehbach, H., Poustra, A., and Murray, N.,** λ replacement vectors carrying polylinker sequences, *J. Mol. Biol.,* 170, 827, 1983.
18. **Slightom, J. L., Blechl, A. E., and Smithies, O.,** Human fetal $^{G}\gamma$ and $^{A}\gamma$ globin genes: complete nucleotide sequences suggest that DNA can be exchanged between these duplicated genes, *Cell,* 21, 627, 1980.
19. **Maniatis, T., Fritsch, E. F., and Sambrook, J.,** *Molecular Cloning. A Laboratory Manual,* Cold Spring Harbor Laboratory Press, Cold Spring Harbor, NY, 1982.
20. **Sambrook, J., Fritsch, E. F., and Maniatis, T.,** *Molecular Cloning. A Laboratory Manual,* 2nd ed., Cold Spring Harbor Laboratory Press, Cold Spring Harbor, NY, 1989.
21. **Chang, L.-Y. E. and Slightom, J. L.,** Isolation and nucleotide sequence analysis of β-type globin pseudogenes from human, gorilla, and chimpanzee, *J. Mol. Biol.,* 180, 767, 1984.
22. **Slightom, J. L., Chang, L. Y. E., Koop, B., and Goodman, M.,** Chimpanzee fetal $^{G}\gamma$- and $^{A}\gamma$-globin nucleotide sequences provides further evidence on gene conversion in Hominine evolution, *Mol. Biol. Evol.,* 2, 370, 1985.
23. **Slightom, J. L., Theisen, T., Koop, B. F., and Goodman, M.,** Orangutan fetal globin genes: nucleotide sequence reveals multiple gene conversion during hominid phylogeny, *J. Biol. Chem.,* 262, 7472, 1987.
24. **Slightom, J. L. and Drong, R. F.,** Procedures for constructing genomic clone banks, in *Plant Molecular Biology Manual,* Gelvin, S. G. and Schilperoort, R. A., Eds., Kluwer Academic Publishers, Dordrecht, Netherlands, 1988, chap. A8.

25. **Lawn, R. M., Fritsch, E. F., Parker, R. C., Blake, G., and Maniatis, T.,** The isolation and characterization of linked δ- and β-globin genes from a cloned library of human DNA, *Cell,* 15, 1157, 1978.
26. **Clark, L. and Carbon, J.,** A colony bank containing synthetic ColE1 hybrid plasmids representative of the entire *E. coli* genome, *Cell,* 9, 91, 1976.
27. **Sun, S. M., Slightom, J. L., and Hall, T. C.,** Intervening sequences in a plant gene — comparison of the partial sequences of cDNA and genomic DNA of French bean phaseolin, *Nature,* 289, 37, 1981.
28. **Tierney, M. L., Bray, E. A., Allen, R. D., Ma, Y., Drong, R. F., Slightom, J. L., and Beachy, R. N.,** Isolation and characterization of a genomic clone encoding β-subunit of β-conglycinin, *Planta,* 172, 356, 1987.
29. **Hong, J. C., Nagao, R. T., and Key, J. L.,** Characterization of a proline-rich cell wall protein gene family of soybean, *J. Biol. Chem.,* 265, 2470, 1990.
30. **Kriz, A. L., Boston, R. S., and Larkins, B. A.,** Structural and transcriptional analysis of DNA sequences flanking genes that encode 19 kilodalton zeins, *Mol. Gen. Genet.,* 207, 90, 1987.
31. **Thompson, G. A., Siemieniak, D. R., Sieu, L. C., Slightom, J. L., and Larkins, B. A.,** Characterization of linked maize 22 KD α-zein genes, *Plant Mol. Biol.,* 18, 827, 1992.
32. **Murray, M. G., Kennard, W. C., Drong, R. F., and Slightom, J. L.,** Use of recombination deficient phage vectors to construct wheat genomic libraries, *Gene,* 30, 237, 1984.
33. **Slightom, J. L., Jouanin, L., Leach, F., Drong, R. F., and Tepfer, D.,** Isolation and identification of TL-DNA/plant junctions in *Convolvulus arvensis* transformed by *Agrobacterium rhizogenes* strain A4, *EMBO J.,* 4, 3069, 1985.
34. **Chlan, C. A., Borroto, K., Kamalay, J. A., and Dure, L., III,** Developmental biochemistry of cottonseed embryogenesis and germination. XIX. Sequences and genomic organization of the α globulin (vicilin) genes of cottonseed, *Plant Mol. Biol.,* 9, 533, 1987.
35. **Rogers, S. O. and Bendich, A.,** Extraction of DNA from plant tissues, in *Plant Molecular Biology Manual,* Gelvin, S. B. and Schilperoort, R. A., Eds. Kluwer Academic Publishers, Dordrecht, Nertherlands, chap. A6.
36. **Strauss, N. A.,** Comparative DNA renaturation kinetics in amphibians, *Proc. Natl. Acad. Sci. U.S.A.,* 68, 799, 1971.
37. **Kamalay, J. C. and Goldberg, R. B.,** Regulation of structural gene expression in tobacco, *Cell,* 19, 935, 1980.
38. **Patanjali, S. R., Parimoo, S., and Weissman, S. M.,** Construction of a uniform-abundance (normalized) cDNA library, *Proc. Natl. Acad. Sci. U.S.A.,* 88, 1943, 1991.
39. **Young, R. A. and Davis, R. W.,** Efficient isolation of genes by using antibody probes, *Proc. Natl. Acad. Sci. U.S.A.,* 80, 1194, 1983.
40. **Young, R. A. and Davis, R. W.,** Yeast RNA polymerase II genes: isolation with antibody probes, *Science,* 222, 778, 1983.
41. **Han, J. H. and Rutter, W. J.,** λgt22, an improved λ vector for the directional cloning of full-length cDNA, *Nucl. Acids Res.,* 15, 6304, 1987.
42. **Murray, M. G. and Thompson, W. F.,** Rapid isolation of high molecular weight plant DNA, *Nucl. Acids Res.,* 8, 4321, 1980.
43. **Saghai-Maroof, M. A., Soliman, K. M., Jorgensen, R. A., and Allard, R. W.,** Ribosomal DNA spacer-length polymorphisms in barley: Mendelian inheritance, chromosomal location, and population dynamics, *Proc. Natl. Acad. Sci. U.S.A.,* 81, 8014, 1984.
44. **Chee, P. P., Fober, K. A., and Slightom, J. L.,** Transformation of soybean (*Glycine max*) by infecting seeds with *Agrobacterium tumefaciens, Plant Physiol.,* 91, 1212, 1989.
45. **Chee, P. P.,** Transformation of *Cucumis sativus* tissues by *Agrobacterium tumefaciens* and the regeneration of transformed plants, *Plant Cell Rep.,* 9, 245, 1990.
46. **Chee, P. P. and Slightom, J. L.,** Transfer of cucumber mosaic virus coat protein gene into the genome of *Cucumis sativus* and analysis of its expression, *J. Am. Soc. Hortic. Sci.,* 116, 1098, 1991.
47. **Chee, P. P., Jones, J. M., and Slightom, J. L.,** Expression of bean storage protein minigene in tobacco seeds: introns are not required for seed specific expression, *J. Physiol. Plant,* 137, 402, 1991
48. **McClean, P., Chee, P. P., Simental, J., Held, B., Drong, R. F., and Slightom, J. L.,** Susceptibility of dry bean (*Phaseolus vulgaris* L.) to *Agrobacterium* infections: transformation of cotyledonary and hypocotyl tissues, *Plant Organ and Tissue Cult.,* 24, 131, 1991.
49. **Okayama, H. and Berg, P.,** High-efficiency cloning of full-length cDNA, *Mol. Cell Biol.,* 2, 161, 1982.
50. **Polites, H. G. and Marotti, K. R.,** A step-wise protocol for cDNA synthesis, *BioTechniques,* 4, 514, 1986.
51. **Hohn, B.,** *In vitro* packaging of λ and cosmid DNA, in *Methods in Enzymology,* 68th ed., Wu, R., Ed., Academic Press, New York, 1979, 299.
52. **Nierman, W. C. and Trypus, C.,** Preservation & stability of bacteriophage λ libraries by freezing in liquid nitrogen, *BioTechniques,* 5, 724, 1987.

# 9

# Transient Analysis of Gene Expression in Plant Cells

*A. N. Nunberg and T. L. Thomas*

## I. INTRODUCTION

Transient expression of genes in plant cells is a powerful tool in plant molecular biology. Though many investigators use this technique to study plant promoter and upstream regulatory elements, it can also be used to study RNA translation[1] and polyadenylation,[2] excision of transposable elements,[3] and many other areas of plant biology. The major advantage of transient assays is time. Studying stably transformed plants may take several weeks to several months, while studying transient assays takes only a matter of days. Expedient assay time is balanced by the limitations of using protoplasts. Protoplasts do not represent normal plant cells, and it is strongly suggested that, if possible, constructs assayed in a transient system should also be assayed in stably transformed plants. This chapter focuses on using transient assays to study promoter and upstream regulatory elements fused to a β-glucuronidase reporter gene, but the principles described here are equally applicable to other problems.

## II. PROTOPLAST PREPARATION

Ideally, protoplasts to be used for a transient assay should be from the same source as the gene of interest. This is not always possible, so alternate sources must be used. When deciding on an alternate source, keep in mind the likelihood of the gene being expressed in the chosen system. Many investigators use plants that are regularly used in transgenic analysis, such as tobacco or petunia. When a source is considered, the type of tissue and physiological state of the tissue or protoplasts should be evaluated. We have investigated the regulation of a carrot gene Dc3,[4] which is induced by abcisic acid (ABA). Initial studies using tobacco mesophyl protoplasts were complicated by high levels of uninduced expression, thought to be a result of ABA biosynthesis during protoplast isolation. The use of hypocotyl protoplasts, which are not a source of ABA, made the investigation much more powerful.

Since tobacco mesophyl protoplasts are the most popular source for transient assays, we have presented a protocol here that is based on the use of these cells. Protocols for *Arabidopsis*,[5] barley,[6] maize,[7] petunia,[8] rice,[9] and soybean[10] are also available. Constabel[11] or Power and Chapman[12] also provide a number of useful protocols.

### A. TOBACCO LEAF PROTOPLAST ISOLATION[13]

1. Grow *Nicotiana tabacum* cv. Xanthi-nc under controlled conditions (25°C). Take fully expanded but young leaves from 50 to 60 day old plants. For some applications, leaves that are not fully expanded may work better.

0-8493-5164-2/93/$0.00+$.50
© 1993 by CRC Press, Inc.

2. Abrade the undersurface of the leaves (10 to 15 cm) with carborundum, and float overnight in a filter-sterilized enzyme solution:
   1.5% cellulase RS (Yakult Honsha, Tokyo, Japan)
   0.1% driselase (Sigma Chemical Co., St. Louis, MO)
   0.5% Macerozyme R-10 (Yakult Honsha, Tokyo, Japan)
   $K_3$ macronutrients[14]
   0.55 *M* mannitol
   pH 5.9
3. Release protoplasts by gently shaking leaf and isolate by floating on 0.55 *M* sucrose (45 × g for 7 min). Collect the protoplasts at the interface and wash two times by sedimentation (100 × g for 4 min) with 0.55 *M* mannitol.

### B. PROTOPLAST QUANTITATION AND VIABILITY

The yield of protoplasts is routinely determined with a hemocytometer. Hemocytometers are essentially microscope slides with a grid etched on the surface. When a cover slip is used, the grid has a known volume. The dimensions of the grid are supplied with the hemocytometer, but most have the dimension of 1 mm × 1 mm × 0.1 mm, for a volume of 0.1 $mm^3$ or $10^{-4}cm^3$. Since 1 ml is equal to 1 $cm^3$, the protoplast concentration can be calculated in the following manner:

$$\text{protoplasts per ml} = \text{average count per triple edged square} \times 10^4$$

Fluorescein diacetate (FDA) is widely used to determine protoplast viability. In living protoplasts, FDA diffuses across the plasma membrane and is cleaved by endogenous enzymes, while the product is retained. Viable protoplasts are visualized following excitation with UV light.

1. Dissolve FDA to 1 mg/ml in acetone and store at –20°C.
2. Add one drop to one of the chambers of the hemocytometer. Quickly wipe the drop with a Kimwipe to spread the FDA over the chamber.
3. Apply the cover slip and add protoplasts to both chambers.
4. Count protoplasts in both chambers and count the number of viable protoplasts by flourescence microscopy.

$$\%\ \text{viabilty} = \text{fluorescing protoplasts/total protoplasts} \times 100$$

## III. DNA TRANSFECTION

### A. PEG METHOD

The PEG method is a relatively simple way of introducing DNA into plant protoplasts. It has the advantage of needing no special equipment, and the relative ease of PEG-mediated transfection makes processing a large number of samples possible. Standard PEG protocols are based on a method used for permanent transformation of tobacco protoplasts.[15] We have found that two PEG solutions work equally well, PEG CMS [16] and PEG C (Rao, A. L. N., personal communication).

Supercoiled DNA is sufficient for transient assays. The DNA need not be linearized, nor is it necessary to add carrier DNA. DNA concentrations can vary from 2 to 50 μg/ml, depending on the construction being used. DNA stored in aliquots at –80°C give the most consistent results. DNA stored at 4°C or in a frost free freezer at –20°C will, over time, yield diminishing expression. The DNA should also be purified at least once by CsCl centrifugation. DNA is sufficiently pure when isolated using PEG precipitation followed by centrifugation through CsCl (Table 1).

**TABLE 1**
**Effects of Storage Conditions and DNA Purity on Transient Expression**

A

| | 4°C | | | −80°C | | |
|---|---|---|---|---|---|---|
| | −ABA | +ABA | Increase | −ABA | +ABA | Increase |
| SF−16 | 0.54 | 2.69 | 5-fold | 0.73 | 7.85 | 11-fold |

A. The plasmid SF-16, containing the ABA inducible promoter of Dc3[4] fused to GUS, was prepared by the PEG method followed by CsCl centrifugation. The column marked 4°C represents expression and induction of the plasmid that was stored at 4°C. The column marked −80°C represents expression and induction of the same plasmid stored at −80°C.

B

| | −CsCl | | | +CsCl | | |
|---|---|---|---|---|---|---|
| | −ABA | +ABA | Increase | −ABA | +ABA | Increase |
| D5 | 0.62 | 2.75 | 4-fold | 0.63 | 7.59 | 12-fold |
| D6 | 0.50 | 2.52 | 5-fold | 0.61 | 2.47 | 4-fold |

B. Two 5′ deletions of SF-16, D5, and D6 were initially prepared by the PEG method. The first column marks expression and induction without CsCl centrifugation of the plasmid, while the second column marks expression after CsCl centrifugation. D5 gives wild-type induction with abcisic acid (ABA) while D6 does not.

*Note:* Activity is expressed as pmole of 4-MU/min/μg protein.

## 1. Materials

**PEG CMS**
40% PEG 3500
0.4 *M* mannitol
0.1 *M* $Ca(NO_3) \cdot 4H_2O$
pH 7 to 9
filter sterilize, store at −20°C

**PEG C**
40% PEG
0.3 *M* $CaCl_2$
0.5%MES
adjust pH to 5.8
filter sterilize, store −20°C

**$W_5$[15]**
154 m*M* NaCl
125 m*M* $CaCl_2 \cdot 2H_2O$
5 m*M* KCl
5 m*M* glucose
pH 5.6 to 6.0

**M/$SP_1$9M[12]**
MS salts
3% sucrose
9% mannitol
2.0 mg/l NAA
0.5 mg/l BAP
pH 5.7

**MaMg[15]**
0.4 to 0.5 *M* mannitol
15 m*M* $MgCl_2$
0.1 % MES
pH 5.6

## 2. PEG Mediated Transfection Protocol

1. Wash protoplasts once in $W_5$, pellet (100 × g, 4 min), and resuspend to a density of $2 \times 10^6$ protoplasts/ml in MaMg.
2. Take at least $2.5 \times 10^5$ protoplasts (250 μl), and place in a 1.5 ml Eppendorf tube.
3. Add DNA in a volume less than 10 μl.
4. Add an equal volume of PEG solution (250 μl), mix gently, and let stand at room temperature for 20 min, mixing every 5 to 10 min.
5. Add to 5 volumes of $W_5$, pellet, and culture.

It is suggested that no less than 250,000 protoplasts per transfection be used since a smaller amount is difficult to work with and may not give satisfactory expression. Culturing media will depend on the source of protoplasts. If protoplasts are isolated from suspension cell cultures, the original culture medium adjusted with osmoticum should be used. If the protoplasts used are from whole plants or adjusted culturing medium cannot be used, an MS based culturing medium can be used. We have used $M/SP_19M$ for sunflower hypocotyl and tobacco mesophyll protoplasts.

## B. ELECTROPORATION

This method uses a voltage potential across a cell to introduce the DNA. Most electroporators deliver a pulse so that the wave produced decays exponentially over time.

There are two main variables in electroporation, electric field (E) and pulse length (expressed here as the RC constant $\tau$). The electric field is the voltage gradient between electrodes, expressed in volts per centimeter. The time constant $\tau$ is equal to RC ($\tau$=RC), where R is the resistance in ohms and C is the capacitance in farads. Resistance refers to the resistance of the sample, while capacitance refers to the capacitance of the discharging circuit. The resistance of the sample is due to the components of the electroporating media, specifically the ion content. Therefore, where possible, use the same electroporating media as described in reported protocols. Most media used contain 150 m*M* NaCl, which appears to be optimal for many plant protoplasts.

When adapting protocols, first determine if the electroporator being used is the same as the protocol being referred to. If it is, simply use the same conditions and settings. Most often though, the electroporator will not be the same, and it will be necessary to determine the appropriate time constant. According to Bio-Rad,[17] a 0.8 ml sample of PBS in a cuvette with a gap width of 0.4 cm has a resistance of approximately 20 ohms. The resistance is affected by the distance between electrodes and the cross-sectional area. For example, a cuvette with a 1 cm gap containing 0.4 ml of PBS has a cross-sectional area of 0.4 $cm^3$/1 cm or 0.4 $cm^2$. The 0.8 ml reference sample has an area of 2 $cm^2$ (0.8 $cm^3$/0.4 cm). Resistance is inversely proportional to area, so the 0.4 ml sample has five times the resistance of the 0.8 ml sample. Resistance is also proportional to the distance between the electrodes, so the 0.4 ml sample has 2.5 times the resistance of the 0.8 ml sample (1 cm/0.4 cm). The total increase between our reference cuvette and example is 12.5-fold, giving a value of 250 ohms. If the protocol is reported using a 50 µf capacitor, the time constant would be

$$(250\ \Omega)(5 \times 10^{-5}\ \text{F}) = 12.5\ \text{msec}$$

For a comparable pulse time, adjust the capacitance of the electroporator being used. In some reports an electrophoresis power supply is used. The charge is either directly applied to the sample or discharged from a capacitor. The latter is easier to duplicate, but in either case, the set voltage is not the voltage delivered to the sample. The problem is further complicated by the different capacitors found in power supplies. Also, different electroporation devices have different internal resistors. Therefore, adjustments made using the above calculations are best used in conjunction with measurements of actual outputs across the cuvette with an oscilloscope.

Supercoiled DNA is generally used for transient electroporation assays. The superhelicity of the DNA does not appear to be as crucial for electroporation as it is for PEG transfection. The range of DNA concentration is the same as with the PEG method, and effective concentrations should be determined experimentally. The DNA should be purified at least once by CsCl centrifugation.

The basic electroporation medium is Hepes buffered saline solution:

30 m*M* Hepes
150 m*M* NaCl
5 m*M* $CaCl_2$

427 m*M* mannitol
pH 7.2

### 1. General Electroporation Protocol

The following protocol is based on that of Fromm et al.[18] and can be used for almost any dicot or monocot protoplast.

1. Wash the protoplasts once with electroporation medium.
2. Resuspend to $1 \times 10^6$ protoplasts/ml in electroporation buffer.
3. Add 1 ml of protoplasts plus DNA to the electroporation cuvette, incubate on ice for 10 min.
4. Electroporate between 500 to 875 V/cm with a time constant of 20 msec.
5. Incubate again on ice 10 min.

Fromm et al.[18] and Hauptman et al.[19] reported that moderate electric fields (500 to 1000 V/cm) with medium time constants (10 to 20 msec) gave the best results. They also found that expression from monocot protoplasts is significantly lower than from dicots. Electroporation protocols have also been reported for carrot,[18,19,26-28] petunia,[19] maize,[18,28,30,31] rice,[30] soybean,[19] tobacco,[18,30,32-34] and wheat.[30,32]

Optimum conditions for electroporation should be determined experimentally. The conditions reported should be tried first; then the electric field should be altered upwards and downwards. In electroporation, optimal DNA delivery is balanced with cell viability. Conditions that result in high efficiency DNA delivery usually lead to substantial cell death. Protoplasts differ in their susceptibility to lysis in an electric field. In general, larger and/or more vacuolated protoplasts are more susceptible than smaller and/or cytoplasmically dense protoplasts. When optimizing the electric field, use increments of 100 V/cm to avoid lysing sensitive protoplasts. Although most electroporation buffers use 150 m*M* NaCl, many differ in pH, cation, and buffer concentration. Most use 5 m*M* $Ca^{+2}$, but pH can range from 5 to 7. Taylor and Larkin[20] found that pH 9 was optimal for electroporation. However, at such high pH, protoplasts fuse, and this may complicate the assay.

Clearly, the PEG method is the easier method for transient assays. Then why electroporate at all? Leon et al.[21] have reported, and we have also observed, that electroporation gives results that are much less variable than the PEG method. The PEG method, however, has the advantage of requiring no special equipment.

## IV. REPORTER GENE ASSAY

Once the DNA is transfected, the next step is to determine the activity of the reporter gene. There are three reporter genes that are commonly used. These are β-glucuronidase (GUS), chloramphenicol acetyl transferase (CAT), and luciferase. In plants, β-galactosidase has not been proven useful. Here, we focus on the GUS assay because of its ease and sensitivity as compared to the CAT assay, and because it does not require the specialized equipment needed in the luciferase assay. For details of the CAT and luciferase assays, see Gorman et al.[22] and De Wet et al.,[23] respectively.

### A. GUS ASSAYS

GUS expression from a strongly expressed constitutive promoter like the cauliflower mosaic virus 35S promoter can be detected as early as 8 h after DNA transfer and for as long as 3 days. Culturing time for optimal expression of such a construction is generally from 24 to 36 h after transfection. Other constructions that we have investigated also give maximal expression between 24 to 36 h after DNA transfection.

**GUS Extraction Buffer**[24]
50 m*M* $NaPO_4$, pH 7.0
10 m*M* β-mercaptoethanol
10 m*M* $Na_2EDTA$
0.1% Sodium Lauryl Sarcosine
0.1% Triton-x-100

**Assay Buffer**
Extraction buffer containing 1m*M* 4-methyl umbelliferyl glucuronide (MUG) (Sigma)

**Stop Buffer**
0.2 *M* $Na_2CO_3$

The following protocol is essentially the same as Bogue et al.[25]

1. Pellet protoplasts from each treatment in a 15 ml falcon tube.
2. Resuspend with a minimum of 300 μl of extraction buffer.
3. Lyse protoplasts by vortexing vigorously.
4. Centrifuge lysate in 1.5 ml Eppendorf tube for 5 min in a microfuge.
5. Dispense 100 μl aliquots into two tubes marked 0 and 60.
6. To the 0 time point add 900 μl of stop buffer.
7. Place both tubes on ice, and add 100 μl of assay buffer.
8. Mix well by vortexing briefly and place the tube marked 60 in a 37°C water bath for 1 h.
9. After 60 min. add 900 μl stop buffer to the tube labeled 60.
10. Determine methyl umbelliferone (MU) concentrations with a fluorometer, excitation at 365 nm and emission at 425 nm.

We employ the one stop method as opposed to taking time points because the activity of GUS is linear for long periods of time.[24] This holds true unless the activity is so high that the substrate is limiting. However, we have found that the one stop method is sufficient even for constructions that are driven by the 35S promoter. Most reports express GUS activity in terms of product per unit time per unit protein, as determined by the Bradford or Lowry method.

This brief overview should allow investigators to develop transient expression assays for their own needs. This technique can be quite valuable, but as stated previously, the data derived must be evaluated cautiously.

## ACKNOWLEDGMENTS

The work from this laboratory was supported by grants from the Texas Advanced Research Program (#010366-038) and Rhône-Poulenc Agrochimie (Lyon, France).

## REFERENCES

1. **Skuzeski, J. M., Nichols, L. M., and Gesteland, R. F.,** Analysis of leaky viral translation termination codons *in vivo* by transient expression of improved beta-glucuronidase vectors, *Plant Mol. Biol.*, 15, 65, 1990.
2. **Guerineau, F., Brooks, L., and Mullineaux, P.,** Effect of deletions in the cauliflower mosaic virus polyadenylation sequence on the choice of the polyadenylation sites in tobacco, *Mol. Gen. Genet.*, 226, 141, 1991.
3. **Houba-Herin, N., Becker, D., Post, A., Larondelle, Y., and Starlinger, P.,** Excision of a Ds-like maize transposable element Ac-delta in a transient assay in petunia is enhanced by a truncated coding region of the transposable element Ac, *Mol. Gen. Genet.*, 224, 17, 1990.

4. **Seffens, W. S., Almoguera, C., Wilde, D. H., Vonder Haar, R. A., and Thomas, T. L.,** Molecular analysis of a phylogenetically conserved carrot gene: developmental and environmental regulation, *Dev. Genet.*, 11, 65, 1990.
5. **Browse, J. and Somerville, C. R.,** Changes in lipid composition during protoplast isolation, *Plant Sci.*, 56, 15, 1988.
6. **Loesch-Fries, L. S. and Hall, T. C.,** Synthesis, accumulation and encapsidation of individual Brome Mosaic Virus RNA components in barley protoplasts, *J. Gen. Vir.*, 47, 323, 1980.
7. **Lyznik, L. A., McGee, D., Tung, P.-Y., Bennezen, J. L., and Hodges, T. K.,** Homologous recombination between plasmid DNA molecules in maize protoplasts, *Mol. Gen. Genet.*, 230, 209, 1991.
8. **Mizrahi, Y., Applewhite, P. B., and Galston, A. W.,** Polyamine binding to proteins in oat and Petunia protoplasts, *Plant Physiol.*, 91, 738, 1989.
9. **Li, Z. and Murai, N.,** Efficient plant regeneration from rice protoplasts in general medium, *Plant Cell Rep.*, 9, 216, 1990.
10. **Mersey, B. G., Griffing, L. R., Rennie, P. J., and Fowke, L. C.,** The isolation of coated vesicles from protoplasts of soybean, *Planta*, 163, 317, 1985.
11. **Constabel, F.,** Isolation and culture of plant protoplasts, in *Plant Tissue Culture Methods*, Wetter C. F., Ed., N.R.C., Ottawa, Ontario, 1982, chap. 6.
12. **Power, J. B. and Chapman, J. V.,** Isolation, culture and genetic manipulation of plant protoplasts, in *Plant Cell Culture: A Practical Approach*, Dixon, R. A., Ed., IRL Press, Washington, D.C., 1985, chap. 3.
13. **Carrington, J. C. and Freed, D. D.,** Cap-independent enhancement of translation by a plant potyvirus 5′ nontranslated region, *J.Vir.*, 64, 1590, 1990.
14. **Kao, K. N., Constabel, F., Michayluk, R., and Gamborg, O. L.,** Plant protoplast fusion and growth of intergenic hybrid cells, *Planta*, 120, 215, 1974.
15. **Negrutiu, I., Shillito, R., Potrykus, I., Biasini, G., and Sala, F.,** Hybrid genes in the analysis of transformation conditions, *Plant Mol. Biol.*, 8, 363, 1987.
16. **Negrutiu, I., De Brouwer, D., Watts, J., Sidorov, V., Dirks, R., and Jacobs, M.,** Fusion of plant protoplasts: a study using auxotrophic mutants of *N. plumbaginifolia*, *Theor. Appl. Genet.*, 72, 279, 1986.
17. **Bio-Rad,** Gene Pulser Transfection Apparatus operating instructions and applications guide, Version 2-89.
18. **Fromm, M., Taylor, L. P., and Walbot, V.,** Expression of genes transferred into monocot and dicot plant cells by electroporation, *Proc. Natl. Acad. Sci. U.S.A.*, 83, 5824, 1985.
19. **Hauptman, R. M., Ozias-Akins, P., Vasil, V., Tabaeizadeh, Z., Rogers, S. G., Horsch, R. B., Vasil, I. K., and Fraley, R. T.,** Transient expression of electroporated DNA in monocotyledonous and dicotyledonous species, *Plant Cell Rep.*, 6, 265, 1987.
20. **Taylor, B. H. and Larkin, P. J.,** Analysis of electroporation efficiency in plant protoplasts, *Aust. J. Bot.*, 1, 52, 1988.
21. **Leon, P., Planckaert, F., and Walbot, V.,** Transient gene expression in protoplasts of *Phaseolus vulgaris* isolated from cell suspension culture, *Plant Physiol.*, 95, 968, 1991.
22. **Gorman, C. M., Moffatt, L. F., and Howard, B. H.,** Recombinant genomes which express chloramphenicol acetyltransferase in mammalian cells, *Mol. Cell Biol.*, 2, 1044, 1982.
23. **De Wet, J. R., Wood, K. V., De Luca, M., Helinski, D. R., and Subramani, S.,** Firefly luciferase gene: structure and expression in mammalian cells, *Mol. Cell Biol.*, 7, 725, 1987.
24. **Jefferson, R. A.**, Assaying chimeric genes in plants: the GUS gene fusion system, *Plant Mol. Biol. Rep.*, 5, 387, 1987.
25. **Bogue, M. A., Vonder Haar, R. A., Nuccio, M. L., Griffing, L. R., and Thomas, T. L.,** Developmentally regulated expression of a sunflower 11S seed protein gene in transgenic tobacco, *Mol. Gen. Genet.*, 222, 49, 1990.
26. **Bates, G. W., Carle, S. A., and Piastuch, W. C.,** Linear DNA introduced into carrot protoplasts by electroporation undergoes ligation and recircularization, *Plant Mol. Biol.*, 14, 899, 1990.
27. **Bower, R. and Birch, R. G.,** Competence for gene transfer by electroporation in a sub-population of protoplasts from uniform carrot cell suspension cultures, *Plant Cell Rep.*, 9, 386, 1990.
28. **Callis, J., Fromm, M., and Walbot, V.,** Expression of mRNA electroporated into plant and animal cells, *Nucl. Acids Res.*, 15, 5823, 1987.
29. **Ecker, J. R. and Davis, R. W.**, Inhibition of gene expression in plant cells by expression of antisense RNA, *Proc. Natl. Acad. Sci. U.S.A.*, 83, 5372, 1986.
30. **Last, D. I., Brettell, R. I. S., Chamberlain, P. A., Chaudhury, A. M., Larkin, P. J., Marsh, E. L., Peacock, W. J., and Dennis, E. S.,** pEMU: an improved promoter for gene expression in cereal cells, *Theor. Appl. Gen.*, 81, 581, 1991.
31. **Planckaert, F. and Walbot, V.,** Transient gene expression after electroporation of protoplasts derived from embryogenic maize callus, *Plant Cell Rep.*, 8, 144, 1989.
32. **Aryan, A. P., An, G., and Okita, T. W.,** Structural and functional analysis of promoter from gliadin, an endosperm-specific storage protein gene of *Triticum aestivum* L., *Mol. Gen. Genet.*, 225, 65, 1991.

33. **Gatz, C. and Quail, P.H.,** Tn10-encoded tet repressor can regulate an operator containing plant promoter, *Proc. Natl. Acad. Sci. U.S.A.*, 85, 1394, 1988.
34. **Luciano, C. S., Rhoads, R. E., and Shaw, J. G.,** Synthesis of potyviral RNA and proteins in tobacco mesophyll protoplasts inoculated by electroporation, *Plant Sci.*, 51, 295, 1987.

# 10

# Techniques for Isolating and Characterizing Plant Transcription Promoters, Enhancers, and Terminators

*Gynheung An and Younghee Kim*

## I. INTRODUCTION

Expression of plant genes is primarily controlled at the transcriptional level. Transcription can be divided into three stages: initiation, elongation, and termination. Initiation begins with the binding of RNA polymerase to the promoter region of each gene. Plant genes fall into three classes defined by their types of promoter. Ribosomal RNA is transcribed by RNA polymerase I, mRNA by RNA polymerase II, and tRNA and other small RNAs by RNA polymerase III. In addition to the RNA polymerases, accessory transcription factors are needed for transcription initiation. Transcription elongation occurs by movement of RNA polymerase along the DNA, extending the RNA chain until RNA synthesis is terminated at the transcription terminator region where the transcription complex dissociates.

Many of the studies of plant promoters have focused on the genes transcribed by RNA polymerase II. Deletion analysis and *in vitro* binding to a nuclear factor have been used to locate the promoter elements, and oligonucleotide-mediated mutagenesis has been used for further characterization of the regulatory elements. As found in other eukaryotic systems, most plant promoters contain the TATA box sequence at about the –30 region. Deletion of the TATA box reduces promoter activity and results in multiple transcription initiation startpoints. The CAAT box sequence is located close to the –80 region of several plant promoters. However, the importance of the CAAT box sequence is not well established for plant promoters. In some cases, the activity of a promoter is significantly increased by the presence of an enhancer, which may be located at a variable distance from the promoter and can function in either orientation.

A hexamer sequence (TGACGT) occurs in most constitutive promoters within a few hundred nucleotides from the transcription start site.[1] The hexamer motifs are often found as repeats, which are separated by six to eight nucleotides.[2] Deletion or point mutation analysis has indicated that the hexamer motif is essential for the transcription activity of the CaMV 35S promoter,[3] *ocs* promoter,[4] and *nos* promoter.[5] Genes for the transcription factor which specifically interacts with the hexamer motif have been isolated from both dicotyledonous and monocotyledonous plant species.[6-8]

One of the most commonly found promoter elements from the environmentally inducible genes is the G-box sequence (CCACGTGG).[1] The conserved sequence motif or its related sequence is located in promoters of various photosynthesis genes, *Arabidopsis* alcohol dehydrogenase gene, abscisic acid-inducible genes, and wound inducible genes. Deletion of the

0-8493-5164-2/93/$0.00+$.50
© 1993 by CRC Press, Inc.

sequence from the wound inducible potato proteinase inhibitor II promoter has resulted in loss of wound inducibility and methyl jasmonate responsivity.[9] Genes coding for the G-box-binding transcription factor have been isolated.[10,11] In addition to the hexamer and G-box motifs, several other regulatory elements have been identified from various plant promoters. Some of the most commonly used schemes for isolation and characterization of plant transcription elements are described below.

## II. ISOLATION OF REGULATORY REGIONS

Several approaches have been used to isolate DNA fragments carrying regulatory elements from plants. Transcriptional regulatory regions can be obtained using a cDNA clone to isolate adjacent regions from genomic DNA. The promoter and terminator regions have been determined by measuring the transcription initiation and termination sites from a genomic clone. Alternatively but not frequently, promoter tagging methods have been used to detect and isolate regulatory regions directly from the plant genome.

### A. ISOLATION OF REGULATORY REGIONS BY cDNA CLONE

Many different cDNA cloning methods are available to isolate a clone starting with mRNA. Unlike genomic clones, cDNA clones lack introns, and therefore they are easier to characterize. The direction of transcription can be obtained by sequencing the clone and searching for an open reading frame which encodes a functional protein corresponding to the starting material. However, it is not necessary to sequence the entire cDNA if sequencing the ends of a cDNA shows the amino terminus of a protein or a poly A tail. The genomic DNA flanking the amino terminus of the cDNA clone should contain the promoter, and the region adjacent to the poly A tail harbors the terminator region of the gene.

Genomic clones carrying a regulatory region can be isolated by screening a genomic library, with the cDNA used as a probe. The procedure for isolation of a genomic clone can be complicated if there is more than one copy of the gene in the plant species. Therefore, it is necessary to determine the copy number of the gene by Southern blot analysis under stringent conditions. To obtain an accurate estimation of the copy number, more than one restriction endonuclease is used to digest the plant DNA. If the DNA is not prepared properly, it is difficult to obtain a complete digestion of plant genomic DNA with most of the restriction enzymes. Compared to plasmid or bacteriophage DNAs, digestion of plant genomic DNA requires a higher amount of restriction enzyme. Precise amounts of enzymes may vary depending on the plant species and restriction enzymes. At least 10 to 20 units of enzyme per microgram of DNA and several hours of overnight digestion are generally required for complete digestion. Since plant DNA is highly methylated,[12] those enzymes which cannot restrict methylated DNA, such as *Eco*RII or *Stu*I, should be avoided. If there is only one copy of the gene, the genomic clone can be isolated with an intact cDNA clone as a probe. If there is more than one copy of the gene, a gene specific probe should be used to avoid accidental cloning of a gene which is related to the starting cDNA. The 3′ untranslated region of a cDNA clone is frequently used for the gene-specific probe.

Once a genomic clone is obtained, restriction mapping and Southern blot analysis are used to locate the amino and carboxy terminal ends of the coding region. The exact locations of the start and end points of its transcript can be obtained by S1 nuclease mapping.[13] In this procedure, the DNA fragment carrying the regulatory element is denatured and incubated with total mRNA prepared from a tissue in which the gene is most abundantly expressed. Hybridization is performed under stringent conditions in which the formation of DNA:RNA hybrids is preferred. DNA that has not formed duplexes is then hydrolyzed with exonuclease S1 or mung bean nuclease, leaving only DNA that is hybridized to mRNA. The size of the protected DNA fragment, which is identical to the length of the mRNA, is measured by gel electrophoresis. The ends of a transcript can also be mapped by alternative approaches.[14]

A restriction fragment carrying the region upstream of the transcription start point or the region downstream of the termination site is considered to contain a regulatory region. The length of the regulatory region is difficult to estimate and may vary significantly from one species and gene to the next. Therefore, it is wise to start with at least a 2 kb fragment for characterization or utilization of the promoter.

### B. ISOLATION OF REGULATORY REGIONS FROM A GENOMIC CLONE

A genomic clone is often obtained without a cDNA clone by screening a library with a heterologous probe, by gene tagging, or by chromosome walking. The regulatory region can be mapped with a heterologous cDNA clone if there is significant homology. If a heterologous cDNA approach is not applicable, DNA sequencing would provide information on the structure of the gene. However, if there are a large number of introns, it is difficult to localize the regulatory region by sequence information alone. Isolation of the cDNA clone corresponding to the genomic clone will facilitate identification of the promoter region.

### C. ISOLATION OF PROMOTER REGIONS BY TAGGING

Promoter tagging vectors have been used for isolation of plant promoters from several plant species using the *Agrobacterium*-mediated Ti-plasmid transformation system. There are two different schemes. In the first approach, a promoterless reporter gene or a reporter gene with a minimal promoter element is placed immediately next to the right border of a tumor-inducing (Ti) plasmid vector, so that, upon transfer into a plant chromosome, the reporter gene is located next to the plant DNA.[15,16] If the insertion occurs at the promoter region of a functional gene, the reporter gene will be activated. The neomycin phosphotransferase (*npt*) gene has been used as a reporter to identify promoter elements, since activation of the gene can be detected easily by selecting the kanamycin resistant transformants. However, using this method, it is difficult to tag a promoter which is active in only a certain developmental stage or which is induced by a specific environmental factor. Therefore, another selectable marker, such as the hygromycin resistant gene, is placed next to the promoterless reporter gene. With this vector, hygromycin resistant transformants are first selected, and activation of the reporter gene by a plant promoter is screened by an enzyme assay. The frequency of fusions to an active promoter element is 5 to 30% depending on the type of organ examined.[15,16] The gene fusion can be rescued from the plant by transformation of the T-DNA-linked plasmid and flanking plant DNA into *E. coli*.

In the second method, plant promoters can be isolated by inserting a random DNA fragment in front of a promoterless reporter gene and screening for activation of the reporter gene. One such vector is pROA97, which contains two markers; one is the *npt* gene, and the other is the GUS gene, with both markers placed under the minimal 35S promoter carrying the TATA box region.[17] Using this type of vector, several promoter fragments were isolated from *Sau*3A segments of *Arabidopsis thaliana* [17] and rice[18] genomic DNA.

## III. CHARACTERIZATION OF PROMOTER AND ENHANCER ELEMENTS

One of the most frequently used methods for characterizing plant regulatory elements is deletion analysis. By this method, a DNA fragment containing the regulatory element is treated with an exonuclease to generate a serial deletion of the promoter region. The deletion mutants are then placed upstream of a reporter gene whose expression can be easily studied by either stable transformation or transient assay. Although this method has been successfully used for studying several plant promoters, some precautions are required, since plant promoters often consist of several regulatory elements whose functions are redundant to each other. Therefore, deletion of one or a few elements does not result in alteration of the promoter characteristics. It is necessary to test smaller DNA fragments for their promoter ability before deletion analysis is conducted. Another way to study complex promoter elements is by making a chimeric promoter between a small segment of the promoter and a minimal promoter whose nature has already been characterized. In this way only a small portion of the promoter is studied at any one time. Combining the results from such studies will eventually reveal the structure and characteristics of a complex promoter.

## A. DELETION ANALYSIS

Deletion mutants can be obtained by removal of a restriction fragment from the regulatory region. However, the frequency of suitable restriction sites within a promoter region is usually not high enough to provide a wide range of deletion mutants, and therefore exonuclease treatments are used to generate additional deletion mutants.

### 1. Nuclease BAL-31

The most popular enzyme used for deletion mutagenesis is the nuclease BAL-31, which is predominantly an exonuclease that simultaneously degrades both the 3′ and 5′ termini of double-stranded DNA without internal scission. Using this nuclease, a series of deletion mutants can be generated from either the 3′ or 5′ end of the promoter fragment.[14] After BAL-31 treatment, a synthetic linker is added at the deletion end point to facilitate characterization and further manipulation of the deletion mutants. One of the problems with this procedure is that the deletion occurs to both the promoter fragment and the cloning vector. Therefore, the DNA sequence adjacent to the end point is different for individual deletion mutants, and it is necessary to subclone the mutated promoter fragments into another vector. Also, the efficiency of linker addition is relatively poor compared to self-ligation. To avoid these problems, we have developed two plasmid vectors, pGA616 and pGA617,[19] from pUC18 and pUC19 by inserting a synthetic linker (GGTACCTCGAGGCCT) containing *Kpn*I, *Xho*I, and *Stu*I sites into the unique *Ssp*I site located about 600 nucleotides from the multiple cloning site. Using the pGA616 or pGA617 vector, a promoter fragment can be deleted without subcloning and linker addition. The DNA sequences adjacent to the deletion end points are identical and contain *Kpn*I and *Stu*I sites, which can be used for further subcloning of the molecules.

The following procedure is used to generate nested deletion mutants using pGA616 or pGA617: A DNA fragment containing the regulatory region is inserted into the multiple cloning site of pGA616 or pGA617 using X-Gal selection. The recombinant plasmid (about 10 μg) is linearized by digestion at a unique restriction enzyme site located at either the 5′ or 3′ end of the insert. It is necessary to make sure that the DNA is completely digested by running about 0.5 μg of the sample on an agarose gel. During the test run, the DNA sample is kept on ice to prevent degradation. An equal volume of 2X BAL-31 buffer (1.2 *M* NaCl, 24 m*M* $CaCl_2$, 24 m*M* $MgCl_2$, 2 m*M* EDTA, and 40 m*M* Tris-HCl, pH 8.0) is added to the DNA solution, and the sample is incubated at 30°C. The deletion reaction is started by addition of 2 to 5 units of BAL-31. Approximately 200 nucleotides per min are removed from both ends. Most commercial preparations of BAL-31 contain a fast and a slow form of the enzyme. The purified slow form, which is a proteolytic degradation product of the fast form, is used to remove short fragments. One tenth of the sample is withdrawn at 1 min intervals and mixed with an equal volume of ice-cold stop solution (0.1 *M* EGTA and 0.6 *M* Sodium acetate, pH 7.0). The enzyme reaction is absolutely dependent on the presence of calcium, and therefore the reaction is stopped by addition of the chelating agent EGTA. The samples are extracted with an equal volume of phenol and chloroform and precipitated with ethanol. The pellet is dissolved in 10 μl TE solution (1 m*M* EDTA and 10 m*M* Tris, pH 7.5), and a small portion of the sample (about 2 μl) is run on an agarose gel to determine the size of deletion. Since BAL-31 digests both ends of the linearized DNA, the length of the deletion at each end is half of the total deletion. The remaining DNA samples are cut with *Stu*I. About 10 to 20% of the BAL-31 product is blunt ended, and the remaining 80 to 90% retains single-stranded tails. Filling in the tails with bacteriophage T4 DNA polymerase or the Klenow fragment of *E. coli* DNA polymerase I will significantly increase the yield. However, the number of deletion mutants obtained by this procedure is usually high enough to skip the filling-in reaction. After extraction with phenol and chloroform, the DNA samples are precipitated with ethanol and dissolved in 20 μl ligation buffer. The DNA is self-ligated at room temperature and transferred into an appropriate *E. coli* host to select for ampicillin resistant transformants. The deletion length of individual promoter mutants is measured by using the restriction enzyme sites present in the synthetic linker. The precise end points can be obtained by sequencing the deletion mutants.

### 2. Exonuclease III

Exonuclease III catalyzes the stepwise removal of 5′ mononucleosides from the recessed or blunt 3′ termini of double-stranded DNA and, in combination with a single-strand nuclease, can also be used to generate nested sets of deletion mutants.[14] Since protruding 3′ termini are resistant to the enzyme activity, exonuclease III can be used for unidirectional deletions. To create mutants, the promoter fragment is inserted into the multiple cloning site of a vector, such as pBluescript plasmid, which carries several unique enzyme sites with either recessed or protruding 3′ termini. The 3′ or 5′ end of the inserted promoter fragment is digested with two restriction enzymes. The enzyme that cleaves nearer the promoter fragment must generate either a blunt end (e.g. *Eco*RV or *Sma*I) or a recessed 3′ terminus (e.g., *Eco*RI or *Hin*dIII), and the other enzyme generates a protruding 3′ terminus (e.g., *Pst*I or *Kpn*I). The linear DNA is then treated with exonuclease III for varying lengths of time. The reaction is stopped by the addition of EDTA. After phenol extraction and ethanol precipitation, the DNA is treated with either nuclease S1 or mung bean nuclease. The ends of the linear DNA can be repaired with T4 DNA polymerase or a Klenow fragment.

### 3. Internal Deletion and Linker-Scanning Mutagenesis

Identification of control elements in the promoter region usually begins with the analysis of sets of 5′ deletions and 3′ deletions. This analysis defines the boundary of the regulatory domain but fails to provide detailed information on the internal organization of the regulatory region. Characterization of the internal structure can be achieved by studying internal deletion or linker-scanning mutants generated by joining 5′ and 3′ deletion mutants via the enzyme site located at the deletion end points. By selecting appropriate deletion end points, the deletion length can be controlled. Although this method is very useful in studying the organization of the regulatory region, the internal deletions occasionally create side effects by altering spacing between the promoter elements.[20] Precise substitution of nucleotide sequences can be obtained by the linker-scanning mutagenesis procedure in which mutants are generated by ligating 5′ and 3′ deletion mutants whose endpoints differ by the number of nucleotides present in the linker.[21] Therefore, the DNA sequence ordinarily present in the wild-type promoter is replaced by the linker sequence without altering the distance within the regulatory region. However, it is a laborious undertaking to obtain a set of linker scanning mutations since a large number of 5′ and 3′ deletion mutants must be obtained and characterized before matching. To overcome this problem, several other approaches have been developed.[14] Among these methods, the oligonucleotide-mediated mutagenesis method is used most commonly for studying plant regulatory elements.

## B. OLIGONUCLEOTIDE-MEDIATED MUTAGENESIS

In contrast to other methods, oligonucleotide-mediated mutagenesis generates specific mutations within a relatively short period of time. Although there are many different mutagenesis methods,[14] they are similar. A DNA fragment is first cloned into a single-stranded phage vector, such as M13 or a phagemid. The M13 vector generally provides a higher yield, especially when a multiple mutation is introduced. Single-stranded DNA is prepared from the recombinant phage and is hybridized with a synthetic oligonucleotide, typically, a 17- to 19-mer with one mismatch in the center. If two or more nucleotides are to be mutated, at least 12 to 15 perfectly matched nucleotides are needed on either side. While the hybridization is usually performed at room temperature, it should be done at 12 to 16°C if the oligonucleotide is rich in A and T. After annealing, the hybridized oligonucleotide is extended using the Klenow fragment and dNTPs, and sealed with DNA ligase and ATP. Finally, an appropriate host cell is transfected with the mutagenized DNA. Recombinant phage containing the mutated DNA sequence are screened using the labeled oligonucleotide as a hybridization probe.

Although this method has been used in many laboratories, screening a mutant and subcloning the mutated fragment into an appropriate vector still require considerable effort. Several schemes have been developed to improve the mutation frequency and the screening procedures.[14] One of the most convenient ways of overcoming these difficulties is to synthesize both strands and then directly replace the target area, thereby saving both the time and effort involved in cloning a target region into a single-stranded phage and recloning the mutated promoter to another vector. Furthermore, the mutation frequency is near 100%, reducing the time required for screening a mutant. In contrast to the single-stranded oligonucleotide method, the length of a synthetic oligonucleotide can be as short as six nucleotides. This strategy is especially useful when several different mutations are to be introduced into the same region.[4] However, mutating a promoter region by the double-stranded synthesis method is possible only when there is a convenient restriction enzyme site at the target area. Usually, unique restriction enzyme sites are generated at the promoter region during the internal deletion mutagenesis. Otherwise, enzyme sites can be generated by the single-stranded vector system as described above. One of the drawbacks of this method is that the insertion of a synthetic oligonucleotide alters the distance between promoter elements, possibly affecting promoter activity.[20] Therefore, an appropriate deletion mutant should be selected for the experiment.

## C. BINDING FACTORS

Promoter elements have also been identified and characterized by studying interactions between the regulatory sequences and nuclear protein factors. Transcription initiation requires several nuclear factors which specifically interact with promoter elements. Since studying regulatory elements with the binding factor does not require deletion mutation analysis, the promoter elements can be quickly identified. Among a variety of *in vitro* techniques, the most popular method is the gel-retardation assay in which a nuclear extract is incubated with DNA fragments carrying native or mutagenized promoter sequences and the protein-DNA complex is detected by gel electrophoresis. Detailed interaction can be further studied by footprinting analysis. Since these analyses are dependent on a crude preparation of nuclear proteins, it is often difficult to study a specific interaction. Of course, cloning genes encoding a trans-acting factor facilitates studying the interaction of a promoter element with a specific transcription factor.

### 1. *In Vitro* Binding

One of the critical steps of *in vitro* binding analysis is the preparation of a good nuclear protein extract. Nuclei are usually isolated from the tissues where the promoter is to be studied.

There are several methods for isolation of nuclei;[22] however, the quality varies significantly among the methods. For *in vitro* binding studies, nuclei should be intact and free from nuclease contamination. The following method works well for plants.[23] The sample is frozen in liquid nitrogen and homogenized in a buffer [0.44 *M* sucrose, 2.5% Ficoll (MW 400,000), 5.0% Dextran 40, 10 m*M* $MgCl_2$, 10 m*M* β-mercaptoethanol, 0.5% Triton X-100, and 25 m*M* Tris-HCl, pH 7.6]. The homogenate is filtered through a single layer of Miracloth. The nuclei are pelleted by low speed centrifugation (150 to 5,000 × g), which leaves almost all other major cell components in the supernatant. After gently suspending in a nuclear resuspension buffer (20% glycerol, 5 m*M* $MgCl_2$, 10 m*M* β-mercaptoethanol, and 50 m*M* Tris-HCl, pH 7.8), the nuclei are further purified by centrifugation at 4,000 × g in a discontinuous Percoll gradient containing 5 ml layers of 40, 60, and 80% Percoll solutions on a 5 ml layer of 2 *M* sucrose. The Percoll solutions contain 0.44 *M* sucrose, 10 m*M* $MgCl_2$, and 25 m*M* Tris-HCl, pH 7.5. The nuclei are banded in the interface between the 80% Percoll and the 2 *M* sucrose. The nuclei are then washed twice with nuclear resuspension buffer and pelleted by low speed centrifugation. The pellet is then resuspended in the nuclear resuspension buffer and is stored at –80°C in the presence of 20% glycerol for up to several months.

Nuclear proteins may be prepared as follows.[24] Nuclei are disrupted by incubation in a nuclear extraction buffer (110 m*M* KCl, 5 m*M* $MgCl_2$, 1 m*M* DTT, 0.5 m*M* phenylmethylsulfonyl fluoride, 0.6 m*M* leupeptin, 7.5 m*M* pepstatin A, and 15 m*M* Hepes-KOH, pH 7.6). One tenth volume of 4 *M* ammonium sulfate is added to the extract and gently mixed at 4°C for 1 h. Chromatin and particulate materials are removed by centrifugation at 130,000 × g for 60 min. Nuclear proteins are then precipitated from the supernatant by the slow addition of 0.3 g/ml freshly ground ammonium sulfate, while stirring, for 30 min at 4°C. Following centrifugation at 10,000 × g for 15 min, the pellet is resuspended in the nuclear extraction buffer and then dialyzed against the nuclear extraction buffer without the protease inhibitors or DTT. Insoluble materials are removed by centrifugation at 12,000 × g for 10 min. Extracts are aliquoted, frozen in liquid nitrogen, and then stored at –80°C.

Obtaining a suitable amount of nuclear protein from certain plant tissues is difficult. If the promoter is active in cultured cells, nuclear proteins can be isolated from suspension cultured cells, which produce a large quantity of nuclear proteins. In the procedure, protoplasts are prepared from the cells with an appropriate enzyme mixture, and nuclear proteins are purified as follows. Protoplasts are suspended in a lysis buffer (18% Ficoll, 0.5 m*M* spermidine, 0.15 m*M* spermine, 0.5 m*M* PMSF, 12 m*M* β-mercaptoethanol, 20 m*M* potassium acetate, 0.6 m*M* leupeptin, 0.15 pepstatin A, and 20 m*M* MES, pH 5.8) and homogenized with a Teflon homogenizer. Homogenates are filtered through a 25 μm nylon filter and centrifuged at low speed (about 100 × g) for 30 min at 4°C. Nuclear proteins may be prepared from the pellet as described above.

The DNA fragments for *in vitro* binding study are prepared by digesting the promoter region with restriction enzymes which produce small segments. The size can vary, but it is best to use smaller fragments (a few hundred bp or shorter) since protein-DNA complexes of this size can be easily separated electrophoretically from unbound DNA. The DNA fragment is end-labeled with either Klenow enzyme or $T_4$ polynucleotide kinase. It is important to obtain a highly-specific labeling since the amount of the transcription factor in the nuclear extract may be limiting.

Binding of a transcription factor to the labeled DNA fragment is studied by gel-shift analysis.[25] About 20,000 cpm of the labeled DNA is incubated with up to 50 μg of the nuclear protein in 20 μl of a binding solution (2 m*M* magnesium acetate, 0.5 m*M* EDTA, 2 m*M* DTT, and 20 m*M* Tris-acetate, pH 7.5) at 28°C for 15 min. Also included is 0.1 to 1 μg of poly-d(I-C) in the binding reaction to reduce nonspecific binding by general DNA binding proteins. The reaction mixture is then run on a 4% nondenaturing polyacrylamide gel with a ratio of polyacrylamide to *bis*-acrylamide of 29.2 to 0.8, to form loose pores so that the binding between the transcription factor and specific DNA fragment is retained during electrophoresis. The gel is dried and exposed to an X-ray film.

It is essential to determine whether the binding has occurred to a specific transcription factor. This is achieved by including about a 10 to 100 molar excess of unlabeled DNA as a competitor in the binding reaction. If the interaction is specific, the unlabeled DNA will compete with the labeled DNA for the binding. In this way, it can be determined whether or not DNA fragments from other promoters also interact with the same transcription factor.[26] DNA fragments prepared from deletion mutants and oligonucleotide-mediated mutants can also be used for studying the promoter elements. Synthetic oligonucleotides containing a specific DNA sequence from within a promoter region can further localize and characterize the promoter elements for the competition study.

### 2. *In Vitro* and *In Vivo* Footprinting

The exact location of where a transcription factor interacts with a promoter element is identified by the footprinting assay. After the *in vitro* binding reaction, DNaseI and $MgCl_2$ (final concentration of 5 m*M*) are added to the mixture to cleave the unbound region of the DNA fragment. The amount of DNaseI is adjusted to introduce about one to a few nicks per DNA fragment. During this treatment, the DNA should remain as a double-stranded form. The bound and free DNAs are separated by running a polyacrylamide gel, and the eluted DNA fragments are denatured and fractionated on a sequencing gel. The analysis visualizes the region where a protein interacts since the protein protects the bound region from the DNaseI reaction. Chemicals such as dimethyl sulfate or 1,10-phenanthroline-copper ion are also used for the footprinting analysis.

The interaction between a promoter element and its regulatory factor can be directly studied by *in vivo* footprinting analysis.[27] In this method, dimethyl sulfate is added to cells at a final concentration of 0.2%. The chemical quickly diffuses into chromosomes and methylates naked DNA. Promoter elements which are actively engaged in transcription are protected. The protected region is determined by genomic sequencing.[28]

### 3. Cloning the Regulatory Gene

Interaction between promoter elements and their regulatory factors is complex since transcription initiation involves the interaction of several nuclear proteins with the promoter. Furthermore, some factors are present in small quantities or are unstable during the preparation. This problem can be overcome by cloning a gene coding for the factor. Regulatory genes have been cloned by screening an expression cDNA library with a radioactively-labeled DNA fragment carrying a specific promoter element under similar conditions used for the *in vitro* binding.[6-8,10,11,29] Plant regulatory genes have also been cloned by transposon tagging[30] or T-DNA insertional mutagenesis.[31] A transcription factor can be produced in *E. coli* and used for studying the interaction with promoter elements.

## IV. CHARACTERIZATION OF TERMINATOR ELEMENTS

Gene expression can also be controlled at the terminator region; e.g., the expression of a reporter gene is dependent upon the type of plant terminator used.[32] However, progress in studying plant terminator elements has been slow since, unlike promoter elements, deletion or mutation of the terminator elements does not result in total loss of transcription. Furthermore, there is no good *in vitro* system to identify a terminator element. However, the deletion analysis and oligonucleotide-mediated mutagenesis methods previously described for studying the promoter elements can also be used for studying terminator elements.

## V. REPORTER GENES

Since there is no good *in vitro* transcription system for plants, promoters need to be introduced into intact plant cells in order to estimate their activity. However, it is difficult to measure the amount of expression from the introduced gene due to the endogenous expression from the original gene present in the chromosome. By inserting a small DNA fragment into the 5′ or 3′ untranslated region of the recombinant gene, the size of transcripts can be distinguished between the introduced and endogenous genes. However, it is difficult to measure routinely mRNA levels unless the gene of interest is strongly expressed. One way to overcome this difficulty is to use a reporter gene. In this method, a DNA fragment carrying the promoter region is fused to a coding region of a foreign reporter gene, and promoter activity is studied by measuring reporter expression. The reporter is normally a promoterless gene whose product is easy to assay and sensitive enough to measure a small change in gene expression. A reporter carrying the minimal promoter sequence, such as the TATA box region of the 35S promoter or *nos* promoter, is useful when a 3′ deletion mutant or a small DNA fragment is studied. There should be no significant level of background or inhibitory activity in plant cells.

One of the most commonly used reporters for promoter study is the chloramphenicol acetyltransferase (*cat*) gene.[33] The assay for the activity of this enzyme monitors the conversion of radioactive chloramphenicol to its acetylated form. Some plant species contain a significant amount of reporter enzyme inhibitory activity, and therefore the plant extract should be inactivated by heating at 60 to 70°C and by including EDTA in the assay buffer. However, this marker cannot be used easily for localization of the cell type that expresses the reporter gene. The *E. coli* β-glucuronidase (GUS) is the most popular reporter gene used for *in situ* localization of promoter activity.[34] The β-galactosidase (*lacZ*) gene has also been used as a reporter, but, due to a high level of endogenous background activity, this reporter is not commonly used in plants.[35] The luciferase gene from firefly[36] and bacteria[37] has also been used for *in situ* localization of plant gene expression. This marker is especially attractive since the assay is nondestructive, allowing the assay to be performed within living tissue. The diphtheria toxin A chain is a self-reporter which visualizes promoter expression by destroying the cells that express the gene.[38]

## VI. TRANSIENT ANALYSIS

Plant regulatory elements have been studied using both transient assay and stable transformation systems. Transient assays are, however, simple and rapid[39] so that promoter elements can be identified and characterized quickly compared to the stable transformation analysis which takes up to several months. In the transient assay system, plasmid DNA containing a mutagenized promoter and a downstream reporter gene is introduced into protoplasts by electroporation or PEG treatment. Depending on the type of promoter to be studied, the source of plant material for protoplast isolation is different. If a promoter, such as those from housekeeping genes, is functional in most plant tissues, protoplasts can be isolated from leaves or suspension cultured cells, which produce a large number of active protoplasts. However, a promoter that is active only in a certain cell type may require protoplasts isolated from a specific tissue. Most of the DNA introduced by electroporation or PEG treatment remains transiently in cells for a few days before being degraded. If the promoter region is recognized by the nuclear transcription factors during this period, the protoplasts will express the reporter gene. The level of transient expression is dependent upon the amount, size, and nature of the DNA. The recombinant plasmid to be used for a transient assay should be as small as possible, since a DNA fragment bigger than 5 kb is not efficiently transferred into protoplasts. For those cells which are difficult to protoplast, the particle gun method may be used to study expression of a promoter in an intact plant or organ. However, this method is not used commonly for characterization of promoter elements, except in monocotyledonous plants, since only a small portion of the plant material takes up the foreign DNA.

## VII. STABLE TRANSFORMATION ANALYSIS

Although the transient assay is a useful way to study regulatory elements, many plant promoters are not expressed efficiently in protoplasts. In some cases, the results obtained from the transient assay need to be re-evaluated by the stable transformation system. The most commonly used method for stable transformation of dicotyledonous plants is the *Agrobacterium*-mediated Ti-plasmid vector.[40] However, cereal plants are normally not susceptible to transformation by this method, and therefore the direct DNA uptake method is used. Several vectors for studying promoter elements have been developed based on the binary Ti-plasmid vector system. One of the promoter-probing vectors is pGA580,[33] which contains, in addition to a broad host replicon that can replicate in both *Agrobacterium* and *E. coli*, the promoterless *cat* reporter gene two border sequences which are required for initiation and termination of T-DNA transfer, and a selectable marker for plant transformation. A promoter fragment is inserted into a cloning site located in front of the reporter gene. In pGA580, the *cat* reporter is located downstream of the *lac* promoter which expresses the *cat* gene in bacteria but not in plants. Therefore, insertion of a foreign DNA can be screened by chloramphenicol sensitivity in *E. coli*. This vector is then moved into *Agrobacterium* using direct DNA transfer or triparental mating methods.[33] *Agrobacterium* carrying the recombinant molecule is used to transform an appropriate host plant by cocultivation methods.

One of the major difficulties in studying promoters by the stable transformation system is that there is a significant variation in the level of the promoter strength among independent transformants. This variation is probably due to an effect from the surrounding genomic DNA. Such position effects can be partially overcome by averaging the results from several independently transformed plants. In this regard, suspension cultured cells are advantageous to use if the promoter is expressed in the cultured cells, since a large number of transformants can readily and rapidly be obtained. Fortunately, the character or pattern of the introduced promoter, as distinguished from its strength, is usually not influenced by the surrounding chromosomal DNA sequences except when a minimal promoter fragment is used. However, it is necessary to assay several transgenic plants to ascertain that the behavior of the introduced promoter is the same as that of the original gene.

## REFERENCES


1. **Weising, K. and Kahl, G.,** Toward an understanding of plant gene regulation: the action of nuclear factors, *J. Biosci.*, 46c, 1, 1991.
2. **Bouchez, D., Tokuhisa, J. G., Llewellyn, D. J., Dennis, E. S., and Ellis, J. G.,** The ocs-element is a component of the promoters of several T-DNA and plant viral genes, *EMBO J.*, 8, 13, 1989.
3. **Lam, E., Benfey, P. N., Gilmartin, P. M., Fang, R.-X., and Chua, N.-H.,** Site-specific mutations alter *in vitro* factor binding and change promoter expression pattern in transgenic plants, *Proc. Natl. Acad. Sci. U.S.A.*, 86, 7890, 1989.
4. **Singh, K., Tokuhisa, J. G., Dennis, E. S., and Peacock, W. J.,** Saturation mutagenesis of the octopine synthase enhancer: correlation of mutant phenotypes with binding of a nuclear protein factor, *Proc. Natl. Acad. Sci. U.S.A.*, 86, 3733, 1989.

5. **Ebert, P. R., Ha, S. B., and An, G.,** Identification of an essential upstream element in the nopaline synthase promoter by stable and transient assays, *Proc. Natl. Acad. Sci. U.S.A.*, 84, 5745, 1987.
6. **Katagiri, F., Lam, E., and Chua, N.-H.,** Two tobacco DNA-binding proteins with homology to the nuclear factor CREB, *Nature*, 340, 727, 1989.
7. **Tabata, T., Takase, H., Takayama, S., Mikami, K., Nakatsuka, A., Kawata, T., Nakayama, T., and Iwabuchi, M.,** A protein that binds to a cis-acting element of wheat histone gene has a leucine zipper motif, *Science*, 245, 965, 1989.
8. **Singh, K., Dennis, E. S., Ellis, J. G., Llewellyn, D. J., Tokuhisa, J. G., Wahleithner, J. A., and Peacock, W. J.,** OCSBF-1, a maize ocs enhancer binding factor: isolation and expression during development, *Plant Cell*, 2, 891, 1990.
9. **Kim, S.-R., Choi, J.-L., Costa, M. A., and An, G.,** Identification of G-box sequence as an essential element for methyl jasmonate response of potato proteinase inhibitor II promoter, *Plant Physiol.*, 99, 627, 1992.
10. **Guiltinan, M. J., Marcotte, W. R., Jr., and Quatrano, R. S.,** A plant leucine zipper protein that recognizes an abscisic acid responsive element, *Science*, 250, 267, 1990.
11. **Oeda, K., Salinas, J., and Chua, N.-H.,** A tobacco bZip transcription activator (TAF-1) binds to a G-box-like motif conserved in plant genes, *EMBO J.*, 10, 1793, 1991.
12. **Gruenbaum, Y., Naveh-Many, T., Cedar, H., and Razin, A.,** Sequence specificity of methylation in higher plant DNA, *Nature*, 292, 860, 1981.
13. **Berk, A. J. and Sharp, P. A.,** Sizing and mapping of early adenovirus mRNAs by gel electrophoresis of S1 endonuclease-digested hybrids, *Cell*, 12, 721, 1977.
14. **Sambrook, J., Fritsch, E. F., and Maniatis, T.,** *Molecular Cloning*, Cold Spring Harbor Laboratory Press, Cold Spring Harbor, New York, 1989.
15. **Koncz, C., Martini, N., Mayerhofer, R., Koncz-Kalman, Z., Korber, H., Redei, G. P., and Schell, J.,** High-frequency T-DNA-mediated gene tagging in plants, *Proc. Natl. Acad. Sci. U.S.A.*, 86, 8467, 1989.
16. **Fobert, P. R., Miki, B. L., and Iyer, V. N.,** Detection of gene regulatory signals in plants revealed by T-DNA-mediated fusions, *Plant Mol. Biol.*, 17, 837, 1991.
17. **Ott, R. W. and Chua, N.-H.,** Enhancer sequences from *Arabidopsis thaliana* obtained by library transformation of *Nicotiana tabacum*, *Mol. Gen. Genet.*, 223, 169, 1990.
18. **Claes, B., Smalle, J., Dekeyser, R., Van Montagu, M., and Caplan, A.,** Organ-dependent regulation of a plant promoter isolated from rice by promoter-trapping in tobacco, *Plant J.*, 1, 15, 1991.
19. **Ha, S.-B. and An, G.,** Identification of upstream regulatory elements involved in the developmental expression of the *Arabidopsis thaliana cab1* gene, *Proc. Natl. Acad. Sci. U.S.A.*, 85, 8017, 1988.
20. **Gilmartin, P. and Chua, N.-H.,** Spacing between GT-1 binding sites within a light-responsive element is critical for transcriptional activity, *Plant Cell*, 2, 447, 1990.
21. **McKnight, S. L. and Kingsbury, R.,** Transcriptional control signals of a eukaryotic protein-coding gene, *Science*, 217, 316, 1982.
22. **Dunham, V. L. and Bryant, J. A.,** Nuclei, in *Isolation of Membranes and Organelles from Plant Cells*, Hall, J. L. and Moore, A. L., Eds., Academic Press, London, 1983, 237.
23. **Luthe, D. S. and Quatrano, R. S.,** Transcription in isolated wheat nuclei, *Plant Physiol.*, 65, 305, 1980.
24. **Green, P. J., Kay, S. A., and Chua, N.-H.,** Sequence-specific interactions of a pea nuclear factor with light-response elements upstream of the rbcS-3A gene, *EMBO J.*, 6, 2543, 1987.
25. **Parker, C. S. and Topol, J.,** A drosophila RNA polymerase II transcription factor contains a promoter-region-specific DNA-binding activity, *Cell*, 36, 357, 1984.
26. **Tabata, T., Nakayama, T., Mikami, K., and Iwabuchi, M.,** HBP-1a and HBP-1b: leucine zipper-type transcription factors of wheat, *EMBO J.*, 10, 1459, 1991.
27. **Schulze-Lefert, P., Dangl, J. L., Becker-Andre, M., Hahlbrock, K., and Schulz, W.,** Inducible *in vivo* DNA footprints define sequences necessary for UV light activation of the parsley chalcone synthase gene, *EMBO J.*, 8, 651, 1989.
28. **Church, G. M. and Gilbert, W.,** Genomic sequencing, *Proc. Natl. Acad. Sci. U.S.A.*, 81, 1991, 1984.
29. **Singh, H., Clerc, R. G., and LeBowitz, J. H.,** Molecular cloning of sequence-specific DNA binding proteins using recognition site probes, *Biotech.*, 7, 252, 1989.
30. **Paz-Ares, J., Wienand, U., Peterson, P. A., and Saedler, H.,** Molecular cloning of the *c* locus of *Zea mays*: a locus regulating the anthocyanin pathway, *EMBO J.*, 5, 829, 1986.
31. **Yanofsky, M. F., Ma, H., Bowman, J. L., Drews, G. N., Feldmann, K. A., and Meyerowitz, E. M.,** The protein encoded by the *Arabidopsis* homeotic gene *agamous* resembles transcription factors, *Nature*, 346, 35, 1990.
32. **An, G., Mitra, A., Choi, H. K., Costa, M. A., An, K., Thornberg, R. W., and Ryan, C. A.,** Functional analysis of the 3′ control region of the potato wound-inducible proteinase inhibitor II gene, *Plant Cell*, 1, 115, 1989.
33. **An, G.,** Binary Ti vectors for plant transformation and promoter analysis, *Meth. Enzymol.*, 153, 292, 1987.

34. **Jefferson, R. A., Kavanagh, T. A., and Bevan, M. W.,** GUS fusions: β-glucuronidase as a sensitive and versatile gene fusion marker in higher plants, *EMBO J.*, 6, 3901, 1987.
35. **Helmer, G., Casadaban, M., Bevan, M., Kayes, L., and Chilton, M.-D.,** A new chimeric gene as a marker for plant transformation: the expression of *Escherichia coli* β-galactosidase in sunflower and tobacco cells, *Bio/Technology*, 2, 520, 1984.
36. **Ow, D. W., Wood, K. V., DeLuca, M., de Wet, J. R., Helinski, D. R., and Howell, S. H.,** Transient and stable expression of the firefly luciferase gene in plant cells and transgenic plants, *Science*, 334, 856, 1986.
37. **Koncz, C., Olsson, O., Langridge, W. H. L., Schell, J., and Szalay, A. A.,** Expression and assembly of functional bacterial luciferase in plants, *Proc. Natl. Acad. Sci. U.S.A.*, 84, 131, 1987.
38. **Czako, M. and An, G.,** Expression of DNA coding region for diphtheria toxin chain A is toxic to plant cells, *Plant Physiol.*, 95, 687, 1991.
39. **Fromm, M., Taylor, L. P., and Walbot, V.,** Expression of genes transferred into monocot and dicot plant cells by electroporation, *Proc. Natl. Acad. Sci. U.S.A.*, 82, 5824, 1985.
40. **Lichtenstein, C. P. and Fuller, S. L.,** Vectors for the genetic engineering of plants, in *Genetic Engineering*, 6, Rigby, P. W. J., Ed., Academic Press, London, 1987, 103.
41. **An, G.,** High efficiency transformation of cultured tobacco cells, *Plant Physiol.*, 79, 568, 1985.

# 11

# Use of *Xenopus* Oocytes to Monitor Plant Gene Expression

*John J. Heikkila*

## I. INTRODUCTION

The *Xenopus* oocyte microinjection system is a valuable tool in the arsenal of molecular biology techniques used to analyze eukaryotic gene expression. This remarkable giant cell has the ability to transcribe injected DNA, translate injected mRNA, and post-translationally modify, compartmentalize, and secrete various types of proteins.[1-8] The following chapter consists of an overview of the use of the *Xenopus* oocyte system in the study of plant gene expression, including a presentation of the methods involved in transcribing and/or translating microinjected nucleic acids.

While the *Xenopus* oocyte microinjection system has been used primarily in the study of animal gene expression, recently it has gained popularity in the analysis of plant gene expression. A summary of plant mRNAs isolated from various tissues, plant viruses as well as synthetic mRNAs made from cloned plant cDNAs, which have been successfully translated in *Xenopus* oocytes is given in Table 1. For example, oocytes injected with barley mRNAs can synthesize and secrete catalytically active barley α-amylase.[16] Also, RNA isolated from elicitor-treated tomato cells has directed the expression of an ethylene-forming enzyme (EFE) when injected into oocytes, as judged by their ability to convert 1-aminocyclopropane-1-carboxylic acid into ethylene.[15] In fact, EFE expressed in oocytes was indistinguishable from tomato cell EFE with respect to its saturation kinetics, iron dependency, sensitivity to inhibitors, stereospecificity, and inducibility by fungal elicitors. Microinjection of plant mRNA synthesized *in vitro* has been employed to evaluate the effect of mutagenized mRNA on translation and post-translational modifications. When artificially mutagenized maize protein mRNA was translated in *Xenopus* oocytes, it was demonstrated that the addition of tryptophan and lysine to zein, which normally lacks these amino acids, does not affect the translation, post-translational modifications, or stability of the mutant protein.[20] This finding suggested that the creation of high lysine corn by genetic engineering was feasible. Recently, synthetic mRNAs made from cDNA clones for carrot and soybean cyclins (proteins which are required for mitosis or meiosis) were shown to cause *Xenopus* oocyte maturation following injection.[17] Thus, certain plant proteins may have comparable functions in animal cells.

In addition to translating plant mRNA, the *Xenopus* oocyte system has been used to analyze the expression of injected plant DNA (Table 1). The regulatory regions for the cauliflower mosaic virus 35S and for nopaline synthase were each fused to the bacterial reporter gene, chloramphenicol acetyl transferase (CAT) and microinjected into the germinal vesicles or nuclei of *Xenopus* oocytes.[23] These DNA constructs actively supported the synthesis of CAT, demonstrating that promoter regions from plant genes can support transcription in *Xenopus* oocytes. Also, a comprehensive study of barley α-amylase gene expression was carried out

0-8493-5164-2/93/$0.00+$.50
© 1993 by CRC Press, Inc.

**TABLE 1**
**Translation of Plant RNA and Expression of Plant DNA in Xenopus Oocytes**

| | Ref. |
|---|---|
| **A) RNA isolated from plant tissue**[a] | |
| Castor bean lectin | Coleman[5] |
| Maize storage proteins | Hurkman et al.[9] |
| | Larkins et al.[10] |
| Common bean storage protein | Matthews et al.[11] |
| Bean phytohemaglutinin | Vitale et al.[12] |
| Barley seed low mol. wt. proteins | Lane et al.[13] |
| Field bean storage globulins | Bassuner et al.[14] |
| French bean storage globulins | Bassuner et al.[14] |
| Pea bean storage globulins | Bassuner et al.[14] |
| Tomato ethylene-forming enzyme | Spanu et al.[15] |
| Barley α-amylase | Simon and Jones[16] |
| **B) Synthetic plant mRNA**[b] | |
| Tomato ethylene-forming enzyme | Spanu et al.[15] |
| Carrot cyclin | Hata et al.[17] |
| Soybean cyclin | Hata et al.[17] |
| 19 kDa maize zein protein | Galili et al.[18] |
| Barley α–amylase | Aoyagi et al.[19] |
| **C) Plant viral RNA** | |
| Tobacco mosaic virus | Knowland[21] |
| Cow pea mosaic virus | Huez et al.[22] |
| Alfalfa mosaic virus | Huez et al.[22] |
| Brome mosaic virus | Huez et al.[22] |
| **D) Plant genomic DNA constructs** | |
| Barley α-amylase | Aoyagi et al.[19] |
| Nopaline synthase[c] | Ballas et al.[23] |
| Cauliflower mosaic virus 35S[c] | Ballas et al.[23] |

[a] Total RNA or poly(A+)RNA.
[b] Sense RNA transcripts using riboprobe technology.
[c] Promoters fused to bacterial chloramphenicol acetyl transferase.

in *Xenopus* oocytes by injecting a genomic DNA clone containing an extensive 5′ regulatory region into oocyte nuclei and assaying for α-amylase gene expression.[19] $S_1$ nuclease mapping demonstrated that the same transcriptional start site of the barley α-amylase gene was used in both oocytes and plants. The α-amylase protein was synthesized in the oocyte and secreted into the medium. The molecular mass, isoelectric point, antigenicity, and enzymatic activity of oocyte-made α-amylase were identical to those of barley α-amylase. Thus, transcription of barley α-amylase genes and translation of the resultant mRNA occurred with high fidelity. However, not all plant genes are necessarily correctly initiated and/or terminated in *Xenopus* oocytes. It has been shown that the expression of some animal genes requires the coinjection of organism- or cell-specific transcription factors. For example, it is possible to achieve *trans*-activation of rat phospoenolpyruvate carboxykinase (*gtp*) gene expression in *Xenopus* oocytes by coinjection of rat liver mRNA.[24] This type of occurrence with plant genes may prove to be invaluable in the isolation of specific plant transcription factors or in the cloning of their cDNA.

## II. SETTING UP THE *XENOPUS* OOCYTE MICROINJECTION SYSTEM

### A. BIOLOGY OF THE *XENOPUS* OOCYTE

The *Xenopus* oocyte, an egg forming cell, is located in the ovary and surrounded by a few thousand follicle cells. The ovary contains oocytes at different stages of oogenesis, with

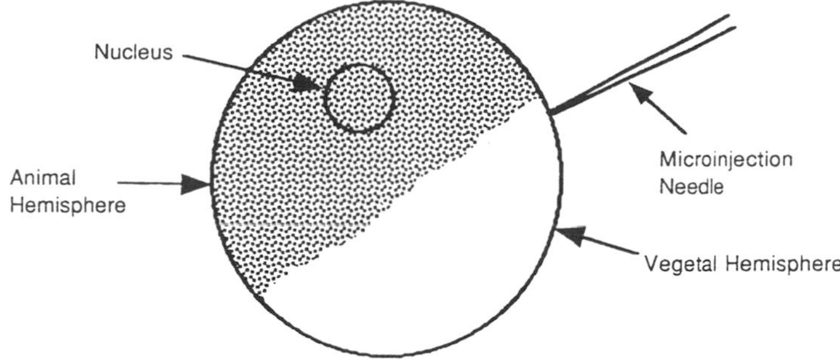


**FIGURE 1.** Diagram of a *Xenopus* oocyte showing the position of a microinjection needle situated for RNA injection. The position of the nucleus in the animal hemisphere is indicated since it is not readily visible due to the heavy pigmentation.

the largest or stage VI oocytes being arrested at meiotic prophase.[6] The stage VI oocytes are normally used for microinjection since they are quite active in RNA and protein synthesis but inactive in DNA synthesis.[6] There are two primary features of these oocytes which make them ideal for microinjection studies. First, the large size of the oocyte (1 to 1.2 mm in diameter) makes it relatively easy to microinject either mRNA into the cytoplasm or DNA into the nucleus (Figure 1). One mature oocyte has approximately the same volume (0.54 µl) and composition as 200,000 somatic cells.[7] Second, the oocyte contains large amounts of the components required for the expression of injected genes, such as RNA polymerases I, II, and III, histones, and ribonucleotide triphosphates.[8] It has been estimated that the oocyte nucleus contains sufficient amounts of the three eukaryotic RNA polymerases to furnish the needs of the developing *Xenopus* embryo at least until the 30,000 cell stage.[4] Each oocyte also has 200,000 more ribosomes and 10,000 more tRNA molecules than are found in a somatic cell.[2] Finally, the oocyte has all of the enzymes required to post-translationally modify and compartmentalize or secrete newly made plant proteins.

## B. *XENOPUS* MAINTENANCE

*Xenopus laevis* adult females can be obtained from a number of suppliers such as Xenopus 1 (716 Northside, Ann Arbor, MI), Carolina Biological Supply Co. (2700 York Rd., Burlington, NC), or Nasco (901 Janesville Ave., Ft. Atkinson, WI). Raising and maintenance of *Xenopus* in the laboratory has been described.[25] If only a few frogs are purchased at any one time, the animals can be kept in large plastic tanks with approximately 4 to 5 l of water per frog at a depth of not more than 10 to 15 cm. These frogs are notorious escape artists, so a lid with air holes should be secured to the top of the tanks. The water must be dechlorinated, free from heavy metals, and maintained at temperatures ranging from 18 to 22°C. The animals are very hardy, but care should be taken not to subject them to drastic changes in water temperature. Since *Xenopus* are fully aquatic, there is no need for a dry surface or resting platform in the tank. The frogs should be fed two to three times per week with either chopped beef heart, Nasco frog brittle, or Purina trout chow. Once one food source has been started do not change it even if the animals are not feeding. Newly arrived animals may not eat for 1 to 2 weeks. When transferring animals from one tank to another or collecting one for oocyte isolation, use a small fish net which is available at local aquarium supply houses or pet stores. Since *Xenopus* have a slippery protective coating of mucous, the best way to hold the animal is to place an index finger between the hind legs and the rest of the hand around the body.

## C. ISOLATION OF THE OOCYTES

An adult female *Xenopus* contains approximately 30,000 oocytes. In a normal experiment, one may use just a few hundred for microinjection. Therefore, it is advantageous to reuse

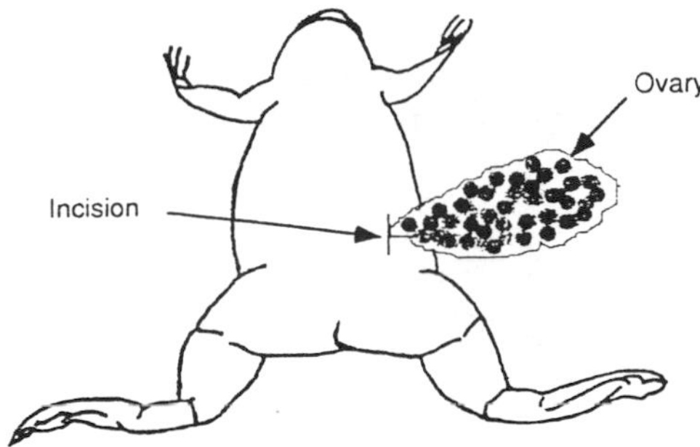


**FIGURE 2.** Diagram illustrating the removal of the ovary from *Xenopus*.

a female for the isolation of oocytes rather than killing it. A large series of experiments may necessitate sacrificing the animal. Persons unfamiliar with the anatomy of the female frog may wish initially to sacrifice an animal for the removal of the oocytes. Anesthetize a female frog by immersing it in 0.5% ethyl *m*-aminobenzoate (Sigma) for approximately 15 to 30 min.[5] Wear gloves as ethyl *m*-aminobenzoate is a potential carcinogen in humans. Once the frog is anesthetized, it is rinsed off with water. Alternatively, one can anesthetize the frog by lowering its body temperature. This is done by immersing it in a bowl of ice for 30 to 45 min.[26,27] Once the animal has ceased moving, the oocytes can be removed.

If the animal is to be sacrificed, place it on a dissecting tray, and decapitate using a sharp scalpel. If the frog is to be reused, then oocytes should be removed surgically employing sterile instruments and technique. Place the frog ventral side up on a bed of ice so that the back but not the head is in contact with the ice. Swab the lower abdomen of the frog with a cotton ball soaked in alcohol. Hold the skin with a pair of forceps and make a 1 to 2 cm incision in the skin using a sharp scalpel or pair of dissecting scissors at the site indicated in Figure 2. A similar cut is made through the underlying muscle. Since the ovary completely fills the abdomen, the oocytes should be visible. Using forceps, pull out a section of ovary, cut it off with a pair of scissors, and transfer it to a Petri dish containing modified Barth's medium (MBS; Table 2). Then push the remaining ovary back into the animal. The muscle is sewn up with dissolving suture material (e.g., Ethicon plain gut 4/0), and the skin is stitched with silk sutures (e.g., Ethicon 3/0 black braided silk). The skin sutures will eventually fall out. Then place the frog on its ventral side in a container with a small amount of water. Make sure the head is elevated by wet paper towels. The frog should recover completely within a few hours.

Using two pairs of fine forceps, remove individual oocytes from the ovarian clumps. This is done by holding the membranous material to which the oocytes are attached with one pair of forceps and gently pulling the oocyte off at its base. Choose the largest or stage VI oocytes, having a diameter of approximately 1.0 to 1.2 mm. Transfer the oocytes, with a wide mouth Pasteur pipette, to a fresh Petri dish containing MBS. Alternatively, oocytes can be released by enzymatic stripping.[5,26] Place clumps of oocytes in a solution containing 2 mg/ml collagenase (Sigma Type II) in MBS, and gently swirl on a rotary shaker for 2 to 4 h at 18 to 24°C. The released oocytes are then washed thoroughly by transferring them through a series of Petri dishes containing MBS. The released oocytes are still surrounded by several layers of follicular cells which can be removed by additional collagenase treatment if necessary.[6] The oocytes can be cultured for up to a week or more with daily changes of medium and removal of dead cells. Dead oocytes can be recognized easily by a color change. For example, the vegetal hemisphere of dead oocytes is quite pale compared to that of live oocytes. After isolating mature oocytes, incubate them overnight prior to injection, and use only the healthy looking ones.

**TABLE 2**
**Composition of Modified Barth's Medium[a]**

| |
|---|
| 88.0 m*M* NaCl |
| 1.0 m*M* KCl |
| 2.4 m*M* $NaHCO_3$ |
| 15.0 m*M* Hepes-NaOH (pH 7.6) |
| 0.82 m*M* $MgSO_4 \cdot 7H_2O$ |
| 0.33 m*M* $CaNO_3 \cdot 4H_2O$ |
| 0.41 m*M* $CaCl_2 \cdot 6H_2O$ |

[a] Using stock solutions for the various ingredients, follow the order given in the table. It may be desirable to make a 10X stock solution which can be filter sterilized and stored for several months at 4°C. Prior to use, add freshly prepared sodium penicillin and streptomycin sulfate to a final concentration of 10 µg/ml. When analyzing translation products which will be secreted, 100 µg/ml gentamicin and 20 U/ml nystatin should be added.

It is difficult to inject oocytes in a Petri dish without stabilizing them in some fashion. One strategy for immobilizing oocytes involves placing them in a small plastic Petri dish which has a piece of polyethylene mesh (Fisher Scientific, Cat. No. 8-670-176) secured to the bottom with a few drops of chloroform. The size of the mesh wells (approximately 750 µm$^2$) is sufficient to seat the oocytes properly.[4,27] Transfer the oocytes to the microinjection support in MBS, and then manipulate them into the correct orientation with a hair loop or a Pasteur pipette which has been flamed to melt the tip into a smooth probe. Prior to injection, the level of MBS can be lowered so that the meniscus is just below the surface of the oocyte.

Another simple technique for immobilizing oocytes involves casting 1% agarose, made up in MBS in small Petri dishes, to a depth of approximately 0.2 to 0.3 cm.[4] Pits in the surface are then created by using a Pasteur pipette to remove plugs of agarose. The Petri dish is filled with MBS, the oocytes are placed in the pits or depressions, and then they are manipulated so that they will be in the proper orientation depending on whether RNA or DNA will be injected.

## D. MICROINJECTION EQUIPMENT

The equipment required for *Xenopus* oocyte microinjection is relatively simple and inexpensive. One requires a stereomicroscope (10 to 40 X) with a light source, micropipettes, and a microinjection system which usually consists of a micromanipulator and syringe. A number of commercial systems for microinjection, capable of accurately delivering volumes in the nanoliter range, are readily available, such as the Pico-Injector (PLI-100; Medical Systems Corp., Greenvale, NY), the Nanoinject (Cat. No. 3-00-203-X; Drummond Scientific Co., Broomall, PA), and the Eppendorf Microinjector (Model 5242; Fremont, CA). Each system comes with detailed instructions and references to studies employing that equipment. Also, the assembly of a microinjection system employing a microprocessor-controlled pipettor has been described.[28,29] Our laboratory has employed the latter system with excellent results in experiments using oocytes as well as fertilized eggs.[30] Additionally, there have been numerous detailed publications outlining microinjection set-ups which can be built very inexpensively.[3-6] Attachment of the microinjection apparatus to the micromanipulator allows the device to be moved in three dimensions. Numerous types are on the market, including ones from Narashige Scientific Instrument Laboratory and Brinkmann Instruments.

Most of the success associated with microinjection of oocytes depends on having a proper sized needle or micropipette. In order to make micropipettes, buy or have access to a micropipette or microelectrode puller (e.g., Harvard Apparatus, Model 50-2005). This device heats a 10 cm hard-glass capillary tube (e.g., Kimax, Article No. 35400; 1.5 to 1.8 mm × 100 mm) evenly in the middle, while at the same time pulling at both ends. When the glass begins to melt, the capillary tube is stretched and eventually pulled apart. The tip of the tube is broken under a microscope using fine forceps to generate an opening of approximately 10 to 20 μm in diameter as judged by a microscope stage micrometer. Micropipettes can be stored by fixing them onto a strip of modeling clay in large Petri dishes so that the tip does not come into contact with either the clay or the walls or bottom of the Petri dish. If necessary, the micropipettes can be sterilized at 200°C for 1 to 2 h prior to use.[5] A microforge has been used by some researchers to generate a sharpened tip that will facilitate microinjection.[5,26] This procedure may not be necessary for routine injection of mRNA.

Once the equipment is assembled, a calibration series should be carried out to ensure that the system can deliver relatively accurate volumes. This can be done by taking up a known amount of radioactive label into the tip of the micropipette and injecting specific amounts into water droplets placed on Parafilm. Analysis of the radioactivity by scintillation counting will allow one to determine the accuracy of the delivery system. The volumes routinely used for microinjection are approximately 50 nl for RNA and 10 to 20 nl for DNA. For experimental reproducibility, it is desirable to use the same needle to inject a set of oocytes.

## III. RNA MICROINJECTION

### A. BACKGROUND

The *Xenopus* oocyte system has many advantages over conventional *in vitro* translation systems. While *in vitro* translation systems respond in a linear fashion for only a short time (90 to 120 min), oocytes can translate mRNA for many hours or days depending on the stability of the mRNA. Gurdon et al.[31] estimated that the half-life of rabbit globin mRNA in *Xenopus* oocytes was more than 1 week. This extensive protein synthetic period permits the translation of very long mRNAs which cannot be translated efficiently in standard *in vitro* systems. The stability of the injected mRNA depends on its sequences as well as on the presence of a 5′ cap and a 3′ poly(A) tail.[2]

*In vitro* systems are also poor in performing post-translational modifications. In contrast, *Xenopus* oocytes possess the ability to post-translationally modify certain proteins. These post-translational modifications include cleavage of protein precursors, phosphorylation, acetylation, and glycosylation.[1,5] Unlike *in vitro* translation systems, the *Xenopus* oocyte system has been used to examine the processes of secretion and compartmentalization of proteins.[1,5] For example, microinjection of barley secretory protein mRNA into oocytes resulted in a translational product which is exported from the oocyte into the medium.[16]

### B. RNA PREPARATION

The injection of polyadenylated mRNA isolated by oligo(dT)-cellulose chromatography or synthetic mRNA is preferable to the injection of total RNA, although the presence of ribosomal RNA and tRNA will not have a deleterious effect on the viability of the oocyte.[5] Purified RNA is usually dissolved in sterile distilled water or injection buffer (88 m*M* NaCl, 15 m*M* Tris-HCl pH 7.5) at a concentration of 0.2 to 2 mg/ml. The final injection volume should not be more than 50 nl since larger volumes have a tendency to leak out of the oocyte. The RNA solution should be spun in a microcentrifuge for 1 to 2 min at 10,000 × g to remove particles that might otherwise clog the injection needle. The RNA solution is placed as a droplet onto a sheet of Parafilm for ease of filling the needle. Since some of the RNA will be degraded by the oocyte upon injection, the translation of the mRNA will not be proportional to the amount injected.[5] It is advantageous to inject a range of RNA concentrations to determine the optimal amount. Using this approach, Spanu et al.[15] determined that 2.5 ng of polyadenylated RNA and 500 ng of total RNA were required to achieve maximal levels of tomato ethylene-forming enzyme. Commercially available rabbit globin mRNA (Bethesda Research Laboratories) can be used as control.

### C. RNA MICROINJECTION PROCEDURE

Transfer five to eight oocytes onto the microinjection support as described above (Section II.C). Injections in the vegetal region or near the "equator" of the oocyte avoid the possibility of hitting and damaging the nucleus. Lower the needle using the micromanipulator at a 20° angle above horizontal, and pierce the oocyte (Figure 1). A steeper angle may result in yolk platelets floating up into the needle. The oocyte will display some resistance and flatten slightly before the needle pops into the cytoplasm. Inject the RNA, and then slowly withdraw the needle. Once all of the oocytes are injected, transfer them carefully to a Petri dish with MBS. Allow the mRNA to translate for 12 to 48 h at 20°C before analyzing the protein product. Maintaining the oocytes at 25°C or higher will decrease their survival. While one can detect the translation products of injected mRNA in a single oocyte, it is advisable to pool ten oocytes in order to average individual differences in oocytes with respect to translational efficiency. With practice, several hundred oocytes can be injected per hour.

### D. RADIOACTIVE LABELING OF THE OOCYTES

Ideally, detection of the mRNA translation product would involve an enzyme or immunological assay. If labeling of the protein is desired, then radioactive amino acids such as $^{35}$S-methionine (1 to 5 mCi/ml of medium; 350 to 800 Ci/mmol) or $^{3}$H-leucine (1 mCi/ml of medium; 50 Ci/mmol) can be added to the medium and incubated for 8 to 10 h.[26] One can also inject label into the oocyte (1 to 2 μCi of $^{35}$S-methionine at 800 to 1200 Ci/mmol or 1 μCi of $^{3}$H leucine at 20 Ci/mmol) followed by incubation for 1 to 6 h. The latter method removes the lag time required for amino acid uptake by the oocyte. If individual oocytes are labeled, this can be performed in microtitre plates,[5] an ideal container when the manufactured protein is secreted into the media. The plastic tops of small vials may also be used for labeling small numbers of oocytes.

### E. PREPARATION OF SOLUBLE PROTEIN EXTRACT

Analysis of the translation product may involve immunoprecipitation, electrophoresis, or measurement of enzyme activity. In order to prepare a soluble extract, oocytes are homogenized on ice in 0.1 *M* NaCl, 1% Triton-X-100, 1 m*M* phenylmethylsulfonyl fluoride, and 20 m*M* Tris-HCl pH 7.6 at 50 μl/oocyte. The homogenate is spun at 10,000 g in a microcentrifuge at 4°C for 5 min before removing the supernatant with a fine pipette and making sure to avoid the overlying pellicle which is composed mostly of yolk. The supernatant can be analyzed by electrophoresis by adding electrophoretic sample buffer, or it can be used directly to immunoprecipitate selected proteins.[5]

## IV. DNA MICROINJECTION

### A. BACKGROUND

The *Xenopus* oocyte system has certain advantages over transfection of eukaryotic cells for studies examining the regulation of gene expression. For example, genomic DNA sequences are microinjected easily into *Xenopus* oocytes in specific reproducible quantities. In contrast, transient transfection assays in eukaryotic cells can result in a variable number of copies of transfected genes per cell as well as variable expression of the introduced gene from cell to cell.[32,33] The most efficient transcription of microinjected DNA in *Xenopus* oocytes occurs when the DNA that is injected into the nucleus is in a circular form.[8] While linear DNA will serve as a template for transcription, it is not as efficient as circular DNA and is gradually degraded by exonuclease activity. In contrast, circular DNA can remain viable for several weeks postinjection, probably due, in part, to the high levels of endogenous ligase activity, which rapidly repairs any single-stranded nicks in the injected DNA. Injected DNA is not replicated and does not integrate into host oocyte chromosome, but is assembled into a normal chromatin structure.[7]

The *Xenopus* oocyte *in vivo* transcription system has proven useful in the analysis of the involvement of *cis*-acting DNA sequences in the transcription of cloned genes. For example, analysis of various deletion mutants of sea urchin histone genes has revealed a number of 5′ and 3′ elements which are essential for or can modulate the expression of these genes.[34] Promoter regions of genes can be fused to reporter genes such as those encoding the bacterial enzyme chloramphenicol acetyl transferase or β-galactosidase. This methodology may be useful for the analysis of plant promoters that do not have a readily assayable product or whose gene product has a counterpart that is active in the *Xenopus* oocyte.

## B. DNA MICROINJECTION PROCEDURE

The microinjection of DNA into the germinal vesicle or nucleus of the oocytes is technically more difficult than microinjection of RNA into oocytes. The main problem is the ability to pierce the germinal vesicle consistently. This takes a bit of practice since the germinal vesicle is located in the animal hemisphere and is not visually apparent due to heavy pigmentation. The approach employed most often involves a "blind" technique.[5] The microinjection support (see Section II.D) is placed in a large Petri dish containing crushed ice. The lower temperature has been shown to reduce mortality following DNA microinjection. Using forceps, orient the oocyte so that the animal pole is directed toward the point of the microinjection needle. Carefully lower the needle using the micromanipulator, and insert the needle approximately one-third of the way from the point of insertion toward the opposite vegetal pole.[6] Care must be taken not to advance the needle too far into the oocyte, or it may pass completely through the nucleus. Microinject approximately 5 to 10 ng of DNA in a total volume of not more than 10 to 20 nl. Beginners may wish to inject dye such as 0.2% trypan blue into the oocyte. Afterwards, the injection efficiency can be tested by dissecting out the nucleus. This is done by transferring the oocytes to a Petri dish containing MBS and then making an incision in the animal hemisphere with the tip of a syringe needle. Gently squeeze the oocyte with a pair of tweezers to expel the blue-stained germinal vesicle from the oocyte. With practice the efficiency of this method can approach 80%.[6] This injection protocol was used successfully by Aoyagi et al.[19] to examine the expression of barley α-amylase genes.

A second approach to nuclear injection involves low speed centrifugation of the oocytes, which causes the nucleus to migrate to near the surface of the animal pole and facilitates injection.[5] The first step in this method involves the preparation of a support as previously described (Section II.D) to fix the individual oocytes in the proper orientation so that the animal hemisphere is uppermost. The Petri dishes are then placed in a flat-bottomed 1-liter centrifuge bucket, attached to a swing-out rotor of a clinical centrifuge, and spun at 450 to 600 × g for 10 min at 20°C. After centrifugation, the nucleus is visible as a pale area in the pigmented animal hemisphere since it has displaced the pigment granules. Transfer the microinjection support onto a larger Petri dish containing ice, and microinject the DNA into the oocyte nuclei. Injections should be carried out within 45 min since the nuclei will tend to migrate back into the animal region. Once again, it is advisable to practice with a solution of 0.2% trypan blue before attempting the procedure with DNA. It should be noted that variability in the efficiency of transcription of injected DNA does exist in oocytes from

**TABLE 3**
**Composition of Buffers Used in the Isolation of RNA from Oocytes**

**A. Extraction Buffer[a,b]**

| Final concentration | per liter |
|---|---|
| 3 *M* lithium chloride | 127.11 g |
| 6 *M* urea | 360.36 g |
| 0.5% sodium dodecyl sulfate | 25 ml of 20% (w/v) stock |
| 70 m*M* β-mercaptoethanol | 4.9 ml of 14.3 *M* solution |
| 10 m*M* sodium acetate | 3 ml of 3 *M* stock |

**B. Resuspension Buffer[b,c]**

| Final concentration | per 100 ml |
|---|---|
| 0.2% sodium dodecyl sulfate | 1 ml of 20% stock |
| 100 m*M* sodium acetate | 3.3 ml of 3 *M* stock |

[a] Filter sterilize buffer; then add sodium dodecyl sulfate and β-mercaptoethanol.
[b] Adjust the pH to 5.0 with acetic acid if necessary.
[c] Autoclave the solution, and then add sodium dodecyl sulfate.

different frogs. Thus, comparisons should be carried out with oocytes from a single batch in which at least ten oocytes have been pooled to average out any differences.

## C. ISOLATION OF OOCYTE RNA

Transcription of injected genes can be monitored by analyzing the RNA transcripts or examining the protein product. The use of techniques such as $S_1$ nuclease mapping or RNase protection analysis will permit the detection of the transcript and whether it was correctly initiated. There are a number of protocols published for the isolation of oocyte RNA.[4,6] One RNA isolation method developed by Auffray and Rougeon[35] and later modified by Mohun et al.[36] has worked effectively in our laboratory and is given below. Homogenize the oocytes at a concentration of 20 to 30 oocytes/ml of cold extraction buffer (Table 3) using a glass homogenizing tube and teflon pestle (five to ten strokes). Pour the homogenate into a sterile 15 ml Corex tube, cover with Parafilm, and leave at 4°C overnight. The extraction buffer will solubilize DNA, but RNA and some protein will precipitate under these conditions. Centrifuge the homogenate in a Sorvall centrifuge in an SA600 rotor at 10,500 rpm for 30 min at 4°C or in an SS34 rotor at 11,500 rpm for 30 min at 4°C. Decant the supernatant and invert the tubes over a clean paper towel for about 15 min to allow all of the buffer to drain. Resuspend the pellet in 4 ml of resuspension buffer (Table 3) by gently vortexing. Extract the solution twice with an equal volume of phenol:chloroform (1:1) and twice with an equal volume of chloroform. Place the final aqueous fraction in a sterile 15 ml Corex tube and add 1/10 volume of 3 *M* sodium acetate (i.e., to a final concentration of 0.3 *M*) and 2 to 2.5 volumes of absolute ethanol. Cover the tubes with Parafilm, mix the contents thoroughly by inverting the tubes, and then place at –20°C overnight. Centrifuge the tubes at 10,000 rpm for 50 min at 4°C in a Sorvall centrifuge using an SA 600 rotor. Drain the tubes, add 4 ml of 70% ethanol, and centrifuge again at 10,000 rpm for 20 min. Pour off the ethanol, and dry the tubes under vacuum for about 15 min. Resuspend the RNA in 200 μl of sterile distilled water. Use 5 μl to determine the RNA concentration with a spectrophotometer. After dividing the RNA into useful aliquots (e.g., 10 μg), lyophilize, and store the material at –80°C.

## V. ADDITIONAL USES OF *XENOPUS* OOCYTES TO STUDY PLANT GENE EXPRESSION

Partial purification of plant mRNA fractions in order to enrich the desired message is desirable prior to preparing a cDNA library. The *Xenopus* oocyte translation system is ideal for the assay of mRNA, particularly when the message is present in low quantities. Fractions enriched for a specific plant mRNA can be detected by assaying for the translation product by two dimensional polyacrylamide gel electrophoresis, immunoblotting, enzyme activity, or some other physical characteristic of the protein. Using this type of protocol, messenger RNA fractions enriched in sequences coding for a portion of a mouse 5-HT receptor were detected following microinjection into *Xenopus* oocytes and voltage clamp analysis.[37] The enriched fractions were used to make a pool of cDNAs. In order to purify the 5-HT receptor cDNA, the pooled cDNAs were used to hybrid-select mRNAs which would synthesize part of the 5-HT receptor. This type of protocol has been reviewed recently by Snutch[38] and is readily applicable to plant mRNA.

Finally, recent work by Tigyi et al.[39] has shown that *Xenopus* oocytes can be used as immunological vectors to produce antibodies to brain cell antigens. In this protocol, *Xenopus* oocytes were used to translate rat brain mRNA. An oocyte membrane fraction, which contained some of the foreign protein, was used to immunize mice which had been rendered immunotolerant to antigens of native oocyte membrane. Immunized mice were then used to generate monoclonal antibodies which reacted specifically to cerebellar pinceau terminals of rat. Undoubtedly, this approach, employing *Xenopus* oocytes as immunological vectors to generate unique antibodies, could also be used to generate antibodies to plant proteins.

## REFERENCES


1. **Heikkila, J. J.,** Expression of cloned genes and translation of messenger RNA in microinjected *Xenopus* oocytes, *Int. J. Biochem.,* 22, 1223, 1990.
2. **Melton, D. A.,** Translation of messenger RNA in injected frog oocytes, in *Methods in Enzymology,* Vol. 152, Berger, S. L. and Kimmel, A. R., Eds., Academic Press, New York, 1987, 288.
3. **Stephens, D. L., Miller, T. J., Silver, L., Zipser, D., and Mertz, J. E.,** Easy-to-use equipment for the accurate microinjection of nanoliter volumes into the nuclei of amphibian oocytes, *Anal. Biochem.,* 114, 299, 1981.
4. **Coleman, A.,** Expression of exogenous DNA in *Xenopus* oocytes, in *Transcription and Translation: A Practical Approach,* Hames, B. D. and Higgins, S. J., Eds., IRL Press, Washington, D.C., 1984, 49.
5. **Coleman, A.,** Translation of eukaryotic messenger RNA in *Xenopus* oocytes, in *Transcription and Translation: A Practical Approach,* Hames, B. D. and Higgins, S. J., Eds., IRL Press, Washington, D.C., 1984, 271.
6. **Gurdon, J. B. and Wickens, M. P.,** The use of *Xenopus* oocytes for the expression of cloned genes, in *Methods in Enzymology,* Vol. 101, Wu, R., Grossman, L., and Moldave, K., Eds., Academic Press, New York, 1983, 370.
7. **Gurdon, J. B. and Wakefield, L.,** Microinjection of amphibian oocytes and eggs for the analysis of transcription, in *Microinjection and Organelle Transplantation Techniques*, Cellis, J. E., et al., Eds., Academic Press, London, 1986, 289.
8. **Gurdon, J. B. and Melton, D. A.,** Gene transfer in amphibian eggs and oocytes, *Annu. Rev. Genet.,* 15, 189, 1981.
9. **Hurkman, W. J., Smith, L. D., Richter, J., and Larkins, B. A.,** Subcellular compartmentalization of maize storage proteins in *Xenopus* oocytes injected with zein messenger RNAs, *J. Cell Biol.,* 89, 292, 1981.
10. **Larkins, B. A., Pedersen, K., Handa, A. K., Hurkman, W. J., and Smith, L. D.,** Synthesis and processing of maize storage proteins in *Xenopus laevis* oocytes, *Proc. Natl. Acad. Sci. U.S.A.,* 76, 6448, 1979.

11. **Matthews, J. A., Brown, J. W. S., and Hall, T. C.,** Phaseolin mRNA is translated to yield glycosylated polypeptides in *Xenopus* oocytes, *Nature,* 294, 175, 1981.
12. **Vitale, A., Sturm, A., and Bollini, R.,** Regulation of processing of a plant glycoprotein in the golgi complex: a comparative study using *Xenopus* oocytes, *Planta,* 169, 108, 1986.
13. **Lane, C. D., Coleman, A., Mohun, T., Morser, J., Champion, J., Kourides, I., Craig, R., Higgins, S., James, T. C., Applebaum, S. W., Ohlsson, R. I., Pancha, E., Houghton, M., Matthews, J., and Miflim, B. J.,** The *Xenopus* oocyte as a surrogate secretory system. The specificity of protein export, *Eur. J. Biochem.,* 111, 225, 1980.
14. **Bassuner, R., Huth, A., Manteuffel, R., and Rapoport, T. A.,** Secretion of plant globulin polypeptides by *Xenopus laevis* oocytes, *Eur. J. Biochem.,* 133, 321, 1983.
15. **Spanu, P., Reinhardt, D., and Boller, T.,** Analysis and cloning of the ethylene-forming enzyme from tomato by functional expression of its mRNA in *Xenopus laevis* oocytes, *EMBO J.,* 10, 2007, 1991.
16. **Simon, P. and Jones, R. L.,** Synthesis and secretion of catalytically active barley α-amylase isoforms by *Xenopus* oocytes injected with barley mRNAs, *Eur. J. Cell Biol.,* 47, 213, 1988.
17. **Hata, S., Kouchi, H., Suzuka, I., and Ishii, T.,** Isolation and characterization of cDNA clones for plant cyclins, *EMBO J.,* 10, 2681, 1991.
18. **Galili, G., Kawata, E. E., Smith, L. D., and Larkins, B. A.,** Role of the 3′-poly(A) sequence in translational regulation of mRNAs in *Xenopus laevis* oocytes, *J. Biol. Chem.,* 263, 5764, 1988.
19. **Aoyagi, K., Sticher, L., Wu, M., and Jones, R. L.,** The expression of barley α-amylase genes in *Xenopus laevis* oocytes, *Planta,* 180, 333, 1990.
20. **Wallace, J. C., Galili, G., Kawata, E. E., Cuellar, R. E., Shotwell, M. A., and Larkins, B. A.,** Aggregation of lysine-containing zeins into protein bodies in *Xenopus* oocytes, *Science,* 240, 662, 1988.
21. **Knowland, J.,** Protein synthesis directed by the RNA from a plant virus in a normal animal cell, *Genetics,* 78, 383, 1974.
22. **Huez, G., Cleuter, Y., Bruck, C., Van Vloten-Doting, L., Goldbach, R., and Verduin, B.,** Translational stability of plant viral RNAs microinjected into living cells: influence of a 3′-poly(A) segment, *Eur. J. Biochem.,* 130, 205, 1983.
23. **Ballas, N., Broido, S., Soreq, H., and Loyter, A.,** Efficient functioning of plant promoters and poly(A) sites in *Xenopus* oocytes, *Nucl. Acids Res.,* 17, 7891, 1989.
24. **Benvenisty, N., Shoshanik, T., Farkash, Y., Soreq, H., and Reshef, L.,** trans-Activation of rat phosphoenolpyruvate carboxykinase (GTP) gene expression by micro-coinjection of rat liver mRNA in *Xenopus laevis* oocytes, *Mol. Cell Biol.,* 9, 5244, 1989.
25. **Wu, M. and Gerhart, J.,** Raising *Xenopus* in the laboratory, in *Methods in Cell Biology,* Vol. 36, Kay, B. K. and Peng, H. B., Eds., Academic Press, New York, 1991, 3.
26. **Kawata, E. E., Galili, G., Smith, L. D., and Larkins, B. A.,** Translation in *Xenopus* oocytes of mRNAs transcribed *in vitro, Plant Mol. Biol. Man.,* B7, 1, 1988.
27. **Kay, B. K.,** Injection of oocytes and embryos, in *Methods in Cell Biology,* Vol. 36, Academic Press, New York, 1991, 663.
28. **Hitchcock, M. J. M. and Friedman, R. M.,** Microinjection of *Xenopus* oocytes: an automated device for volume control in the nanoliter range, *Anal. Biochem.,* 109, 338, 1980.
29. **Hitchcock, M. J. M., Ginns, E. I., and Marcus-Sekura, C. J.,** Microinjection into *Xenopus* oocytes: equipment, in *Methods in Enzymology,* Vol. 152, Berger, S. L. and Kimmel, A. R., Eds., Academic Press, New York, 1987, 276.
30. **Krone, P. H. and Heikkila, J. J.,** Expression of microinjected hsp 70/CAT and hsp 30/CAT chimeric genes in developing *Xenopus laevis* embryos, *Development,* 106, 271, 1989.
31. **Gurdon, J. B., Lingrel, J., and Marbaix, G.,** Message stability in injected frog oocytes: long life of mammalian α and β globin messages, *J. Mol. Biol.,* 80, 539, 1973.
32. **Domen, J., Van Leen, R. W., Lubsen, N. H., and Schoenmakers, J. G. G.,** A vital staining method for measuring the efficiency of transfection of eukaryotic cells, *Anal. Biochem.,* 155, 379, 1986.
33. **Spandidos, D. A. and Wilkie, N. M.,** Expression of exogenous DNA in mammalian cells, in *Transcription and Translation: A Practical Approach,* Hames, B. D. and Higgins. S. J., Eds., IRL Press, Oxford, 1984, 1.
34. **Grosschedl, R. and Birnstiel, M. L.,** Spacer DNA sequences upstream of the TATAAATA sequence are essential for promotion of H2A histone gene transcription *in vivo, Proc. Natl. Acad. Sci. U.S.A.,* 77, 7102, 1980.
35. **Auffray, C. and Rougeon, F.,** Purification of the mouse immunoglobulin heavy chain messenger RNAs from total myeloma tumour RNA, *Eur. J. Biochem.,* 107, 303, 1980.
36. **Mohun, T. J., Brennan, S., Dathan, N., Fairman, S., and Gurdon, J. B.,** Cell-type specific activation of actin genes in the early amphibian embryo, *Nature,* 311, 716, 1984.

37. **Lubbert, H., Hoffman, B. J., Snutch, T. P., Van Dyke, T., Levine, A. J., Hartig, P. R., Lester, H. A., and Davidson, N.,** cDNA cloning of a serotonin $5HT_{1C}$ receptor by electrophysiological assays of mRNA-injected *Xenopus* oocytes, *Proc. Natl. Acad. Sci. U.S.A.*, 84, 4332, 1987.
38. **Snutch, T. P.,** The use of *Xenopus* oocytes to probe synaptic communication, *Trends Neurosci.*, 11, 250, 1988.
39. **Tigyi, G., Matute, C., and Miledi, R.,** Monoclonal antibodies to cerebellar pinceau terminals obtained after immunization with brain mRNA-injected *Xenopus* oocytes, *Proc. Natl. Acad. Sci. U.S.A.*, 87, 528, 1990.

# 12

# *In Situ* Localization of Specific mRNAs in Plant Tissues

*Heather I. McKhann and Ann M. Hirsch*

## I. INTRODUCTION

*In situ* hybridization permits the localization of specific mRNA transcripts at the cellular level by hybridizing a specific probe to tissue sections in which the mRNA is preserved. Thus, *in situ* hybridization is a powerful method for studying gene expression during development. In addition to specific cellular localization, it offers the advantage over other methods, such as northern analysis, of increased sensitivity: mRNAs of low abundance can be detected even in a few cells. Moreover, genes that are expressed only at certain developmental stages and/or in specific cell types can be studied. For example, we have used *in situ* hybridization techniques successfully to study the spatial patterns of expression of nodulin genes in *Rhizobium meliloti*-induced alfalfa root nodules (Figure 1A–D).

*In situ* hybridization, first described in 1969,[1,2] was used initially to map ribosomal sequences in *Xenopus* oocytes. Subsequently, it was used to map repetitive sequences of *Drosophila* polytene chromosomes[3,4] and later to map single copy genes to animal[5,6] and plant[7-9] chromosomes. Technological advances have now made it possible to localize RNA in cytological preparations. *In situ* hybridization has been widely used in animal systems, and in more recent years, has been applied to plant tissues (see Table 1). Because *in situ* hybridization has been in use for a much longer time in animal systems, many excellent review articles, which may be useful for researchers working on plant tissues, are available.[10-12] We cite these references when comparable information for plant tissues is not available. For example, there is a very useful troubleshooting guide written by McCabe and Pfaff,[12] and there are also a number of papers describing the use of oligonucleotide probes.[13,14] Furthermore, many of the newest techniques have been developed in animal systems (e.g., *in situ* transcription[11]) and are only beginning to be applied to plants.[15]

## II. EXPERIMENTAL CONSIDERATIONS

The protocols given here are the ones used in our laboratory. However, many variations on the procedures exist, particularly with regard to how the tissue is sectioned and the type of probe used. The procedure presented here for hybridization with $^{35}S$-labeled probes is derived largely from that of Cox and Goldberg,[16] while that for using digoxigenin-labeled probes was developed in our lab[17] using the Boehringer Mannheim protocol for blot hybridization as a basis. These references can be consulted for additional details. We will outline the main steps in our protocol and discuss some of the factors involved in deciding which technique is most appropriate for a given experiment. Because it may be necessary to modify

0-8493-5164-2/93/$0.00+$.50
© 1993 by CRC Press, Inc.

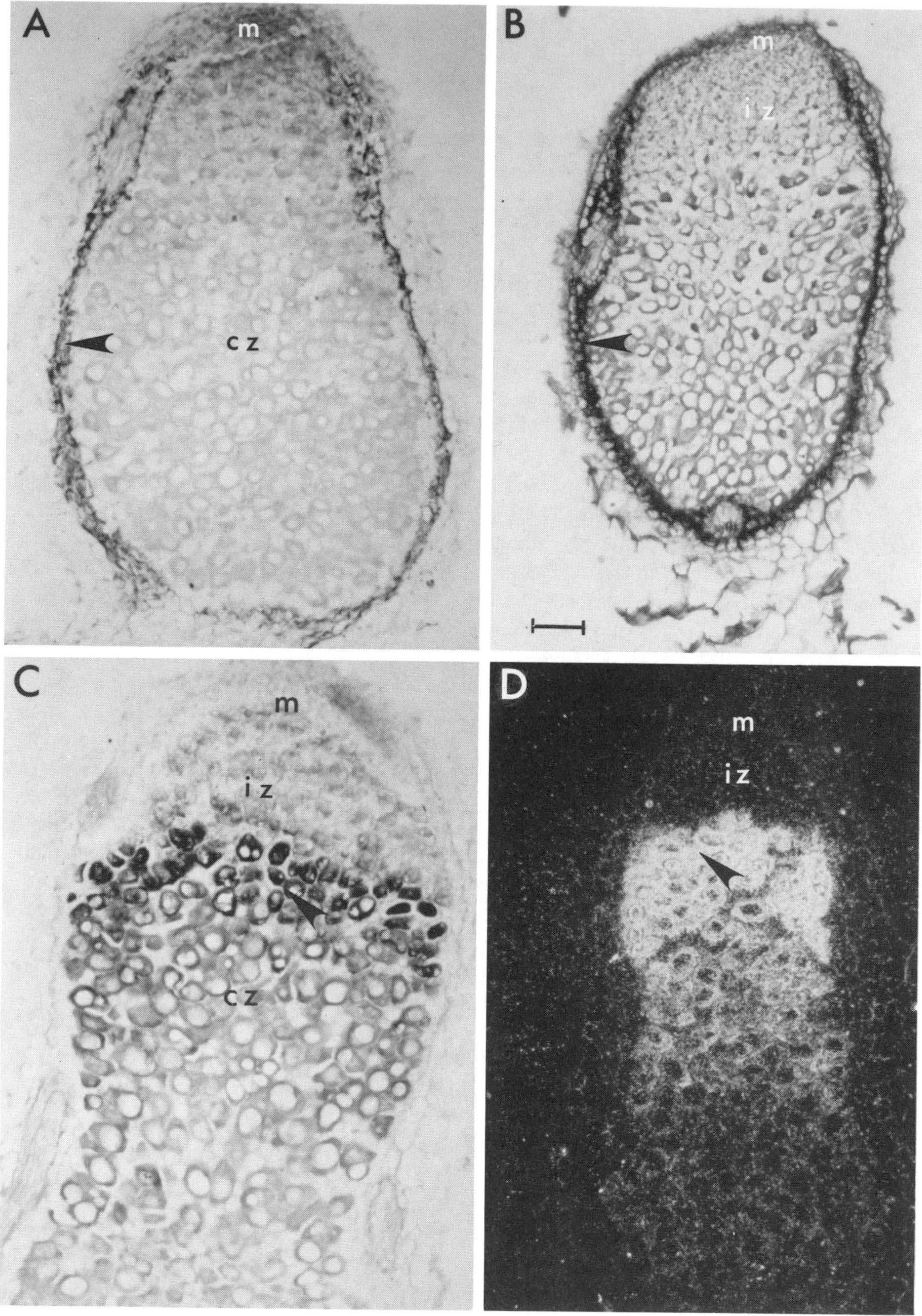


**FIGURE 1.** Comparison of digoxigenin and $^{35}$UTP labeling using comparable tissue and probes. All photographs are X85. Bar = 100 µm. Full caption accompanies color plate, which can be found between pages 268 and 269. (Parts A and C from *Plant Mol. Biol. Rep.* 8(4): 237–248 (1990). With permission.)

**TABLE 1**
**Some Applications of *In Situ* Hybridization in Plants to Date**

| Gene or gene product | Plant | Localization | Comments | Ref. |
|---|---|---|---|---|
| T-DNA | *Crespis capillaris* | Chromosome | Biotinylated or $^3H$-labeled DNA | 7, 8 |
| Stylar S2 Glycoprotein | *Nicotiana alata* | Flowers, style | $^{32}P$-labeled cDNA probe | 41–43 |
| Legumin | Pea | Seedlings | Biotinylated cDNA probe | 44–46 |
| PEPC, RuBISCO | Maize | Leaves, mesophyll | $^3H$-labeled RNA probe | 47 |
| Seed storage protein | *Arabidopsis* | Embryos | $^3H$-labeled DNA, RNA $^{35}S$-labeled DNA | 25 |
| Floral-specific | Tomato | Anther, pistils | | 20 |
| Pyruvate Pi dikinase | *Triticum aestivum* | Seeds | | 48 |
| β-Conglycinin | Soybean, transformed tobacco | Seeds, leaves | | 49 |
| Chalcone synthase | Parsley | Chromosome | | 50 |
| rRNA | *Vigna unguiculata* | Epicotyls | $^{32}P$-labeled cDNA probe | 22, 43 |
| β-Glucanases | *Hordeum vulgare* | Germinated grains | $^{32}P$-labeled cDNA | 51 |
| Self-incompatibility genes | *Brassica oleracea* | Stigma | $^3H$-labeled cDNA | 52 |
| Chalcone synthase | Parsley | Leaf epidermis | $^3H$-labeled cDNA | 53 |
| Isocitrate lyase, proteinases | *Brassica napus* | Seedlings | | 54, 55 |
| Gametogenesis specific genes | *Lilium* | Flowers | Biotinylated DNA | 56 |
| Kunitz trypsin inhibitor gene | Soybean, transformed tobacco | Seeds, embryos | | 57 |
| PEPC, RuBISCO | Maize | Leaves, mesophyll, bundle sheath | Sulfonated cDNA probes | 58 |
| AGAMOUS | *Arabidopsis* | Flowers | | 59, 60 |
| Anther-specific mRNAs | Tobacco | Anthers | | 61 |
| ENOD12 | Pea | Nodules | | 62 |
| ENOD2 | Pea, soybean | Nodules | | 63 |
| ENOD2 | Alfalfa | Nodules | Digoxigenin-labeled RNA probe | 64 |
| *floricaula* | *Antirrhinum majus* | Inflorescences | Digoxigenin-labeled RNA probe | 30 |
| rbcS | *Lemna gibba* | Fronds, roots | | 65 |
| Auxin-responsive transcripts GH3, SAURs | Soybean | Seedlings, flowers | | 66 |
| Chalcone synthase | *Antirrhinum majus* | Flower petals | $^3H$-labeled RNA probe | 18 |
| *myb* | *Antirrhinum* | Flowers | | 67 |

the protocol according to the application, we have noted a number of variations that other researchers have used.

## A. TISSUE PREPARATION

Successful *in situ* hybridization depends on obtaining optimally-fixed, intact sections of plant material so that morphological integrity is maintained and the mRNA is preserved. Once living tissue is excised, various enzymatic changes, which alter the tissue and ultimately cause cell death, take place. To prevent distortions and loss of tissue integrity, plant and animal tissue is quickly killed or "fixed" either by chemicals or by freezing. Depending on the type of tissue collected (nodules, roots, leaves, stems), slightly different procedures are required. We routinely fix material in formaldehyde-acetic acid-alcohol (FAA), embed it in paraffin, and section it. We have also used other aldehyde fixatives such as 3.5% glutaraldehyde and 1.5% paraformaldehyde in 0.1 *M* phosphate buffer (pH 7.2). For an alternative method of fixing and embedding plant material, see Jackson.[18] Some workers have found cryosectioning preferable to paraffin embedding and sectioning.[19,20] Each method has its advantages. Paraffin sections are easily obtained once a series of time-consuming steps have been completed. They also can be stored over long periods of time. On the other hand, cryostat sections are difficult to store but are more easily prepared initially. For additional information on cryostat sectioning, see Watkins.[21] Cryostat sectioning protocols designed specifically for plant material[17,19] and for floral tissue are also available.[20] McFadden has used plastic-embedded plant tissue for *in situ* hybridization.[22] Following fixation, embedding, and sectioning, sections are placed onto microscope slides that have been acid-washed and dipped in poly-L-lysine to aid in their adherence to the slide.

## B. PRETREATMENT OF SLIDES

Prior to hybridization, the sections, now affixed to microscope slides, are put through a series of treatments designed to increase the accessibility of mRNAs to the probe and to reduce background hybridization. These treatments may include incubation of the sections in HCl or proteases. Heat may also be used. The protocol herein employs dilute HCl treatment (to disrupt RNA secondary structure and dissociate polysomes[19] followed by incubation with proteinase K (to remove proteins and to increase the hybridization signal). The sections are then acetylated with acetic anhydride to reduce any nonspecific binding of the probe caused by positive charges on the tissue and the poly-L-lysine-coated slides.[23]

## C. HYBRIDIZATION

During the hybridization step, the sections are incubated with a specific, labeled probe, so that a duplex that can be subsequently visualized is formed. This is the critical step in the procedure, and there are several considerations to keep in mind.

### 1. Type of Probe

DNA, RNA, and oligonucleotide probes have all been used successfully. The type of probe used will depend on what type of transcript you wish to detect, although RNA probes, as described here, are the most widely used. RNA probes offer the advantage that both antisense and sense strands can be synthesized, the sense strand being used as a control for nonspecific hybridization. For a more detailed discussion of RNA probes, see Cox et al.[24] DNA probes offer the advantage that they are easier to prepare and do not depend on the fragment being cloned into a transcription vector (see Meyerowitz[25] for probe preparation). However, we have had limited success with DNA probes with our tissue material. In certain cases (e.g., when looking at specific expression of gene family members) oligonucleotide probes may be the best choice. Another consideration is the length of the probe. Very small probes form less stable hybrids, and therefore specific binding of the probe to the mRNA is decreased. Longer probes, on the other hand, result in increased background. The optimal probe length is 100 to 250 bp, so probes that are longer are digested (DNA) or hydrolyzed (RNA) to give an average length that is within this range.

## 2. Choice of Label

Radioactive or nonradioactive probes can be used for *in situ* hybridization. Factors to consider when choosing the type of label include the resolution offered and the time needed to detect the label. Thus, although $^{32}P$ and $^{3}H$ have been used for *in situ* hybridization (see Table 1), neither is ideal. The high energy of $^{32}P$ permits brief exposure times, but resolution is poor. On the other hand, tritium's low energy gives excellent resolution, but at least several months are required for exposure of the emulsion. $^{35}S$ offers a compromise between these two radioisotopes, offering good resolution in a reasonable amount of time. However, when using $^{35}S$ to make radioactive probes, it is necessary to add DTT[26-28] or β-mercaptoethanol[29] to all solutions to prevent oxidation of the thiol group. Recently, NEN has started marketing $^{33}P$, an isotope with characteristics intermediate between $^{32}P$ and $^{35}S$, which may have application to *in situ* hybridization. However, this isotope is currently too expensive for most routine applications.

Although the use of radioactive probes is an effective method of performing *in situ* localization, the time needed for exposing the photographic emulsion varies from several days to several weeks. A number of techniques, which have been used successfully on plant tissues, have been developed for producing nonradioactive labeled probes. One of these is biotin-labeling followed by detection with a fluorescent or enzymatic reagent or with colloidal gold. Another method employs digoxigenin-labeled probes. This technique offers a relatively quick and sensitive means of detecting mRNA *in situ*, and probes can be prepared using the Genius™ Kit (Boehringer Mannheim). Digoxigenin, a plant-derived product, can be used to label DNA, RNA, and oligonucleotides. The resulting hybrid is visualized by enzyme-linked immunoassay using an antidigoxigenin antibody conjugated to alkaline phosphatase. When 5-bromo-4-chloro-3-indolyl phosphate (X-phosphate) and nitroblue tetrazolium (NBT) are added to the digoxigenin-antidigoxigenin conjugate, alkaline phosphatase activity results in an insoluble blue precipitate that marks the position of the digoxigenin-labeled hybrid.

We have developed a protocol for *in situ* hybridization using probes made with the Genius Kit[17] and present this method here (see also Coen et al.[30]). Use of digoxigenin-labeled probes allows one to hybridize two different probes to tissue sections simultaneously. For example, Young[31] has used a digoxigenin-labeled probe together with a $^{35}S$-labeled probe to detect two different transcripts in rat brains simultaneously. Others[32] have used biotin- and digoxigenin-labeled probes that are detected by different antibody conjugates. In plants, Leitch et al.[33] have used biotin- and digoxigenin-labeled probes that are visualized with different fluorescent molecules to detect repetitive DNA sequences in rye, while McFadden et al.[34] have developed an *in situ* hybridization protocol for double-labeling using electron microscopy. Another approach that has been used in plants is to incubate roots in $^{3}H$-thymidine and then do *in situ* hybridization on sectioned root tissue with a digoxigenin-labeled histone probe.[35] Nuclei actively involved in DNA synthesis are labeled with both $^{3}H$ and digoxigenin.

## 3. Hybridization Conditions

The temperature, probe concentration, and length of hybridization affect the specificity and sensitivity of the hybridization signal. Hybridizations are normally carried out between 37 and 50°C, with 42°C being a common starting point. For a discussion of how to determine the optimal hybridization conditions, see Altar et al.,[11] Cox et al.,[24] and Angerer et al.[36,37]

**4. Washing Conditions**

Washing removes nonhybridized probe and results in a decreased background. The stringency of the washes affects the amount of background and is determined by the temperature, the salt concentration, and the presence of formamide in the wash solutions. Background can be decreased further by doing an RNase A digestion of the nonhybridized probe when using RNA probes.

### D. VISUALIZATION OF HYBRIDIZATION

The method of visualization of the hybridized probe will depend on how the probe was labeled. $^{35}S$ labeled probes require that the slides be dipped in photographic emulsion and left to expose over some period of time, usually not less than 3 days. Digoxigenin-labeled probes are visualized via a color reaction which may occur within as little as 10 min after addition of the substrate. When using RNA probes, specific hybridization of the antisense probe can be compared to the hybridization with the sense probe, which provides a control for nonspecific binding of the probe.

## III. EXPERIMENTAL PROCEDURE FOR TISSUE PREPARATION

### A. EQUIPMENT AND SOLUTIONS NEEDED

RNA is extremely labile, therefore it is *very important* that clean working conditions are maintained. Wear gloves at all times during the procedure to avoid RNase contamination. All solutions and equipment must be kept RNase-free! See Appendix for how to prepare RNase-free solutions and glassware.

- Vials for fixing tissue
- Vacuum desiccator
- Aluminum weighing dishes or other containers for embedding tissue
- Oven, 60°C, for melting paraffin and embedding tissue
- Warming table
- Alcohol lamp
- Dissecting needles
- Glass staining dishes with glass slide racks
- Poly-L-lysine coated microscope slides (see Appendix)
- Rotary microtome
- Wooden or plastic chucks
- Camel's hair brushes
- Fixative:

| FAA | | Glutaraldehyde-paraformaldehyde |
|---|---|---|
| 95% ethanol | 50 ml | 3 to 4% glutaraldehyde |
| glacial acetic acid | 5 ml | 1.5% paraformaldehyde |
| formaldehyde | 10 ml | in 0.1 M $PO_4$-buffer |
| water | 35 ml | (pH 7.2) |

- Melted, filtered, paraffin (Paraplast)
- 50% ethanol
- Phosphate buffer (pH 7.2) (for glutaraldehyde fixation only)
- 5, 10, 20, 30, 40, and 50% ethanol series (for glutaraldehyde fixation only)
- Tertiary butyl alcohol (TBA) series (see text)
- Absolute tertiary butyl alcohol
- DEP-DI (DEP-treated distilled water)
- Xylenes

## B. FIXATION OF PLANT MATERIAL

The following steps are the same for both the radioactive and nonradioactive hybridization protocols.

Several factors are critical in obtaining good preservation. They are (1) adequate and rapid access or **penetration** of the fixative to the tissues; (2) the appropriate **osmotic potential** of the fixative; (3) the appropriate **pH** of the fixative. We use either FAA (formaldehyde-acetic acid alcohol) or a mixture of glutaraldehyde-paraformaldehyde. Although FAA is a very harsh fixative and not balanced to cellular pH or osmoticum, it penetrates rapidly. Also, plant tissues can be stored in FAA almost indefinitely, although we have not used tissues for *in situ* hybridization that have been stored in FAA longer than 6 months. Because paraffin embedding itself results in considerable tissue distortion, the type of fixative used prior to paraffin embedding may not be critical as long as the tissue is properly fixed. The structural details of highly vacuolate cells are lost in paraffin-embedded material. However, small cells with dense cytoplasm usually retain some structural integrity. For further details on tissue fixation, see O'Brien and McCully.[38]

For fixation, the material should be of a manageable size. Smaller pieces of tissue allow greater penetration of the fixative, dehydrating solutions, and embedding material.

After subdividing, place the material as well as a label stating the identity or code number for the tissue in small vials or in stoppered bottles with 10 to 20 times the volume of fixing fluid per volume of tissue. Air bubbles interfere with proper fixation, so they should be removed by placing the unstoppered vials in an aspirator or vacuum dessicator under vacuum.

1. Aspirate for short intervals of time until material sinks to the bottom or just under the surface.
   Meristematic tissue should be aspirated for a **very** short interval.
2. Tap the specimen bottle gently to aid in the removal of air bubbles.
3. Very buoyant material should be placed in a tall thin vial and held under the surface with a cheesecloth plug.
4. Most pieces are submerged after vacuuming; usually any floating piece will sink if pushed under the surface.
5. Be sure to aspirate the same type of tissue together; i.e., do not mix highly vacuolate material with meristematic tissue.
6. If a specimen is too large or especially recalcitrant, leave the tissue overnight under vacuum, making certain that enough liquid is present in the vial to cover the tissue.

Overfixation should be avoided. We fix with FAA for 3 h to overnight and with glutaraldehyde-paraformaldehyde for 1.5 to 3 h depending on the size of the tissue. Material fixed in FAA can be fixed at room temperature and can be stored in the fixative for a longer time. **Exhibit caution with all fixatives!** They kill human as well as plant tissue.

After fixation in FAA, the tissue is washed twice in 50% ethanol, each time for 30 min, before going on with the dehydration steps (see below). Material fixed in glutaraldehyde-paraformaldehyde is fixed at 4°C and then rinsed 2X in $PO_4$-buffer (pH 7.2). Each rinse is 15 to 30 min, and then the tissue is dehydrated in increasing percentages of ethanol from 5 to 50%, 15 to 30 min each step.

## C. DEHYDRATION

After fixing, the water must be removed from the tissue before it is embedded in a water-insoluble matrix such as paraffin. Dehydration is accomplished by exposing the tissue to higher concentrations of ethanol or acetone. (We use ethanol because we use the TBA series as an intermediate before infiltrating the tissue with paraffin.) Tertiary butyl alcohol is routinely used for paraffin material because paraffin is miscible with TBA.

From the 50% ethanol, the tissue is dehydrated according to the TBA dehydration series:[39]

| Step # | 95% ethanol (%) | 100% ethanol (%) | TBA (%) | Water (%) |
|---|---|---|---|---|
| 1 | 50 | — | 10 | 40 |
| 2 | 50 | — | 20 | 30 |
| 3 | 50 | — | 35 | 15 |
| 4 | 40 | — | 55 | 5 |
| 5 | — | 25 | 75 | — |

Dehydrate in TBA steps 1 to 5 for a minimum time of 30 min. Larger tissue pieces will take longer. After TBA step 5, transfer the tissue to absolute TBA. (Vials should be filled one third with TBA.) Caution! TBA solidifies at room temperature (25.5°C), so place the vials with absolute TBA on top of an oven or in another warm place. Follow with three changes of absolute TBA, keeping the tissue in one change of absolute TBA overnight.

## D. INFILTRATION

1. Material should now be transferred to aluminum weighing dishes and filled 1:1 with 100% TBA and molten paraffin.
2. Place the dishes in a warm oven (ca. 60°C) overnight. As the TBA evaporates, the paraffin will gradually infiltrate the tissue.
3. On the following day, pour off any remaining mixture, and add fresh melted paraffin. (We filter the paraffin to make certain that it is particle-free).
4. Make one change with additional paraffin, and let the paraffin infiltrate the tissue overnight.
5. Make two to three more changes of molten paraffin (every 4 h). The material is now ready to embed.

## E. EMBEDDING

1. Have tools warm (use an alcohol lamp).
2. Quickly transfer the aluminum dishes to a warming table and orient the tissue with warmed needles.
3. When oriented, gradually move the dishes to a cooler area of the warming table.
4. Allow the paraffin to solidify and cool completely at room temperature before transferring the embedded tissue to 4°C.
5. Store at 4°C until use.
6. When ready to use, remove the tissue segment from the paraffin wafer by scoring the wax. Gently break the wafer. Carefully cut out the tissue segment and remove excess paraffin with a razor blade.
7. Mount the block on a wooden or plastic support using molten paraffin as an adhesive.
8. Trim the block into either a trapezoid or a rectangle.
9. Section on a rotary microtome at 7 to 10 μm.

## F. SECTIONING

Prior to sectioning, prepare poly-L-lysine-coated slides (see Appendix). These may be prepared in advance in large numbers and stored at 4°C.

1. Once a trimmed specimen and knife have been mounted in the microtome, sectioning can begin. (Position the specimen **before** putting the knife into place!)

2. The block face of the specimen can either have a rectangular or a trapezoidal form, but in either case, make sure that the block is exactly parallel to the knife edge. Thus, as each section is cut, it will press evenly upon the trailing edge of the previous section and detach it cleanly from the knife edge with a minimum of wrinkling. Check that there is no grease, dust, or other contaminants on the block face or knife edge, as these may damage the tissue or the edge of the block. Check, too, that all clamps holding the specimen and knife are tightened.
3. Move the wheel of the microtome with a steady, even stroke. Hold the camel's hair brush in one hand to lift the ribbons of sections as they come off the knife.
4. Spread the ribbons in an appropriate container (e.g., an empty photographic paper box) or place them directly onto the slides (see below).
5. Examine the sections under the dissecting microscope to determine which sections you wish to affix to a slide. ( On one slide, we usually position a length of ribbon which fits underneath a 24 × 60 mm coverslip.)
6. Prior to affixing them to the slides, sections can be stored overnight or longer in the storage box at 4°C.
7. Affixing the sections to the slides;
   a. Place several drops of DEP-DI on the slide.
   b. With a brush, place the sections on the flooded area, and very gently pull on their edges with dissecting needles to align them on the slide.
   c. Place the slide on a warming table (40 to 50°C) until any wrinkles in the ribbon are removed. (Again, the ribbon may be stretched slightly with needles.)
   d. Remove excess liquid with "Kimwipes", and dry the slides overnight on the warming table in a dust-free area.
   e. Transfer the slides to a slide box and keep them at 4°C until use. Slides can also be deparaffinized at this point by putting them in a slide rack and subjecting them to two changes of xylene (in a fume hood), 30 min each. Remove the slide tray to paper towels. After the xylene has evaporated, the sections may be examined with the microscope to determine if they are suitable for hybridization. The slides can be stored at 4°C indefinitely.

### G. NOTES

Poor hybridization may be caused by inaccessibility of tissue mRNA. To prevent this, decrease the fixation time. Underfixation, on the other hand, can lead to diffusion of tissue mRNA as well as poor tissue preservation. The optimal time must be determined empirically for each tissue type. Loss of tissue sections during the hybridization steps may be due to poorly coated slides (see Appendix for preparation of poly-L-lysine coated slides).

## IV. EXPERIMENTAL PROCEDURE FOR HYBRIDIZATION WITH $^{35}$S-LABELED RNA PROBE

### A. INTRODUCTION

To generate an RNA probe, the corresponding cDNA fragment must be cloned into a vector which contains RNA polymerase promoters (e.g., Riboprobe Gemini system (pGEM), Promega or pBluescript (pBS), Stratagene). These companies also sell transcription kits which include instructions for making radiolabeled transcripts.

The fragment can be cloned into the vector with different ends, or if it has the same ends, it can be ligated into the middle of the multiple cloning site and subsequently linearized with restriction enzymes that recognize adjacent sites. Mapping or sequencing the plasmid will determine the orientation of the insert. A restriction enzyme is then used that will cut the DNA on one or the other side of the insert, generating linearized plasmid from which transcripts are generated (see Figure 2).

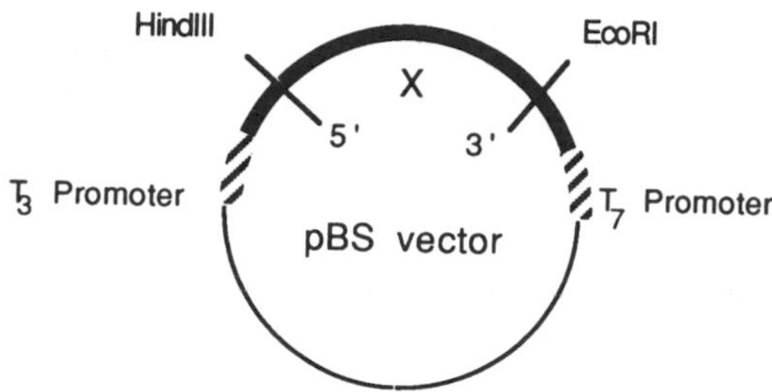


**FIGURE 2.** In this example, the enzyme *Hin*dIII cleaves the plasmid at the 5′ end of the gene X. Subsequent treatment with T7 RNA polymerase will generate RNA antisense transcripts to the gene X that can be used as *in situ* probes for the transcribed gene. Linearization of the plasmid with *Eco*RI and then using T3 RNA polymerase will provide a sense probe which is the negative control. Note: This is true only if the insert X is not cloned into the *Hin*dIII site or into the *Eco*RI site.

## B. PREPARATION OF RNA PROBE

### 1. Equipment and Solutions Needed

All solutions should be RNase-free once the plasmid has been linearized.

- DEP-treated microfuge tubes (see Appendix for DEP-treatment protocol)
- DEP-treated pipette tips
- Microfuge
- 37°C water bath
- 60°C water bath
- Restriction enzyme(s)
- Agarose (low melting point)
- TBE buffer
- Phenol/chloroform/isoamyl alcohol (PCIA) (25:24:1; TE-saturated)
- 3 *M* Na Acetate (pH 5.2)
- Absolute ethanol (Store at –20°C)
- 70% ethanol (Store at –20°C)
- TE
- DEP-DI
- 5X transcript buffer*
- DTT (0.75 *M*)*
- rATP (10 m*M*)*
- rCTP (10 m*M*)*
- rGTP (10 m*M*)*
- RNAsin (optional)*
- $^{35}$S-UTP (For example, Amersham $^{35}$S-UTP, 40 mCi/ml, >800 mCi/mmol. We purchase the T7/SP6 grade.)
- SP6, T7 (or T3) RNA polymerase*
- RNase-free DNase (Boehringer Mannheim, Cat. #1119915)
- Sephadex G-50 columns
- Yeast tRNA (see Appendix for preparation)
- Formamide (Ultra-pure "DNA-Grade", RNase-free formamide [Boehringer Mannheim])
- 1 *M* DTT
- 0.1 *M* $NaHCO_3$ (pH 10.2)
- 5% acetic acid
- 10 *M* $NH_4$ Acetate

The components marked * are provided in most transcription kits (Stratagene, Promega).

## 2. Linearization

The following steps are the same for making both radioactive and nonradioactive probes.

1. Linearize the plasmid with the appropriate restriction enzyme in a 50 μl volume. A clean plasmid preparation is needed; however, it is possible to use "miniprep" DNA if extra steps are taken to remove contaminating proteins (see notes). We usually linearize 5 μg at a time and then store the linearized plasmid at –20°C. An excess of enzyme and longer digestion times may be needed to achieve complete linearization.
2. Check the linearity of the preparation using electrophoresis by running a 1% agarose gel in TBE buffer with ethidium bromide (2.5 μl/25 ml from a stock of 5 mg/ml) to visualize nucleic acids. It may be useful to run the digest on a low melting point agarose gel and extract the DNA from the agarose. This ensures that only the linearized plasmid is used for subsequent steps.
3. Extract the DNA twice with an equal volume of PCIA.
4. Precipitate DNA by adding

| | |
|---|---|
| 3 *M* Na Acetate (pH 5.2) | 1/10 vol |
| 100% ethanol | 2 vol |

   Store the eppendorf tube at –20°C overnight or for 1 h at –80°C. (DNA may be safely stored at –20°C at this point.)
5. Centrifuge 30 min at 15000 × g (4°C), wash pellet with ice cold 70% ethanol, and then dry the pellet.
6. Resuspend DNA to 1 μg/μl in TE buffer (RNAse-free). Linearized template may be stored at –20°C.

It is important to have clean, completely linearized DNA at this step. Nonlinearized plasmid will generate "run-around" transcripts instead of "run-off" transcripts. Raikhel et al.[19] repeat the digestion of the plasmid to be transcribed after repeated extractions with PCIA. They follow the digestion with a proteinase K incubation to remove any proteins that may interfere with transcription. The proteinase K digestion may also precede the PCIA extraction.[16] In addition to sense transcript controls, some researchers recommend the use of another portion of the same gene as a probe, especially when analyzing multigene families.[12,40] Both portions of the gene should give the same result.

## 3. Preparation of $^{35}$S-Labeled Transcript

If using a kit, follow the instructions provided with the kit, e.g., Riboprobe Gemini II Core system (Promega, Cat. # P1270) or Stratagene RNA transcription kit (Cat. # 200340).

1. Add to a microcentrifuge tube (in the following order) at room temperature:

| | |
|---|---|
| DEP-DI | 9.0 μl |
| 5X transcript buffer | 5.0 μl |
| DTT (0.75 *M*) | 1.0 μl |
| rATP (10 m*M*) | 1.0 μl |
| rCTP (10 m*M*) | 1.0 μl |
| rGTP (10 m*M*) | 1.0 μl |
| DNA template (1 μg) | 1.0 μl |
| RNAsin (optional) | 1.0 μl |
| $^{35}$S-UTP | 5.5 μl |
| SP6, T7 (or T3) RNA polymerase | 0.5 μl |
| **total volume** | 25.0 μl |

   The mixture is radioactive from this point on. Take appropriate precautions.

2. Incubate 1 h at 37°C.
3. Add 1 μl of 1 mg/ml RNase-free DNase (10 U); incubate 30 min at 37°C.
4. Add 24 μl DEP-DI to make 50 μl.

Cox and Goldberg[16] add a proteinase K digestion following DNase treatment.

## 4. Column Purification

This procedure removes unincorporated $^{35}$S-nucleotides from the probe solution so that an increased autoradiograph signal-to-noise ratio is achieved. An alternative procedure is to do an ethanol precipitation.[19,40]

Use Nick Columns, Sephadex G-50, "DNA-Grade," Pharmacia (Cat. #170855-02).

1. Cleaning the column:
   a. Remove the top cap and the excess liquid of the column and rinse once with DEP-DI.
   b. Remove the bottom cap, fill the column with DEP-DI, and let it run through.
   c. Fill the column now with DEP-treated TE and let it drip dry. Repeat.
2. Running the sample:
   a. Load the sample (50 μl) onto the Nick column.
   b. Add 400 μl TE, collect the 400 μl in a microcentrifuge tube labeled "waste," and discard properly.
   c. Place a fresh tube under the column, then add 400 μl TE, and collect the probe.
   d. Discard the Nick column containing free NTPs into radioactive waste. At least half of the radioactivity should be in your probe when compared to the column. If it is not, it is likely that the transcription did not work.
3. Extract twice with an equal volume of PCIA.
4. Precipitate the RNA with

| | |
|---|---|
| 3 *M* Na Acetate | 30 μl |
| yeast tRNA (50 mg/ml) | 1 μl |
| 2.5 volumes ethanol | 750 μl |

5. Precipitate RNA at –70°C for 1 h or overnight at –20°C.
6. Centrifuge 30 min at 10000 × g (4°C), wash pellet with cold 70% ethanol, and dry.

At this point, the investigator can either resuspend the RNA in 20 μl formamide/10 m*M* DTT or partially hydrolyse the RNA, depending on the size of the transcript generated.

Raikhel et al.[19] recommend back extraction of the organic phase during PCIA extraction for increased recovery of RNA transcripts.

## 5. Hydrolysis of the RNA Transcript

The following steps are the same for both radioactive and nonradioactive probes.

The alkaline hydrolysis step creates smaller pieces of RNA which penetrate the tissue better. The time of hydrolysis obeys the relationship below.[24] However, the appropriate hydrolysis time can be determined empirically:

$$t = \frac{L_o - L_f}{kL_oL_f} \tag{1}$$

$t$ = time in minutes
$L_o$ = initial length in kb
$L_f$ = desired length in kb
$k$ = 0.11 cuts/kilobase/min

1. Resuspend RNA in 50 µl 0.1 *M* $NaHCO_3$ (pH 10.2), 10 m*M* DTT.
2. Incubate for the time determined at 60°C to hydrolyse RNA to an average length of 100 to 200 bases. Fifteen minutes appears to work well for any probe between 300 and 1000 nucleotides.
3. Neutralize the reaction solution and precipitate the RNA by adding

| | |
|---|---|
| 5% acetic acid | 5 µl |
| 10 *M* $NH_4$ acetate | 15 µl |
| absolute ethanol | 175 µl |

4. Leave 1 h at –70°C or overnight at –20°C.
5. Centrifuge 30 min at 10,000 × g (4°C), wash pellet with ice cold 70% ethanol, and dry.
6. Resuspend RNA in 20 µl formamide, 10 m*M* DTT. Probe can be stored ≤ two months at –70°C.
7. Determine the cpm in your probe by removing 1 µl and adding it to 99 µl water. Remove 10 µl and count it in a scintillation counter. This gives you the cpm per 0.1 µl. If you are not going to use the probe immediately, it should be counted again before use so that the appropriate amount is added to the hybridization buffer.

Low incorporation of labeled nucleotide into the transcripts can be due to a number of factors, including precipitation of the DNA by the spermidine in the transcription buffer (Be sure to do this reaction at room temperature.); excess NaCl (Precipitate DNA with sodium acetate, not sodium chloride.); RNase contamination (Use RNase inhibitor in reaction.); or inactive enzyme (Use control template as a test.). It may be useful in certain cases to generate unlabeled transcripts. This can be done by using 1 µl of 10 m*M* UTP and omitting the $^{35}$S-UTP. Raikhel et al.[19] give a procedure for analyzing the RNA transcripts, including determination of specificity and size determination of hydrolyzed transcripts. Consult this reference for further information. For troubleshooting transcripts of the wrong size, consult the Promega technical manual on *in vitro* transcription systems or consult other sources. If no silver grains are seen after the autoradiography is completed, one possible cause is message degradation. Check the integrity of the message by gel electrophoresis.

## C. PROBING TISSUE SECTIONS WITH $^{35}$S-LABELED RNA PROBES

### 1. Equipment and Solutions Needed

- Glass slide-staining dishes and glass slide racks for dipping slides into solutions. These items can be DEP-treated and baked at 180°C to remove RNases (baking time 5 h or overnight; overnight is advised). Be sure to let the staining dishes cool down gradually. They are not Pyrex glass and crack easily.
- Metal holder for transferring slide racks
- Coverslips
- Stir bar, DEP-treated and baked within a slide staining dish
- Magnetic stirrer
- Slide box
- Incubator set to desired hybridization temperature
- 0.2 *M* HCl
- 2X SSPE or 2X SSC (see Appendix for 20X stock)
- Proteinase K stock (20 mg/ml)
- Proteinase K buffer (see Appendix)

- DEP-DI
- 100 m*M* triethanolamine
- Acetic anhydride (reagent grade)
- 30, 70, 95, 100% ethanol
- Prehybridization buffer (see Appendix)
- Hybridization buffer (see Appendix)
- Wash solutions:

| For Method A | | For Method B |
|---|---|---|
| 3 M dithiothreitol (DTT) stock | or | β-mercaptoethanol |
| 4X SSC | or | wash solution A (see Appendix) |
| 2X SSC | or | wash solution B (see Appendix) |
| 0.1 or 0.5X SSC | or | wash solution C (see Appendix) |
| | | 2X SSC |

- RNase A stock (50 mg/ml)
- NTE (see Appendix)
- 30, 50, 70, 95% ethanol containing a final concentration of 0.3 *M* $NH_4$ Acetate

## 2. Pretreatment of Slides

The following steps are the same for radioactive and nonradioactive probes.
The next series of steps take about 3.5 h (up to the hybridization step).

1. Remove paraffin with two treatments of xylene, 30 min each, if it has not been done so previously.
2. Acid treatment:
   a. Incubate slides in 0.2 *M* HCl for 20 min at room temperature.
   b. Rinse twice in 2X SSPE or 2X SSC, 5 min each time.
3. Proteinase K treatment:
   a. Incubate slides for 30 min at room temperature in 200 ml of 1.0 μg/ml proteinase K dissolved in 1X proteinase K buffer (10 μl proteinase K stock/200 ml buffer). The best timing should be determined empirically.
   b. Rinse 2X at room temperature in DEP-DI for 5 min each.
4. Acetic anhydride treatment (in fume hood):
   a. Dip slides in 100 m*M* triethanolamine in DEP-DI for 5 min. (**Do not** heat sterilize triethanolamine.)
   b. Add acetic anhydride to 0.25% (v/v: 500 μl/200 ml) to fresh triethanolamine, then incubate slides for 10 min with vigorous stirring.
   c. Rinse slides twice in 2X SSPE or 2X SSC for 5 min each.
   d. Rinse slides twice in DEP-DI for 2 min each.
5. Dehydrate slides 2 min each in the following ethanol series: 30%, 70%, 95%, 100%.
6. Air dry slides in a dust-free area or under vacuum for at least 1 h.

These steps are critical for obtaining optimal hybridization. Overexposure to proteolytic enzymes can cause the loss of tissue mRNA and poor tissue morphology. Too short an exposure to proteolytic enzymes can lead to inaccessibility of the message. We have found that the times reported here work well. Raikhel et al.[19] reported that a heat treatment step prior to proteinase K treatment helps retain the tissue sections on the slides. Cox and Goldberg[16] incubate the tissue sections in 1% bovine serum albumin in 10 m*M* Tris-HCl (pH 8.0) to block the positive charges of the poly-L-lysine and omit the acid treatment. Omission of the acetic anhydride step can lead to high background due to nonspecific binding of the probe to tissue protein or to heterologous mRNA.

## 3. Hybridization of Tissue Sections

### *a. Calculations*

Volume of $^{35}$S-labeled RNA transcript needed:

$$\mu\text{l transcript} = \frac{1}{(\text{cpm}/\mu\text{l})} \times \frac{5\times10^5\,\text{cpm}}{\text{slide}} \times (\#\,\text{slides}+1) \tag{2}$$

Total volume $^{35}$S-RNA transcript for all slides:

$$\text{total } \mu\text{l volume } = (\#\text{ of slides } +1) \times \frac{22\ \mu\text{l}}{\text{slide}} \tag{3}$$

Volume of transcript dilution buffer:

$$\mu\text{l } 50\%\text{ formamide, } 10\text{ m}M\text{ DTT} = (0.2\times\text{ Equation } 3) - \text{Equation } 2 \tag{4}$$

Volume of hybridization buffer:

$$\mu\text{l hybridization buffer} = 0.8\times\text{total volume } ^{35}\text{S} - \text{RNA transcript} \tag{5}$$

### *b. Prehybridization*

1. Apply 100 to 200 µl prewarmed hybridization buffer to each slide, cover sections with coverslip, and incubate for 2 h at 42°C in a humid chamber. Adding more prehybridization buffer allows coverslips to come off more easily, preventing the loss of tissue sections. (A plastic slide box with moistened filter paper inside works well as an incubation chamber. The slides should sit horizontally in the box.)
2. After 2 h, let the coverslips slide off, and go directly to "Hybridization".

### *c. Hybridization*

1. Dilute RNA transcript (Equation 2 above) with volume determined by Equation 4 (transcript dilution buffer; 42% formamide (ultrapure), 10 m*M* DTT).
2. Mix diluted RNA transcript with the hybridization buffer (Equation 5).
3. Apply 22 µl of the probe in prewarmed hybridization buffer directly onto the sections for each 22 × 22 mm coverslip. Increase the amount of probe proportionally for larger coverslips (i.e., 100 µl for 60 × 24 mm).
4. Cover the sections with coverslips. (Baking the cover slips at 180°C is strongly advised.)
5. Incubate slides at 42°C overnight in a humid chamber. (The slides should sit horizontally in the box.)

We use the following rule of thumb: apply $10^6$ cpm of probe in 100 µl hybridization buffer to each slide, and use a large (60 × 24 mm) coverslip. Because there will be hybridization buffer remaining from the prehybridization step, the molar concentration of applied probe might vary slightly among slides. Raikhel et al.[19] heat the hybridization buffer containing the probe to 80°C for 30 sec prior to applying to the slides, presumably to remove any secondary structure. The optimal time and temperature of hydridization can be calculated (e.g., see Cox et al.[24]) or determined empirically. Cox and Goldberg[16] recommend temperatures between 45 and 50°C for homologous RNA probes. At these temperatures, they find it useful to place the slides in mineral oil for hybridization to prevent drying of the hybridization buffer. Others[19] seal the coverslips with rubber cement. We have found that temperatures between 37 and 42°C are sufficient when hybridization is done in a plastic slide box lined on the bottom with saturated Whatman paper.

Low signal may be caused by low copy number of the tissue mRNA. Detection may be aided by increasing the probe concentration, increasing the hybridization time, using the optimal hybridization temperature, or increasing the length of exposure time. Probe length may also affect the signal strength. The size of the probe following hydrolysis can be checked by gel electrophoresis.[19]

### d. Washing Tissue Sections

The washing conditions needed to remove the excess probe vary depending on the nature of the probe. Thus, two methods are given for wash conditions. The second method provides higher stringency at a lower temperature by using formamide and may be more suitable when background is a problem. A third variation is found in Raikhel et al.,[19] and this paper should be consulted for details.

**Note added in proof:** We have developed a wash protocol which combines the use of a single formamide wash with SSC washes, making it a less expensive, yet higher stringency protocol than wash protocol B described in the text.

1. Let coverslips come off in 4 × SSC, 10 m*M* DTT.
2. Wash in three changes of 4 × SSC, 10 m*M* DTT, 10 min each.
3. Wash in NTE at room temperature, 15 min.
4. Incubate for 30 min in 20 μg/ml final concentration RNase A (made in NTE) at 37°C.
5. Wash for 10 min in 2 × SSC, 10 m*M* DTT at 37°C.
6. Wash for 30 min in 2 × SSC, 50% formamide, 10 m*M* DTT at 65°C.
7. Wash for 10 min in 2 × SSC, 10 m*M* DTT at 65°C.
8. Wash 15 min in 0.1 × SSC, 10 m*M* DTT at 65°C.
9. Rinse for 2 min in 0.1 × SSC at room temperature.
10. Dehydrate (see main protocol).

*Notes:* Prewarm wash solutions (except RNase) to correct temperature. Omit DTT if isotope is not $^{35}S$. β-mercaptoethanol can be used instead of DTT.

Method A (Based on Cox and Goldberg[16])

1. Place slides in glass rack, and soak in 4X SSC, 5 m*M* DTT at room temperature to remove cover slips.
2. Rinse slides in fresh 4X SSC, 5 m*M* DTT for 5 min.
3. Incubate slides in RNase A solution (50 μg/ml in NTE) at 37°C for 30 min.
4. Wash in NTE at 42°C for 30 min; add DTT to a final concentration of 5 m*M* just before washing.
5. Wash twice in 2X SSC, 5 m*M* DTT at room temperature, 30 min each wash (with stirring).
6. Incubate slides with prewarmed 0.1X SSC, 5 m*M* DTT at 60°C for 1 h (with stirring).

Method B (Based on Zeller and Rogers[40])

1. Let coverslips slide off by dipping slides into wash solution C (see Appendix). Use fresh solution for each probe. Store slides in glass rack in a holding tank with wash solution C.

2. The following series of washes are then done:
   a. 2 × 15 min in wash solution A at 50°C.
   b. 2 × 15 min in wash solution B at 50°C.
   c. 5 min in wash solution C at room temperature.
   d. 2 × 5 min in 2X SSC at room temperature.
   e. 35 min in 50 μg/ml RNaseA in NTE at room temperature.
   f. 2 × 30 min in wash solution C at 50°C.
   g. 2 × 30 min in wash solution A at 50°C.
   h. 2 × 5 min in 2X SSC at room temperature.

High background may be caused by nonspecific binding of the probe to tissue protein or to heterologous mRNA. Increasing the stringency of the washes may alleviate this problem. Also, a control can be included in which sections are treated with RNase prior to hybridization with the antisense probe. This indicates whether nonspecific binding of the probe to cellular components has occurred.

#### *e. Dehydration*

1. Dehydrate the sections (2 min each step): 50% ethanol/0.3 *M* $NH_4$ acetate, 70% ethanol/0.3 *M* $NH_4$ acetate; 95% ethanol/0.3 *M* $NH_4$ acetate; 100% ethanol.
2. Air dry slides for 2 h in a dust free area.

## D. AUTORADIOGRAPHY OF MICROSCOPE SLIDES

All steps can be performed under a red safelight (15W bulb, Kodak safelight filter #2). However, we generally coat the slides in complete darkness.

For $^{35}S$ probes, use Kodak NTB-2 emulsion. Warm emulsion at 40 to 50°C ≥ 10 min in a water bath. (Heat expansion may result in the cap of the emulsion container coming off, causing some of the emulsion to flow out from under the cap. This usually does not cause a light leak into the container itself.) Aliquot the emulsion into slide mailers, and mix with an equal volume of distilled water. The easiest way to do this is to fill the mailers with water first, and then add the emulsion in the dark with a syringe marked in a way that you can recognize the mark in the dark. Wrap the mailers individually in several layers of foil. The emulsion can be stored for 2 to 3 months at 4°C. Make sure the mailers you use are tight. DO NOT FREEZE THE EMULSION.

### 1. Equipment and Solutions Needed

- Kodak NTB-2 emulsion (Kodak Cat. # 165-4433)
- Slide mailers (Ted Pella, Cat. #22518)
- 42°C waterbath
- Forceps
- Test tube rack
- Large, light-tight box
- Glass staining dishes (3)
- Glass slide rack
- Small slide storage boxes (e.g., VWR, Cat. # 48444-003), containing "Kimwipes" or other absorbent paper and a small amount of Drierite or other desiccant wrapped in cheesecloth taped to one side
- Kodak D-19 developer (Kodak Cat. #146-4593)
- Kodak Rapid Fixer

### 2. Coating the Slides

1. Remove a slide mailer containing emulsion from the refrigerator, and warm to 40 to 50°C. The emulsion should be left undisturbed for about 15 min to allow air bubbles to rise.
2. Dip the slides in the emulsion, and place the slides upright in a test tube rack to dry. Bench paper, paper towels, or "Kimwipes" placed in the rack will keep the slides vertical and absorb excess emulsion.
3. Transfer the entire rack to a light-tight area or cabinet, and let it dry undisturbed for 2 to 4 h.
4. After drying, transfer the slides to an opaque slide box, tape the sides of the box, wrap it in several layers of foil, and expose the slides for 3 days to several weeks at 4°C.

Cox and Goldberg[16] dilute the emulsion 1:1 with 600 m*M* $NH_4$ acetate which helps to precipitate the RNA hybrids. We recommend coating a blank slide with emulsion to determine the amount of background exposure.

### 3. Developing the Emulsion

All steps can be performed under a red safelight. However, we also do these steps in the dark.

1. Warm slides to room temperature.
2. Place slides in a slide rack, and develop in Kodak D-19 for 3 to 5 min at 16°C. Agitate gently every 30 sec.
3. Rinse in distilled water for 30 sec.
4. Fix in Kodak Rapid Fixer for 2 to 3 min at 16°C.
5. Wash in distilled water for $\geq$ 15 min.

Some workers suggest using Kodak Fixer instead of Rapid Fixer. High background may be caused by overexposure or overdevelopment. It may also be caused by exposure of the emulsion to light or isotope emission. Also check the expiration date on the emulsion. A blank slide control will help you determine that the emulsion is good. Photodevelopment may also lead to a lack of silver grains; be sure that the emulsion and photographic chemicals are good. A protocol for removing fogged emulsion appears in *BioTechniques* 13:498–499 (1992).

## E. STAINING AUTORADIOGRAPHED SLIDES

### 1. Equipment and Solutions Needed

- Safranin or toluidine blue
- Coplin jars
- 30, 50, 70, 95, 100% ethanol
- Eukitt (Calibrated Instruments, Inc.) or other coverslip mounting medium

### 2. Staining

1. Plant tissue can be stained in either safranin (0.5% in 5% ethanol) or toluidine blue (0.05% in water or 1% borax (pH 9.0), w/v). Toluidine blue is a polychromatic stain that gives good results without using a counterstain. However, it is not colorfast and may fade after several months or years.
2. Dehydrate sections in a graded ethanol series — 30%, 50%, 70%, 95%, 100% — for up to 2 min at each step.
3. After slides are dry, mount cover glasses using a mounting medium.

# V. EXPERIMENTAL PROCEDURE FOR HYBRIDIZATION WITH DIGOXIGENIN-LABELED PROBES

## A. PREPARATION OF PROBE

### 1. Equipment and Solutions Needed

- DEP-treated microfuge tubes
- DEP-treated pipet tips
- Genius RNA Labeling Kit (Boehringer Mannheim Cat. #1175025)
- 0.2 *M* EDTA
- 4 *M* LiCl
- Absolute ethanol
- 70% ethanol
- DEP-DI

### 2. Linearization of the DNA Template

The plasmid is linearized as described for $^{35}$S-labeled probe.

### 3. Preparation of Digoxigenin-Labeled Probes Using the Genius RNA-Labeling Kit

1. To a microfuge tube, add in order:
   1 µg of linearized plasmid
   2 µl NTP labeling mixture (vial #7 in RNA-labeling kit)
   2 µl 10X transcription buffer (vial #8 in RNA-labeling kit)
   1 µl RNase inhibitor (vial #10 in RNA-labeling kit)
   Bring to 18 µl with DEP-DI, then add:
   2 µl RNA polymerase (either T7 or SP6 polymerase, vials #11 or #12 in RNA-labeling kit)
2. Mix gently and incubate for 2 h at 37°C.
3. Add 2 µl RNase-free DNase (vial #9 in RNA-labeling kit)
4. Incubate 15 to 30 min at 37°C.
5. Add 2 µl of 0.2 *M* EDTA (pH 8.0) to stop the reaction.
6. Precipitate RNA with 2.5 µl 4 *M* LiCl, then 75 µl absolute ethanol.
7. Keep at –80°C for 1 h or –20°C overnight.
8. Centrifuge for 20 min at 4°C, wash pellet with ice cold 70% ethanol, and dry.
9. RNA can now be hydrolyzed to smaller pieces, following the procedure given for $^{35}$S-labeled probes. (See also *Boehringer Mannheim Biochemica* ,Vol. 8, No. 6, Nov. 1991.)
10. Resuspend pellet in 100 µl DEP-DI with 1 µl RNase inhibitor for 30 min at 37°C.
11. Store at –20°C until use.

## B. HYBRIDIZATION

Tissue preparation and pretreatment of slides are done as described for $^{35}$S-labeled probes.

### 1. Equipment and Solutions Needed

- Prehybridization buffer (see Appendix)
- Hybridization buffer (see Appendix)
- Coverslips
- Slide box
- Incubator set to desired hybridization temperature
- 2X SSC

- 1X SSC
- 0.5X SSC

### 2. Pretreatment of Slides

Follow Steps 1 through 6 under pretreatment of slides for $^{35}S$-labeled probes.

### 3. Prehybridization

1. Apply 100 to 200 µl prewarmed prehybridization solution to each slide. (Using more prehybridization buffer allows the coverslip to come off more easily, preventing loss of tissue sections.)
2. Cover the slides with baked coverslips.
3. Incubate the slides 2 h or more in a humid chamber at room temperature to 42°C.

### 4. Hybridization

1. Carefully remove the coverslips, and allow the prehybridization solution to drip off. Do not allow the sections to dry.
2. Apply 100 µl of prewarmed hybridization buffer containing the labeled RNA.
3. Cover the sections with fresh, baked coverslips.
4. Seal the chamber so that it is airtight and incubate overnight. Do not exceed 47°C.

### 5. Washes

1. Allow the coverslips to come off in 2X SSC.
2. Wash the slides twice with 2X SSC at room temperature, each time 1 h.
3. Wash the slides once in 1X SSC for 1 h at room temperature.
4. Wash the slides once in 0.5X SSC for 30 min at room temperature.

## C. IMMUNOLOGICAL REACTION USING GENIUS™ NUCLEIC ACID DETECTION KIT

### 1. Equipment and Solutions Needed

- Glass staining dishes with racks, metal holders
- Slide box
- Coverslips
- Genius Nucleic Acid Detection Kit (Boehringer Mannheim, Cat. #1175041)
- Buffer #1 (see Appendix)
- Buffer #2 (see Appendix)
- Buffer #3 (see Appendix)
- Antibody buffer (see Appendix)
- Bovine serum albumin
- Triton X-100
- Color solution (make fresh; see Appendix)
- 30, 70, 95, 100% ethanol
- Eukitt or other coverslip mounting medium

### 2. Procedure

1. Wash slides with buffer #1 for 5 min at room temperature.
2. Incubate with 2% BSA, 0.3% Triton-X in buffer #1 for 30 min at room temperature, 200 µl per slide. Cover with coverslips and keep slides in a moist chamber.

3. Remove coverslips and add dilute antibody conjugate (see discussion) to slides, 200 μl per slide.
4. Cover with a fresh coverslip, and incubate in a moist chamber for 2 h at room temperature.
5. Wash slides twice with buffer #1, each time 15 min, at room temperature with gentle shaking.
6. Wash slides once with buffer #2, 2 min at room temperature with gentle shaking.
7. Apply freshly made color solution to slides.
8. Cover with coverslips, and incubate in a moist chamber away from light. Exposure to light causes an increase in background color.
9. Monitor the progress of the reaction by microscopy. The reaction may take anywhere from 10 min to several days. As a rule, shorter incubation times are better, since background color is reduced.
10. When the reaction is complete, stop it with buffer #3 for 5 min at room temperature.
11. Dip slides in an increasing ethanol series, 30 sec each dip. DEP-DI, 30% ethanol, 70% ethanol, 95% ethanol, 100% ethanol.
12. After the slides have dried, mount a coverslip on the slide with Eukitt or other mounting medium.

The concentration of the probe should be 1 to 10 ng per slide per 100 μl. We usually add 20 μl of the probe to 1 ml of hybridization buffer. Coen et al.[30] used 4% of each labeling reaction in 40 μl of hybridization buffer per slide and hybridized at 50°C overnight. They also modified the washing procedure: several washes in 2X SSC with 50% formamide at 50°C were done, followed by a wash in NTE buffer (see Appendix). This was followed by RNase A treatment in this buffer at 37°C for 30 min. The slides were then washed for 1 h in 2X SSC with 50% formamide and then washed several times in 130 m*M* NaCl, 10 m*M* sodium phosphate (pH 7.0). Slides were stored in this buffer for 1 to 3 days at 4°C prior to the immunological reaction.

We usually dilute the antibody 1:500, although higher dilutions may give less background.[31] The antibody is diluted in buffer #1 containing 1% BSA and 0.3% Triton X-100, but Young[31] used buffer #1 containing 3% normal goat serum, and we have used buffer #1 alone successfully. The length of the antibody incubation step may also affect signal strength and background. Young[31] found a 5-h incubation to be optimal. Coen et al.[30] also modified the detection procedure. Slides were incubated with gentle shaking in blocking agent (provided in the detection kit) in 100 m*M* Tris-HCl, 150 m*M* NaCl (pH 7.5), followed by 45 min in 1% BSA, 0.3% Triton X-100 in buffer #1. The slides were then incubated for 1 h in dilute antibody conjugate (1:2500) and then given four 20 min washes in buffer #1 with no antibody, followed by a brief wash in buffer #2. Using this protocol, Coen et al.[30] found that 1 to 2 days of incubation with the color solution were required.

## VI. APPENDIX

### A. PREPARATION OF POLY-L-LYSINE COATED SLIDES

1. Place slides in a glass rack, transfer the rack to a glass staining dish containing Chromerge, and soak slides there overnight.
2. Rinse 1 h under running distilled water.
3. Drain the water from the dish, and bake the slides overnight at 180°C.
4. Cool them gradually.
5. Immerse slides in a 100 μg/ml poly-L-lysine (Sigma) solution in 10 m*M* Tris-HCl (pH 8.0) for at least 20 min at room temperature.

6. Air dry overnight in a dust-free area.
7. Store the slides in a slide box at 4°C.

## B. DEP TREATMENT OF GLASS- AND PLASTICWARE

To DEP treat glass- and plasticware, a plastic dishpan containing about 5 l of water is prepared and approximately 5 ml of diethylpyrocarbonate (DEP) is added. Material to be treated is immersed completely and left to soak about 10 min. The material is then placed into DEP-treated containers or covered with foil as appropriate and autoclaved (plastics) or baked overnight at 180°C.

## C. SOLUTIONS

### 1. General Solutions

DEP-DI: 50 µl of DEP (diethylpyrocarbonate (Sigma)) stock in 100 ml distilled water, mix well, let sit for 5 to 10 min, and autoclave

20X SSPE:

| | | *grams/liter* |
|---|---|---|
| NaCl | 3.6 *M* | 210.4 g |
| $Na_2HPO_4$ | 0.2 *M* | 25 g |
| NaOH | 75 m*M* | 3 g |
| EDTA | 20 m*M* | 7.4 g |
| | | (pH 7.7) |
| | $H_2O$ | to 1 liter |

add 500 µl DEP and autoclave

20X SSC:

| | | *grams/liter* |
|---|---|---|
| NaCl | 3.0 *M* | 175.32 g |
| $Na_3$ Citrate | 0.3 *M* | 88.23 g |
| | | (pH 7.0) |
| $H_2O$ | | to 1 liter |

add 500 µl DEP and autoclave

DEP treatment of buffers: same as for water; see above

**Exceptions are the following**: Tris, NaOH, and SDS, which have to be made with DEP-DI.

Yeast tRNA: Dissolve in TE, extract 5X with phenol/chloroform
To aqueous phase add:
- 1/4 vol 10 *M* $NH_4$ acetate
- 2.5 vol ethanol

Precipitate 1 h at –20°C
Centrifuge, wash with ice cold 70% ethanol, dry
Resuspend at 50 mg/ml
Store at –20°C.

Poly(A): Dissolve in TE, extract 5X with phenol/chloroform
To aqueous phase add:
- 1/4 vol 10 *M* $NH_4$ Acetate
- 2.5 vol ethanol

Precipitate 1 h at –20°C
Centrifuge, wash with ice cold 70% ethanol, dry
Resuspend at 10 mg/ml
Store at –20°C.

5X transcription buffer: 200 m*M* Tris-HCl (pH 7.5)
30 m*M* $MgCl_2$
10 m*M* spermidine
50 m*M* NaCl
DEP-treat and autoclave all ingredients except spermidine. Filter-sterilize the spermidine stock.

2X SSPE: Dilute 20X SSPE 1:10 with water, add DEP to 0.05%, autoclave

0.2 *M* HCl: Add 3.4 ml conc. HCl to 196.6 ml sterile DEP-DI

Proteinase K buffer:

| | | *Stock per 200 ml* |
|---|---|---|
| Tris-HCl, pH 7.5 | 100 m*M* | 20 ml 1*M* Tris (pH 7.5) |
| EDTA | 50 m*M* | 20 ml 0.5 *M* EDTA (pH 8) |
| DEP-DI | | 160 ml |

Proteinase K: Stock: 20 mg/ml in DEP-treated water. Store at –20°C
Add 10 µl to 200 ml proteinase K buffer just before use

100 m*M* triethanolamine: Add 3.71 g triethanolamine to 200 ml sterile DEP-DI

## 2. Solutions for Radioactive Hybridization

Pre/hybridization buffer:[1]

| | |
|---|---|
| 20X SSPE | 100 µl |
| formamide (ultrapure)[2] | 500 µl |
| 50% Dextran sulfate | 200 µl |
| Yeast tRNA (50 mg/ml) | 25 µl |
| 3 *M* DTT | 3.3 µl |
| Poly(A) (10 mg/ml) | 62.5 µl |
| DEP-DI | 109.2 µl |

[1] For hybridization, remember to adjust volume of hybridization buffer to account for the volume of added probe.
[2] Ultra-pure "DNA-Grade", RNase-free formamide (Boehringer Mannheim).

RNase A (Sigma): Dissolve RNase A at 10 mg/ml in 10 m*M* Tris-HCl (pH 7.5), 15 m*M* NaCl.

Heat to 100°C for 15 min
Allow to cool slowly to room temperature
Aliquot and store at –20°C

NTE buffer:

| | | *Stock per 200 ml* |
|---|---|---|
| NaCl | 500 m*M* | 20 ml 5 *M* NaCl |
| Tris-HCl | 10 m*M* | 2 ml 1 *M* Tris (pH 8) |
| EDTA | 1 m*M* | 0.4 ml 0.5 *M* EDTA (pH 8) |
| DI | | 177.6 ml |

Wash solution A: 50% formamide, 2X SSC, 20 m*M* β-mercapto ethanol

Wash solution B: 50% formamide, 2X SSC, 20 m*M* β-mercaptoethanol, 0.5% Triton X-100

Wash solution C: 2X SSC, 20 m*M* β-mercaptoethanol (555 µl/400 ml)

### 3. Solutions for Nonradioactive Hybridization

Pre/hybridization buffer:[1] can also use buffer as for radioactive probes, omitting the DTT.

| | *Stock per ml* |
|---|---|
| 4X SSC | 200 μl 20X SSC |
| 50% formamide (ultrapure)[2] | 500 μl |
| 1X Denhardt's solution | 20 μl 50X Denhardt's |
| 5% dextran sulfate | 200 μl 50% dextran sulfate |
| 0.5 mg/ml salmon sperm DNA | 50 μl 10 mg/ml stock |
| 0.25 mg/ml yeast tRNA | 25 μl 50 mg/ml stock |

[1] For hybridization, remember to adjust the volume of the hybridization buffer to account for the volume of added probe.

[2] Ultra-pure "DNA-Grade", RNase-free formamide (Boehringer Mannheim).

Buffer #1: 100 m*M* Tris-HCl
150 m*M* NaCl (pH 7.5)

Buffer #2: 100 m*M* Tris-HCl
100 m*M* NaCl
50 m*M* $MgCl_2$ (pH 9.5)

Buffer #3: 10 m*M* Tris-HCl
1 m*M* EDTA (pH 8.0)

Color solution: for 10 ml
45 μl NBT solution
35 μl X-phosphate
2.4 mg levamisole in buffer #2

(Levamisole inhibits endogenous nonintestinal phosphatase activity.)

## ACKNOWLEDGMENTS

We are especially grateful to Dr. Marian Löbler and to Birgit Bochenek for their help in perfecting the protocols described from our laboratory. Agway, Inc. (Syracuse, N.Y.) is thanked for their generous gift of *Medicago sativa* cv. Iroquois, and we thank F.M. Ausubel for *Rhizobium meliloti* strain Rm1021.

This work was supported by DCB-9021587 from the National Science Foundation. Heather I. McKhann was supported by a California Biotechnology training grant, University of California.

## REFERENCES

1. **John, H. A., Birnstiel, M. L., and Jones, K. W.,** RNA-DNA hybrids at the cytological level, *Nature*, 223, 582, 1969.
2. **Gall, J. G. and Pardue, M. L.,** Formation and detection of RNA-DNA hybrid molecules in cytological preparations, *Proc. Natl. Acad. Sci. U.S.A.*, 63, 378, 1969.
3. **Pardue, M. L. and Gall, J. G.,** Nucleic acid hybridisation to the DNA of cytological preparations, *Meth. Cell Biol.*, 10, 1, 1975.
4. **Szabo, P., Elder, R., Steffensen, D. M., and Uhlenbeck, O. C.,** Quantitative *in situ* hybridization of ribosomal RNA species to polytene chromosomes of *Drosophila melanogaster*, *J. Mol. Biol.*, 115, 539, 1977.

5. **Gerhard, D. S., Kawasaki, E. S., Bancroft, F. C., and Szabo, P.,** Localization of a unique gene by direct hybridization *in situ*, *Proc. Natl. Acad. Sci. U.S.A.*, 78, 3755, 1981.
6. **Harper, M. E., Ullrich, A., and Saunders, G. F.,** Localization of the human insulin gene to the distal end of the short arm of chromosome 11, *Proc. Natl. Acad. Sci. U.S.A.*, 78, 4458, 1981.
7. **Ambros, P. F., Matzke, M. A., and Matzke, A. J. M.,** Localization of *Agrobacterium rhizogenes* T-DNA in plant chromosomes by *in situ* hybridization, *EMBO J.*, 5, 2493, 1986.
8. **Ambros, P. F., Matzke, M. A., and Matzke, A. J. M.,** Detection of a 17 kb unique sequence (T-DNA) in plant chromosomes by *in situ* hybridization, *Chromosoma*, 94, 11, 1986.
9. **Mouras, A., Saul, M. W., Essad, S., and Potrykus, I.,** Localization by *in situ* hybridization of a low copy number chimaeric resistance gene introduced into plants by direct gene transfer, *Mol. Gen. Genet.*, 207, 204, 1987.
10. **Pardue, M. L.,** *In situ* hybridisation, in *Nucleic Acid Hybridisation: A Practical Approach*, Hames, B. D. and Higgins, S. J., Eds., IRL Press, Oxford, 1988, 179.
11. **Altar, C. A., Ryan, S., Abood, M., and Eberwine, J. H.,** *In situ* hybridization: standard procedures and novel approaches, in *Methods in Neurosciences*, Vol. 1, Conn, P. M., Ed., Academic Press, San Diego, 1989, 238.
12. **McCabe, J. T. and Pfaff, D. W.,** *In situ* hybridization: a methodological guide, in *Methods in Neurosciences*, Vol. 1, Conn, P. M., Ed., Academic Press, San Diego, 1989, 98.
13. **Baldino, F. and Lewis, M. E.,** Nonradioactive *in situ* hybridization histochemistry with digoxigenin-deoxyuridine 5′-triphosphate-labeled oligonucleotides, in *Methods in Neurosciences*, Vol. 1, Conn, P. M., Ed., Academic Press, San Diego, 1989, 282.
14. **Penschow, J. D., Haralambidis, J., Pownall, S., and Coghlan, J. P.,** Location of gene expression in tissue sections by hybridization histochemistry using oligodeoxyribonucleotide probes, in *Methods in Neurosciences*, Vol. 1, Conn, P. M., Ed., Academic Press, 1989, 222.
15. **Raikhel, N. V., Bednarek, S. Y., and Wilkins, T. A.,** Cell-type-specific expression of a wheat-germ agglutinin gene in embryos and young seedlings of *Triticum aestivum*, *Planta*, 176, 406, 1988.
16. **Cox, K. H. and Goldberg, R. B.,** Analysis of plant gene expression, in *Plant Molecular Biology: A Practical Approach*, Shaw, C. H., Ed., IRL Press, Oxford, 1988, 1
17. **Bochenek, B. and Hirsch, A. M.,** *In situ* hybridization of nodulin mRNAs in root nodules using non-radioactive probes, *Plant Mol. Biol. Rep.*, 8, 237, 1990.
18. **Jackson, D.,** *In-situ* hybridisation in plants., in *Molecular Plant Pathology: A Practical Approach*, Bowles, D. J., Gurr S. J., and McPherson, M., Eds., Oxford University Press, Oxford, 1991, 163.
19. **Raikhel, N. V., Bednarek, S. Y., and Lerner, D. R.,** *In situ* RNA hybridization in plant tissues, in *Plant Molecular Biology Manual*, Kluwer Academic Publishers, Dordrecht, Netherlands, 1989, B9.
20. **Smith, A. G., Hinchee, M., and Horsch, R.,** Cell and tissue specific expression localized by *in situ* RNA hybridization in floral tissues, *Plant Mol. Biol. Rep.*, 5, 237, 1987.
21. **Watkins, S.,** Cryosectioning, in *Current Protocols in Molecular Biology*, Ausubel, F. M., Brent, R., Kingston, R. E., Moore, D. D., Seidman, J. G., Smith J. A., and Struhl, K., Eds., Greene Publishing, New York, 1989, 14.2.1.
22. **McFadden, G. I.,** *In situ* hybridisation in plants: from macroscopic to ultrastructural resolution, *Cell Biol. Int. Rep.*, 13, 3, 1989.
23. **Hayashi, S., Gillam, I. C., Delaney, A. D., and Tener, G. M.,** Acetylation of chromosome squashes of *Drosophila melanogaster* decreases the background in autoradiographs from hybridization with $^{125}$I-labeled RNA, *J. Histochem. Cytochem.*, 26, 677, 1978.
24. **Cox, K. H., DeLeon, D. V., Angerer, L. M., and Angerer, R. C.,** Detection of mRNAs in sea urchin embryos by *in situ* hybridization using asymmetric RNA probes, *Dev. Biol.*, 101, 485, 1984.
25. **Meyerowitz, E. M.,** *In situ* hybridization to RNA in plant tissue, *Plant Mol. Biol. Rep.*, 5, 242, 1987.
26. **McCabe, J. T., Morrell, J. I., Richter, D., and Pfaff, D. W.,** Localization of neuroendocrinologically relevant RNA in brain by *in situ* hybridization, in *Front. Neuroendocrinol.*, 9, Ganong, W. F. and Martini, L., Eds., Raven Press, New York, 1986, 145.
27. **Lawrence, J. B., Singer, R. H., Villnave, C. A., Stein, J. L., and Stein, G. S.,** Intracellular distribution of histone mRNAs in human fibroblasts studied by *in situ* hybridization, *Proc. Natl. Acad. Sci. U.S.A.*, 85, 463, 1988.
28. **Williamson, D. J.,** Specificity of riboprobes for intracellular RNA in hybridization histochemistry, *J. Histochem. Cytochem.*, 36, 811, 1988.
29. **Bandtlow, C. E., Heumann, R., Schwab, M. E., and Thoenen, H.,** Cellular localization of nerve growth factor synthesis by *in situ* hybridization, *EMBO J.*, 6, 891, 1987.
30. **Coen, E. S., Romero, J. M., Doyle, S., Elliott, R., Murphy, G., and Carpenter, R.,** *Floricaula*: a homeotic gene required for flower development in *Antirrhinum majus*, *Cell*, 63, 1311, 1990.
31. **Young, W. S.,** Simultaneous use of digoxigenin- and radiolabeled oligodeoxyribonucleotide probes for hybridization histochemistry, *Neuropeptides*, 13, 271, 1989.

32. **Herrington, C. S., Bhatt, B., Burns, J., Graham, A. K., and McGee, J. O.,** Simultaneous non-isotopic *in situ* hybridization using biotinylated and digoxigenin-labelled probes, *J. Pathol.*, 157, 163, 1989.
33. **Leitch, I. J., Leitch, A. R., and Heslop-Harrison, J. S.,** Physical mapping of plant DNA sequences by simultaneous *in situ* hybridization of two differently labelled fluorescent probes, *Genome*, 34, 329, 1991.
34. **McFadden, G., Bönig, I., and Clarke, A.,** Double label *in situ* hybridization for electron microscopy, *Trans. R. Microsc. Soc.*, 1, 683, 1990.
35. **Tanimoto, E. Y., Rost, T. L., and Comai, L.,** Histone H2A expression during S-phase: histological co-localization of H2A mRNA and DNA synthesis in pea root tips, in *Cell Cycle Regulation in Plant Growth and Development*, Francis, D. and Ormrod, J. C., Eds., Kluwer Academic Publishers, Dordrecht, Netherlands, in press.
36. **Angerer, R. C., Cox, K. H., and Angerer, L. M.,** *In situ* hybridization to cellular RNAs, in *Genetic Engineering*, Vol. 7, Setlow, J. K. and Hollaender, A., Eds., Plenum Press, New York, 1985, 43.
37. **Angerer, L. M., Cox, K. H., and Angerer, R. C.,** Demonstration of tissue-specific gene expression by *in situ* hybridization, in *Methods in Enzymology*, Vol. 152, Berger, S. L. and Kimmel, A. R., Eds., Academic Press, New York, 1987, 649.
38. **O'Brien, T. P. and McCully, M. E.,** *The Study of Plant Structure: Principles and Selected Methods*, Termarcarphi Pty., Ltd., Melbourne, Australia, 1981.
39. **Sass, J. E.,** *Botanical Microtechnique*, Iowa State University Press, Ames, Iowa, 1958.
40. **Zeller, R. and Rogers, M.,** *In situ* hybridization to cellular RNA, in *Current Protocols in Molecular Biology*, Ausubel, F. M., Brent, R., Kingston, R. E., Moore, D. D., Seidman, J. G., Smith J. A., and Struhl, K., Eds., Greene Publishing, New York, 1989, 14.3.1.
41. **Anderson, M. A., Cornish, E. C., Mau, S.-L., Williams, E. G., Hoggart, R., Atkinson, A., Bönig, I., Grego, B., Simpson, R., Roche, P. J., Haley, J. D., Penschow, J. D., Niall, H. D., Tregear, G. W., Coghlan, J. P., Crawford, R. J., and Clarke, A. E.,** Cloning of cDNA for a stylar glycoprotein associated with expression of self-incompatibility in *Nicotiana alata*, *Nature*, 321, 38, 1986.
42. **Cornish, E. C., Pettitt, J. M., Bönig, I., and Clarke, A. E.,** Developmentally controlled expression of a gene associated with self-incompatibility in *Nicotiana alata*, *Nature*, 326, 99, 1987.
43. **McFadden, G. I., Bönig, I., Cornish, E. C., and Clarke, A. E.,** A simple fixation and embedding method for use in hybridization histochemistry on plant tissues, *Histochem. J.*, 20, 575, 1988.
44. **Harris, N. and Croy, R. R. D.,** Localization of mRNA for pea legumin: *in situ* hybridization using a biotinylated cDNA probe, *Protoplasma*, 130, 57, 1986.
45. **Harris, N., Grindley, H., Mulchrone, J., and Croy, R. R. D.,** Correlated *in situ* hybridisation and immunochemical studies of legumin storage protein deposition in pea (*Pisum sativum* L.), *Cell Biol. Int. Rep.*, 13, 23, 1989.
46. **Harris, N., Mulcrone, J., and Grindley, H.,** Tissue preparation techniques for *in situ* hybridisation studies of storage-protein gene expression during pea seed development, in *In Situ Hybridisation: Application to Developmental Biology and Medicine*, Harris N. and Wilkinson, D. G., Eds., Cambridge University Press, Cambridge, 1990, 175.
47. **Martineau, B., and Taylor, W. C.,** Cell-specific photosynthetic gene expression in maize determined using cell separation techniques and hybridization *in situ*, *Plant Physiol.*, 82, 613, 1986.
48. **Aoyagi, K. and Chua, N.-H.,** Cell-specific expression of pyruvate, Pi dikinase, *Plant Physiol.*, 86, 364, 1988.
49. **Barker, S. J., Harada, J. J., and Goldberg, R. B.,** Cellular localization of soybean storage protein mRNA in transformed tobacco seeds, *Proc. Natl. Acad. Sci., U.S.A.*, 85, 458, 1988.
50. **Huang, P.-L., Hahlbroch, K., and Somssich, I. E.,** Detection of a single-copy gene on plant chromosomes by *in situ* hybridization, *Mol. Gen. Genet.*, 211, 143, 1988.
51. **McFadden, G. I., Ahluwalia, B., Clarke, A. E., and Fincer, G. B.,** Expression sites and developmental regulation of genes encoding (1->3, 1->4)-β-glucanases in germinated barley, *Planta*, 173, 500, 1988.
52. **Nasrallah, J. B., Yu, S.-M., and Nasrallah, M. E.,** Self-incompatibility genes of *Brassica oleracea*: expression, isolation, and structure, *Proc. Natl. Acad. Sci. U.S.A.*, 85, 5551, 1988.
53. **Schmelzer, E., Jahnen, W., and Hahlbrock, K.,** *In situ* localization of light-induced chalcone synthase mRNA, chalcone synthase, and flavonoid end products in epidermal cells of parsley leaves, *Proc. Natl. Acad. Sci. U.S.A.*, 85, 2989, 1988.
54. **Comai, L., Dietrich, R. A., Maslyar, D. J., Baden, C. S., and Harada, J. J.,** Coordinate expression of transcriptionally regulated isocitrate lyase and malate synthase genes in *Brassica napus* L., *Plant Cell*, 1, 293, 1989.
55. **Dietrich, R. A., Maslyar, D. J., Heupel, R. C., and Harada, J. J.,** Spatial patterns of gene expression in *Brassica napus* seedlings: identification of a cortex-specific gene and localization of mRNAs encoding isocitrate lyase and a polypeptide homologous to proteinases, *Plant Cell*, 1, 73, 1989.
56. **Jones, K. G., Crossley, S. J., and Dickinson, H. G.,** Investigation of gene expression during plant gametogenesis by *in situ* hybridisation, in *In Situ Hybridisation: Application to Developmental Biology and Medicine*, Harris, N. and Wilkinson, D. G., Eds., Cambridge University Press, Cambridge, 1990, 189.

57. **Perez-Grau, L. and Goldberg, R. B.,** Soybean seed protein genes are regulated spatially during embryogenesis, *Plant Cell*, 1, 1095, 1989.
58. **Perrot-Rechenmann, C., Joannes, M., Squalli, D., and Lebacq, P.,** Detection of phosphoenolpyruvate and ribulose 1,5-bisphosphate carboxylase transcripts in maize leaves by *in situ* hybridization with sulfonated cDNA probes, *J. Histochem. Cytochem.*, 37, 423, 1989.
59. **Bowman, J. L., Drews, G. N., and Meyerowitz, E. M.,** Expression of the *Arabidopsis* floral homeotic gene AGAMOUS is restricted to specific cell types late in flower development, *Plant Cell*, 3, 749, 1991.
60. **Drews, G. N., Bowman, J. L., and Meyerowitz, E. M.,** Negative regulation of the *Arabidopsis* homeotic gene AGAMOUS by the APETALA2 product, *Cell*, 65, 991, 1991.
61. **Koltunow, A. M., Truettner, J., Cox, K. H., Wallroth, M., and Goldberg, R. B.,** Different temporal and spatial gene expression patterns occur during anther development, *Plant Cell*, 2, 1201, 1990.
62. **Scheres, B., van de Wiel, C., Zalensky, A., Horvath, B., Spaink, H., Van Eck, H., Zwartkruis, F., Wolters, A.-M., Gloudemans, T., Van Kammen, A., and Bisseling, T.,** The ENOD12 gene product is involved in the infection process during the pea-*Rhizobium* interaction, *Cell*, 60, 281, 1990.
63. **van de Wiel, C., Scheres, B., Franssen, H. J., van Lierop, M., van Lammeren, A., van Kammen, A., and Bisseling, T.,** The early nodulin transcript ENOD2 is located in the nodule parenchyma (inner cortex) of pea and soybean nodules, *EMBO J.*, 9, 1, 1990.
64. **van de Wiel, C., Norris, J. H., Bochenek, B., Dickstein, R., Bisseling, T., and Hirsch, A. M.,** Nodulin gene expression and ENOD2 localization in effective, nitrogen-fixing and ineffective, bacteria-free nodules of alfalfa, *Plant Cell*, 2, 1009, 1990.
65. **Silverthorne, J. and Tobin, E.,** Post-transcriptional regulation of organ-specific expression of individual *rbcS* mRNAs in *Lemna gibba*, *Plant Cell*, 2, 1181, 1990.
66. **Gee, M. A., Hagen, G., and Guilfoyle, T. J.,** Tissue-specific and organ-specific expression of soybean auxin-responsive transcripts GH3 and SAURs, *Plant Cell*, 3, 419, 1991.
67. **Jackson, D., Culianez, F., Prescott, A. G., Roberts, K., and Martin, C.,** Expression patterns of *myb* genes from *Antirrhinum* flowers, *Plant Cell*, 3, 115, 1991.

# 13

# Immunological Methods for Assessing Protein Expression in Plants

*Erwin B. Dumbroff and Shimon Gepstein*

## I. INTRODUCTION

Within the past two decades, immunological techniques have become indispensable tools in the physiological, biochemical, and molecular disciplines of plant science. Their principal attraction resides in the high specificity of the immunological reaction, which allows accurate recognition of an antigenic substance even in the presence of contaminating antigens that are not of immediate interest. The methodology is now routinely used for the rapid purification, visualization, and quantification of proteins, polysaccharides, and even small molecules to which antibodies were induced by conjugating the molecule to a large immunogenic carrier protein.

The introduction of a purified immunogen into an animal system triggers rapid mutational and recombinatorial mechanisms that alter specific DNA sequences in B lymphocytes. These B cells then differentiate and generate a family of antibodies (i.e., a polyclonal antiserum) that has an affinity for a specific antigen (i.e., for the immunogen itself) and for any hapten that was part of the immunogenic complex. The phenomenon appears to be unlimited in its capacity to respond, in a highly specific way, to an infinite number of immunogens of natural or synthetic origin.[1] The antibodies produced are glycoproteins, specifically immunoglobulins, that constitute the major proteinaceous component of blood serum. IgG is the principal immunoglobulin and comprises up to 80% of all antibodies in the serum with IgM, IgA, IgD, and IgE present in varying amounts.

A molecule of IgG consists of two identical heavy and two identical light polypeptide chains with variable amino acid sequences at the amino-terminal ends of each chain. The sequences in each of the variable regions form the antigen binding sites, and they respond to an antigenic stimulus by changes in sequence that account, in large measure, for antibody specificity. The amino acid sequences in the variable regions of one light (L) and one heavy (H) chain interact to form one variable domain for each H and L pair, which makes IgG bivalent and provides two separate sites for binding antigens.[2-4]

The high affinity binding between antigen and antibody results from a combination of hydrophobic, ionic, and Van der Waals forces, but covalent bonds are not involved. The points of attachment on the antigen are referred to as epitopes or antigenic determinants, many of which occur on the surface of a protein molecule and account for the family of polyclonal antibodies produced in response to immunization with a particular protein.[4] When a small molecule is attached to a carrier protein as a hapten, a new epitope is formed, and a portion of the family of antibodies induced by the carrier protein will have an affinity for the hapten itself. This, of course, is the basis of immunodetection of small, nonproteinaceous compounds

0-8493-5164-2/93/$0.00+$.50
© 1993 by CRC Press, Inc.

like plant hormones and of small, weakly-immunogenic proteins like ubiquitin. The occurrence of the same or even similar epitopes on different proteins is the basis of antibody cross reactivity, but for the most part, even small changes in an epitope prevent antibody recognition.[2,3]

Labeling antibodies with radioisotopes, light-emitting compounds, or colored reagents provides an investigator with a sensitive means of detecting and measuring the amount of antigen in a sample. Although most immunological techniques are used in a quantitative or semiquantitative sense, they also have important applications in preparative work for isolating large amounts of a given antigen or antibody from crude mixtures using the techniques of immunoaffinity chromatography.[5,6]

## II. PRODUCTION OF ANTIBODIES

### A. PROTEIN PURIFICATION

Procedures commonly used for the production of monospecific polyclonal antibodies, i.e., antibodies that recognize only one species of protein, usually require injection of microgram to milligram quantities of a highly purified sample. It is essential, therefore, to employ one or more stringent purification techniques that have the potential to yield protein homogeneity. Unfortunately, however, monospecificity is not always achieved since only small amounts of some contaminating proteins are sufficient to induce an immunogenic response.[7] If the antigen of interest can be purchased commercially, so much the better, but homogeneity should be tested and ensured by additional purification when necessary.

Proteins are usually extracted from fresh-frozen or lyophilized plant tissue in a suitable, near-neutral buffer containing a cocktail of protease inhibitors (e.g., 5 m*M* 6-amino-n-hexanoic acid + 1 m*M* benzamidine hydrochloride + 1 m*M* phenylmethysulfonyl fluoride) and 1% (w/v) polyvinylpolypyrrolidone to adsorb polyphenolic compounds that are readily oxidized and then bind and crosslink sample proteins.[7-9] A degree of purification equivalent to that achieved by 2-D polyacrylamide gel electrophoresis (PAGE), using isoelectric focusing in the first dimension and separation by molecular size in the second, is usually satisfactory for the production of highly specific antisera. However, protein yields from 2-D gels are generally low. An alternative approach is to use a Rotofor Isoelectric Focusing Cell (Bio-Rad, Richmond, CA.) for initial purification of relatively large amounts of crude extract, followed by one-dimensional separation on standard size polyacrylamide gels. Initial set-up costs are moderate, but the method is no more complex than 2-D gel electrophoresis and yields milligram quantities of highly purified protein within a matter of 1 to 2 days. HPLC with a modern chromatofocusing column can also be used for both analytical and preparative separations of molecular species that have similar charge characteristics. The mechanism of separation is based on the isoelectric pH of proteins, and in that respect, it is similar to isoelectric focusing (IEF) in a Rotofor.[10] For laboratories or research groups that routinely require large amounts of highly purified protein for antibody production and other purposes, the cost of a free-flow electrophoresis system (Elphor VAP 22, Dr. Weber GmbH, D-8045, Ismaning, Germany) that facilitates both analytical and preparative-scale IEF plus separation by size may be justified. Protein purification may also be carried out using the more classical procedures of ammonium sulfate fractionation, size-exclusion chromatography (e.g., on Sephadex, Sephacryl, Biogel) and ion-exchange chromatography (e.g., on DEAE- or CM-cellulose) on low pressure columns or by HPLC.[7,11]

One-dimensional PAGE is often used as the last step in protein purification, i.e., before injecting the immunogen into a laboratory animal. Sodium dodecyl sulfate (SDS) is usually added to the running buffer to dissociate the proteins into their constituent polypeptides and thereby facilitate their resolution on the gel. Detailed protocols for running SDS-PAGE are provided in a number of publications[2,12-16] and are also provided with the purchase of commercial gel electrophoresis units.

Between 0.1 and 10 mg of partially purified protein can be strip-loaded on a standard, 16 × 20-cm slab gel using a 1.5 mm 2-D comb. Somewhat smaller amounts of protein, ranging between 0.05 and 1.0 mg, are usually loaded on a minigel (7 cm × 8 cm × 1.5 mm), but in both cases, the amounts selected should give tight bands that are well separated from contaminating polypeptides. The concentration of acrylamide:bisacrylamide used to prepare a gel can be varied to optimize separation of the target protein (A concentration of 16% is suitable for proteins between 10 and 30 kD, 12% for proteins in the range of 50 kD, and 8% for proteins with molecular weights of approximately 80 kD and over.[12] Minigels are available precast from commercial sources (e.g., Bio-Rad), and though expensive, they provide convenience, reproducibility, and considerable savings in time.

After electrophoretic separation, narrow strips cut from each side of a gel can be stained and then used to locate proteins of interest. If preferred, the whole gel can be lightly stained with Coomassie Brilliant Blue R-250 (0.05% w/v in water) for about 10 min and destained in distilled water for several hours. Staining a gel with 0.3 *M* $CuCl_2$ for 5 min and then rinsing in water provides a negative image in which the protein bands remain clear and the portions of gel devoid of protein are opaque.[2] Elution of the protein from the gels is not required before injection into rabbits or other large laboratory animals since polyacrylamide constitutes an effective adjuvant that increases the immunological response.

One-dimensional SDS-PAGE can resolve proteins that differ by as little as 1% in size,[7] and contaminating proteins of the same molecular weight usually can be eliminated by a careful choice of purification methods preceding the final electrophoretic step. However, the presence of more than one band in an SDS gel following an initial, nondenaturing purification step may not be indicative of contaminant polypeptides since many proteins have more than one subunit. Moreover, during 2-D electrophoresis the presence of several spots, separated during IEF but showing no difference in molecular size when run in the second dimension by SDS-PAGE, may represent different products of a multigene family and not contaminating proteins.[8]

### B. IMMUNIZATION

A number of animal species are routinely used to generate antisera, but rabbits provide a convenient compromise, given their ease of handling and care and their relatively large yield of serum.[2] If possible, two to four animals should be injected with the same immunogen since the intensity of the immunoresponse can vary markedly from animal to animal. The protein bands previously cut from the gels are broken and 1 to 3 ml of gel are mixed with 1.5 ml of distilled water or appropriate buffer plus 1.0 to 1.5 ml of complete Freund's adjuvant (Difco, Surrey, England; Gibco/BRL, Gaithersburg, MD) in a glass homogenizer or in a small mortar and pestle and then drawn, several times, into a glass, Luer-lock syringe through an 18-gauge needle until the mixture is well emulsified. The formulation of the mixture should be adjusted so that fluid does not leak from the syringe. If highly purified protein, free of gel, is available, the same procedure is followed, but the volumes of water and adjuvant are reduced to about 1 ml each, and the mixture is injected with a 22-gauge needle. Freund's Complete Adjuvant (FCA) is used for the first injection, but Freund's Incomplete Adjuvant (FIA) is used for all subsequent injections to avoid side effects and harm to the animal. The adjuvant consists of a mixture of water in mineral oil, and FCA contains *Mycobacterium tuberculosis* to stimulate the initial immune response.

The upper portion of a rabbit's back is shaved to remove the long hair, and 200 to 400 μl of the gel-protein-adjuvant suspension is injected subcutaneously (never intravenously!) at each of four to five well-spaced sites along either side of the upper back using a total of

100 to 500 μg of plant protein. Be sure to maintain sterile conditions and to check animal-care regulations for maximum allowable volume for any one series of injections. Each site is swabbed with 70% ethanol before injection, and the needle is inserted, bevel side up, into a fold of skin held between thumb and forefinger until the needle tip can be moved freely from side to side. After withdrawing the needle, gently pinch the hole to prevent loss of inoculum. A second series of injections, using 0.5 to 1.0 times the original amount of immunogen mixed with FIA, is generally started 2 to 4 weeks later. The rabbits can be bled 10 to 15 days after this first boost, collecting 20 to 30 ml of blood each time from a 4-kg rabbit at 3- to 4-week intervals. A third series of injections may be required after 4 to 6 weeks to maintain or perhaps to achieve high antibody titer, but in most instances this second boost is not required. If high serum titers are not achieved after the second boost, i.e., after the third complete series of injections, the rabbit should be replaced.

Immunization of rabbits with a protein bound to nitrocellulose paper (NCP) can provide a useful variant of the previous procedure, particularly when only small amounts of immunogen are available for injection. After transfer of polypeptides from a gel to NCP (Western Blotting, Sections III.A and B), the blot, i.e., the NCP plus protein, is stained with Ponceau S red (0.1% w/v in 1% v/v acetic acid)[2,17] or 1% fast green (w/v in water) for several minutes.[18] The strip of interest is excised, destained in water, cut into fine pieces, dried, and solubilized in 0.5 ml of dimethylsulfoxide (DMSO) for 30 min before adding 0.5 ml of Freund's adjuvant.[2,19] After thorough mixing, the suspension is injected by following the same procedures used for the polyacrylamide gel fractions. Depending on the immunogenicity of the protein, this method facilitates effective immunization with as little as 5 to 20 μg of sample. The NCP apparently increases the availability of the protein and may provide some degree of protection from proteolysis.[20] Although we have had our best success by dissolving the NCP-protein complex in DMSO prior to injection, techniques have also been described for direct injection of a fine powder of the NCP complex that has been macerated and suspended in phosphate-buffered saline (PBS, 0.01 *M* $Na_2HPO_4/NaH_2PO_4$, 0.15 *M* NaCl, pH 7.3) without DMSO or Freund's adjuvant.[2,19,21] Use of these methods in conjunction with 2-D electrophoresis does facilitate the production of highly specific antibodies with only small amounts of protein. Nevertheless, we have had mixed results with the approach, and as reported by others,[2] the intensity of the immunological response tends to vary with the kind of protein used.

The immunogenicity of small polypeptides, less than 5 kD, and other molecules that do not induce a strong immunogenic response can be increased substantially by their coupling to a large carrier protein such as bovine serum albumin or bovine gamma globulin using glutaraldehyde or other cross-linking reagents. The detailed methodology required for a specific target molecule is usually determined by preliminary experiments, but basic procedures for conjugation are described in several texts.[2,22,23] Specific examples are also available for ubiquitin, a highly conserved protein with 76 amino acids found in both plants and animals.[24-26]

## C. BLEEDING, SERUM PREPARATION, AND STORAGE

The test rabbit should be placed in a restraining cage or wrapped in a large towel or blanket with its head sticking out and gently but firmly held by one person while a second collects a volume of blood. Do not put the rabbit directly on a smooth surface; they tend to panic! Use of a tranquilizer may not be required if the animal is kept warm and is gently stroked during bleeding, but subcutaneous injection of 0.55 to 1.1 mg of acepromazine per kilogram of weight 20 min before bleeding will calm the animal and tends to increase blood flow. Shave an area around the marginal vein on the convex surface of the ear to be cut, about one-third the distance from the tip. To elicit blood flow after cutting, paint the center artery (not the marginal vein) with a 40% (v/v) solution of d-limonene in 95% ethanol. Xylene is an effective and more traditional vasodilator, but it can cause discomfort and some damage to the rabbit's skin.[27] After about 3 min, clean the area of the cut with 70% ethanol, and make a 5 mm incision at a 45° to 90° angle into but not through the marginal vein using a new, sterilized razor blade or scalpel. A clean, fine cut is essential to minimize clot formation. Place thumb with light pressure on the marginal vein above the cut, i.e.,toward the rabbit's head, to retard return flow, and collect about 20 to 30 ml of blood in a glass (not plastic) beaker or tube within 5 to 10 min. Wipe the cut with cotton or gauze to remove clots, and keep the ear slightly bent to maintain flow. After collection, remove the d-limonene with 70% ethanol, and apply pressure to the cut with gauze. If flow does not cease within 30 to 60 sec, keep the gauze in place by wetting with glycerol or secure with a paperclip. Observe the rabbit for 5 min before returning it to its cage and for 5 to 10 min thereafter to ensure that bleeding has stopped. Be sure to remove the paper clip and gauze.

Hold the blood at room temperature for 1 h to induce clotting. Dislodge the clot from the sides of the holding beaker by ringing with a glass rod or pipette and place in a refrigerator overnight. Pipette the serum into a tube and spin in a clinical centrifuge for 10 min. Decant and then divide the sample into small aliquots in Eppendorf vials and store at –20°C or lower. Clean serum should be clear and range in color from a light tan to pink. If repeated freezing and thawing are avoided, the antiserum can be stored for several years. Diluted antiserum, i.e., a working solution, can be stored at 4°C for several weeks or longer if 0.02% (w/v) $NaN_3$ is added as an antimicrobial agent.

When facilities for raising chickens are available, chicken eggs from immunized hens provide a particularly convenient source of large amounts of IgG. The immunoglobulin content of egg yolks is higher than the amount present in a hen's serum, and there is, of course, a significant advantage in the ease of collecting eggs and extracting their yolks compared to the effort involved in bleeding rabbits.[28]

## D. ANTIBODY PURIFICATION

Several methods are available to concentrate and purify antibodies from crude serum; however all have drawbacks including possible damage to the antibodies plus the additional time and cost of the procedures. Moreover, purification of antisera is usually not required when sufficient quantities of a highly purified immunogen are used to induce the immunoresponse. In those instances when it is necessary to reduce background and cross-reactivity and to ensure a higher degree of immunospecificity toward the target protein, purification by affinity chromatography is a well-documented and effective procedure.[2,5-7,13,29] It involves coupling sufficient amounts of a purified antigen to a suitable matrix in a column, e.g., Affi-Gels (Bio-Rad) or CNBr-activated Sepharose CL-4B (Pharmacia, Piscataway, NJ), adding the antiserum, elution of contaminating material with a neutral buffer, and finally desorption and collection of the purified antibodies. The desorption step is critical since the acidic or basic eluents often required to dislodge the antibodies may also cause denaturation. However, even distilled water can be an effective eluent, so preliminary tests should be run to determine elution conditions that are sufficiently mild but still consistent with good antibody recovery.[2,7]

As an alternative to the use of affinity columns, purification of small amounts of antisera can be achieved by immobilizing the antigen on 8-cm squares of nitrocellulose membrane (0.2 μm). The membrane is incubated for 4 h at room temperature in Tris-HCl (25 m*M* Tris, 140 m*M* NaCl to pH 7.6 with HCl) containing 0.2 to 0.5% (w/v) of the target protein, followed by several 10 min washes in Tris-HCl and subsequent rinses in PBS (pH 7.3). The protein is fixed on the membrane with 0.2% glutaraldehyde (w/v) in PBS on a slow shaker for 1 h, rinsed in PBS, and then incubated for 12 h in 10 ml of Tris-HCl plus 1 ml of crude antiserum. Antibodies bound to the membrane are eluted by shaking in 10 ml of 0.1 *M* glycine-HCl (pH 2.8) for 5 min. The decanted solution is quickly neutralized with 1.5 ml of 1 *M* potassium phosphate buffer (pH 9.3) before overnight dialysis at 4°C against 2.5 m*M* potassium phosphate (pH 7.4), 15 m*M* NaCl, and 0.5 m*M* $Na_2EDTA$. One ml aliquots of the dialyzed sample are added to sterile, 1.5 ml Eppendorf tubes, lyophilized, and stored at –20°C.[30-33]

When the primary purpose of purification is simply elimination of background contaminants from the antiserum and concentration of the IgG fraction, passing the antiserum through commercially prepared columns of agarose coated with either Protein A (Bio-Rad) or Protein G (Pierce, Rockford, IL) provides a simpler, faster, and more convenient approach than affinity binding to the target antigen. Proteins A and G are both derived from bacterial cell walls and selectively bind IgG. The added cost and the previously noted limitations associated with the desorption step are the principal disadvantages attendant to the use of these columns.

## E. POLYCLONAL VERSUS MONOCLONAL ANTIBODIES

Monoclonal antibodies are the components of a population of identical antibodies that all recognize (are specific for) one epitope on an antigen. They are produced by hybridoma cells, which are fusion products of cancerous B lymphocytes, i.e., myeloma cells and normal spleen cells obtained from mice immunized with a specific antigen. After initial selection of the hybridoma clones, antibody production is maintained in continuous culture, or the hybridomas are injected into mice to induce tumors that secrete large quantities of the antibody into ascites fluid, which is collected periodically and may either be used immediately or concentrated and stored.

Descriptions of the protocols necessary for producing monoclonal antibodies are described in detail elsewhere.[2,3,22,34,35] Nevertheless, it is appropriate to note here that a decision to use monoclonal or polyclonal antibodies in an immunological study involves the evaluation of a set of relatively clear-cut advantages and disadvantages before the choice is made.[3,13,22] Monoclonal antibodies offer significantly higher specificity than polyclonals, and large amounts are readily and continuously produced once cultures are established. They are particularly useful for quantitative detection of antigens that are routinely assayed in large numbers of samples, and a sustained supply of antibodies from the same clone of hybridoma cells affords significant confidence in test results. When available, a bank of different monoclonal antibodies generated against a particular protein may also provide important information concerning protein structure.[36]

On the negative side, monoclonal antibodies are technically more difficult and more expensive to produce, and the hybridoma cells can also destabilize and degrade, necessitating regeneration of the hybridoma cultures.[13,22] The epitope recognized by a monoclonal antibody can also occur on other, unrelated contaminant molecules, thus increasing the likelihood of cross-reactivity and consequently the need to purify, at least partially, the target antigen.

Monospecific polyclonal antibodies recognize multiple epitopes on an antigen and have widespread application in exploratory work, particularly in physiological studies of seed germination and of plant stress and senescence in which it is desirable to detect degradation products of target molecules. They also have a distinct advantage for screening gene expression libraries since they do recognize multiple determinants.[37]

The availability of stringent methods of protein purification that can provide essentially pure immunogen has made preparation of highly discriminating polyclonal antisera a fairly routine matter. Nevertheless, the quantities of a given batch of antibodies are quite limited, necessitating frequent renewal of supplies and testing of new antisera for specificity and avidity.

A large number of monoclonal and polyclonal antibodies are now available commercially (Amersham, Arlington Heights, IL; Boehringer Mannheim, Indianapolis, IN; Pierce; Jackson Immuno Research, West Grove, PA), but antibodies specific to plant proteins are still rare, although polyclonals can be ordered on an ad hoc basis from Ribi (Hamilton, Montana).

# III. WESTERN BLOTS

## A. DESCRIPTION

The potential to visualize and identify several proteins in a crude or partially purified extract from biological tissues defines the principal advantages of the Western blot procedure. Proteins separated by SDS-PAGE are transferred electrophoretically from a polyacrylamide gel to a suitable electroblot membrane, and one or more of the blotted proteins are subsequently identified using appropriate antibodies conjugated to a sensitive detection system. Nonspecific binding of the antibodies to unoccupied sites on the electroblot is minimized by first immersing the membrane in a blocking solution containing a foreign protein and/or a detergent. The membrane is then flooded with primary antibodies that bind the target protein on the blot. At this point, the bound antibodies are usually detected by a reporter enzyme conjugated to a secondary antibody directed against rabbit IgG. The target protein/primary antibody/secondary antibody complex is then visualized by adding the necessary chromogenic substrate for the reporter enzyme on the secondary antibody. Densitometric scans can provide estimates of the relative amounts of protein in each of the observed bands. Depending on the detection system used, the immunoblots can be stored dry in a cool, dark cabinet for several months.

There are many variations in the details of the basic Western blot procedure, particularly with regard to the choice of detection systems used to visualize the target proteins. Specific directions for transferring proteins from the gel to the membrane are usually provided by commercial suppliers of immunoblot equipment, and the principles of several different detection methods are also clearly described, with color illustrations, in catalogues from several supply houses (Amersham, Bio-Rad, Boehringer Mannheim, and Pierce). The following protocol for Western blot analysis of proteins separated by SDS-PAGE on minigels has been used routinely in our laboratories, and it represents a convenient and effective method for use with plant-derived proteins.

## B. THE ELECTROBLOT

1. Prepare concentrated (10X) transfer buffer[38] containing 30.3 g Tris, 144.0 g glycine, and distilled, deionized water to 1 l. The pH will range between 8.2 and 8.4; do not adjust. Store at 4°C.
2. Prepare 1X transfer buffer (25 m*M* Tris, 192 m*M* glycine, 20% MeOH) containing 100 ml of 10X buffer, 200 ml of absolute MeOH, and water to 1 l. Keep at 4°C before use.
3. Cut Whatman 3MM chromatography paper or equivalent into 7.5 ×10 cm sections and nitrocellulose paper (NCP, 0.2 μm) membrane into 7 × 9 cm sections. To avoid introducing contaminant polypeptides, always wear disposable gloves when handling NCP and throughout the immunoblot procedure.
4. Separate proteins on a minigel apparatus following the manufacturer's instructions. Include commercial, prestained protein standards on the gel to monitor subsequent transfer efficiency to the electroblot membrane and to provide an approximation (±10%) of molecular weights of the sample proteins. (For a more precise estimation of molecular weights see Section III.G.4.)
5. After separation of proteins on the minigel, soak the gel and the precut NCP membrane in 1X transfer buffer for at least 15 min. Cut a small corner from the lower right-hand side of both the gel and the membrane as directional markers in subsequent procedures.
6. Prepare a "gel-membrane sandwich" by placing the sandwich holder (the cassette from a commercial trans-blot apparatus) in a glass tray 3/4 full of 1X transfer buffer. Place the cathode (–) side of the cassette on the bottom of the tray, and construct the sandwich

under the surface of the buffer, layering each component from bottom to top as follows: (a) the foam or fibrous support pad, usually supplied with the apparatus, or for example a Scotch-Brite scouring pad, (b) one precut piece of 3MM chromatography paper, (c) the gel, (d) NCP membrane, (e) a second piece of precut 3MM chromatography paper, and (f) a second fiber support pad. Make sure that the gel and NCP are tightly appressed. Press the submerged sandwich with your fingers, or roll with a short glass rod or test tube to exclude all air bubbles, ensuring complete transfer and maximum resolution. Close and latch the cassette.

7. If several gels (up to four) are to be transferred in one sandwich, use the following sequence: (a) pad, (b) 3MM paper, (c) first gel, (d) NCP membrane, (e) two pieces of 3MM paper, (f) second gel, (g) NCP membrane, (h) two pieces of 3MM paper, (i) third gel, (j) NCP membrane, (k) two pieces of 3MM paper, and (l) fiber or foam support pad. Exclude all air bubbles and close cassette cover.
8. Place an empty transfer tank on a stirplate in a cold chamber at 4°C.
9. Fill the transfer tank to about 90% of total volume with cold, 1X transfer buffer. Add a medium-sized magnetic stir bar.
10. Place the prepared sandwich in the transfer tank with the gel side of the cassette (i.e., the bottom) closest to the cathode (–) and the NCP side facing the anode (+). Fill the tank with transfer buffer to a point just covering the anode on the tank. Cover the tank, and connect the electrodes from the transfer tank to the commercial power unit (red to red and black to black).
11. We prefer the constant current mode of operation with complete transfer from a 1-mm thick gel attained within 1 h at a current of 0.4 A at 4°C. The current can be raised to 0.5 A for high molecular weight proteins. Currents of 0.1 to 0.15 A are used for overnight runs at 4°C.
12. After transfer, turn off the power, unplug the cables and carefully remove the cassette, placing it in a dish containing clean transfer buffer.
13. Carefully open the cassette, but only enough to determine if the prestained standards have been completely transferred to the NCP. If not, close the cassette, and return it to the transfer tank to complete the process. Although transfer buffer can be reused, often three to four times before it turns slightly yellow, fresh buffer is preferred for each run.

## C. BLOCKING EXPOSED SITES

1. When transfer is complete and the cassette has been returned to the tray of transfer buffer (Step B.12), carefully remove the NCP from the sandwich and place it gel-side-up, in a small glass tray with 10 ml of blocking buffer (Tris-buffered saline + 1% BSA, see below). Shake gently for 1 h at room temperature to block any unoccupied sites on the NCP membrane.[39]
2. Concentrated, 10X, Tris-buffered saline (TBS) contains 30.3 g Tris and 81.8 g NaCl per liter of solution adjusted to pH 7.5 with HCl. Store at 4°C.
3. To prepare 1X TBS (25 m*M* Tris-HCl, 140 m*M* NaCl), dilute 100 ml of the concentrated stock solution to 1 l. Blocking buffer contains 1X TBS plus 10 g BSA (Fraction V, 98 to 99% pure or ELISA grade) per liter of solution and can be frozen in plastic containers to prolong storage life.

## D. BINDING WITH PRIMARY ANTIBODIES

1. After 1 to 2 h of incubation in blocking buffer, add $NaN_3$ to 0.02% and primary antiserum (e.g., 1 to 100 μl) for a total dilution of 100 to 10,000X depending on the

antibody titer. Cover and place at slow speed on a shaker at room temperature for 2 to 12 h or overnight. The buffer with antibodies can often be used several times, particularly when the antibodies are in short supply. If necessary, the antibodies can be immunopurified on Protein A or Protein G, as described in Section II.D. Screening tests (Section IV.C) of antibody versus antigen concentrations can be used to determine the optimum dilution of antiserum in a particular study.

2. Wash the NCP 3X for 10 min each with 10 ml of fresh blocking buffer to remove unbound antibodies.
3. Up to three different antibodies can sometimes be used to probe well-separated proteins on a single membrane once the likelihood of cross-reactivity between the antibodies and the three proteins has been excluded and after determining that any secondary bands that might be present are not located at the primary reaction site of another antibody. Alternatively, a blotted membrane can be cut into two or more strips and each strip probed independently with one antiserum.

## E. PROBING WITH ENZYME-LINKED SECONDARY ANTIBODIES

1. Transfer the NCP to 10 ml of 1X TBS without BSA. Add 3 μl of secondary antibody (goat anti-rabbit IgG, H and L) conjugated to alkaline phosphatase and shake at slow speed for 1 h at room temperature; then wash with fresh TBS for 10 min.
2. Wash once for 10 min in 10 ml of TBS made 5 m*M* with $Na_2EDTA$ plus 1% (v/v) Triton X-100 at pH 6.5[8] to ensure low background. (Caution: see Section III.G.5.) Prepare 100 ml of stock solution and store at 4°C.
3. Wash 2X, for 10 min each, with plain TBS.

## F. DETECTION OF ANTIBODY MARKERS, ADDITION OF SUBSTRATE

1. Prepare alkaline-phosphatase buffer: 100 m*M* Tris, 100 m*M* NaCl, 5 m*M* $MgCl_2$.[2] Dissolve 12.11 g of Tris, 5.844 g NaCl, 1.017g $MgCl_2$ per liter of solution, adjust to pH 9.5 with HCl. Store at 4°C.
2. Prepare *p*-nitroblue tetrazolium chloride (NBT) solution. Put 50 mg NBT in a brown, screw-top vial, add 1 ml of 70% (v/v) n,n-dimethylformamide (DMF), mix, and store at 4°C.[2]
3. Prepare solution of 5-bromo-4-chloro-3-indolylphosphate toluidine salt (BCIP). Put 50 mg BCIP in a brown, screw-top vial, add 1 ml of 100% DMF, mix, and store at 4°C.[2,8]
4. Put 10 ml of alkaline phosphatase buffer in a small dish just large enough to hold the NCP membrane with bound antibodies. Add 66 μl of NBT solution and 33 μl of the BCIP to yield final concentrations of 0.3 ml $\mu l^{-1}$ and 0.15 mg $\mu l^{-1}$ respectively. Place the NCP membrane in the dish and mix by shaking gently for not more than 2 to 3 min until light, purplish-brown bands of a suitable intensity appear.
5. Decant the reaction mixture, and cover the NCP with stop buffer (20 m*M* Tris, 5 m*M* EDTA) containing 1.21 g Tris and 931 mg $Na_2EDTA$ in 400 ml of solution. Adjust to pH 8.0, make to 500 ml with water, and store at 4°C.
6. After 5 to 10 min, remove NCP, and place on a paper towel to partially dry. Transfer NCP to waxed paper, allow to air-dry, and if desired, photograph.*
7. A scanning densitometer used in the reflectance mode can provide quantitative estimates of protein levels in each band when compared to standards or controls included on the same immunoblot.

* High backgrounds on immunoblots derive from several sources, principally from use of impure antibodies and/or reagents, incomplete blocking and washing of the transfer membrane, and failure to wear gloves during the full course of the immunoblot procedure.

## G. VARIATIONS IN THE GENERAL METHOD

### 1. Electroblot Membranes

Overall performance and relative cost make standard NCP the membrane of choice in many laboratories. However, break-resistant (supported) NCP, though somewhat more expensive, retains the essential characteristics of the pure nitrocellulose. Membranes with somewhat different characteristics are also in routine use and are readily available commercially. PVDF (polyvinylidene difluoride) is slightly more expensive than NCP, but it has the advantages of high mechanical strength, a higher protein binding capacity, and compatibility with many organic solutes. Its chemical stability makes it suitable for direct use in protein sequence analysis of the small amounts of purified protein isolated by PAGE.[40] Activated nylon membrane is also flexible and chemically resistant, with a protein binding capacity several-fold higher than NCP or PVDF.[41] However, nylon membrane is expensive, and it requires stringent blocking with BSA or nonfat dry milk to avoid high background. Nevertheless, it provides optimal results in terms of a high signal-to-noise ratio when used in conjunction with chemiluminescent detection.[41]

Plain photocopy paper (Xerox 4024 DP 20-lb) has been suggested as an alternative, inexpensive immunoblot membrane that provides quantitative detection over a wider range of protein concentrations than PVDF.[42] High background is commonly observed, but it can be subtracted using image analysis after a densitometric scan. Although the method is well described, its efficacy with a wide range of proteins has not been established.

### 2. Buffers

Although Tris-glycine is usually recommended for electroblot transfer, we have also had excellent results using 10 m*M* CAPS buffer [3-(cyclohexylamino)-1-propanesulfonic acid] with 10% methanol (v/v) at pH 11.[40] CAPS is less expensive than Tris-glycine, and it yields a shorter transfer time and can be used with NCP or PVDF membranes. To prepare stock buffer (10X), dissolve 22.13 g of CAPS in 900 ml of water, titrate to pH 11 with 2 *M* NaOH and add water to 1 l. Store at 4°C. For the working buffer, mix 100 ml of the 10X stock with 100 ml of methanol and 800 ml of water.

The inclusion of methanol in transfer buffers is not without drawbacks. It reduces pore size in gels and may thus interfere with the transfer of proteins larger than 175 kD. However, it also reduces protein loss during subsequent washing steps by removing some SDS from the detergent-protein complex, thereby increasing affinity between the proteins and the transfer membrane.[41] Methanol is not required for transfers to the positively charged nylon membrane or when blotting from gels that do not contain SDS.

### 3. Detection Systems

Several choices are available for the detection of proteins on transfer membranes. The alkaline phosphatase system (described in Sections III.E and F) has proven highly satisfactory for most of our work, but other approaches, usually more expensive, do provide specific advantages. For example, the standard alkaline phosphatase reaction can be amplified by using a biotinylated secondary antibody and taking advantage of the unusually high affinity between biotin and avidin or microbially derived streptavidin. When several molecules of alkaline phosphatase are conjugated to each avidin or streptavidin molecule, the amount of enzyme bound to the biotinylated secondary antibody and thus to a primary antibody is increased, and the standard alkaline phosphatase detection signal is enhanced. Particular care should be taken with the blocking step, since background signal can also be increased, but use of streptavidin instead of avidin may reduce some of the nonspecific binding.

Amplification of the alkaline phosphatase reaction can also be achieved with commercial kits that use enhanced chemiluminescence as opposed to color development to detect the blotted protein.[43] Light is emitted from an alkaline phosphatase-activated substrate, and the processed membrane is exposed to instant photographic or X-ray film, generally for one to several minutes, to provide a permanent, fade-proof record that is convenient for densitometric analysis and is easily filed. Light emission is usually stable for several hours which permits multiple exposures or a delay in processing. NCP, PVDF, or nylon membranes can be used, but nylon has the distinct advantage of requiring a two- to four-fold reduction in the amounts of biotinylated antibody and streptavidin-alkaline phosphatase conjugate used in processing.

Horseradish peroxidase can be substituted for alkaline phosphatase in the standard and amplified systems, but sensitivity of detection is lower and color fades with time.

It should be emphasized that whatever detection system is chosen, inadequate blocking of exposed binding sites on an electroblot and use of excessive amounts of secondary antibody with marker enzyme will yield high backgrounds on the immunoblot.

### 4. Molecular Weight Determinations

As noted in Section III.B.4, prestained protein standards used to monitor transfer efficiency will only provide rough approximations of the molecular weights of sample proteins immobilized on a membrane. The amount of dye bound to the standards alters their molecular weights and also varies among the molecules of each particular protein, causing their bands to broaden during migration in the gel.[41] For more precise determination of molecular weights, it is convenient to use biotinylated protein standards since their mobilities in gels and their molecular weights are not altered appreciably by the small biotin molecules bound to their surfaces. As a result, their bands remain tight, and they are intensely stained and sharply defined when the immunoblots are processed using the streptavidin-alkaline phosphatase conjugate and BCIP/NBT as substrate as described in Section III.F.

Alternatively, duplicate samples can be run on two halves of one gel and a mixture of nonbiotinylated, unstained, commercial, molecular-weight standards added to marker lanes on each half of the gel. After electrotransfer, the membrane is cut, one half is probed with primary and secondary antibodies, and the other half is stained for total protein. Fast green (1% w/v) or a prestained protein standard can be added to a center lane as a cutting guide. The same steps are followed to compare duplicate, uncut membranes. Molecular weights of the immunoblotted proteins are determined from a plot of relative mobilities ($R_f$) of the standard proteins over the logarithms of their molecular weights. $R_f$ values are determined as distances moved by each protein divided by the distance moved by a tracking dye (bromphenol blue, 0.1% w/v) added to the protein samples before SDS-PAGE. Instructions are usually available from suppliers of commercial kits (e.g., Sigma, St. Louis, MO).

The traditional stains used to detect proteins in gels, Coomassie blue and silver, are not suitable for staining blotted membranes. They commonly reduce the signal-to-noise ratio due to high background, and the membranes are often distorted by methanol included in the staining solutions. However, total protein on membranes can be conveniently and effectively stained with one of the following methods: (a) India ink,[2] by rinsing the blot in PBS (pH 7.3) containing 0.3% (v/v) Tween 20, then incubating in PBS containing 1 $\mu l\ ml^{-1}$ India ink and 0.3% Tween 20 for 2 to 20 h, with final rinsing in PBS; (b) Ponceau S as described in Section II.B; (c) the highly sensitive colloidal gold assay; and (d) biotinylation of all proteins on the blot with subsequent color development and detection using streptavidin-alkaline phosphatase with BCIP/NBT substrate as described in Sections III.F. and III.G.3. Commercial kits with instructions are available for methods (c) and (d).

### 5. Blocking Solutions

Blocking exposed sites on blotted membranes is one of the most crucial and perhaps underrated steps in Western blot analysis. Thus, it is essential to use a blocking agent that is compatible with all other conditions in the test, and if high backgrounds are observed, a more concentrated solution or a different blocking agent should be tried. BSA (1 to 5% w/v,

Fraction V, 98 to 99% pure) and reconstituted nonfat dry milk (1 to 5% w/v) are good choices that are compatible with many test systems, but other proteins (e.g., ovalbumin, gelatin) and even nonionic detergents (e.g., Tween 20, Triton X-100) are favored in some laboratories. Although dry milk is the least expensive choice, its composition can vary, and milk carbohydrates may bind to primary antibodies that recognize carbohydrate haptens on glycoproteins.[44] Nonionic detergents can be used alone (0.05 to 1.0% v/v) or combined with a protein in a blocking solution. Low concentrations of blocking agents are preferred if background on a membrane remains low since high concentrations of nonionic detergents can wash proteins off a blot, and proteinaceous blocking agents may conceivably bind and mask some sample proteins. In all cases, the membrane and blocking solution should be rocked or shaken gently throughout the blocking procedure. Protein-based solutions are made fresh or stored for a few days, at most, at 4°C. The addition of metabolic inhibitors, such as sodium azide (0.02% w/v) or thimersol (0.02% w/v) will extend storage time, but the azide may inhibit the horseradish peroxidase reaction if the blocking buffer is also used in the binding step.

#### 6. Semi-Dry Electroblots

Semi-dry electroblotting is a relatively new technique that provides several advantages over more conventional transfers in liquid-filled tanks.[45,46] The procedure is carried out using either standard-size or minigels. From one to six blots are run at one time by separating each gel-membrane sandwich with dialysis membrane to prevent cross-contamination. The amount of buffer required is about half the volume used in a mini-tank, with run times of only 15 to 60 min and no cooling requirement. Prestained protein markers can be included to monitor transfer efficiency using NCP, PVDF, or nylon membranes. The gels are oriented horizontally, with each sandwich composed from bottom (anode) to top (cathode) of one to three sheets of buffer-saturated filter paper, prewet transfer membrane, slab gel, and again one to three sheets of buffer saturated filter paper. Single sheets of prewet dialysis membrane are placed between each sandwich but never between a sandwich and the anode or cathode plates.

Several types of semidry electrotransfer units are now available commercially (American Bionetics, Hayward, CA; Bio-Rad; E & K Scientific, Saratoga, CA.), but it is fair to say that procedures, and to some extent equipment, are still being refined. We have tested the method with good results, but certain problems can arise; e.g., extended blotting times lead to buffer depletion from the filter paper resulting in incomplete transfer of high molecular weight proteins from the gel. In contrast, some low molecular weight proteins may pass through the membrane without binding, but this problem can be avoided by using a membrane with a smaller pore size. Both problems apparently derive from reduced flexibility in the choice of running conditions, resulting principally from the absence of a large volume of transfer buffer.

## IV. IMMUNOASSAYS AND SCREENING TESTS

### A. DESCRIPTION

Immunoassays and screening tests are rapid and comparatively simple microscale binding assays in which an immune complex is formed between antigen and antibody in the wells of microtitration plates or on one of the membranes already described for Western blot analysis. The methods employed are sensitive, specific, and relatively inexpensive. They allow qualitative and quantitative screening for specific kinds of proteins and polypeptides and for smaller compounds, such as hormones or sugars, to which antibodies were originally induced by conjugation of the small molecule to an immunogenic carrier protein. All of the assays involve measuring one of the components of the antibody-antigen complex using radiotracer labeling or, for our purposes, one of the enzyme-linked chromogenic detection systems described in Section III.F.

Although a wide range of immunoassays and screening methods have been developed, they can conveniently be divided into three broad categories, (a) antigen capture assays, (b) double-antibody sandwich assays, and (c) antibody capture assays. The protocols for these immunological tests are fairly straightforward, and details are included in several texts[2,22,47,48] and are usually provided by several commercial suppliers of immunological products (Amersham, Bio-Rad, Boehringer Mannheim, Gibco/BRL, Pierce, Schleicher and Schuell, Keene, NH).

### B. ENZYME-LINKED IMMUNOSORBENT ASSAYS

Detection in these assays (ELISAs) is dependent on the use of one of several possible reactions that can yield quantitation in the picogram range after preparation of a suitable standard curve. Alkaline phosphatase and horseradish peroxidase are the most commonly used enzyme tracers, facilitating tests that are safer, simpler, less expensive, and often more sensitive than radioimmunoassays, in which antigen or antibody are usually radiolabeled with a gamma emitter.

As described below, ELISA tests can assume three different formats.

#### 1. Competitive Antigen Capture

This method is commonly used for the analysis of plant hormones and other low molecular-weight compounds that can be bound as haptens to a commercial protein and the complex used to elicit production of antibodies against the hapten. The immunoassay involves the following general steps with examples of possible antigens and suitable volumes, concentrations, and buffers noted in parentheses:

1. Coat wells of a microtitration plate with monoclonal or purified, monospecific polyclonal antibodies [e.g., 100 μl of diluted IgG specific for abscisic acid (ABA) conjugated to BSA; IgG is diluted in 50 m*M* carbonate buffer to 5 ng $\mu l^{-1}$, pH 9.6; incubate 2 h at room temperature or 4°C overnight; wash with PBS, pH 7.3; add blocking buffer of PBS, 1 to 3% BSA, 0.02% $NaN_3$, pH 7.3, and incubate 2 to 4 h].
2. Add (i) a specific amount of the tracer (100 μl), i.e., the antigen standard conjugated to a reporter enzyme (ABA conjugated to alkaline phosphatase) and (ii) either unlabeled antigen from a standard solution (100 μl, 1 to 2 pmol ABA) or from a sample containing an unknown amount of unlabeled antigen. Use blocking solution for all dilutions.
3. Mix to initiate competition between the tracer antigen and the unlabeled antigen for the antibodies in the wells.
4. Incubate (2 to 4 h, room temperature), decant unbound tracer, and wash wells with buffer (PBS).
5. Add substrate for reporter enzyme on tracer [e.g., for alkaline phosphatase add 50 μl *p*-nitrophenyl phosphate at 1 mg $ml^{-1}$ in 10 m*M* diethanolamine, 0.5 m*M* $MgCl_2$, pH 9.6, incubate 20 to 30 min, stop reaction with 50 μl of 2 *M* NaOH or 1 m*M* EDTA, read absorbance at 405 nm with an ELISA plate reader]. Color is inversely proportional to amount of antigen in sample, a result of competition between antigen in sample and antigen in tracer.
6. Construct a sigmoid standard curve of amount of tracer bound divided by total amount of tracer added (%) versus the log of increasing antigen concentration; convert to a linear form using a logit transformation.

Detailed protocols for competitive ELISA techniques are described in several publications[49-51] and are also included in the kits supplied by Idetek, Inc. (San Bruno, CA) for the analysis of plant hormones. Other sources evaluate and discuss the principles and advantages of the method.[52-54]

### 2. Double Antibody Sandwich, Antigen Capture

Although a purified antigen is preferred in any immunoassay to minimize cross-reactivity and background, this type of ELISA is commonly used with partially purified or even crude extracts, i.e., when a good supply of highly purified antigen is not available. It is a noncompetitive method in which two antibodies bind to the same antigen but not to each other. Monoclonal antibodies from two different species can be used, or one population of affinity purified polyclonals will work. One population of monoclonal antibodies is free, and the other is conjugated to a reporter enzyme, or if polyclonal antibodies are used, a portion of that population is enzyme labeled. A three-antibody indirect approach is also described in Step 2 below.

1. Coat wells of microtitration plate with the free antibody; incubate and wash with appropriate blocking buffer (see Section IV.B.1.1).
2. Add aliquot (100 μl, 5 ng $\mu l^{-1}$) of test antigen to each well, mix, and rinse.
3. Add solution of second antibody conjugated to alkaline phosphatase; incubate and wash with buffer. Alternatively, an indirect assay can be used in which the second set of antibodies, from a different animal, are also unlabeled, but a third set of commercial, enzyme-linked antibodies is used that is specific for IgG in the second set of antibodies, e.g., rabbit anti-mouse IgG conjugated to alkaline phosphatase. Although the three-antibody approach provides an indirect measure of antigen present, it avoids the necessity of preparing enzyme-antibody conjugates for each different antigen.
4. Incubate; then wash wells with appropriate buffer.
5. Add enzyme substrate, incubate, and stop reaction (see Section IV.B.1.5).
6. Read absorbance at 405 nm with an ELISA plate reader.
7. Construct a standard curve. Intensity of color developed is proportional to the amount of antigen present.

Detailed protocols for the method, including amounts, concentrations and volumes, are given by Pratt et al.,[36] Gaastra,[47] and Clarke et al.[48]

### 3. Antibody Capture, Indirect Detection

The antigen of interest is bound to the walls of a microtitration plate followed by binding of the primary antibody (polyclonal or monoclonal), which is then detected by a secondary, enzyme-linked, commercial polyclonal antibody that acts as the tracer. This is convenient to use when sufficient amounts of highly purified antigen are available to coat the wells so that background does not obscure detection. In some cases, high purity antigen may not be required if both sets of antibodies are highly specific. The general steps are as follows:

1. Coat wells with purified tissue extract or standard antigen dissolved in appropriate coating buffer (e.g., 100 μl of antigen at 10 ng $\mu l^{-1}$ in PBS).
2. Incubate (2 to 4 h, room temperature); wash with buffer (PBS), then blocking buffer (see Section IV.B.1.1).
3. Add specific antiserum in blocking buffer, incubate, and wash with blocking buffer (see Section IV.B.1.1).
4. Add commercial, enzyme-linked secondary antibody specific for IgG in primary antibody (see Section IV.B.2.3).
5. Add enzyme substrate, incubate, and stop reaction (see Section IV.B.1.5).
6. Read absorbance at 405 nm, and construct standard curve. Color response is proportional to amount of antigen-bound antiserum present.

Detailed protocols are provided by Clarke et al.[48] and by Penrose and Glick.[55]

### C. DOT BLOTS AND SLOT BLOTS

Dot and slot blots are similar to the ELISA antibody capture assay described in Section IV.B.3, but the plastic microtitration plates are replaced by electroblot transfer membranes, which can bind more protein per unit of surface area. The sample dots or slots are read with a scanning densitometer at a wavelength appropriate for the chromogenic probe. Useful applications of these microblots include evaluation of antibody titers from different animals, rapid screening of large numbers of hybridomas for monoclonal antibodies, monitoring protein effluent from a separation column, antibody dilution assays to determine optimum antibody to antigen ratios, and preliminary screening assays of crude protein extracts.

The procedure involves spotting microliter amounts of antigen on NCP, PVDF, or nylon transfer membrane and maintaining the same precautions for handling membranes noted in Section III. For convenience and for precise quantitation of applied antigen, the membrane is usually mounted in a microfiltration apparatus that has a vacuum manifold and a template that facilitates application of a large number of equally-spaced samples. The samples are applied with standard or multichannel pipettes in volumes ranging from a few microliters to 500 μl, containing 100 pg to 10 μg of protein on spots 1 to 3 mm in diameter. Although samples can be applied without microfiltration, sample loading is usually reduced, and efficient interaction between spotted antigen and subsequent reactants may be limited. Nevertheless, unfiltered dot blots are useful for monitoring the presence, absence, or direction of change in antigen titer in crude or purified samples.

For precise quantitation using the microfiltration procedure, the dot-blot template is replaced with a slot-blot template in which each sample is applied in a rectangular format that is convenient for densitometric scanning. The samples can be visualized using a chromogenic marker enzyme (e.g., alkaline phosphatase with BCIP/NBT substrate) or colloidal gold conjugated to a commercial secondary antibody. Specific details are provided in several sources[2,22,41] and in protocols from commercial suppliers of immunological products.

## ACKNOWLEDGMENTS

We thank Mrs. Patricia Dumbroff and Dr. Sibdas Ghosh for their many helpful suggestions and for their critical review of this manuscript.

## REFERENCES

1. **French, D. L., Laskov, R., and Scharff, M. D.,** The role of somatic hypermutation in the generation of antibody diversity, *Science*, 244, 1152, 1989.
2. **Harlow, E. and Lane, D.,** *Antibodies, A Laboratory Manual*, Cold Spring Harbor Laboratory, New York, 1988.
3. **Wilson, K. and Goulding, K. H., Eds.,** *A Biologist's Guide to Principles and Techniques of Practical Biochemistry*, 3rd ed., Edward Arnold, Great Britain, 1986, chap. 3, 4.
4. **Darnell, J., Lodish, H., and Baltimore, D.,** *Molecular Cell Biology*, 2nd ed., Sci. Am. Books, Freeman and Co., New York, 1990, chap. 25.
5. **Dean, P. D. G., Johnson, W. S., and Middle, F. A., Eds.,** *Affinity Chromatography, A Practical Approach*, IRL Press, Oxford, 1985.
6. **Anonymous,** *Affinity Chromatography, Principles and Methods*, Handbook, Pharmacia Fine Chemicals, Ljungforetagen AB, Orebro, Sweden, 1983, chap. 5.
7. **Scopes, R. K.,** *Protein Purification, Principles and Practice*, 2nd ed., Springer-Verlag, New York, 1987.
8. **Ghosh, S., Gepstein, S., Glick, B. R., Heikkila, J. J., and Dumbroff, E. B.,** Thermal regulation of phosphoenolpyruvate carboxylase and ribulose-1,5-bisphosphate carboxylase in $C_3$ and $C_4$ plants native to hot and temperate climates, *Plant Physiol.*, 90, 1298, 1989.

9. **Gegenheimer, P.,** Preparation of extracts from plants, in *Guide to Protein Purification, Methods in Enzymology*, Vol. 182, Deutscher, M. P., Ed., Academic Press, San Diego, 1990, chap. 14.
10. **Giri, L.,** Chromatofocusing, in *Guide to Protein Purification, Methods in Enzymology*, Vol. 182, Deutscher, M. P., Ed., Academic Press, San Diego, 1990, chap. 31.
11. **Deutcher, M. P., Ed.,** *Guide to Protein Purification, Methods in Enzymology*, Vol. 182, Academic Press, New York, 1990.
12. **Davis, L. G., Dibner, M. D., and Battey, J. F.,** *Basic Methods in Molecular Biology*, Elsevier, New York, 1986, chap. 19.
13. **Dunbar, B. S.,** *Two-Dimensional Electrophoresis and Immunological Techniques*, Plenum Press, New York, 1987, chap. 2, 3, 8, Appendix 11, 12.
14. **Garfin, D. E.,** One-dimensional gel electrophoresis, in *Guide to Protein Purification, Methods in Enzymology*, Vol. 182, Deutscher, M. P., Ed., Academic Press, San Diego, 1990, chap. 33.
15. **Markwell, J.,** Electrophoretic analysis of photosynthetic pigment-protein complexes, in *Photosynthesis Energy Transduction, A Practical Approach*, Hipkins, M. F. and Baker, N. R., Eds., IRL Press, Oxford, 1986, chap. 3.
16. **Smith, B. J.,** SDS polyacrylamide gel electrophoresis of proteins, in *Methods in Molecular Biology, Proteins*, Vol. 1, Walker, J. M., Ed., Humana Press, New Jersey, 1984, chap. 6.
17. **Harrington, M. G.,** Elution of protein from gels, in *Guide to Protein Purification, Methods in Enzymology*, Vol. 182, Deutscher, M. P., Ed., Academic Press, San Diego, 1990, chap. 37.
18. **Merril, C. R.,** Gel-staining techniques, in *Guide to Protein Purification, Methods in Enzymology*, Vol. 182, Deutscher, M. P., Ed., Academic Press, San Diego, 1990, chap. 36.
19. **Knudsen, K. A.,** Proteins transferred to nitrocellulose for use as immunogens, *Anal. Biochem.*, 147, 285, 1985.
20. **Kilberg, M.,** Immunization using S&S NC nitrocellulose membranes, in *Sequences*, Schleicher & Schuell, 1988, 26, 1.
21. **Diano, M., Le Bivic, A., and Hirn, M.,** A method for the production of highly specific polyclonal antibodies, *Anal. Biochem.*, 166, 224, 1987.
22. **Tijssen, P.,** *Practice and Theory of Enzyme Immunoassays, Laboratory Techniques in Biochemistry and Molecular Biology*, Vol. 15, Burdon, R. H. and van Knippenberg, P. H., Eds., Elsevier, The Netherlands, 1985, chap. 5, 13, 14, 16.
23. **Johnstone, A. and Thorpe, R.,** *Immunochemistry in Practice*, Blackwell, Oxford, 1982, chap. 11, 12.
24. **Hershko, A., Eytan, E., and Ciechanover, A.,** Immunological analysis of the turnover of ubiquitin-protein conjugates in intact cells, *J. Biol. Chem.*, 257, 13964, 1982.
25. **Haas, A. L. and Bright, P. M.,** The immunological detection and quantitation of intracellular ubiquitin-protein conjugates, *J. Biol. Chem.*, 260, 12464, 1985.
26. **Shimogawara, K. and Muto, S.,** Heat shock induced change in protein ubiquitination in *Chlamydomonas*, *Plant Cell Physiol.*, 30, 9, 1989.
27. **Lacy, M. J., Kent, C. R., and Voss, E. W., Jr.,** d-Limonene: an effective vasodilator for use in collecting rabbit blood, *Lab. Anim. Sci. Am. Assoc. Lab. Anim. Sci.*, 485, 1987.
28. **Jensenius, J. C., Andersen, I., Hau, J., Crone, M., and Koch, C.,** Eggs: conveniently packaged antibodies. Methods for purification of yolk IgG, *J. Immunol. Meth.*, 46, 63, 1981.
29. **Choi, J. H., Liu, Liang-Shi, Borkird, C., and Sung, Z. R.,** Cloning of genes developmentally regulated during plant embryogenesis, *Proc. Natl. Acad. Sci., U.S.A.*, 84, 1906, 1987.
30. **Olmstead, J. B.,** Affinity purification of antibodies from diazotized paper blots of heterogenous protein samples, *J. Biol. Chem.*, 256, 11955, 1981.
31. **Gershoni, J. M. and Palade, G. E.,** Review: protein blotting: principles and applications, *Anal. Biochem.*, 131, 1, 1983.
32. **Khayat, E.,** unpublished data, 1989.
33. **Smith, D. E. and Fisher, P. A.,** Identification, developmental regulation, and response to heat shock of two antigenically related forms of a major nuclear envelope protein in *Drosophila* embryos: application of an improved method for affinity purification of antibodies using polypeptides immobilized on nitrocellulose blots, *J. Cell Biol.*, 99, 20, 1984.
34. **Gatz, R. L., Young, B. A., Facklam, T. J., and Scantland, D. A.,** Monoclonal antibodies: emerging product concepts for agriculture and food, *Bio/Technology*, 33, 1983.
35. **Pratt, L. H.,** Phytochrome immunochemistry, in *Techniques in Photomorphogenesis*, Smith, H. and Holmes, M. G., Eds., Academic Press, London, 1984, 201.
36. **Pratt, L. H., Senger, H., and Galland, P.,** Phytochrome and other photoreceptors, *Meth. Plant Biochem.*, 4, 185, 1990.
37. **Dunbar, B. S. and Schwoebel, E. D.,** Preparation of polyclonal antibodies, in *Guide to Protein Purification, Methods in Enzymology*, Vol. 182, Deutscher, M. P., Ed., Academic Press, San Diego, 1990, chap. 49.

38. **Towbin, J., Staehlin, T., and Gordon, J.,** Electrophoretic transfer of proteins from polyacrylamide gels to nitrocellulose sheets: procedure and some applications, *Proc. Natl. Acad. Sci. U.S.A.*, 76, 4350, 1979.
39. **Nelson, N.,** Structure and synthesis of chloroplast ATPase, *Meth. Enzymol.*, 97, 510, 1983.
40. **Matsudaira, P.,** Sequence from picomole quantities of proteins electroblotted onto polyvinylidene difluoride membranes, *J. Biol. Chem.*, 262, 10035, 1987.
41. **Anonymous,** Protein blotting, a guide to transfer and detection, Bulletin 1721, Bio-Rad, Richmond, CA.
42. **Yom, Heng-Cherl and Bremel, R. D.,** Xerographic paper as a transfer medium for Western blots: quantification of bovine αS1-casein by Western blot, *Anal. Biochem.*, 200, 249, 1992.
43. **Durrant, I.,** Light-based detection of biomolecules, *Nature*, 346, 297, 1990.
44. **Johnson, D. A., Gautsch, J. W., Sportsman, J. R., and Elder, J. H.,** Improved method for utilizing nonfat dry milk for analysis of proteins and nucleic acids transferred to nitrocellulose, *Gene Anal. Technol.*, 1, 3, 1984.
45. **Lin, W. and Kasamatsu, H.,** On the transfer of polypeptides from gels to nitrocellulose membranes, *Anal. Biochem.*, 128, 302, 1983.
46. **Kyhse-Anderson, J.,** Electroblotting of multiple gels: a simple apparatus without buffer tank for rapid transfer of proteins from polyacrylamide to nitrocellulose, *J. Biochem. Biophys. Meth.*, 10, 203, 1984.
47. **Gaastra, W.,** Enzyme-linked immunosorbent assay (ELISA), in *Methods in Molecular Biology*, Vol. 1, Walker, J. M., Ed., Humana Press, Clifton, NJ, 1984, chap. 38.
48. **Clark, M. F., Lister, R. M., and Bar-Joseph, M.,** ELISA techniques, in *Methods for Plant Molecular Biology*, Weissbach, A. and Weissbach, H., Eds., Academic Press, San Diego, 1988, chap. 32.
49. **Weiler, E. W., Jourdan, P. S., and Conrad, W.,** Levels of indole-3-acetic acid in intact and decapitated coleoptiles as determined by a specific and highly sensitive solid-phase enzyme immunoassay, *Planta*, 153, 561, 1981.
50. **Weiler, E. W.,** Enzyme-immunoassay for cis (+)– abscisic acid, *Physiol. Plant.*, 54, 510, 1982.
51. **Daie, J. and Wyse, R.,** Adaptation of the enzyme-linked immunosorbent assay (ELISA) to the quantitative analysis of abscisic acid, *Anal. Biochem.*, 119, 365, 1982.
52. **Hedden, P.,** Gibberellins, in *Principles and Practice of Plant Hormone Analysis*, Vol. 1, Rivier, L. and Crozier, A., Eds., Academic Press, London, 1987, chap. 2.
53. **Neill, S. J. and Horgan, R.,** Abscisic acid and related compounds, in *Principles and Practice of Plant Hormone Analysis*, Vol. 1, Rivier, L. and Crozier, A., Eds., Academic Press, London, 1987, chap. 3.
54. **Sandberg, G., Crozier, A., and Ernsten, A.,** Indole-3-acetic acid and related compounds, in *Principles and Practice of Plant Hormone Analysis*, Vol. 2, Rivier, L. and Crozier, A., Eds., Academic Press, London, 1987, chap. 4.
55. **Penrose, D. M. and Glick, B. R.,** Production of antibodies against sorghum leaf phosphoenolpyruvate carboxylase monomer and their use in monitoring phosphoenolpyruvate carboxylase levels in sorghum tissues, *Biochem. Cell Biol.*, 64, 1234, 1986.

# 14

# *In Situ* Immunocytochemical Localization of Plant Proteins

*Nicole Benhamou and Alain Asselin*

## I. INTRODUCTION

Plant proteins are key components in a number of biological functions such as cell growth and development, metabolic regulation, transport, self-incompatibility, and resistance to stress. In considering the broad spectrum of research devoted to plant proteins, it has become apparent that visualizing the spatial distribution of these molecules could complement our understanding of their functions. It has been more than two decades since immunocytochemical studies of extracellular antigens were first applied to plant tissues.[1] Since then, they have proved extremely useful for identifying and determining the precise location of a wide range of molecules in plant cells.[2-3] There are now several areas in which immunocytochemistry is being explored as an essential complement to chemical, biochemical, and molecular investigations. A growing body of evidence indicates that research on particular topics such as membrane structure, cell wall elongation, photosynthesis, hormonal regulation, storage processes, and resistance to stress has greatly benefited from the information provided by the *in situ* localization of plant antigens.[4-7]

Although plants exhibit some of the features of immunity in that they are capable of discriminating self from nonself, they do not possess an immune system similar to that found in vertebrates. Thus, antigen-antibody interactions are restricted to defense reactions in vertebrates, and the analogy in plants may be somewhat related to the production of inhibitory phenols and phytoalexins as well as to the synthesis of lectins with agglutinating properties. The discovery that many plant proteins and glycoproteins are powerful antigens has led to the concept that antibodies, raised in experimental animals against purified plant molecules, could be valuable tools for serological, immunological, and immunocytochemical purposes.[8] The considerable improvement of methods for plant protein purification, together with the production of more and more antisera with specific binding properties, explains the current interest in plant immunocytochemistry.[2-3]

Since immunoglobulins are not electron-opaque, antibody binding sites cannot be directly visualized under the electron microscope without the use of an electron-dense marker. Among the various markers (e.g., peroxidase and ferritin particles) that have been used, colloidal gold particles have undoubtedly led to the best immunocytochemical results in terms of high resolution labeling at the electron microscope level.[9] Because of its particulate nature and its high electron density, colloidal gold has rapidly acquired increasing applicability and relevance in immunocytochemistry.[3] Antigen-antibody interactions can be visualized according to different procedures. Antibodies may be applied in a pre-embedding approach prior to or after tissue fixation, or in a post-embedding approach after tissue sectioning. The latter

0-8493-5164-2/93/$0.00+$.50
© 1993 by CRC Press, Inc.

method includes a one step procedure in which the specific antibody is directly conjugated to colloidal gold, and a two step, indirect procedure in which the specific antibody binds to its target antigen and is then detected by either gold-labeled protein A[9] or gold-labeled anti-antibody.[10] This indirect method offers some advantages over the other alternatives. First, the antibody has several determinants to which labeled-protein A or anti-antibody may bind, thus leading to an amplification of the response, and second, both labeled compounds are commercially available and can be stored for several months without loss of reactivity.

Despite the increasing availability of highly specific antibodies and the development of accurate methods of *in situ* investigation, plant immunocytochemistry still poses some difficulties.[11] Thus, tissue processing leading to satisfactory preservation of cellular structures may, in some cases, mask or even destroy antigenic sites. Similarly, tissue fixation with aldehydes may represent a limitation for the *in situ* detection of soluble proteins such as enzymes or pathogenesis-related (PR) proteins.[12] Finally, access of the antibody to internal receptor sites may be restricted by the plant cell wall, which constitutes a potential barrier especially when preembedding techniques are used.[8,11]

Several reviews have recently been published that address the breadth of applications of immunocytochemistry in plant biology.[2,3,8] It is not our purpose in this chapter to review the wide variety of applications that have been reported in the literature. Instead, our objectives are to outline the techniques developed in the authors' laboratories for the production and screening of antibodies used as immunocytochemical reagents for the *in situ* localization of some proteins in plant tissues. Particular emphasis will be given to the applications of immunocytochemistry to plants for the localization of those proteins that are induced upon pathogenic attack and that are thought to be involved in plant disease resistance. If one considers that current interest seems to be focused on the possibility of tranforming plants in such a way that they express constitutively or inducibly "foreign" resistance genes, it is obvious that information regarding the location and spatial distribution of the gene products may be highly informative in the understanding of the mechanism underlying induced plant resistance.

## II. IMMUNOCYTOCHEMICAL LABELING OF PLANT PROTEINS: PRINCIPLES, GENERAL PRACTICE, AND LIMITATIONS

Immunocytochemistry for the localization of plant proteins is identical in principle to that used for the identification of any other type of molecule. The success of the approach is entirely dependent upon highly specific antibodies, good preservation of protein antigenicity and cellular structure, and low degree of nonspecific interaction between the immunological probes and the embedded plant tissue.[13] A number of procedures using different types of fixatives and/or embedding resins have been reported. However, it appears that the conditions yielding optimal results in terms of ultrastructure preservation and appropriate retention of protein antigenicity have to be worked out separately in each case.

### A. PRINCIPLES OF PLANT PROTEIN IMMUNOCYTOCHEMISTRY

Immunocytochemical techniques are based on the binding of antibody molecules to specific sites on the surface of an antigen against which they were raised in a mammalian species. Provided the antibody molecules have sufficient access to their antigens, the antibody-antigen interaction may be visualized under the electron microscope through the use of gold-labeled secondary antibody or protein A.[9] This post-embedding technique is widely applied in many laboratories for the *in situ* localization of various intracellular substances.

Protein A from *Staphylococcus aureus* consists of a single polypeptide of 42 kDa which interacts with immunoglobulins (especially immunoglobulin G, IgG) in the Fc region. One molecule of protein A contains four homologous Fc-binding sites and thus can interact with two IgG molecules.[9] Because of its high affinity for IgGs, protein A has been widely used as a secondary reagent in immunocytochemical studies using polyclonal antisera from rabbits.

Gold-conjugated secondary antibodies are often used as an alternative to protein A in the two step post-embedding labeling technique. Usually these antibodies are raised in goats against immunoglobulins from rabbits, mice, or other animal species. Labeled-secondary antibodies are commercially available and supplied in various gold particle sizes. They are highly stable at –4°C and can be stored for several months. Both protein A and secondary antibodies have been used successfully in studies dealing with protein localization in plant tissues.[14-16] However, secondary antibodies may be preferred because of their higher affinity binding properties for primary antibody molecules.

Although immunocytochemical techniques are apparently easy to perform, several parameters need to be considered carefully when dealing with plant tissues. Among the most important, one can cite (1) isolation and purification of proteins used as antigens for immunization, (2) antibody preparation and its specificity against the plant antigen, (3) tissue processing that allows both cellular preservation and retention of protein antigenicity, (4) antibody access to intracellular molecules in tissue sections, and (5) efficient and sensitive labeling technique that avoids nonspecific binding to the embedding resin. At present, there is no single technique of choice, and the approach must be adapted according to the antigen to be identified. As an example, difficulties may be encountered in localizing soluble extracellular antigens.[13] Because of their high solubility combined with their low molecular masses (e.g., PR proteins), these molecules may readily diffuse during tissue processing, thus giving rise to nonspecific labeling patterns.[12] Alternative methods of tissue processing such as the use of cryofixed tissues[17] or microwave energy fixation[12] must be worked out for obtaining a realistic image of the distribution of soluble proteins in plant tissues.

## B. TISSUE PROCESSING

The influence of fixation on the preservation of antigenic sites may vary according to the plant material used or even the antigen to be detected. Similarly, embedding resins may have to be selected to suit particular plant tissues. More generally, fixation and/or embedding techniques should be adjusted in order to obtain optimal degrees of antigen-antibody binding on tissue sections.

### 1. Standard Fixation Protocol

The aim of fixation is to immobilize cell structures in a state as close as possible to the native state. In this context, fixation appears necessary to halt diffusion of compounds into and out of cells and to reinforce the plant tissue against the effect of other reagents during tissue processing.[18] The ideal fixative should penetrate rapidly into the tissue and create stable cross-links between reactive groups of the protein molecules, resulting in satisfactory ultrastructural preservation. However, a compromise must be made so that protein cross-linking is minimized in such a way that antigenic determinants are still available.

A variety of aldehyde-based methods have been suggested, and again the appropriate fixation approach must be determined for each antigen to be detected. Among the aldehydes, glutaraldehyde is currently used for plant tissue fixation.[19] Being a dialdehyde, this fixative offers the advantage of increasing cross-linking between protein chains, allowing good ultrastructure preservation. However, this property may, in turn, reduce antibody recognition of proteins. Concentrations of fixatives, as well as times of fixation and conditions such as temperature, should be worked out to optimize antigenicity preservation while allowing sufficient tissue integrity. An alternative to glutaraldehyde is paraformaldehyde, which can be used alone at varying concentrations or in combination with glutaraldehyde.[19] Being a monoaldehyde, this fixative leads to a low degree of tissue preservation but allows better retention of antigenicity. Another class of fixatives, known to precipitate proteins rather than creating cross-links between their reactive groups, may also be used for immunocytochemical purposes. This is the case of picric acid, although this reagent has seldom been reported to be efficient.

Post-fixation with osmium tetroxide, usually recommended for basic electron microscope investigations, is most often avoided in fixation protocols for immunocytochemistry because of its adverse effect on antigenic sites.[3] The masking effect of this fixative may cause considerable loss of antibody recognition, resulting in false labeling patterns. However, osmium tetroxide may be removed from sections of post-fixed tissues by treatment with saturated sodium metaperiodate.[20] Because periodate may alter the structure of most glycoconjugates, it is not clear whether or not use of this reagent is appropriate prior to immunocytochemical studies. In fact, the best technique leading to optimal results consists of using 2 to 3% (v/v) glutaraldehyde in 0.1 *M* sodium cacodylate buffer, pH 7.2, for 3 h at room temperature or for longer periods at 4°C, and omitting post-fixation with osmium tetroxide.

## 2. Microwave Fixation Protocol

Fixation and immobilization of low molecular mass soluble proteins pose a number of problems for their localization *in planta*. The difficulty in maintaining sufficient protein antigenicity is, in the case of soluble proteins, compounded by the problem of retaining them in their original location. Because of their low molecular masses and their extreme solubility in aqueous media, these proteins readily diffuse from one cell compartment to another. Hosokawa and Ohashi[21] and Benhamou et al.[12] have pointed out the difficulty of localizing pathogenesis-related (PR) proteins in intercellular spaces using plant tissues fixed under conventional procedures. Various means have been devised for retaining maximal amounts of soluble antigens while providing optimal tissue preservation. Freeze-drying and freeze substitution, as well as the use of frozen sections,[17] have proved to be useful alternatives for preserving soluble protein molecules. Another approach that has been introduced recently in plant cytology is the microwave energy fixation method.[12] This procedure, based on the use of microwave irradiation during aldehyde fixation, was found effective for both preserving cellular structures and retaining soluble proteins in specific cell compartments. In a recent report, Benhamou et al.[12] described the experimental design and conditions yielding optimal preservation of tissue ultrastructure and protein antigenicity. The authors demonstrated that irradiation exposure time and final fixative temperature were important factors for optimal tissue integrity. It was found that maximal preservation was obtained with 15 to 20 s of microwave irradiation and a final fixative solution temperature of 37 to 42°C. It is thought, although not yet proven, that rapid diffusion of the fixative into tissues (15 to 20 s) is enhanced by the thermal effect of microwave irradiation. By reducing the incubation time for tissue fixation from 2 to 3 h to 15 to 20 s, microwave energy fixation offers the advantage of immobilizing large amounts of soluble antigens while preserving sufficiently the structural integrity of plant tissues. This approach also has the advantage of being inexpensive, considering that microwaves are readily available in most laboratories. It is a valuable alternative to freeze-substitution which requires costly instrumentation.

## 3. Embedding Procedures

Two types of resins that give the best structural preservation and beam stability of sections under the electron microscope are currently used. These are the epoxy resins such as Epon, Spurr, and Araldite and the hydrophilic acrylics, including L.R. White and Lowicryl.[22] Epoxy resins generally exhibit low water absorption and several hydrophobic groups in their structures that may be the cause of reduced antigenic preservation in some cases. However, good immunocytochemical results have been reported with Epon-embedded plant tissues.[6,7] The acrylic resin, L.R. White, has become popular for plant immunocytochemical studies.[3,5] This resin, which combines a high hydrophobic character with an electron beam stable cross-link, is usually heat-cured at 50°C using an aromatic tertiary amine accelerator. Lowicryl K4M, a mixture of acrylates and methacrylates, is highly hydrophilic and usually exhibits a very good retention of antigenicity.[23] This resin is cured at low temperatures by exposure to ultraviolet. Although Lowicryl is undoubtedly one of the best resins for antigenic preservation, it has some limitations that frequently make its use with plant tissues inadequate.

### C. PRODUCTION AND SCREENING OF ANTIBODY PROBES

The usefulness of the immunogold labeling technique relies on the availability of a specific antibody. Polyclonal and monoclonal antibodies can be used. Polyclonal antibodies are usually raised in rabbits against purified proteins, while monoclonal antibodies are produced in antibody-secreting hybridoma cell lines. The comparative advantages and limitations of both types of antibodies have been discussed in detail in previous reviews.[3] Practically, high titer monospecific polyclonal antiodies are very useful, easier to prepare, and cheaper than monoclonal antibodies that require time-consuming and expensive procedures.[3]

The titer and specificity of the antibodies have to be thoroughly determined. The specificity of antibodies is now routinely assessed by using dot immunoblots or immunoblots after polyacrylamide gel electrophoresis (PAGE).[24,25] In addition, proper controls have to be included, such as the concurrent analysis of the reactivity of pre-immune serum.

The analysis of antibody specificity by dot immunoblots is valuable if a pure protein is available. Protein purity is often determined by PAGE analysis followed by silver staining of the protein.[26] Staining with Coomassie blue (R-250 or G-250) is much less sensitive.[27] In some cases, protein purity can also be assessed by homogeneity of N-terminal sequences of purified proteins. Analysis by PAGE can include either denaturing (sodium dodecyl sulfate, SDS) or native gels. Native PAGE is usually performed at pH 4.3 or 8.9, depending on whether the protein to be analyzed is basic or acidic. Two-dimensional gels can also be very informative,[25,28] although in most cases, one-dimensional SDS-PAGE is used. It is important to realize that such gels cannot separate proteins with the same molecular mass. On the other hand, native PAGE systems are especially suited for separating proteins with the same molecular mass but with a different overall charge. However, it is possible that two distinct proteins migrate as one spot, even if a 2-D gel system is used. In such cases, affinity electrophoresis can be valuable if one protein has a differential affinity for a given substrate. For example, β-1,3-glucanases have been separated from other plant proteins by affinity electrophoresis in the presence of laminarin.[29] If the purified protein has detectable enzymatic activity toward a substrate, detection of its activity after PAGE can provide useful complementary information.[29] Thus, plant lysozymes, chitinases, β-1,3-glucanases, and chitosanases are easily and rapidly detected by activity staining after PAGE separation of proteins.[28,30-32]

Analysis of the specificity of the antibody toward the plant protein under study is of crucial importance and should be performed rigorously by several complementary approaches, as previously suggested for the analysis of the purity of the protein antigen. The specificity of the antibody should be tested not only against the purified protein but also against whole tissue components. Analysis with immunoblots after PAGE separation of tissue proteins under various conditions should be attempted. It is noteworthy that results from immunoblots following SDS-PAGE may be difficult to interpret because of the fully denatured state of the protein separated under reducing conditions. For example, a tobacco β-1,3-glucanase showed immunological cross-reactivity with bovine carbonic anhydrase when proteins were denatured and reduced, while under native conditions, no such cross-reactivity was detected.[25] *In planta*, this enzyme occurs in its native state, and the fully denatured and reduced state of proteins is not necessarily a proper reflection of the protein state even in embedded tissues.[33]

There are technical means for increasing the specificity of some antibodies. The isolation of antibodies bound to protein antigens separated by PAGE techniques is feasible by dissociating the antigen-antibody complex at acidic pH.[34] There are also alternative methods for purifying antibodies such as antigen affinity chromatography.[13]

## D. POST-EMBEDDING LABELING TECHNIQUES

Post-embedding labeling techniques have been more widely applied because they allow direct access of antibodies to their binding sites. The general principles underlying these techniques have been reported in several excellent reviews.[3,35]

### 1. Colloidal Gold

Since its introduction in immunocytochemistry by Faulk and Taylor,[36] colloidal gold has acquired increased applicability and relevance as a particulate, electron-dense marker. In the last ten years, a number of reviews have been devoted to methods for the preparation and stabilization of colloidal gold suspensions.[37,38] All of these methods are based on the reduction of tetrachloroauric acid with a suitable agent. One of the most commonly used reducing agents is sodium citrate. Depending on the amount of sodium citrate added, colloidal gold with particle sizes ranging from 15 to 150 nm can be obtained.[39] Success in the preparation of colloidal gold is highly dependent on glassware cleanliness.

### 2. Gold-Labeled Protein A

Protein A-gold complexes are easily prepared and can be stored at 4°C for up to one year without loss of binding properties to immunoglobulins. Protein A is complexed to colloidal gold at pH 6.9 and used at pH 7.0 to 7.4 for section labeling.[9] Usually, a 20- to 30-fold dilution of the stock solution obtained after ultracentrifugation and resuspension of the pellet in 0.5 ml of phosphate buffered saline (PBS), pH 7.4, containing 0.01% (w/v) polyethylene glycol 20000 (PEG 20000) is used for immunocytochemical labeling.

### 3. Gold-Labeled Second Antibody

Gold-labeled second antibodies are widely used as an alternative to protein A. They are also highly stable and can be stored at –20°C for several months. Secondary antibody-gold conjugates are easily obtained from several companies in a wide range of particle sizes.

### 4. Labeling Protocol

Ultrathin sections of embedded-plant tissue are usually collected on nickel or gold grids, which can be stored prior to further treatment. The use of copper grids for immunocytochemical purposes is not recommended because of the corrosive effect of buffers (e.g., Tris) and other chemicals on this metal.[19] Although some variations may occur in immunogold labeling protocols, they all rely on the same basic principles. Taken together, these principles aim at minimizing nonspecific antibody attachment while amplifying antigen-antibody interactions. The main features of these basic principles are

1. Blocking of nonspecific primary antibody binding sites on tissue sections by incubating the grids with a saline buffer containing a "blocking" agent such as bovine serum albumin (BSA), ovalbumin, or gelatin;
2. Blocking of nonspecific secondary antibody binding sites on tissue sections by incubating the grids with diluted nonimmune serum of the species in which these antibodies were raised;
3. Determining the proper dilution of primary antibody which gives intense specific signal and low background attachment. Antibodies are diluted in saline buffer containing 1% (w/v) BSA or ovalbumin. In some cases, a detergent such as Tween 20 (0.05%, v/v) can be added to the buffer.

A typical immunogold labeling procedure, currently performed in the authors' laboratories, is as follows:

1. Incubate the sections with 1% (w/v) BSA in PBS, pH 7.4, for 5 min at room temperature in a moist chamber.
2. Transfer the sections, without rinsing, onto drops of nonimmune goat serum diluted 1:10 in PBS-BSA for 30 min, at room temperature.
3. Incubate the sections, without rinsing, with the primary antibody at the appropriate dilution in PBS-BSA for 2 h, at 37°C.
4. Thoroughly wash the grids carrying sections with Tris-buffered saline (TBS), pH 8.2, containing 1% (w/v) BSA, and remove the excess buffer with filter paper.
5. Incubate the sections with gold-conjugated goat antiserum directed against rabbit (or mouse) immunoglobulins, diluted 1:20 in TBS-BSA, pH 8.2, for 60 min, at room temperature.
6. Thoroughly wash the sections with PBS, pH 7.4, and rinse with distilled water.
7. Stain the sections with uranyl acetate and lead citrate prior to examination under the electron microscope.

When the protein A-gold approach is used in place of gold-conjugated secondary antibody, the incubation with nonimmune goat serum is omitted.

The reliability and specificity of any immunogold labeling protocol have to be assessed by control tests which should give negative results or a reasonably low level of background. The most important controls are

1. The use of nonimmune serum instead of primary antibody;
2. The use of primary antibody previously incubated with its corresponding antigen;
3. The omission of the primary antibody step.

### E. ALTERNATIVE TO THE ANTIBODY APPROACH FOR THE *IN SITU* LOCALIZATION OF SOME PROTEINS

*In situ* localization of proteins in plant and animal tissues is currently performed by means of antibodies. However, Benhamou et al.[40] reported recently that, in particular cases, the antibody approach could be successfully replaced by the use of macromolecules bearing specific binding sites for the protein to be detected. This is of special interest because of the novelty of the approach used; e.g., the localization of a fungal endopolygalacturonase can be performed in bean tissues using a polygalacturonase inhibiting protein (PGIP) from bean.[40] The purified PGIP, bearing specific binding sites for an endopolygalacturonase produced by the fungus *Colletotrichum lindemuthianum*,[41] was successfully complexed to colloidal gold at pH 9.2 and used in a one-step post-embedding procedure. High resolution results were obtained, and the background signal, sometimes linked to the number of steps in the antibody labeling procedure, was avoided. This simple approach may be extended to the localization of other proteins such as enzymes or lectins, provided highly purified molecules with strong binding properties for the protein to be detected are available.

## III. IMMUNOCYTOCHEMICAL LOCALIZATION OF SPECIFIC PROTEINS IN PLANT TISSUES

In the last decade, despite some problems encountered with tissue preparation and antibody binding, plant immunocytochemistry has been successfully applied to a wide variety of plant tissues for the localization of specific molecules that could not be detected *in situ* by other

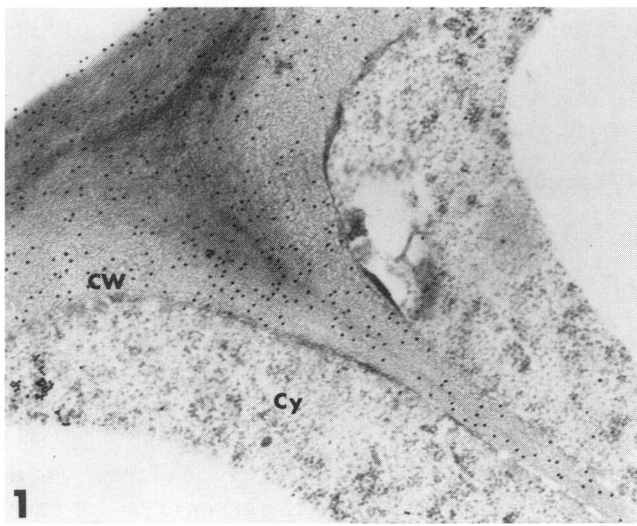


**FIGURE 1.** Immunogold labeling of lysozyme in wheat coleoptile cells. Sections were incubated with polyclonal antibody against wheat germ lysozyme followed by 15 nm Protein A-gold. (Cy) cytoplasm, (CW) cell wall. (Magnification × 36,000.)

means.[3,8] Immunocytochemical studies of plant tissues or cells not only allow the accurate localization of a defined protein but also can provide unique information on (1) cellular compartments containing the detected protein, (2) spatial and temporal changes in protein distribution during growth, development, or pathological processes, (3) accumulation sites of newly-synthesized proteins in response to stress, (4) implication of the protein under study in biological functions such as storage, transport, and/or defense mechanisms, and (5) protein synthesis and accumulation in transgenic plants constitutively expressing "foreign" genes. The potential value of immunocytochemistry in plant studies will be illustrated by a few selected examples, keeping in mind that many other applications are possible.[2,3,8]

## A. *IN SITU* LOCALIZATION OF A WHEAT GERM LYSOZYME

Plant lysozymes have recently attracted attention because of their strong chitinolytic activity, in addition to their ability to hydrolyze peptidoglycan in bacterial cell walls.[30] Taken together, these properties have led to the suggestion that these hydrolytic enzymes are involved in plant resistance to fungal and/or bacterial attack. Polyclonal antibodies, raised in rabbits against a highly purified lysozyme from wheat germ, were used in conjunction with gold-complexed protein A for *in situ* localization of the enzyme in wheat embryo and coleoptile tissues fixed with glutaraldehyde and embedded in Epon 812.[42] Clearly, the enzyme accumulated in the wall matrix (Figure 1). Since the cell wall represents the first barrier encountered by pathogens, it is likely that the association of lysozyme with cell walls participates in the overall defense strategy and protects wheat embryos against microbial attack.

## B. *IN SITU* LOCALIZATION OF INVERTASE IN *FUSARIUM*-INFECTED TOMATO ROOT TISSUES

Invertase, the enzyme that hydrolyzes sucrose into glucose and fructose, normally occurs in developing sink organs where unloading of sucrose from the phloem takes place. However, this enzyme, known to be involved in native carbohydrate metabolism of healthy plants, markedly increases upon fungal or bacterial infection, as demonstrated by biochemical studies and mRNA investigations of infected cells.[43] To understand the relationship between fungal

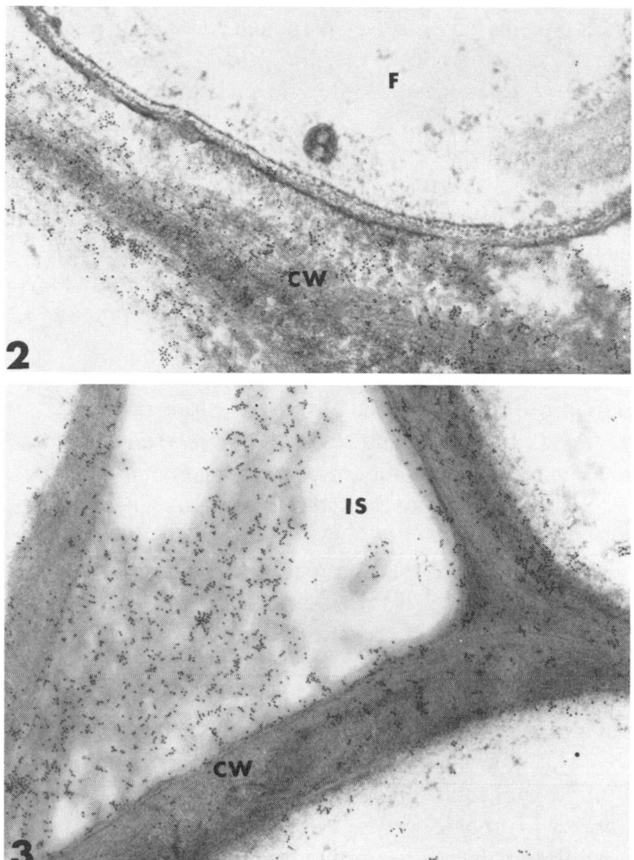


**FIGURES 2–3.** Immunogold localization of invertase in tomato root cells infected by *Fusarium oxysporum* f. sp. *radicis-lycopersici*. Sections were incubated with polyclonal antibody against deglycosylated invertase followed by 5 nm goat anti-rabbit gold. (CW) cell wall, (F) fungal cell, (IS) intercellular space. 2. (Magnification × 54,000.) 3. (Magnification × 45,000.)

infection and invertase accumulation, Benhamou et al.[16] used immunogold labeling to study the spatial and temporal distribution of this enzyme in tomato plants either susceptible or resistant to *Fusarium oxysporum* f.sp. *radicis-lycopersici* (FORL) (Figure 2). This *in situ* localization of invertase brought new insights into the possible implication of this enzyme in plant disease resistance. The demonstration that the basal level of enzyme increased earlier and to a larger extent in resistant than in susceptible plants led the authors to suggest that the induction of invertase upon fungal attack was part of the plant defense strategy. The finding that invertase accumulated in wall appositions (papillae), known to be physical barriers formed in response to infection, and in intercellular spaces (Figure 3) brought additional support for a putative function in resistance. The induction of invertase in infected plants may be a signal that converts colonized tissues or cells into sinks where sucrose is unloaded and carbohydrates are rapidly mobilized, as a response to the marked rise in respiratory activity. One may expect that invertase increase in infected plant tissues provides the carbon sources required for the establishment of defense responses.[16]

## C. *IN SITU* LOCALIZATION OF PATHOGENESIS-RELATED (PR) PROTEINS

An attack by pathogens elicits a complex plant defense response in which numerous genes are activated, resulting in the synthesis and accumulation of a variety of proteins.[44] Among these, the so-called PR proteins have received particular attention in relation to their biochemical properties, and their gene expression at the mRNA level. These newly synthesized proteins have been grouped into five families according to their physico-chemical properties. Two families have been shown to consist of PR proteins with chitinase and β-1,3-glucanase activities.[45] Understandably, much attention has been paid to these hydrolytic enzymes because of their believed antimicrobial activity as shown by fungal growth inhibition *in vitro*.[46] If one considers that chitin and β-1,3-glucans are major structural components of the walls of most fungi, then it is conceivable that newly synthesized plant chitinases and β-1,3-glucanases may play a key role in plant defense mechanisms. In this case again, immunocytochemical studies of their localization and distribution in infected plant tissues provide a simple means of delineating their potential functions *in vivo*.

### 1. Immunogold Localization of Chitinase and β-1,3-Glucanase

Specific antisera raised against a tomato chitinase (26 kDa) and a tobacco β-1,3-glucanase (33 kDa) were used as immunological probes for localizing both hydrolases in FORL-infected tomato root tissues.[6,47] The specificity of the antisera was assessed on immunoblots after SDS-PAGE.[6]

Using the immunogold labeling procedure described above, Benhamou et al.[6, 47] studied changes in the spatial and temporal distribution of chitinase and β-1,3-glucanase in both resistant and susceptible FORL-infected plants. Several results emerged from these time-course studies:

1. Both enzymes occurred at the fungus cell surface, thus supporting the view of an antifungal activity (Figures 4, 4′, 5). However, the preferential association of chitinase with altered fungal wall areas indicated that chitinase activity was likely preceded by the action of other hydrolytic enzymes.
2. Both enzymes accumulated earlier in cell walls of resistant than susceptible tomato plants.
3. In susceptible tomato plants, the accumulation of chitinase and β-1,3-glucanase was largely correlated with pathogen distribution, since it was confined to infected cells only.
4. In resistant tomato plants, β-1,3-glucanase was widely distributed in all tissues including noninfected ones, whereas chitinase accumulation was restricted to outer, infected tissues (e.g., epidermis, cortex).
5. The presence of β-1,3-glucanase at strategic sites such as wall appositions provided additional support for its implication in the overall plant defense strategy.
6. Taken together, the observations generated from these immunocytochemical studies suggested that induction of β-1,3-glucanase was an early event, likely associated with protection against fungal invasion, whereas chitinase production rather reflected a punctual response probably stimulated by elicitors released from fungal cell walls digested by the β-1,3-glucanase.

These examples of specific protein localization in plant tissues highlight the value of immunocytochemistry as a means for delineating precisely the distribution pattern of the proteins under study *in planta*, and for providing unique information regarding their biological functions *in vivo*.

### 2. Immunogold Localization of PR-1 Proteins

In addition to chitinases and β-1,3-glucanases, plants respond to pathogenic stresses by *de novo* synthesis of low molecular mass, soluble proteins, the so-called PR-1 proteins.[48] Although these proteins have been extensively studied in terms of physicochemical properties,

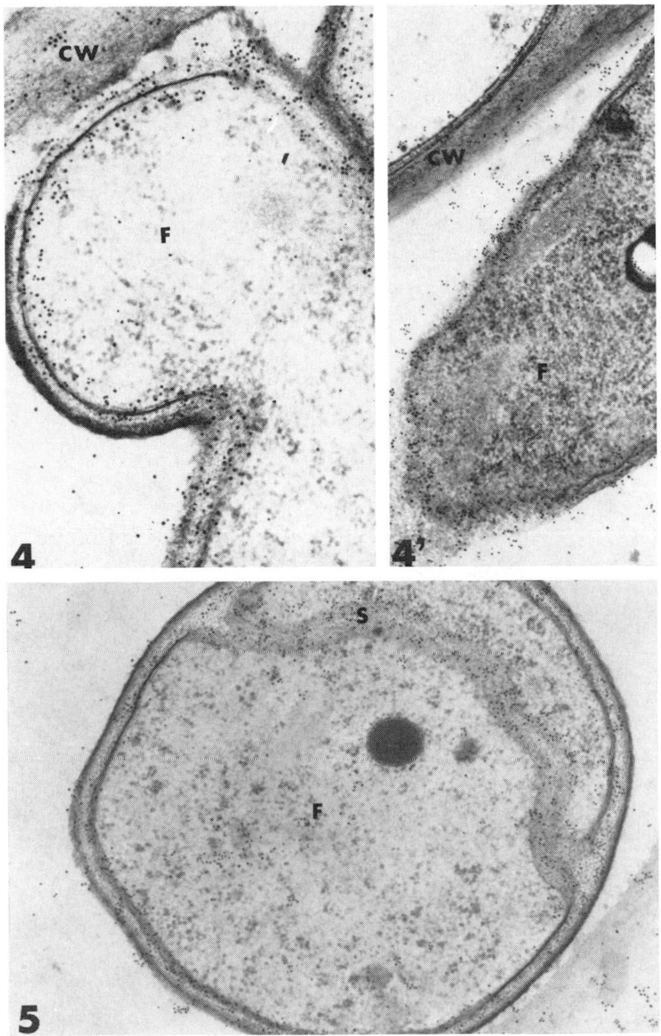


**FIGURES 4–5.** Immunogold localization of chitinase and β-1,3-glucanase in *Fusarium*-infected tomato root tissues. Sections were incubated with antiserum raised against either chitinase or β-1,3-glucanase followed by 5 nm goat anti-rabbit gold. (CW) cell wall, (F) fungus, (S) septum. 4. (Magnification × 27,000.) 4′ (Magnification × 54,000.) 5. (Magnification × 54,000.)

amino acid sequence, and gene expression, their biological functions are still unknown. In this context, a true image of their distribution in infected plant tissues could facilitate an understanding of their biological roles. However, a major difficulty encountered in immunogold studies of PR-1 localization is related to the rapid diffusion of these proteins from one cell compartment to another during tissue processing.[21] Due to their low molecular masses and their high solubility in aqueous reagents, these molecules do not necessarily remain at the same site during standard fixation procedures, and this may lead to incorrect patterns of protein distribution. This problem has been somewhat circumvented by the microwave energy fixation method introduced by Benhamou et al. for localizing PR-1 proteins in tobacco leaf tissues reacting hypersensitively to tobacco mosaic virus (TMV) infection.[12] The authors showed that microwave fixation of tobacco plant tissues in conjunction with an aldehyde mixture containing 0.1% glutaraldehyde and 2% paraformaldehyde was suitable for yielding good morphological preservation and satisfactory retention of PR-1 proteins in their original

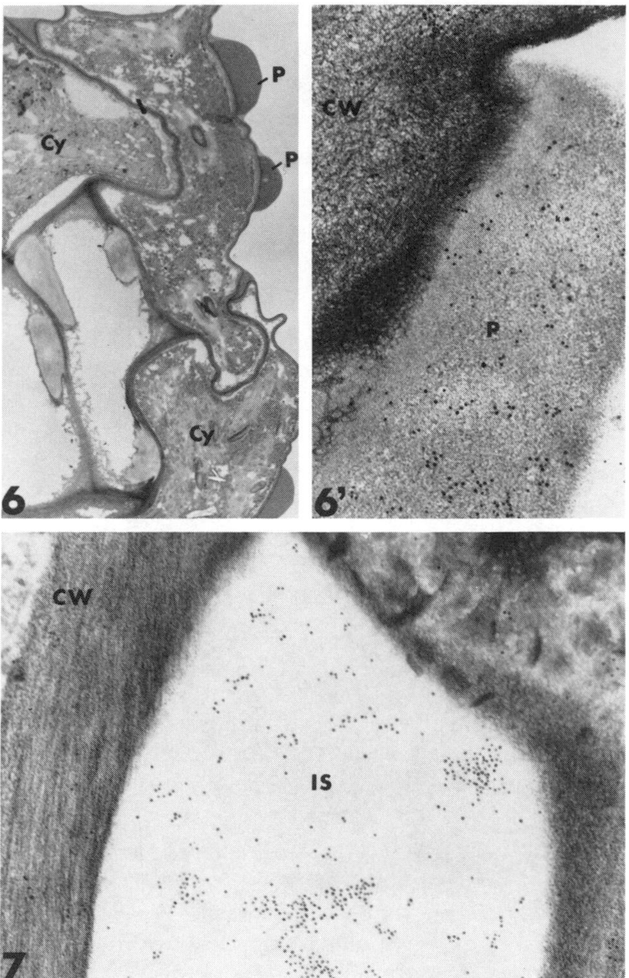


**FIGURES 6–7.** Immunogold localization of PR-1 proteins in tobacco leaf tissue reacting hypersensitively to tobacco mosaic virus infection. Tissues were fixed by the microwave energy fixation method. Sections were incubated with anti-PR-1 antibodies followed by 5 nm goat anti-rabbit gold. (Cy) cytoplasm, (CW) cell wall, (IS) intercellular space, (P) hemispherical protuberance. 6. (Magnification × 9,000.) 6′ (Magnification × 72,000.) 7. (Magnification × 72,000.)

location. The significant decrease in time fixation (15 to 20 s instead of 2 to 3 h) may explain the considerable reduction of protein diffusion.

In addition to being localized predominantly in intercellular spaces (Figure 7), PR-1 proteins were also found to be associated with hemispherical protusions formed in response to infection in cells surrounding infection sites (Figures 6, 6′). Whether or not PR-1 proteins act as structural components to reinforce the architecture of physical barriers, thought to prevent virus diffusion, remains to be determined.

## D. *IN SITU* LOCALIZATION OF HYDROXYPROLINE-RICH GLYCOPROTEINS IN DIFFERENT PLANT TISSUES

Hydroxyproline-rich glycoproteins (HRGPs), which contain dityrosine residues in addition to hydroxyproline, generally constitute less than 10% of the protein content of normal plant cell walls.[49] These HRGPs, termed extensin due to their believed structural function during cell wall elongation, markedly increase during a challenge by pathogens.[50] Enrichment of cell wall bound HRGPs upon pathogenic attack has been related to increased resistance of plant cell walls to pathogen ingress. The demonstration that an increase in translatable cytosine-rich RNAs occurred at the onset of *in vivo* HRGP accumulation in *Colletotrichum*-infected melon plants[51] was used to support the idea that HRGP accumulation is part of the inducible plant defense system. However, to learn more about the exact functions of these glycoproteins, it was essential to determine where these glycoproteins accumulated in infected tissues. A detailed picture of the spatial and temporal distribution of HRGPs in susceptible and resistant tomato plants infected by FORL has been reported recently.[10]

Antibodies raised against deglycosylated melon HRGPs were used in that study. Cross-reaction between this antiserum and tomato HRGPs was checked by Western blotting. Because antisera raised against glycoproteins may contain immunoglobulins that bind to the oligosaccharide side-chains of the molecule, it is recommended that the protein be deglycosylated prior to antibody production. An alternative is to remove from the antiserum immunoglobulins that bind to oligosaccharides.

A time-course study of HRGP accumulation in Epon-embedded tomato root tissues showed that these molecules accumulated to a greater extent in resistant than in susceptible plants (Figure 8). Their generalized accumulation in cell walls of resistant tissues, as well as their association with physical barriers (Figure 9) and intercellular spaces, provided support for the concept that HRGPs may play an important role in protection against fungal penetration.

Using the same approach, HRGPs have also been localized successfully in hypersensitive tobacco plants,[52] where they accumulated predominantly in physical barriers. Finally, these glycoproteins have been found to be associated with peribacteroid membranes in bean root nodule cells infected with *Rhizobium leguminosarum* bv. *phaseoli*[7] (Figure 10). It is believed that HRGPs may contribute to the strength of the peribacteroid membrane and thereby prevent direct contact of the bacteroids with the host cytoplasm.[7] The observations derived from all studies dealing with HRGP localization in plant tissues tend to indicate that these glycoproteins have a mechanical function and probably participate in defense by contributing with other compounds such as callose and lignin to the formation of more protective barriers.[7,10,52]

### E. *IN SITU* LOCALIZATION OF SOME OTHER PROTEINS IN PLANT TISSUES

Recent advances in methods of tissue preparation have made it possible to investigate the major components of the cytoskeletal system. As an example, monoclonal antibodies were used to identify myosin heavy chains in flowering plants and to localize them in a green alga.[53] Similarly, actin has been detected in coleoptile cells.[54]

Immunocytochemical methods have been used widely for localizing storage proteins in seeds.[5] To date, major seed storage proteins such as legumin, vicilin, cruciferin, and phaseolin have been successfully visualized in their respective cell compartments using various embedding resins including glycolmethacrylate, L.R. White, and Spurr or frozen sections.[3] More recently, a number of "foreign" seed proteins have been identified in transgenic tobacco plants by immunogold labeling on L.R. White-embedded tissue sections.[55]

A number of enzymes involved in several physiological processes such as nitrogen assimilation,[56] starch biosynthesis,[57] and photorespiration[58] have been detected by the immunogold approach. Similarly, localization of lectins and agglutinins[11] has been accomplished using either polyclonal or monoclonal antibodies.

The structure and function of plant coated vesicles have also been studied by immunocytochemistry.[59] It is assumed that coated vesicles in plant cells are involved in transport processes from the Golgi apparatus to the plasma membrane.

The applications of immunocytochemistry to plant tissues are too numerous to be presented here, but many of them have been reviewed elsewhere.[60]

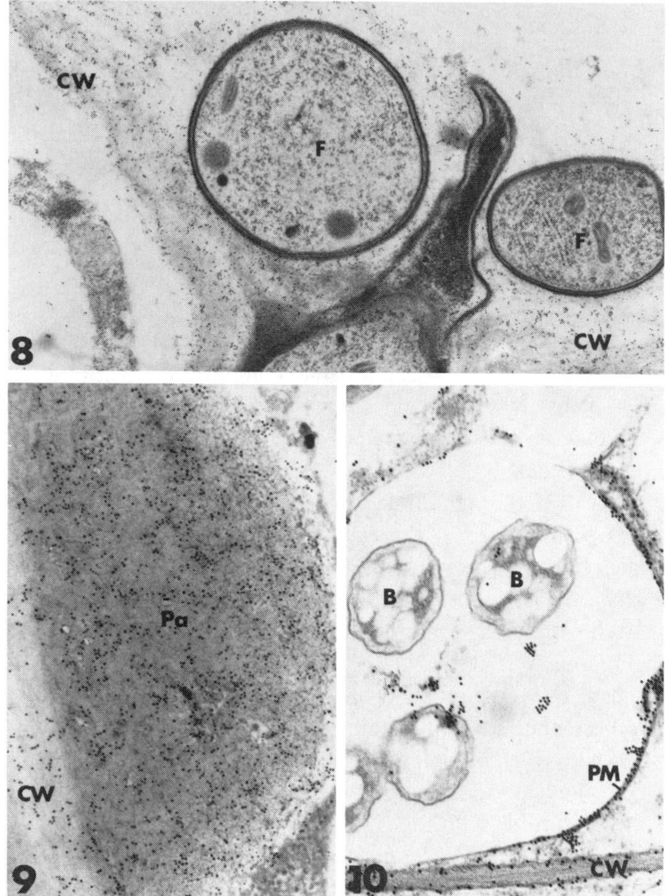


**FIGURES 8–10.** Immunogold localization of hydroxyproline-rich glycoproteins in *Fusarium*-infected tomato root tissues (Figures 8, 9) and in bean root nodule cells (Figure 10). Sections were incubated with antibody raised against deglycosylated HRGPs followed by 5 or 15 nm goat anti-rabbit gold. (B) bacteroid, (CW) cell wall, (F) fungus, (Pa) papilla, (PM) peribacteroid membrane. 8. (Magnification × 21,600.) 9. (Magnification × 54,000.) 10. (Magnification × 36,000.)

## IV. CONCLUSION AND FUTURE PROSPECTS

The recent advances in the isolation and purification of numerous plant proteins and in the preparation of highly specific antibodies have led to the consideration that immunocytochemistry could be a useful tool for studying the precise location of these proteins in their respective cell compartments and helping to elucidate their functions. However, the use of such techniques to localize accurately specific proteins raises some questions about their general applicability to all plant tissues. Several lines of evidence indicate that the key factors for success are antigenicity preservation and accessibility in the tissue section. It seems that each antigen requires a "homemade" tissue preparation technique leading to optimal preservation for accurate localization. Although aldehyde fixation of plant tissues is not always compatible with optimal retention of antigenicity, the use of cross-linking fixatives is preferable from a structural point of view. Thus, conditions which yield the best results have to be worked out in each case. Problems of preserving antigens may be, in some cases, circumvented by using rapid freeze-fixation and freeze substitution. This method proved valuable for preserving cytoskeletal components, such as actin microfilaments.[61] A number of procedures have been devised to enhance retention of antigenicity, such as the use of rapid freeze-fixation in conjunction with embedding, either at low temperature or in an acrylic resin, such as polyvinyl pyrollidone.[3] Satisfactory results have also been obtained with the microwave energy fixation technique, which is rapid and does not require costly instrumentation.[12]

Although some problems still exist in immunocytochemistry, this approach has already acquired increased relevance and applicability and has contributed to a better understanding of some interactions in which proteins or glycoproteins are involved. Together with the *in situ* hybridization techniques that have been developed recently for localizing nucleic acids, immunocytochemical methods will undoubtedly provide new insights in various areas of plant biology.

## REFERENCES


1. **Knox, R. B., Heslop-Harrison, J., and Reed, C. E.,** Localization of antigens associated with the pollen grain wall by immunofluorescence, *Nature*, 225, 1066, 1970.
2. **Herman, E. M.,** Immunocytochemical localization of macromolecules with the electron microscope, *Annu. Rev. Plant Physiol. Plant Mol. Biol.*, 39, 139, 1988.
3. **Vandenbosh, K. A.,** Immunogold labeling, in *Electron Microscopy of Plant Cells*, Hall, J. L. and Hawes, C., Eds., Academic Press, New York, 1991, 181.
4. **Sonnewald, U., Studer, D., Rocha-Sosa, M., and Willmitzer, L.,** Immunocytochemical localization of patatin, the major glycoprotein in potato (*Solanum tuberosum* L.) tubers, *Planta*, 178, 176, 1989.
5. **Murphy, D. J., Cummins, I., and Ryan, A. J.,** Immunocytochemical and biochemical study of the biosynthesis and mobilization of the major seed storage proteins of *Brassica napus*, *Plant Physiol. Biochem.*, 27, 647, 1989.
6. **Benhamou, N., Grenier, A., Asselin, A., and Legrand, M.,** Immunogold localization of β-1,3-glucanases in two plants infected by vascular wilt fungi, *Plant Cell*, 1, 1209, 1989.
7. **Benhamou, N., Lafontaine, P. J., Mazau, D., and Esquerré-Tugayé, M. T.,** Differential accumulation of hydroxyproline-rich glycoproteins in bean root nodule cells infected with a wild-type or a $C_4$-dicarboxylic acid mutant of *Rhizobium leguminosarum bv. phaseoli*, *Planta*, 184, 457, 1991.
8. **Knox, R. B.,** Methods for locating and identifying antigens in plant tissues, in *Techniques in Immunocytochemistry*, Vol. 1, Bullock, G. R. and Petrusz, P., Eds., Academic Press, New York, 1982, 205.
9. **Bendayan, M.,** Protein A-gold and enzyme-gold: two novel affinity techniques for ultrastructural localization of macromolecules, *Int. Congr. Electron Microsc.*, 2, 427, 1982.
10. **Benhamou, N., Mazau, D., and Esquerré-Tugayé, M. T.,** Immunocytochemical localization of hydroxyproline-rich glycoproteins in tomato root cells infected by *Fusarium oxysporum* f. sp. *radicis-lycopersici*: study of a compatible interaction, *Phytopathology*, 80, 173, 1990.
11. **Raikhel, N. V., Mishkind, M., and Palevitz, B. A.,** Immunocytochemistry in plants with colloidal gold conjugates, *Protoplasma*, 121, 25, 1984.
12. **Benhamou, N., Noël, S., Grenier, J., and Asselin, A.,** Microwave energy fixation of plant tissues: an alternative approach that provides excellent preservation of ultrastructure and antigenicity, *J. Electron Microsc. Tech.*, 17, 81, 1991.
13. **Knox, R. B., Vithange, H. I. M. V., and Howlett, B. J.,** Botanical immunocytochemistry: a review with special references to pollen antigens and allergens, *Histochem. J.*, 12, 246, 1980.
14. **Craig, S. and Millerd, A.,** Pea seed storage proteins: immunocytochemical localization with protein A-gold by electron microscopy, *Protoplasma*, 105, 333, 1981.
15. **Vaughn, K. C. and Campbell, W. H.,** Immunogold localization of nitrate reductase in maize leaves, *Plant Physiol.*, 88, 1354, 1988.
16. **Benhamou, N., Grenier, J., and Chrispeels, M. J.,** Accumulation of β-fructosidase in the cell walls of tomato roots following infection by a fungal wilt pathogen, *Plant Physiol.*, 97, 739, 1991.
17. **Tokuyasu, K. T.,** Immunocytochemistry on ultrathin frozen sections, *Histochem. J.*, 12, 381, 1980.
18. **Brandtzaeg, P.,** Tissue preparation methods for immunocytochemistry, in *Techniques in Immunocytochemistry*, Vol. 1, Bullock, G. R. and Petruzsz, P., Eds., Academic Press, New York, 1982, 2.
19. **Roland, J. P. and Vian, B.,** General preparation and staining of thin sections, in *Electron Microscopy of Plant Cells*, Hall, J. L. and Hawes, C., Eds., Academic Press, New York, 1991, 2.

20. **Bendayan, M. and Zollinger, M.,** Ultrastructural localization of antigenic sites on osmium-fixed tissues applying the protein A-gold approach, *J. Histochem. Cytochem.*, 31, 101, 1983.
21. **Hosokawa, D. and Ohashi, Y.,** Immunocytochemical localization of pathogenesis-related proteins secreted into the intercellular space of salicylate-treated tobacco leaves, *Plant Cell Physiol.*, 29, 1035, 1988.
22. **Causton, B. E.,** Choice of resins for electron immunocytochemistry, in *Immunolabelling for Electron Microscopy*, Polak, J. M. and Varndell, M., Eds., Elsevier Press, Amsterdam, 1984, 29.
23. **Roth, J., Bendayan, M., Carleman, E., Villiger, W., and Garavito, M.,** Enhancement of structural preservation and immunocytochemical staining in low temperature embedded pancreatic tissue, *J. Histochem. Cytochem.*, 29, 663, 1981.
24. **Benhamou, N., Parent, J. G., Garzon, S., Asselin, A., Ouellette, G. B., and Joly, J. R.,** Use of monoclonal antibody against poly [ I ]: poly [ C ] for detecting mycoviruses and potential applications to potato spindle tuber viroid and animal reoviruses, *Can. J. Plant Pathol.*, 9, 106, 1987.
25. **Parent, J. G., Hogue, R., and Asselin, A.,** Serological relationships between pathogenesis-related leaf proteins from four *Nicotiana* species, *Solanum tuberosum*, and *Chenopodium amaranticolor*, *Can. J. Bot.*, 66, 199, 1988.
26. **Grenier, J., Benhamou, N., and Asselin, A.,** Colloidal gold-complexed chitosanase: a new probe for ultrastructural localization of chitosan in fungi, *J. Gen. Microbiol.*, 137, 2007, 1991.
27. **Morrissey, J. H.,** Silver stain for proteins in polyacrylamide gels: a modified procedure with enhanced uniform sensitivity, *Anal. Biochem.*, 117, 307, 1981.
28. **Trudel, J., Audy, P., and Asselin, A.,** Electrophoretic forms of chitinase activity in Xanthi-nc tobacco, healthy and infected with tobacco mosaic virus, *Mol. Plant-Microbe Interact.*, 2, 315, 1989.
29. **Côté, F., Letarte, J., Grenier, J., Trudel, J., and Asselin, A.,** Detection of ß-1,3-glucanase activity after native polyacrylamide gel electrophoresis: application to tobacco pathogenesis-related proteins, *Electrophoresis*, 10, 527, 1989.
30. **Audy, P., Grenier, J., and Asselin, A.,** Lysozyme activity in animal extracts after sodium dodecyl sulfate-polyacrylamide gel electrophoresis, *Comp. Biochem. Physiol.*, 92B, 523, 1989.
31. **Grenier, J. and Asselin, A.,** Some pathogenesis-related proteins are chitosanases with lytic activity against fungal spores, *Mol. Plant-Microbe Interact.*, 3, 401, 1990.
32. **Côté, F., El Ouakfaoui, S., and Asselin, A.,** Detection of β-glucanase activity on various β-1,3- and β-1,4-glucans after native and denaturing polyacrylamide gel electrophoresis, *Electrophoresis*, 12, 69, 1991.
33. **Cusak, M. and Pierpoint, W. S.,** Similarities between sweet protein thaumatin and a pathogenesis-related protein from tobacco, *Phytochemistry*, 27, 3817, 1988.
34. **Fortin, M. G., Parent, J. G., and Asselin, A.,** Comparative study of two groups of b proteins (pathogenesis-related) from the intercellular fluid of *Nicotiana* leaf tissue infected by tobacco mosaic virus, *Can. J. Bot.*, 63, 932, 1985.
35. **Roth, J.,** The colloidal gold marker system for light and electron microscopic cytochemistry, in *Techniques in Immunocytochemistry*, Vol. 2, Bullock, G. R. and Petrusz, P., Eds., Academic Press, New York, 1983, 217.
36. **Faulk, W. P. and Taylor, C. M.,** An immunocolloidal method for the electron microscope, *Immunocytochemistry*, 8, 1081, 1971.
37. **De Mey, J.,** The preparation of gold probes, in *Immunocytochemistry: Modern Techniques and Applications*, Polak, J. M. and Van Noorden, S., Eds., John Wright & Sons, Bristol, 1986, 3.
38. **Horisberger, M.,** Electron-opaque markers: a review, in *Immunolabelling for Electron Microscopy*, Polak, J. M. and Varndell, I. M., Eds., Elsevier Press, Amsterdam, 1984, 17.
39. **Frens, G.,** Controlled nucleation for the regulation of particle size in monodisperse gold solutions, *Nature Phys. Sci.*, 241, 10, 1973.
40. **Benhamou, N., Lafitte, C., Barthe, J. P., and Esquerré-Tugayé, M. T.,** Cell-surface interactions between bean leaf cells and *Colletotrichum lindemuthianum*: cytochemical aspects of pectin breakdown and fungal endopolygalacturonase accumulation, *Plant Physiol.*, 97, 234, 1991.
41. **Cervone, F., De Lorenzo, G., Dagra, L., and Salvi, G.,** Interaction of fungal polygalacturonase with plant proteins in relation to specificity and regulation of plant defense response, in *Microbe-Plant Pathogenic Interactions*, Lugtemberg, B., Ed., NATO ASI Series, Vol. H4, Springer-Verlag, Berlin, 1986, 253.
42. **Audy, P., Benhamou, N., Trudel, J., and Asselin, A.,** Immunocytochemical localization of a wheat germ lysozyme in wheat embryo and coleoptile cells and cytochemical study of its interaction with the cell wall, *Plant Physiol.*, 88, 1317, 1988.
43. **Sturm, A. and Chrispeels, M. J.,** cDNA cloning of carrot extracellular β-fructosidase and its expression in response to wounding and bacterial infection, *Plant Cell*, 2, 1107, 1990.
44. **Bell, A. A.,** Biochemical mechanisms of disease resistance, *Annu. Rev. Plant Physiol.*, 32, 21, 1981.
45. **Boller, T.,** Induction of hydrolases as a defense reaction against pathogens, in *Cellular and Molecular Biology of Plant Stress*, Key, J. L. and Kosuge, T., Eds., Alan R. Liss, New York, 1985, 247.

46. **Mauch, F., Mauch-Mani, B., and Boller, T.,** Antifungal hydrolases in pea tissue: inhibition of fungal growth by combination of chitinase and b-1,3-glucanase, *Plant Physiol.*, 88, 936, 1988.
47. **Benhamou, N., Joosten, M. H. A. J., and DeWit, P. J. G. M.,** Subcellular localization of chitinase and of its potential substrate in tomato root cells infected by *Fusarium oxysporum* f. sp. *radicis-lycopersici*, *Plant Physiol.*, 92, 1108, 1990.
48. **Carr, J. P. and Klessig, D. F.,** The pathogenesis-related proteins in plants, in *Genetic Engineering Principles and Methods*, Vol. 11, Setlow, J. K., Ed., Plenum Press, New York, 1989, 65.
49. **Lamport, D. T. A. and Catt, J. W.,** Glycoproteins and enzymes of the cell wall, in *Encyclopedia of Plant Physiology*, *Plant Carbohydrates II, Extracellular Carbohydrates*, Tanner, W. and Loewus, F. A., Eds., New Series, Vol. 13B, Springer-Verlag, Berlin, 1981.
50. **Esquerré-Tugayé, M. T. and Lamport, D. T. A.,** Cell surfaces in plant-microorganism interactions. I. A structural investigation of cell wall hydroxyproline-rich glycoproteins which accumulate in fungus-infected plants, *Plant Physiol.*, 64, 314, 1979.
51. **Rumeau, D., Mazau, D., and Esquerré-Tugayé, M. T.,** Cytosine-rich RNAs from infected melon plants and their *in vitro* translation products, *Physiol. Mol. Plant Pathol.*, 31, 305, 1987.
52. **Benhamou, N., Mazau, D., Esquerré-Tugayé, M. T., and Asselin, A.,** Immunogold localization of hydroxyproline-rich glycoproteins in necrotic tissue of *Nicotiana tabacum* L. cv. Xanthi-nc. infected by tobacco mosaic virus, *Physiol. Mol. Plant Pathol.*, 36, 129, 1990.
53. **Quiao, L., Grolig, F., Jablonsky, P. P., and Wiliamson, R. E.,** Myosin heavy chains: detection by immunoblotting in higher plants and localization by immunofluorescence in the alga *Chara*, in *Cell Biology International Report*, Vol. 13, Vigil, E. L. and Hawes, C., Eds., Academic Press, New York, 1989, 107.
54. **Parthasarathy, M. V.,** F-actin architecture in coleoptile epidermal cells, *Eur. J. Cell Biol.*, 39, 1, 1985.
55. **Greenwood, J. S. and Chrispeels, M. J.,** Immunocytochemical localization of phaseolin and phytohemagglutinin in the endoplasmic reticulum and Golgi complex of developing bean cotyledons, *Planta*, 164, 295, 1985.
56. **Thalouarn, P., Rey, L., Hirel, B., Renaudin, S., and Fer, A.,** Activity and immunocytochemical localization of glutamine synthetase in *Latharaea clandestina* L., *Protoplasma*, 141, 95, 1987.
57. **Kim, W. T., Franceschi, V. R., Okita, T. W., Robinson, N. L., Morell, M., and Preiss, J.,** Immunocytochemical localization of ADP glucose pyrophosphorylase in developing potato tuber cells, *Plant Physiol.*, 91, 217, 1989.
58. **Mangeney, E., Hawthornthwaite, A. M., Codd, G. A., and Gibbs, S. P.,** Immunocytochemical of phosphoribulose kinase in the cyanelles of *Cyanophora paradoxa* and *Glaucocystis nostochinearum*, *Plant Physiol.*, 84, 1028, 1987.
59. **Hawes, C., Coleman, J., Evans, D., and Cole, L.,** Recent advances in the study of plant coated vesicles, in *Cell Biology International Report*, Vol. 13, Vigil, E. L. and Hawes, C., Eds., Academic Press, New York, 1989, 119.
60. **Vigil, E. L. and Hawes, C.,** *Cytochemical and Immunological Approaches to Cell Biology*, Academic Press, New York, 1989, 1.
61. **Lancelle, S. A. and Hepler, P.K.,** Immunogold labeling of actin on sections of freeze-substituted plant cells, *Protoplasma*, 150, 72, 1989.

# 15

# Accessing Computer Software for Molecular Biology

*J. J. Pasternak*

## I. GETTING STARTED

More and more, computer applications are becoming a necessary adjunct to molecular biological research. Unlike a few years ago when research software was awkward, highly specialized, and limited to a few operating systems, today it is relevant, easy to use, and readily available for DOS, MAC, VAX, and UNIX machines. Part of the impetus to develop newer and more effective programs has come from the rapid accumulation of nucleic acid and protein sequence data with the accompanying need to analyze this information. Now there is an array of programs that extend from multipurpose packages to "stand-alone", single task applications. Most but not all of the multifaceted programs are commercial products that are designed for thorough sequence analyses of nucleic acids and proteins, including restriction enzyme site analysis, open reading frame determination, reverse translation, codon usage determination, multiple alignments, sequence editing, publication-ready graphics, hydropathy plots, searches of sequence databases for similar sequences, etc. These packages range in features and price from about $200 (US) to $3500 (US) and are designed to operate for the most part with personal computers, although comprehensive programs for mainframe computers are also available.

In addition to commercial products, there are high quality applications that are available at no (freeware) or minimal (shareware) cost. The major emphasis of this chapter is to describe how to access this software. Although reviews of software computer programs can be informative, unfortunately they also become outdated quickly. Consequently, existing programs will not be described here.

For a very active research group that is involved routinely in DNA sequencing and sequence analyses, the commercial packages are quite appropriate, although costly. These programs are designed for use with DOS and MAC personal computer systems. There are also some packages that can be used on mainframe computers. If a personal computer system is preferred, then the Macintosh™ operating system should be given careful consideration. Specifically, an excellent self-contained, off-line computer analysis system would include a Mac II series computer: a math coprocessor, 5 megabytes of RAM, and an 80 to 100 megabyte hard drive. If similarity searches are to be carried out often, a CD-ROM reader should be added. The more substantial, full-featured commercial packages for the Mac computer system are MacVector (IBI, PO Box 9558, New Haven, CT 06535, USA); GeneWorks (Intelligenetics, 700 E. El Camino Real, Mountain View, CA 44040, USA); MacMolly Tetra (Soft Gene GMBH, Offenbacher Str.5, D-1000, Berlin, Germany) and LaserGene (DNASTAR, Inc., Corporate Headquarters, 1228 Smith Park St., Madison, WI 53715, USA). Each package

0-8493-5164-2/93/$0.00+$.50
© 1993 by CRC Press, Inc.

should be tested thoroughly. One advantage of the Mac-based systems is that anyone in the laboratory with a minimum of computer expertise can learn to run the modules easily. With commercial packages, there is usually excellent technical support, and upgrades should not be particularly costly to registered owners. Although off-line systems can be self-contained, connection to a mainframe computer either directly (e.g., ethernet) or by modem is important. Such connections will enable the network(s) of the mainframe computer at your institution to be used to contact molecular biology newsgroups and to obtain additional computer applications.

If an off-line molecular biology software system is neither financially feasible nor necessary, connection to a mainframe computer is sufficient for conducting most, if not all of the kinds of, molecular biology analyses that are available for computer use. Use of a mainframe, of course, does not preclude using a personal computer for some analyses. For an on-line system, a personal computer, modem, and communications software with a large capture buffer will suffice. To make the connection with the mainframe computer at your institution, contact the computer system manager for details. The ever-invaluable computer system manager (a.k.a. sysop; *sys*tem *op*erator) will establish an account for you, give you some idea of the relevant commands that are used by the operating system of the mainframe computer, and provide the telephone number for making the connection with a modem. As well, you must learn how to send and receive electronic mail (e-mail). With this arrangement, you will be able to run programs on the mainframe computer. As well, those programs that require less memory and running time can be used on your personal computer. The documentation that accompanies the freeware and shareware molecular biology programs is usually extremely helpful; therefore, the programs can be run without technical support. To reiterate, run your major applications on the mainframe computer because of the speed and memory capacity. You will probably be charged for both the time spent on the mainframe computer and disk storage. However, at most institutions, these costs tend to be very modest.

It is hoped that the reader will be encouraged to enter the "electronic world of molecular biology." Certainly, not all the software that you will try will be to your liking; but, some applications will be so appropriate that you will wonder how you survived without them. And, more importantly, some applications might enable you to develop lines of research that, hitherto, you were reluctant to pursue.

## II. OBTAINING SOFTWARE BY ELECTRONIC MAIL

One of the services provided by the EMBL data library includes a network file server through which free molecular biology software can be obtained by request using electronic mail. The software collection includes programs that will run on DOS, MAC, VAX, and UNIX operating systems. To access the file server, send a specific command line in the body of an e-mail message to NETSERV@EMBL-HEIDELBERG.DE. For example, to get the HELP documentation for MAC software, send the message HELP MAC_SOFTWARE. The response by e-mail will include information about the file server, how to get MAC software, how to handle the MAC files, and a list of the MAC software that is available from the EMBL File Server. By substituting DOS, VAX, or UNIX for MAC in a separate command line, with one command per line in the message, you will receive the HELP documents for each of these software directories.

To get a specific program, for example, the MAC-based phylogenetic tree drawing program TREEDRAW, send the command line GET MAC_SOFTWARE:TREEDRAW.HQX to the EMBL File Server. The program will be sent to you by e-mail. For the file server to comply with your request, you must use the correct filename (TREEDRAW) and extension (.HQX). The appropriate filenames with extensions are available from the directory listing in the HELP documents. Use the same format for the GET command to obtain DOS, VAX, or UNIX software.

To facilitate electronic transfer, programs are compressed (compacted, archived, packed) and then encoded ("binhexed", "uuencoded") to an ASCII file. To create an executable application, you will need the appropriate programs to (1) decode and then (2) uncompress the received file. For each type of operating system there are programs to do these jobs. The HELP documentation explains the procedures for each operating system. Parenthetically, the extensions .HQX and .UUE denote, as abbreviations, the programs that should be used for decoding. The extension .HQX indicates that a MAC program should be decoded by the program BinHex 4.0 (or equivalent) and the .UUE extension indicates that the program uudecode should be used for decoding purposes. The EMBL File Server uses DOS-, VAX-, and UNIX-based versions of uudecode for decoding.

The documentation obtained from the EMBL File Server is sufficient to instruct a first-time user in decoding and uncompressing files. Two examples, one for UNIX programs and one for MAC programs, will be outlined here. The procedures for creating executable programs for either VAX or DOS operating systems are very similar but not identical to those used for UNIX-based programs. Consult the HELP documentation for information about decoding and uncompressing VAX and DOS files.

For a UNIX program that you want to run on a mainframe computer, transfer the program from your mail box to one of your directories. Remove the e-mail header from the program, and try the resident uudecode program to see if it will decode the file. The command at the prompt is *uudecode filename.uue*. If the version of uudecode on your machine does not work, get uud.c from the EMBL File Server (i.e., GET UNIX_SOFTWARE:UUD.C). Remove the header from uud.c, and compile the program (i.e., *cc uud.c -o uud*). If the program does not compile properly, show the error message to your sysop, and more than likely s/he will modify the source code for you or at least tell you how to edit the source code so that the program will compile properly. Once compiled, use the program uud to uudecode the UNIX file (i.e., *uud filename.uue*). This procedure will produce a file with a double extension; viz., filename.tar.Z. With the command *uncompress filename.tar.Z*, the first step of the unpacking process will be carried out, and a file named filename.tar will be produced. To finish the unpacking, use the command *tar -xvf filename.tar*. Each command, as usual, should be implemented with a carriage return. The uncompress and tar programs that are resident on your UNIX machine should not give you any problems.

In some cases, the EMBL File Server sends a set of "split" files because the program is too large to send as one e-mail message. For UNIX programs, for example, if the extension in the directory is .uaa, you can expect more than one e-mail message; i.e., the program will be sent to you in parts. If the program comprises two parts, you will receive two files: filename.uaa and filename.uab. The file server will send you the complete package with the single command GET UNIX_SOFTWARE:FILENAME.UAA. For decoding split UNIX files, put them in one directory, and run the uud program from EMBL (i.e., *uud filename.uaa*), and the parts will be joined and decoded. Then, continue with "uncompressing" and "un-tar-ing" as you would with a single file.

If the program that you've requested is for a personal computer, you will have to download it from the mainframe computer. There are at least three ways of downloading programs. First, the file transfer program Kermit can be used for this task, although Kermit is very slow. Second, if you are networked, files can be downloaded rapidly. Third, you can display the contents of the file on your monitor and, from the capture buffer of your communications software, save it to your hard disk. When you use certain decoding programs with a personal computer, the e-mail header from the file must be deleted. Use a word processor to do this, and save the file as a text file. The decoding and unpacking software for MAC and DOS programs is available from the EMBL File Server.

For single-file MAC programs, you can decode directly with StuffIt, BinHex 4.0, Compactor, BinHqx 1.02 or DeHQX v. 2.0.0. With the 'decode BinHex file . . .' command in each of the decoding programs, double click on the filename with the .hqx extension which will create a file with a .sea extension. The .sea file will self-extract when it is double-clicked. The extension .sea denotes a *s*elf-*e*xtracting *a*rchive. For split files, use a word processor to remove the headers and join the parts, in the correct order, by "cutting, copying, and pasting," and save as a text file. Then, decode the single file. Alternatively, use DeHQX v. 2.0.0, which will ignore the headers, join the parts, and decode them automatically. If you cannot implement an EMBL file, contact SOFTWARE@EMBL-HEIDELBERG.DE by e-mail, and explain your problem(s).

The University of Houston Gene-Server also has a software collection that is accessible by e-mail. However, unlike the EMBL File Server the commands to Gene-Server must be on the subject line of the e-mail message and not in the body of the message. To learn more about Gene-Server, send the message HELP on the Subject line to gene-server@bchs.uh.edu from Internet or to gene-server%bchs.uh.edu@CUNYVM from BITNET. To get a listing of the MAC software holdings, send the message on the Subject line, SEND MAC INDEX. For information about UNIX, VMS, and DOS software, substitute UNIX, VMS, or DOS for MAC. To obtain a file, the message SEND MAC MULTIDNA.HQX is sufficient. The same procedures for downloading (if required) and then decoding and uncompressing will have to be followed with these files as with the EMBL files to create executable programs. Parenthetically, for MAC files that after decoding have the double extension filename.sit.seg1, filename.sit.seg2, etc., use the "join" option in UnStuffIt to make a single .sit file, and then 'unstuff' with the programs UnStuffIt or SitExpand.

## III. OBTAINING SOFTWARE BY ANONYMOUS FTP

If your mainframe computer is on Internet with a TCP/IP connection, you will be able to contact many sites that have software and retrieve programs that may be useful to you. This procedure is called anonymous ftp (*f*ile *t*ransfer *p*rotocol). Some of the major molecular biology software sites that can be accessed by anonymous ftp are embl-heidelberg.de (192.54.41.20), genbank.bio.net (134.172.1.160), ftp.bio.indiana.edu (129.79.224.5250), menudo.uh.edu (129.7.1.6), and nic.funet.fi (128.214.6.100). The numbers in parentheses are the numerical addresses and are preceded by the 'name of site' addresses. To start an anonymous ftp session, type in at the prompt on your mainframe computer *ftp* followed by either the site-name address or its numerical address (e.g., *ftp embl-heidelberg.de* or *ftp 192.54.41.20*). Each command should be implemented by a carriage return. Here the notation <CR> represents a carriage return. When the ftp prompt (ftp>) asks for NAME, type in *anonymous* <CR>. At the request for PASSWORD, type in *guest* <CR>. With some systems, the password may be either your surname or your e-mail address. You will be given a second chance if you use the wrong password the first time. Now you are in the ftp site! Often the biggest stumbling block to carrying out an anonymous ftp software acquisition is locating the directory that contains the file(s) that you want. If possible, which may not always be the case, it is advisable to know the directory location before you begin a session, because ftp site managers have distinctive filing systems, and in many instances, directories are included within directories almost *ad infinitum*. Let us assume for this introduction to "anonymous ftp-ing" that you neither know the directory location nor the precise filename of the program that you want to retrieve and examine.

After you have logged into a ftp site, at the ftp prompt (ftp>), type in *dir* <CR>. This command will show you all the directories at this level. If, for example, you are interested in a MAC program, the directory mac, if it appears, would be a good place to start your search. At the ftp prompt, type in *cd mac* <CR>. The change directory (cd) command takes you to the mac directory. If you are interested in retrieving the phylogenetic tree drawing program TREEDRAW, at the ftp prompt type in *ls tr** <CR>. This command will list all the programs in the mac directory that have tr as the first two letters. For a complete listing of a directory, type in *ls* <CR>. If TREEDRAW is there, you can get it by typing in at the ftp prompt *get treedraw.hqx* <CR>. Use the filename and extension exactly as presented in the directory listing. To "see" the progress of the file transfer procedure before you type in the get command, type in *hash* <CR> at the ftp prompt. During the transfer process, a hash mark (#) will appear on your screen for every 1024 bytes that are sent. To obtain a set of related multiple files, use the command mget. For example, the TREEDRAW package may have both treedraw.hqx and treedraw.readme files. To receive both files with one command, type in *mget treedraw.** <CR>. If you want to look around the ftp site, return to the previous level with command *cd* .. <CR> and change to the next directory that may be of interest to you.

If the directory path to a specific file is known, the cd command with this path should take you into the directory that has the program you want to get. For example, the English version of mumac (MacMul) is at ftp.bio.indiana.edu with the directory location science/mac/multivar. After logging in to ftp.bio.indiana.edu, type in *cd /science/mac/multivar* <CR>. You will be taken directly to the multivar directory. Alternatively, you can go from directory to directory with single-step cd commands.

If your personal computer is networked to a mainframe computer that is on Internet, then with the appropriate software, which is readily available, you can download directly to your computer from an ftp site. Your sysop or the local PC or MAC gurus can advise you about this capability. If your personal computer is not networked, then download the transferred file with a standard routine (e.g., Kermit, Zmodem, etc.). Of course, if you retrieved a file for use on the mainframe computer, you must transfer it to an appropriate directory. In all cases, the file(s) retrieved by anonymous ftp will have to be decoded and uncompressed. Separate document files (e.g., read.me files) can be read directly as text files on a mainframe computer or by using a word processor on a personal computer.

If the files obtained by anonymous ftp have the same extensions as those from the EMBL software collection prior to decoding (e.g., .hqx, .uue) and after (e.g., .sea, .tar.Z, .tar, .zoo), use the same programs and procedures for decoding and uncompressing as described in the EMBL File Server HELP documents. If MAC software after decoding has the extension .sit, use the programs UnStuffIt or SitExpand to unpack it. If the extension is .cpt, then use Compactor Pro or Compactor to uncompress it. The extension designation dictates what decoding and uncompressing programs must be used. Although there is no universal standard for encoding and packing programs for any one operating system, most ftp sites tend to use the same protocols. If you retrieve a program that can neither be decoded nor unpacked, either ask the sysop for advice or seek out the MAC or DOS experts among your colleagues. One or the other will undoubtedly have the programs that you will need. As well, most computer science departments maintain freeware and shareware collections, which invariably have the full range of decoding and unpacking programs for MAC, DOS, UNIX, or VAX software. Occasionally, a file that you obtained by anonymous ftp is corrupted and is therefore not functional. In those instances you should repeat the transfer process.

The anonymous ftp system enables software to be retrieved easily. What is presented here is a basic, "getting-started" primer and does not include all of the commands and features of this system. File transfer by anonymous ftp is not fast; therefore, whenever possible, retrieve files in off-hours, i.e., after 9 p.m. and before 9 a.m. relative to the local time at the ftp site.

## IV. OBTAINING SOFTWARE BY "SNAIL" MAIL

In some cases, you may have to obtain copies of software applications directly from the developer. Formatting disks, copying programs, and sending out floppy disks is a time-consuming and tedious process. If you can avoid requesting programs by "floppy", do so. If you have no choice, follow the instructions of the developer fully. Generally, send properly formatted diskettes with blank labels in a diskette mailing case with a full return address label or shipping envelope. Include a few extra diskettes as a modest payment for the developer's time and effort. In the main, the programs will come as executables; sometimes a source code is also included which will enable you to compile the program for use with other operating systems.

## V. ASKING ARCHIE

Browsing through an ftp site and looking for a particular file or searching different ftp sites for a specific program can be frustrating and time-consuming. To overcome these problems, a first-rate computer service has been established at the McGill School of Computer Science at McGill University in Montreal. Basically, they have assembled the directory contents from many different ftp sites, including menudo.uh.edu and nic.funet.fi, and by searching a database comprised of these directory holdings, the site(s) and directory location(s) of a specific program are identified and conveyed to you. The system is called ARCHIE. There are two ways that ARCHIE can work for you. First, you can communicate directly with ARCHIE if your mainframe computer is on Internet. Second, you can send ARCHIE an e-mail message, and in return ARCHIE will send you the output of its database search by e-mail.

To contact ARCHIE by Internet, type in *telnet quiche.cs.mcgill.ca* <CR> at the mainframe computer prompt. Numerical addresses are 132.206.2.3 or 132.206.51.1. Assume, for example, that you want ARCHIE to tell you the site(s) and directory location(s) of the program mumac, a multivariate analysis program for the Macintosh. After making connection with ARCHIE, at the logon prompt, type in *archie* <CR>; then at the archie prompt, type in *prog mumac* <CR>. ARCHIE will search its database for mumac and print out to the screen all the sites and directory locations where it can be found. With this example, ARCHIE found mumac, with its six parts, both at the ftp site menudo.uh.edu located in the /pub/gene-server/mac directory path and at the ftp site nic.funet.fi located in the /pub/sci/molbio/mac directory path. In addition, the exact filename designation for each part is listed. With this information, you can retrieve mumac from either ftp site, and you will know how to navigate through the directories at these sites to get mumac.

ARCHIE's inventory is not all-inclusive nor is every ftp site represented in the database. Certain program names are ubiquitous, and you may have a number of hits that are not relevant to your specific search. As well, ARCHIE has other features. By typing in *help* <CR> at the archie prompt during a session, you will find more detailed information about ARCHIE's capabilities displayed.

If your mainframe computer is not on Internet, ARCHIE can be contacted by e-mail. For example, after sending an e-mail message to archie@quiche.cs.mcgill.ca with the command prog mumac in the body of the message, you will receive the results of the database search by e-mail. For more information about the e-mail mode for an ARCHIE session, either type in *help email* <CR> at the archie prompt or send the message help email to ARCHIE at its e-mail address.

For multiple requests by e-mail, the program names following the command word prog can be placed on the same line. In this case, the results of all the searches will be sent to you as one file. However, if the prog command is used on successive lines, you will get a separate reply for each line. In some cases, if the file is greater than 45 kbytes, ARCHIE may mail you a reply that will have to be "uudecoded" and "uncompressed".

## VI. KEEPING INFORMED

The BIOSCI bulletin board network is an excellent way for a molecular biologist to stay informed about a range of scientific topics, including software, methods, protein analysis, linkage studies, molecular evolution, and *Arabidopsis* research. In addition, if you have any specific queries about available software or other issues, by posting a message to the appropriate bulletin board (newsgroup), you can expect a number of responses in a short period of time.

If the computer system at your institution has USENET, you can subscribe directly to BIOSCI bulletin boards through USENET NEWS. For example, to subscribe to the software newsgroup, enter the USENET NEWS mode by typing *rn* <CR> or, on some systems, *vnews* <CR> at the prompt on the mainframe computer; then type *g bionet.software* <CR>. In this way, you will be able to read all the messages sent to the BIO-SOFTWARE bulletin board. There are other bulletin boards that may be of interest, for example, METHODS-AND-REAGENTS (USENET name: bionet.molbio.methods-reagents), ARABIDOPSIS (USENET name: bionet.genome.arabidopsis), MOLECULAR-EVOLUTION (USENET name : bionet.molbio-evolution), and BIONEWS (USENET name : bionet.general). Use the *g* command in USENET NEWS and the USENET name to subscribe to these and any of the other BIOSCI bulletin boards.

To read the messages that have been sent to a bulletin board, type in *rn* (or *vnews*) <CR> at the mainframe computer prompt. Once in the USENET NEWS mode, the commands are straightforward. Type *h* before entering a newsgroup, and a set of commands for handling newsgroups (newsgroup selection commands) will be displayed. At this level, at the query "read now?", typing *n* means next newsgroup, typing *q* means quit USENET NEWS, and typing *y* means enter this newsgroup. Once inside a newsgroup, by typing *h* you will get on-screen help with the commands to facilitate navigating within a newsgroup (paging commands). Briefly, within a newsgroup at the query "what next?", typing *n* means go forward to next unread message, pressing space bar at the 'MORE' prompt means continue displaying more of the message, typing *p* means go to previous unread message, and typing *q* means quit this newsgroup. If you do not wish to continue reading a message, type *n* at the "MORE" prompt. To unsubscribe to a USENET newsgroup, type *u* "USENET name" at the newsgroup level.

If you do not have access to USENET NEWS but are on Internet, BITNET, EARN, NETNORTH, HEANET, or JANET, you can subscribe to a BIOSCI newsgroup by e-mail. There are four BIOSCI nodes which are geographically based that receive subscriptions and other queries. For North and South America, the node address is biosci@genbank.bio.net through Internet or BITNET. For the United Kingdom, the node address is biosci@uk.ac.daresbury through JANET. For Scandinavia and Continental Europe, the node address is biosci@bmc.uu.se through Internet. For Ireland and continental Europe, the node address is LISTSERV@IRLEARN through EARN/BITNET. For the first three nodes, the e-mail message requesting a subscription, help, or cancellation (sign-off) can be in plain language. To subscribe (or sign-off) to the BIOSCI node in Ireland, a specifically formatted message must be sent because it will be handled solely by computer. The form of the message to LISTSERV@IRLEARN for subscribing to the software newsgroup is SUBSCRIBE BIO+SOFT John Smith, where "John Smith" is, of course, your full name. For more information about the BIOSCI Bulletin Board Network, send a plain language message to any of the three "plain language" nodes noted above or to BIOSCI@IRLEARN.BITNET.

If your mainframe computer has USENET, the postnews software can be used to post a message to a BIOSCI bulletin board. To do this, type in *pnews* <CR>, and follow the menu prompts. Briefly, answer the NEWSGROUP prompt with the USENET name of the newsgroup (e.g., bionet.software); at the DISTRIBUTION prompt, type in WORLD; type in a short descriptive title at the appropriate prompt, and finally with a prepared message, type its filename at the "PREPARED FILE TO INCLUDE [NONE]" prompt. If you do not have a prepared file, a carriage return will invoke the postnews editor. Implement each command with a carriage return.

Alternatively, you can post a message to a newsgroup by e-mail. There are four BIOSCI nodes set up to receive e-mail messages that are destined for posting. To distinguish between subscription, sign-off, and help messages, the messages that are directed to a newsgroup for posting are identified by a specific mailing address name for each newsgroup. For example, the mailing address name for the BIO-SOFTWARE newsgroup is BIO-SOFT. The address nodes are the same as those that receive subscriptions and cancellations, i.e., the 'plain language' nodes. If you are located in North or South America and you want to post a query to the BIO-SOFTWARE newsgroup, the e-mail message should be sent to BIO-SOFT@GENBANK.BIO.NET. Some other mailing address names are BIONEWS, GEN-LINK, GENE-ORG, METHODS, MOL-EVOL, PROTEINS, and ARAB-GEN.

# 16

# Sequence Similarity Searches, Multiple Sequence Alignments, and Molecular Tree Building

*J. J. Pasternak*

## I. INTRODUCTION

In this chapter, a practical guide for carrying out sequence similarity searches of databases, multiple sequence alignments, and molecular tree building is presented. Background information is available in a number of articles and books.[1-5] The services and programs that have been developed for computer analysis of sequence data are designed for the convenience of the user. However, access to these systems requires not only being properly connected electronically, but having some understanding of how to tell either a program or a service what you want to achieve. Initially, a first-time user may not have a clear understanding of the criteria that have been used to generate the data by various analytical procedures. With some experience, however, the output of the various programs becomes comprehensible. Generally, the on-line services have good documentation and provide references to key articles in the literature.

For most sequence computer analyses, connection to a mainframe computer is useful. If you have access to a personal microcomputer equipped with search software and up-to-date sequence databases, similarity searches can also be conducted readily. If such an "off-line" system is not readily available, an on-line system is easy to establish. The first and most important step is to contact the computer system manager at your institution about setting up a mainframe computer account. You should ask what network(s) (e.g., Internet, BITNET, etc.) are available and if there are gateways to other networks. Also, learn how to send and receive electronic mail (e-mail) messages. For similarity searches of databases, on-line computer costs are meager. For multiple sequence alignments and tree building, you can expect modest costs for both computer use and disk storage. You should plan to use the mainframe computer for multiple sequence alignments and some tree-building tasks, since the large amount of required memory is usually beyond the range of most personal microcomputers. Also, mainframe computers do most analyses very rapidly.

To work on a mainframe computer, you will need to establish electronic contact. Check with your system manager for information about connecting to the mainframe by means of a modem from your personal microcomputer. If such a connection is available (i.e., a telephone number), you will need a modem (probably 2400 baud) and a communications software package with a large capture buffer.

0-8493-5164-2/93/$0.00+$.50
© 1993 by CRC Press, Inc.

```
SEARCH
>Arabidopsis APRT
MATEDVQDPRIAKIASSIRVIPDFPKPGIMFQDITTLLLDTEA
FKDTIALFVDRYKDKGISVVAGVEARGFIFGPPIALAIGAKFVPMRKPKK
LPGKVISEEYSLEYGTDTIEMHVGAVEPGERAIIDDLIATGGTLAAAIR
LLERVGVKIVECACVIELPELKGKEKLGETSLFVLVKSAA*
END SEARCH
```

**FIGURE 1.** Example of an electronic mail message to NBRF-PIR for a FASTA search of the PIR database for similarities to the query sequence *Arabidopsis thaliana* adenosine phosphoribosyltransferase (APRT).

## II. SEQUENCE SIMILARITY SEARCHES

The major nucleic acid sequence databases are GenBank® and the EMBL data library, while for proteins there are the PIR and SWISS-PROT sequence databases. Each of the databases, their services, e-mail addresses, and, in some cases, how to obtain information about using the resources have been described.[6-10]

The FASTA algorithm[11] is used by NBRF-PIR (National Biomedical Research Foundation-Protein Information Resource) to search for sequence similarities in both protein and nucleic acid databases. PIR has established a network fileserver that handles database retrievals, sequence searches, and sequence submissions by e-mail messages. Information about conducting a search can be obtained by sending an e-mail message containing the words SEARCH HELP in the body of the message to FILESERV@GUNBRF.BITNET.

The message to instruct PIR to carry out a sequence similarity search must be in one of four specific formats. Correct formatting of this mail message is essential if the search is to be carried out at all. Until you are familiar with the parameters of the FASTA search process, use the default values. An example of one of the valid formats for an e-mail message to PIR to instruct it to carry out a database search with a protein query sequence is shown in Figure 1. In this case, the default settings are accepted. A properly formatted message with a query sequence can be prepared easily using a word processor on a microcomputer and then sending the message as a text file to your directory on the mainframe computer. From there, the file can be mailed to FILESERV@GUNBRF.BITNET. The output data from the search will be sent to you by e-mail. To interpret the output data, the 'SEARCH HELP' documentation and original articles should be consulted.

A portion of the output of a FASTA search of the NBRF-PIR protein library with the query sequence adenine phosphoribosyltransferase (APRT) from the plant *Arabidopsis thaliana* is shown in Figure 2. Briefly, the FASTA algorithm finds local similarities based on identities during the first pass of the query sequence through the database and calculates a score for these segments which is printed in the **initn** column. Once identified, the regions of local similarities are scored again for both identities and conserved changes (**init1**), and finally each alignment is optimized by using gaps (**opt**). In the run with the *A. thaliana* APRT sequence, the query sequence "found" APRT sequences from mouse, human (two entries), *E. coli*, and fruit fly; other sequences were also selected. The question, then, is which of the selected sequences are truly related to the *A. thaliana* APRT sequence. As noted above, the articles written by the developers of the FASTA search algorithm should be consulted[11,12] for a full understanding of the results. Generally, if there is an increase in the scores from init1 to opt or, as is more often the case, if the initn, init1, and opt scores are all high, then the query sequence has probably identified a biologically related sequence. As a rule of thumb, although not statistically definitive, if the value of the init1 score for a particular sequence minus the mean init1 score for all comparisons divided by the standard deviation of the mean init1 score (i.e., Z score) is >6, there are grounds for considering that the selected sequence is related to the query sequence. Here, for the transcobalamin I precursor sequence, the Z score calculation is 56–21.8/6.12=5.59. Thus, unless there were some compelling additional evidence, it is unlikely that the transcobalamin I precursor protein is related to the APRT sequence from *A. thaliana*. A program (RDF2 in earlier versions of FASTA and RSS in FASTA v. 1.6) is available as part of the FASTA package for determining whether two sequences are similar to an extent greater than expected by chance alone. RSS and its predecessor RDF2 can be implemented readily on a personal microcomputer and used for pairwise comparisons to determine whether the two sequences are significantly similar. Healthy skepticism should be exercised when trying to establish a relationship between sequences with marginal similarity scores.

NBRF-PIR also uses the QUICKSEARCH routine to detect sequence similarities between a submitted query sequence and the sequences of the databases. Details about QUICKSEARCH and how to use it are presented in the 'SEARCH HELP' document.

GenBank® has initiated an e-mail server for similarity searches with the GenBank®, PIR, and SWISS-PROT databases using the BLAST (Basic Local Alignment Search Tool) program.[13] This program is very fast. A number of networks, including Internet, BITNET, EARN, NETNORTH, and JANET access GenBank®. The mail message requesting a BLAST search must be formatted in a specific way. A HELP document can be obtained by sending an e-mail message with the word HELP in its body to BLAST@GENBANK.BIO.NET. A typical message for directing a search of the PIR database, using the default settings, is shown in Figure 3. Each line of the message must be less than 80 characters. The results of the search will be sent to you by e-mail. Careful interpretation of the output should be based on the criteria discussed by the authors of the program.[13,14]

A portion of the output of a BLAST search run with the APRT protein sequence from *Arabidopsis thaliana* as the query sequence is shown in Figure 4. High scores were obtained with APRT sequences from *E. coli*, mouse, human (two entries), and fruit fly. The EXPECT-value denotes the expected number of alignments that would yield an identical or greater score. The Poisson P-value denotes the probability that the score of the actual query sequence will occur with a random query sequence that was the same length and residue composition when run against a database of the same size. The extremely low EXPECT- and Poisson P-values for the selected APRT sequences denote significant similarity among these sequences. The search also detected, with lower scores, a smaller segment of potential similarity from the mouse, human, and fruit fly APRT sequences, which aligns with *A. thaliana* APRT. In these cases, the EXPECT-values are low and the Poisson P-values are very low, indicating that the sequence alignments may be significant despite the modest scores. By contrast, the EXPECT- and Poisson P-values of the other sequences from the output file (e.g., human uridine monophosphate synthase) are about or greater than 0.001. With BLAST, using the default settings, these values should not be construed as significant.

## III. MULTIPLE SEQUENCE ALIGNMENT

Often when a query sequence identifies significant similarities from sequences within a database, it is necessary to align these sequences not only for molecular evolution studies but possibly to identify conserved domains that may have specific biological functions. Computationally, multiple sequence alignment is not a trivial exercise.[3,4] Manual alignments can be difficult to carry out and are often highly subjective. Computer-generated sequence alignments are, if the same settings are retained, reproducible and relatively quick to create. The problem of subjectivity is not completely overcome with computer-assisted alignments because the weighting values for mismatches, gaps, insertions, and deletions can be manipulated by the researcher. Notwithstanding the limitations, computer-aided alignments are reasonably accurate for most applications.

```
 fasta 1.4c [May, 1990] searches a sequence data bank
Please cite: W.R. Pearson & D.J. Lipman PNAS (1988) 85:2444-2448
>Arabidopsis APRT
MATEDVQDPRIAKIASSIRVIPDFPKPGIMFQDITTLLLDTEA
FKDTIALFVDRYKDKGISVVAGVEARGFIFGPPIALAIGAKFVPMRKPKK
LPGKVISEEYSLEYGTDTIEMHVGAVEPGERAIIIDDLIATGGTLAAAIR
LLERVGVKIVECACVIELPELKGKEKLGETSLFVLVKSAA*
 >Arabidopsis APRT MATEDVQDPRIAKIASSIRVIPDFPKPGIMF : 140 aa
 vs
 NBRF Protein Library (PIR1 PIR2 & PIR3) library
 searching DB$PIR1:.SEQ library
 searching DB$PIR2:.SEQ library
 searching DB$PIR3:.SEQ library


      initn   init1
<  2    57     57:=============================
   4     2      2:=
   6    23     23:============
   8    98     98:=================================================
  10   359    359:==================================================
  12   957    957:==================================================
  14  1414   1414:==================================================
  16  3934   3934:==================================================
  18  3437   3437:==================================================
  20  4969   4969:==================================================
  22  4222   4222:==================================================
  24  3985   3985:==================================================
  26  3493   3493:==================================================
  28  2204   2273:==================================================
  30  1450   1556:==================================================
  32   944   1048:==================================================
  34   631    663:==================================================
  36   493    474:==================================================
  38   303    269:==================================================
  40   230    166:==================================================
  42   132     73:-------------------------------------+++++++++++++
  44    76     46:-----------------------+++++++++++++++
  46    65     33:-----------------++++++++++++++++
  48    51     32:----------------++++++++++
  50    23      9:-----+++++++
  52    29     15:--------+++++++
  54    23      5:---+++++++++
  56     9      6:---++
  58     1      0:+
  60     1      0:+
  62     4      3:==
  64     3      1:-+
  66     0      0:
  68     0      0:
  70     0      0:
  72     0      0:
  74     0      0:
  76     0      0:
  78     0      0:
  80     0      0:
> 80     5      5:===
```

**FIGURE 2.** Partial results from a FASTA search of the PIR database with the query sequence *Arabidopsis thaliana* APRT.

A number of heuristic methods have been devised to achieve some semblance of precision of alignment with sets of similar sequences. To develop a dataset of sequences that is to be aligned, it is necessary either to retrieve the sequences from a database or to type in the sequences from the original articles. The former strategy is much preferred, although it is advisable to carry out a search of the literature because not all published sequences are submitted automatically to the databases.

The outputs from FASTA and BLAST provide accession numbers for all sequences that are possibly similar to the query sequence. For example, in the FASTA run with *A. thaliana* APRT (Figure 2), the accession number of the mouse APRT in the PIR library is RTMSA. This sequence is in database 1 of the three PIR databases. In addition, with the PIR Network

```
9608345 residues in 33627 sequences
 mean initn score:  21.9 (6.40)
 mean init1 score:  21.8 (6.12)
 5572 scores better than 27 saved, ktup: 2, fact: 8  scan time:  0:04:55
The best scores are:                                           initn init1   opt
PIR1:RTMSA Adenine phosphoribosyltransferase (EC 2.4.2.7) - Mous 310   310   344
PIR2:A28021 Adenine phosphoribosyltransferase (EC 2.4.2.7) - Hum 306   306   343
PIR1:RTHUA Adenine phosphoribosyltransferase (EC 2.4.2.7) - Huma 306   306   343
PIR1:RTECA Adenine phosphoribosyltransferase (EC 2.4.2.7) - Esch 295   295   388
PIR2:A29596 Adenine phosphoribosyltransferase (EC 2.4.2.7) - Fru 231   231   286
PIR2:A34227 Transcobalamin I precursor - Human                    64    56    64
PIR3:S12617 *Glycinamide ribonucleotide synthetase-aminoimidazol  64    47    49
PIR2:JQ0169 Coat protein - Melon necrotic spot virus              63    63    64
PIR3:S07575 *120kD surface-exposed protein - Rickettsia ricketts  62    62    69
PIR1:WMBEX6 UL6 protein - Herpes simplex virus type 1 (strain 17  62    43    48
PIR3:A30408 *Ribose-phosphate pyrophosphokinase (EC 2.7.6.1) - S  61    61    71
PIR1:KIECRY Ribose-phosphate pyrophosphokinase (EC 2.7.6.1) - Es  61    61    71
PIR2:S04518 Anthranilate synthase multifunctional protein - Emer  60    52    58
PIR3:A38337 *Amidophosphoribosyltransferase (EC 2.4.2.14) - Chic  58    50    75
PIR3:JA0073 *Coat protein precursor - Potato virus Y (fragment)   56    56    58
PIR1:RRNZB3 Polymerase-associated nucleocapsid phosphoprotein -   56    35    37
PIR2:S08660 Dihydrofolate reductase (EC 1.5.1.3)/thymidylate syn  56    56    74
PIR2:A25562 Sucrose alpha-glucosidase (EC 3.2.1.48) - Bacillus s  56    39    40
PIR3:S15756 *Dihydrofolate reductase/thymidylate synthase - Leis  56    56    74
PIR3:JQ0948 *A5-protein - African clawed frog                     56    48    51

RTMSA Adenine phosphoribosyltransferase (EC 2.4.2.7) - Mouse     310   310   344
  43.6% identity in 140 aa overlap

                                                   10        20        30
Arabid                                     FKDTIALFVDRYKDK---GISVVAGVEARGFIFGPP
                                           :...: :.... :..     :. .X:...:::.:::.
RTMSA  EPELKLVARRIRVFPDFPIPGVLFRDISPLLKDPDSFRASIRLLASHLKSTHSGKIDYIAGLDSRGFLFGPS
                10        20        30        40        50        60        70

          40        50        60        70        80        90       100
Arabid IALAIGAKFVPMRKPKKLPCKVISEEYSLEYCTDTIEMHVGAVEPGERAIIIDDLIATGGTLAAAIRLLERV
       .: ..:.  : .::. :::: ..:..:::::....:.. .:.:::.:..:.:::.:::::. ::  ::...
RTMSA  LAQELGVGCVLIRKQGKLPGPTVSASYSLEYGKAELEIQKDALEPGQRVVIVDDLLATGGTMFAACDLLHQL
         80        90       100       110       120       130       140

      110       120       130       140
Arabid GVKIVECACVIELPELKGKEKLGETSLFVLVKSAA
        ...:::....::..:::.:.:X ...: :..
RTMSA  RAEVVECVSLVELTSLKGRERLGPIPFFSLLQYD
      150       160       170       180

A28021 Adenine phosphoribosyltransferase (EC 2.4.2.7) - Human    306   306   343
  42.1% identity in 140 aa overlap

                                                   10        20        30
Arabid                                     FKDTIALFVDRYKDKG---ISVVAGVEARGFIFGPP
                                           :...:.:.. . :..     :. .X:...:::.:::.
A28021 DSELQLVEQRIRSFPDFPTPGVVFRDISPVLKDPASFRAAIGLLARHLKATHGGRIDYIAGLDSRGFLFGPS
                10        20        30        40        50        60        70

          40        50        60        70        80        90       100
Arabid IALAIGAKFVPMRKPKKLPGKVISEEYSLEYGTDTIEMHVGAVEPGERAIIIDDLIATGGTLAAAIRLLERV
       .: ..:   : .::. :::: .. ..:::::::....:.. .:.:::.:....:::.::::::..::  ::.:.
A28021 LAQELGLGCVLIRKRGKLPGPTLWASYSLEYGKAELEIQKDALEPGQRVVVVDDLLATGGTMNAACELLGRL
         80        90       100       110       120       130       140

      110       120       130       140
Arabid GVKIVECACVIELPELKGKEKLGETSLFVLVKSAA
        ....::....::..:::.::X. ...: :..
A28021 QAEVLECVSLVELTSLKGREKLAPVPFFSLLQYE
      150       160       170       180

RTHUA Adenine phosphoribosyltransferase (EC 2.4.2.7) - Human     306   306   343
  42.1% identity in 140 aa overlap

                                                   10        20        30
Arabid                                     FKDTIALFVDRYKDKG---ISVVAGVEARGFIFGPP
                                           :...:.:.. . :..     :. .X:...:::.:::.
RTHUA  DSELQLVEQRIRSFPDFPTPGVVFRDISPVLKDPASFRAAIGLLARHLKATHGGRIDYIAGLDSRGFLFGPS
                10        20        30        40        50        60        70
```

**FIGURE 2** continued.

```
              40        50        60        70        80        90       100
Arabid IALAIGAKFVPMRKPKKLPGKVISEEYSLEYGTDTIEMHVGAVEPGERAIIIDDLIATGGTLAAAIRLLERV
       .: ..:   : .::. :::: .. ..:::::::...:.. .:.:::.:...:::.::::::..::  ::.:.
RTHUA  LAQELGLGCVLIRKRGKLPGPTLWASYSLEYGKAELEIQKDALEPGQRVVVVDDLLATGGTMNAACELLGRL
              80        90       100       110       120       130       140

          110       120       130       140
Arabid GVKIVECACVIELPELKGKEKLGETSLFVLVKSAA
        ....::....::..:::.::X.  ...: :..
RTHUA  QAEVLECVSLVELTSLKGREKLAPVPFFSLLQYE
          150       160       170

RTECA Adenine phosphoribosyltransferase (EC 2.4.2.7) - Escherich 295   295   388
  48.5% identity in 132 aa overlap

                                                       10        20        30
Arabid                                          FKDTIALFVDRYKDKGISVVAGVEARGFIFGPPIAL
                                                .   .:.:.:.X::.  ::.  :.:.:::::.::.:.::
RTECA  AQQLEYLKNSIKSIQDYPKPGILFRDVTSLLEDPKAYALSIDLLVERYKNAGITKVVGTEARGFLFGAPVAL
           10        20        30        40        50        60        70

         40        50        60        70        80        90       100
Arabid AIGAKFVPMRKPKKLPGKVISEEYSLEYGTDTIEMHVGAVEPGERAIIIDDLIATGGTLAAAIRLLERVGVK
       ..:. :::.::: ::: ..:::.:.::::: .:.::.:..::.....:::.::::X..:...:. ::.: .
RTECA  GLGVGFVPVRKPGKLPRETISETYDLEYGTDQLEIHVDAIKPGDKVLVVDDLLATGGTIEATVKLIRRLGGE
         80        90       100       110       120       130       140

      110       120       130       140
Arabid IVECACVIELPELKGKEKLGETSLFVLVKSAA
       ... : .:.: .: :...:.. ..
RTECA  VADAAFIINLFDLGGEQRLEKQGITSYSLVPFPGH
      150       160       170       180

A29596 Adenine phosphoribosyltransferase (EC 2.4.2.7) - Fruit fl 231   231   286
  40.3% identity in 134 aa overlap

                                                 10        20        30        40
Arabid                                    FKDTIALFVDRYKDKG--ISVVAGVEARGFIFGPPIALAIGA
                                                :.::. ....    ....:...X::.:.  :: ..:
A29596 VKSKIGEYPNFPKEGILFRDIFGALTDPKACVYLRDLLVDHIRESAPEAEIIVGLDSRGFLFNLLIATELGL
           20        30        40        50        60        70        80

                 50        60        70        80        90       100       110
Arabid KFVPMRKPKKLPGKVISEEYSLEYGTDTIEMHVGAVEPGERAIIIDDLIATGGTLAAAIRLLERVGVKIVEC
         .:.::  ::.:.:.: ::.::::.::.:.. .:..::.....:::.::::.:.::. :. .::  .:X.
A29596 GCAPIRKKGKLAGEVVSVEYKLEYGSDTFELQKSAIKPGQKVVVVDDLLATGGSLVAATELIRKVGGVVVES
         90       100       110       120       130       140       150

               120       130       140
Arabid -ACVIELPELKGKEKLGETSLFVLVKSAA
        . :.:: .:.:...: ....  :.:
A29596 LVVVMELVGLEGRKRL-DGKVHSLIKY
       160       170       180

A34227 Transcobalamin I precursor - Human                          64    56    64
  19.4% identity in 98 aa overlap

                                                      10        20        30
Arabid                                        FKDTIALFVDRYKDKGISVVAGVEARGFIFGPPIALAIG
                                                .. X: ..:.. ..    . :.... ::.......:
A34227 HLTDKLENKFQAEIENMEAHNGTPLTNYYQLSLDVLALCLFNGNYSTAEVVNHFTPENKNYYFGSQFSVDTG
          120       130       140       150       160       170       180

     40        50          60        70        80         90       100
Arabid AK-FVPMRKPKK--LPGKVISEEYSLEYGTDTIEMHVGAVEPGERAI-IIDDLIATGGTLAAAIRLLERVGV
       X.  ...   ::  . :.. ..: ::.   .  ..  :. . .....  .:.. ..:....:
A34227 AMAVLALTCVKKSLINGQIKADEGSLKNISIYTKSLVEKILSEKKENGLIGNTFSTGEAMQALFVSSDYYNE
      190       200       210       220       230       240       250       260

       110       120       130       140
Arabid KIVECACVIELPELKGKEKLGETSLFVLVKSAA

A34227 NDWNCQQTLNTVLTEISQGAFSNPNAAAQVLPALMGKTFLDINKDSSCVSASGNFNISADEPITVTPPDSQS
              270       280       290       300       310       320       330
```

**FIGURE 2** continued.

```
S12617 *Glycinamide ribonucleotide synthetase-aminoimidazole rib  64    47    49
  28.6% identity in 63 aa overlap

              20        30        40        50        60        70        80
Arabid KDKGISVVAGVEARGFIFGPPIALAIGAKFVPMRKPKKLPGKVISEEYSLEYGTDTIEMHVGAVEPGERAII
                                           X::  ::... . ::  . .. . :  ..:.: :..
S12617 GSRTEFDSAVDRVLEEFSVELICLAGFMRILSGPFVKKWEGKILNIHPSLLPSFKGANAHKLVLEAGVRVTG
          870       880       890       900       910       920       930

           90       100       110       120       130       140
Arabid IDDLIATGGTLAAAIRLLERVGVKIVECACVIELPELKGKEKLGETSLFVLVKSAA
        .  ......  :.:: . : : ::X .
S12617 CTVHFVAEEVDAGAIIFQEAVPVKIGDTVETLSERVKEAEHRAFPAALQLVASGAVQVGEAGKICWK
        940       950       960       970     980       990      1000
```

**FIGURE 2** continued.

```
BLASTPROGRAM blastp
DATALIB pir
BEGIN
>Arabidopsis APRT
MATEDVQDPRIAKIASSIRVIPDFPKPGIMFQDITTLLLDTEAFKDTI
ALFVDRYKDKGISVVAGVEARGFIFGPPIALAIGAKFVPMRKPKKL
PGKVISEEYSLEYGTDTIEMHVGAVEPGERAIIDDLIATGGTLAAA
IRLLERVGVKIVECACVIELPELKGKEKLGETSLFVLVKSAA
```

**FIGURE 3.** Example of an electronic mail message to GenBank® for a BLAST search of the PIR database for similarities to the query sequence *Arabidopsis thaliana* APRT.

Server, it is possible to obtain accession numbers for sequences within the databases by using various "search" commands. Some of these commands include KEYWORD, TITLE, FEATURE, SUPERFAMILY, TAXONOMY, and HOST. For specific information about these commands send an e-mail message with the words HELP KEYWORD, HELP TITLE, etc., with each complete command on a separate line, to FILESERV@GUNBRF.BITNET. If, for example, you were interested in obtaining database accession members for actin nucleic acid and protein sequences that have been determined for members of the genus *Arabidopsis*, the e-mail message TITLE ACTIN ARABIDOPSIS would be a good starting point. If you requested KEYWORD ACTIN, you would receive by return e-mail a complete listing of all the actin entries, as well as entries that contain the letters *actin* (e.g., pro*lactin*) in the PIR, GenBank®, and EMBL databases. With specialized searches, you can assemble the accession numbers for most of the sequences that should comprise your dataset.

Once you have the accession numbers, the sequences must be retrieved. An electronic mail message to RETRIEVE@GENBANK.BIO.NET with an accession number on each line in the body of the message will result in each sequence being returned in a separate mail message from GenBank®. In this case, accession numbers from the EMBL, GenBank®, and Swiss-Prot databases are recognized, but PIR entries are not.

The PIR File Server uses a slightly different approach for retrieving sequences. The GET command in the form GET EMBL:ACCESSION NUMBER in the body of an e-mail message, with one command per line, to FILESERV@GUNBRF.BITNET will retrieve sequences from the EMBL database library. Of course, you would use the actual accession number in the GET command line. The database identifiers are GB, EMBL, and PIR for GenBank®, EMBL, and PIR, respectively. To obtain sequences from PIR, the form of the GET command should be GET PIRx:ACCESSION NUMBER where x is 1, 2, or 3 depending on the PIR database containing the sequence that is being sought. With the PIR system, it is unwise to ask for more than ten sequences per message because of the possible excessive size of the return e-mail file. To reiterate, with the PIR File Server, retrieval of a sequence is accomplished by noting the database source and the corresponding accession number with a GET command.

```
Here are your search results from the GenBank BLAST e-mail server.

If you use BLAST as a research tool, we ask that this reference be
cited in your paper.

        S. F. Altschul, W. Gish, W. Miller, E. W. Myers and
        D. J. Lipman (1990) J. Mol. Biol.  215, 403-410.


Database versions currently in use on GOS:

Database        Version Used
--------        ------------
GenBank              68 plus new data through 26 August 1991.
NBRF/PIR             28
SWISS-PROT           18

Please report suspected bugs to us at blast-req@genbank.bio.net

*****************************************************************



Initiating /usr/local/bin/blastp query on database /blast/db/pir
Query= >Arabidopsis APRT



>RTECA Adenine phosphoribosyltransferase - Escherichia coli
        Length = 183

  Score = 439, Expect = 2.7e-55, Poisson P = 2.7e-55, length = 155

Query:     16 SSIRVIPDFPKPGIMFQDITTLLLDTEAFKDTIALFVDRYKDKGISVVAGVEARGFIFGP 75
              +SI+ I D+PKPGI+F+D+T+LL D  A+  +I L V+RYK+ GI+ V G EARGF+FG+
Sbjct:     13 NSIKSIQDYPKPGILFRDVTSLLEDPKAYALSIDLLVERYKNAGITKVVGTEARGFLFGA 72

Query:     76 PIALAIGAKFVPMRKPKKLPGKVISEEYSLEYGTDTIEMHVGAVEPGERAIIIDDLIATG 135
              P+AL++G  FVP+RKP KLP   ISE Y LEYGTD +E+HV A+ PG++ +++DDL+ATG
Sbjct:     73 PVALGLGVGFVPVRKPGKLPRETISETYDLEYGTDQLEIHVDAIKPGDKVLVVDDLLATG 132

Query:    136 GTLAAAIRLLERVGVKIVECACVIELPELKGKEKL 170
              GT+ A+++L+ R+G  + + A +I+L +L G ++L
Sbjct:    133 GTIEATVKLIRRLGGEVADAAFIINLFDLGGEQRL 167


>RTMSA Adenine phosphoribosyltransferase - Mouse #EC-number
        Length = 180

  Score = 291, Expect = 4.7e-34, Poisson P = 4.7e-34, length = 120

Query:     60 ISVVAGVEARGFIFGPPIALAIGAKFVPMRKPKKLPGKVISEEYSLEYGTDTIEMHVGAV 119
              I  +AG+++RGF+FGP++A  +G   V +RK  KLPG  +S  YSLEYG   +E++  A+
Sbjct:     58 IDYIAGLDSRGFLFGPSLAQELGVGCVLIRKQGKLPGPTVSASYSLEYGKAELEIQKDAL 117

Query:    120 EPGERAIIIDDLIATGGTLAAAIRLLERVGVKIVECACVIELPELKGKEKLGETSLFVLV 179
              EPG+R +I+DDL+ATGGT+ AA  LL ++   +VEC  ++EL  LKG+E+LG   +F L+
Sbjct:    118 EPGQRVVIVDDLLATGGTMFAACDLLHQLRAEVVECVSLVELTSLKGRERLGPIPFFSLL 177
```

**FIGURE 4.** Partial results from a BLAST search of the PIR databases with the query sequence *Arabidopsis thaliana* APRT.

The EMBL Network File Server retrieves sequences from the EMBL, GenBank®, and SWISS-PROT libraries. To get a nucleic acid sequence from the EMBL database, send the message GET nuc:EMBL ACCESSION NUMBER to NETSERV@EMBL-HEIDELBERG.DE. A GET nuc:GB ACCESSION NUMBER message will retrieve a se-

```
 Score = 120, Expect = 1.6e-09, Poisson P = 2.9e-23, length = 49

Query:     8 DPRIAKIASSIRVIPDFPKPGIMFQDITTLLLDTEAFKDTIALFVDRYK 56
             +P +  +A  IRV PDFP PG++F+DI+ LL D ++F+ +I L   + K
Sbjct:     3 EPELKLVARRIRVFPDFPIPGVLFRDISPLLKDPDSFRASIRLLASHLK 51


>RTHUA Adenine phosphoribosyltransferase - Human #EC-number
          Length = 179

 Score = 282, Expect = 9.2e-33, Poisson P = 9.1e-33, length = 120

Query:    60 ISVVAGVEARGFIFGPPIALAIGAKFVPMRKPKKLPGKVISEEYSLEYGTDTIEMHVGAV 119
             I  +AG+++RGF+FGP++A  +G   V +RK  KLPG  +   YSLEYG  +E++  A+
Sbjct:    57 IDYIAGLDSRGFLFGPSLAQELGLGCVLIRKRGKLPGPTLWASYSLEYGKAELEIQKDAL 116

Query:   120 EPGERAIIIDDLIATGGTLAAAIRLLERVGVKIVECACVIELPELKGKEKLGETSLFVLV 179
             EPG+R +++DDL+ATGGT+ AA  LL R+   ++EC  ++EL  LKG+EKL+  + F L+
Sbjct:   117 EPGQRVVVVDDLLATGGTMNAACELLGRLQAEVLECVSLVELTSLKGREKLAPVPFFSLL 176

 Score = 100, Expect = 1.2e-06, Poisson P = 1.6e-17, length = 33

Query:    18 IRVIPDFPKPGIMFQDITTLLLDTEAFKDTIAL 50
             IR  PDFP PG++F+DI+ +L D  +F+ +I+L
Sbjct:    12 IRSFPDFPTPGVVFRDISPVLKDPASFRAAIGL 44


>A28021 Adenine phosphoribosyltransferase - Human #EC-number
          Length = 180

 Score = 282, Expect = 9.2e-33, Poisson P = 9.1e-33, length = 120

Query:    60 ISVVAGVEARGFIFGPPIALAIGAKFVPMRKPKKLPGKVISEEYSLEYGTDTIEMHVGAV 119
             I  +AG+++RGF+FGP++A  +G   V +RK  KLPG  +   YSLEYG  +E++  A+
Sbjct:    58 IDYIAGLDSRGFLFGPSLAQELGLGCVLIRKRGKLPGPTLWASYSLEYGKAELEIQKDAL 117

Query:   120 EPGERAIIIDDLIATGGTLAAAIRLLERVGVKIVECACVIELPELKGKEKLGETSLFVLV 179
             EPG+R +++DDL+ATGGT+ AA  LL R+   ++EC  ++EL  LKG+EKL+  + F L+
Sbjct:   118 EPGQRVVVVDDLLATGGTMNAACELLGRLQAEVLECVSLVELTSLKGREKLAPVPFFSLL 177

 Score = 100, Expect = 1.2e-06, Poisson P = 1.6e-17, length = 33

Query:    18 IRVIPDFPKPGIMFQDITTLLLDTEAFKDTIAL 50
             IR  PDFP PG++F+DI+ +L D  +F+ +I+L
Sbjct:    13 IRSFPDFPTPGVVFRDISPVLKDPASFRAAIGL 45


>A29596 Adenine phosphoribosyltransferase - Fruit fly
          Length = 183

 Score = 224, Expect = 1.9e-24, Poisson P = 1.9e-24, length = 98

Query:    62 VVAGVEARGFIFGPPIALAIGAKFVPMRKPKKLPGKVISEEYSLEYGTDTIEMHVGAVEP 121
             ++ G+++RGF+F   IA  +G   P+RK  KL+G V+S EY LEYG+DT E++ +A+ P
Sbjct:    64 IIVGLDSRGFLFNLLIATELGLGCAPIRKKGKLAGEVVSVEYKLEYGSDTFELQKSAIKP 123

Query:   122 GERAIIIDDLIATGGTLAAAIRLLERVGVKIVECACVI 159
```

**FIGURE 4** continued.

quence from the GenBank® database. Remember, there should only be one GET command per line in the body of the e-mail message. For protein sequences, only accession numbers from the SWISS-PROT library can be used. To retrieve a protein sequence, the command GET prot:SWISS-PROT ACCESSION NUMBER should be used.

For aligning multiple sequences by computer, the sequences of the dataset have to conform to a specific format dictated by the program that you choose to use. The documentation that

```
             G++ +++DDL+ATGG+L AA  L+ +VG  +VE   V+
Sbjct:   124 GQKVVVVDDLLATGGSLVAATELIRKVGGVVVESLVVV 161

 Score = 60, Expect = 0.65, Poisson P = 4.8e-06, length = 13

Query:    22 PDFPKPGIMFQDI 34
             P+FPK GI+F+DI
Sbjct:    22 PNFPKEGILFRDI 34


>A30148 Uridine monophosphate synthase - Human
        Length = 480

 Score = 80, Expect = 0.00088, Poisson P = 0.00088, length = 38

Query:   117 GAVEPGERAIIIDDLIATGGTLAAAIRLLERVGVKIVE 154
             G+++PGE  +II+D++++G+++  ++ +L++ G+K+ +
Sbjct:   111 GTINPGETCLIIEDVVTSGSSVLETVEVLQKEGLKVTD 148


>S08691 Dihydrofolate reductase/thymidylate synthase -
        Length = 520

 Score = 68, Expect = 0.046, Poisson P = 0.045, length = 37

Query:   111 TIEMHVGAVEPGERAIIIDDLIATGGTLAAAIRLLER 147
             T+E ++++   G+RA   +D++   G LA A+RLL R
Sbjct:   107 TVEELLAPLPEGQRAAAAQDVVVVNGGLAEALRLLAR 143


>RDLNTS Dihydrofolate reductase/thymidylate synthase -
        Length = 520

 Score = 68, Expect = 0.046, Poisson P = 0.045, length = 37

Query:   111 TIEMHVGAVEPGERAIIIDDLIATGGTLAAAIRLLER 147
             T+E ++++   G+RA   +D++   G LA A+RLL R
Sbjct:   107 TVEELLAPLPEGQRAAAAQDVVVVNGGLAEALRLLAR 143


>S08660 Dihydrofolate reductase/thymidylate synthase -
        Length = 520

 Score = 64, Expect = 0.17, Poisson P = 0.16, length = 37

Query:   111 TIEMHVGAVEPGERAIIIDDLIATGGTLAAAIRLLER 147
             T+E ++++     RA   +D++   G LAAA+RLL R
Sbjct:   107 TVEELLAPLPEEKRAAAAQDIVVVNGGLAAAVRLLAR 143


>RTHYG Hypoxanthine phosphoribosyltransferase - Chinese
        Length = 217

 Score = 59, Expect = 0.91, Poisson P = 0.60, length = 38

Query:   122 GERAIIIDDLIATGGTLAAAIRLLERVGVKIVECACVI 159
             G   +I++D+I TG T+ + + L+ R  +K+V  A ++
Sbjct:   126 GKNVLIVEDIIDTGKTMQTLLSLVKRYNLKMVKVASLL 163
```

**FIGURE 4** continued.

accompanies each program describes how the input file must be organized. In all instances, it is necessary to edit the files that contain sequences that were retrieved from the databases. Using a word processing program on your personal microcomputer is a relatively easy way to prepare the input files for most multiple sequence alignment programs. You can either save the e-mail messages that contain the retrieved sequences from the capture buffer of your communications program to disk or download the files from the mainframe to your personal computer using the program Kermit. Alternatively, although "editors" on mainframes tend to be primitive, they can be useful for preparing a text file that only contains the sequence information with no comments or numbers, etc. for each member of the dataset.

It is preferable, in most cases when many sequences are to be aligned, to use a mainframe computer to take advantage of the relatively short computation times and the extended memory. To set up the programs on your mainframe computer, you will need the source codes. Usually, the source codes can be obtained by e-mail from the authors. As well, many of the programs can be acquired from software collections maintained by various file servers.

After obtaining the source code, an executable form of the program must be installed on your mainframe computer. The documentation that accompanies the program will explain how it should be implemented. Initially, you will have to compile the source code. Your system manager can be helpful with amending the source code, if necessary, to make it compile properly on your system. Currently, a number of multiple sequence alignment programs are bundled with tree-building software. However, in most cases, these programs will create a separate output file with the aligned sequences. Inevitably, these output files will have to be edited with a word processor to remove extraneous information. Here, only a few of the available multiple sequence alignment programs will be briefly described. For more details about aligning nucleic acid and protein sequences, the volume edited by R. F. Doolittle[3] should be consulted.

The progressive sequence alignment (PSA) procedure[15] provides a means of aligning either nucleic acid or protein sequences. However, to use the PSA programs, some knowledge of tree-building is required. The documentation provided with the programs is excellent, and the authors have presented a detailed discussion of their procedures.[15] Briefly, the process for PSA multiple alignments entails

1. Formatting each individual sequence of the dataset with the program FORMAT (the input file for FORMAT is a text file of the sequence only with no comments),
2. Assembling each of formatted sequence files into one file,
3. With the program SCORE, determining the scores from pairwise alignments and constructing an initial branching order,
4. Arranging subclusters of sequences identified by SCORE into files with the program PREALIGN,
5. Creating a file with the single formatted and prealigned clusters in the order dictated by the branching order determined by SCORE,
6. Progressively aligning the sequences of the "ordered" file with either the program ALIGN or TREE.

The program ALIGN only does multiple sequence alignments, whereas TREE also computes the final branching order and branch lengths for the molecular tree. With one of the output files from either ALIGN or TREE, the program MULPUB will create a file of the sequence of the dataset aligned in an aesthetically pleasing form (Figure 5).

With less than ten sequences, the PSA procedure is not too cumbersome; however, with many sequences (>20) the programs SCORE, TREE, and ALIGN tend to be slow and should be run as batch jobs. Your computer system manager will explain how you can direct the program to do the analysis in this mode.

The program CLUSTALV has a multiple sequence alignment function that works well with large numbers of either nucleic acid or protein sequences. This program is available from NETSERV@EMBL-HEIDELBERG.DE for DOS, UNIX, VAX, or MAC operating systems. For example, to get the UNIX-based version, send the e-mail message GET UNIX_SOFTWARE:CLUSTALV.UAA to the EMBL Network File Server. You will receive

```
            *                       *  *  * *    * *         *
mouse       MSEPE     LKLVARRIRVFPDFPIPGVLFRDISPLLKDPDSFRASIRL
hamster     MAESE     LQLVAQRSAVSPTSPSPGVLFRDISPLLKDPASFRASIRL
human       MADSE     LQLVEQRIRSFPDFPTPGVVFRDISPVLKDPASFRAAIGL
arabidop    MATEDVQDPRIAKIASSIRVIPDFPKPGIMFQDITTLLLDTEAFKDTIAL
E. coli     MTATAQQ   LEYLKNSIKSIQDYPKPGILFRDVTSLLEDPKAYALSIDL
Drosophila  MSPSISAEDKLDYVKSKIGEYPNFPKEGILFRDIFGALTDPKACVYLRDL

                             *   *** *    *   *      **  **
mouse       LASHLKSTHSGKIDYIAGLDSRGFLFGPSLAQELGVGCVLIRKQGKLPGP
hamster     LASHLKSTHGGKIDYIAGLDSRGFLFGPSLAQELGLGCVLIRKRGKLPGP
human       LARHLKATHGGRIDYIAGLDSRGFLFGPSLAQELGLGCVLIRKRGKLPGP
arabidop    FVDRYKDK   GISVVAGVEARGFIFGPPIALAIGAKFVPMRKPKKLPGK
E. coli     LVERYKNA   GITKVVGTEARGFLFGAPVALGLGVGFVPVRKPGKLPRE
Drosophila  LVDHIRES APEAEIIVGLDSRGFLFNLLIATELGLGCAPIRKKGKLAGE

                 * ****    *    *  **      *** ****   *   *
mouse       TVSASYSLEYGKAELEIQKDALEPGQRVVIVDDLLATGGTMFAACDLLHE
hamster     TVSASYALEYGKAELEIQKDALEPGQKVVVVDDLLATGGTMCAACELLGQ
human       TLWASYSLEYGKAELEIQKDALEPGQRVVVVDDLLATGGTMNAACELLGR
arabidop    VISEEYSLEYGTDTIEMHVGAVEPGERAIIIDDLIATGGTLAAAIRLLER
E. coli     TISETYDLEYGTDQLEIHVDAIKPGDKVLVVDDLLATGGTIEATVKLIRR
Drosophila  VVSVEYKLEYGSDTFELQKSAIKPGQKVVVVDDLLATGGSLVAATELIRK

                          *  * *   *         *
mouse       LRAEVVEC VSLVELTSLKGRERLGPIPF  FSLLEYD
hamster     LQAEVVEC VSLVELTSLKGREKLGSVPF  FSLLQYE
human       LQAEVLEC VSLVELTSLKGREKLAPVPF  FSLLQYE
arabidop    VGVKIVEC ACVIELPELKGKEKLGETSL  FVLVKSAA
E. coli     LGGEVADA AFIINLFDLGGEQRLEKQGITSYSLVPFPGH
Drosophila  VGGVVVESLVVVMELVGLEGRKRLDGKV   HSLIKY
```

**FIGURE 5.** The output of the program MULPUB from Feng and Doolittle[15] with six APRT protein sequences. The asterisks denote completely conserved residues.

the program in two parts by return e-mail. The program has been encoded and compacted to facilitate electronic transfer. By sending the e-mail message HELP UNIX_SOFTWARE to NETSERV@EMBL-HEIDELBERG.DE, you will receive information about decoding and unpacking the UNIX programs. You may have to obtain the decoding program UUD.C from EMBL if your resident undecoding program, i.e., uudecode, does not extract all the CLUSTALV files. To do this, send the message GET UNIX_SOFTWARE:UUD.C to NETSERV@EMBL-HEIDELBERG.DE. Before compiling UUD.C, either delete or 'comment out' the mail header. If UUD.C does not compile, then contact your system manager. The required change(s) in the source code will probably be made quickly and easily by an expert. The documentation with the program will describe how to install CLUSTALV on your computer and how to use it.

The program TreeAlign[16] aligns sets of related nucleic acid or protein sequences. With a specifically formatted input sequence file and a simple set of instructions in a file called par.dat, the program ALIGN (*sensu* Hein) will produce an output file with the extension .ali. This file contains the multiply aligned sequences. For further use, it will probably be necessary to edit by means of a word processor the ".ali" file to isolate only the aligned sequences free of extraneous data. Because TreeAlign does more than align multiple sequences, with large data sets (>10 sequences) it is a slow running program and should be processed as a batch job. The program TreeAlign is available by means of 'anonymous ftp' from the ftp address ftp.bio.indiana.edu where it is in the 'align' directory. TreeAlign is also in various software collections that send programs by e-mail.

The Genetic Data Environment (GDE) program uses the CLUSTALV algorithm to align multiply either nucleic acid or protein sequences. The GDE set of programs is available by anonymous ftp from the address golgi.harvard.edu where the path to its location is pub/GDE_Release_1.0. The DOC directory, which is in the GDE_Release_1.0 directory at this ftp site, contains the documentation files for the GDE program. Currently, GDE is configured for the SUN family of workstations, including Sun 3 and Sun 4 (Sparcstation) systems using OpenWindows 2.0 or MIT X11R4. A version for the Silicon Graphics Iris system is available at the same ftp address as GDE for the SUNsystems, but it is located in the 'contrib' directory which is within the 'pub' directory.

A set of programs[17] written in the programming language C for multiple sequence alignments of proteins that is fast and can conduct a number of aligning operations including creating pretty alignments is available from the authors[17] by e-mail from VINGRON@EMBL.BITNET or from the EMBL software collection. Finally, the program Multiple Alignment Construction and Analysis Workbench (MACAW) is designed to locate blocks of sequence segments that are highly similar and, as part of its operations, it can multiply align up to sixteen protein sequences.[18] MACAW is designed to run on a DOS personal computer that has 2 megabytes of memory, mouse control, and graphics display with Microsoft Windows v. 3.0.

## IV. MOLECULAR TREE-BUILDING PROCEDURES

Multiple sequence alignments provide data for inferring molecular trees that may reflect the historical relationships among a set of sequences.[3] The approach taken here will be to describe how such trees can be constructed using distance matrix, parsimony, and "simultaneous alignment and phylogeny" strategies. There are a number of other tree-building procedures that use aligned sequences,[3,19] but these will not be discussed.

Briefly, distance matrix methods use some measure of similarity (or dissimilarity) of the aligned sequences of the dataset. On the basis of pairwise scores, a matrix is formulated which, in turn, is used by a tree-building algorithm to create an inferred topology of relatedness of the sequences of the dataset. Parsimony methods use sequences of the dataset, attempt to infer the most likely ancestral ("common", "shared") sequences, and reconstruct trees with the minimum number of presumed changes leading to each branch of the tree.

### A. DISTANCE MATRIX TREE-BUILDING USING MULTIPLE SEQUENCE ALIGNMENTS

Feng and Doolittle[15] have created a suite of computer programs, which are available from them, for aligning nucleic acid and protein sequences and determining a similarity score for each pairwise alignment. With these values in the form of a matrix, which is one of the output files of the program SCORE, a file can be formulated that can be utilized by the Neighbor-Joining (NJ) program.[20] The source codes and a PCDOS form of the NJ programs are available from Dr. L. Jin (Genetics Centers, Graduate School of Biomedical Sciences, The University of Texas Health Science Center at Houston, P.O. Box 20344, Houston, Texas, 77225, USA). The documentation accompanying the NJ package clearly explains how to prepare the input file for the program NJTREE. The output file of the program NJTREE is used by the NJDRAW program to create both an ASCII file that prints out on a dot matrix printer the derived tree and a HP Graphics language file that can be used to plot the tree on a Hewlett-Packard plotter. The NJ programs are very fast and can be used either on a mainframe or a PCDOS computer. The output file from the program NJTREE provides a step-by-step record of the tree-building process including the branch lengths for all nodes. The output files from NJDRAW do not always record all of the branch lengths, especially if the branches are very short although the NJTREE output file provides these values.

It is also possible, albeit using more computer time, to use the Feng and Doolittle programs to create Fitch-Margolish distance matrix trees. Briefly, the branching order that is determined by the program SCORE is used to ascertain which sequences are "clustered". These nested sequences must be "prealigned" using the program PREALIGN. Then with the order of sequences dictated by the results from SCORE, an input file for the program TREE is created. With TREE, the sequences are multiply aligned, and a final branching order with branch lengths is calculated.

The plotting capabilities of the Feng and Doolittle package are limited. Consequently, it may be necessary to create the final tree with a drawing program. The program TREEDRAW is an excellent tree drawing program for use with Macintosh™ computers. This program can be obtained by anonymous ftp from ftp.bio.indiana.edu located in molbio/mac directory or by sending an e-mail message GET MAC_SOFTWARE:TREEDRAW.HQX to NETSERV@EMBL-HEIDELBERG.DE. From either source, the program must be decoded and uncompressed before it can be used. To carry out these steps, you will need the program BinHex 4.0 (or a compatible alternative) to carry out the ASCII conversion and the program STUFFIT 1.6 (or an equivalent alternative, e.g., SitExpand) to uncompress the application. 'unBinHexing' precedes 'unstuffing'. TREEDRAW is a HyperCard application and has on-board documentation. Unless you are adept at working with the New Hampshire Standard for representing trees, then for large trees it is best to build up the tree by successive branches, because if you make a mistake with a bracket, colon, or comma, the program will crash, and you will have to reboot and start over again.

The program CLUSTALV, as noted above, carries out multiple sequence alignments. It also creates an input file for NJ analysis. There is an option for calculating the confidence limits of the nodes in the tree. Source code for CLUSTALV is available for compiling from the EMBL Net Server, and the programs can be implemented for both mainframe and personal computers. Final, "pretty" trees may have to be drawn with TREEDRAW or a similar tree drawing program.

The GDE suite of programs has a tree-building function. In this case, the De Soete algorithm[21] has been implemented to carry out least squares additive tree analysis. A tree drawing utility (TreeTool) is also part of this package.

## B. SIMULTANEOUS ALIGNMENT AND PHYLOGENY

Hein[16] has devised a program that aligns either nucleic acid or protein sequences and uses a fast approximation algorithm to generate a parsimonious tree. The package is easy to implement on a Silicon Graphics (IRIX [UNIX]) computer and other mainframe computers. The main program ALIGN should be run as a batch job. There are two output files produced by ALIGN. One contains the aligned sequences and other relevant data. The other file has the tree drawn with branch lengths and nodes demarcated. This "tree" file is not useful for the presentation of a final tree. Furthermore, the package does not have an on-board tree drawing program; consequently, TREEDRAW is a handy adjunct program. Figure 6 shows a phylogenetic tree of six APRT protein sequences drawn using TREEDRAW and a standard computer paint/draw program. The documentation that comes with TreeAlign explains how to install the programs, prepare the proper input file, and interpret the output data. In its present form with protein sequences, ALIGN will handle up to 38 sequences of about 500 amino acids each.

## C. Parsimony Tree-Building

The PHYLIP 3.4 suite of programs is distributed by Dr. J. Felsenstein (Department of Genetics SK-50, University of Washington, Seattle, WA, 98195, USA). Both source code and PCDOS executable versions can be obtained by anonymous ftp at evolution.genetics.edu in the 'pub' directory. PHYLIP is an acronym for *phylo*genetic *i*nference *p*ackage. The

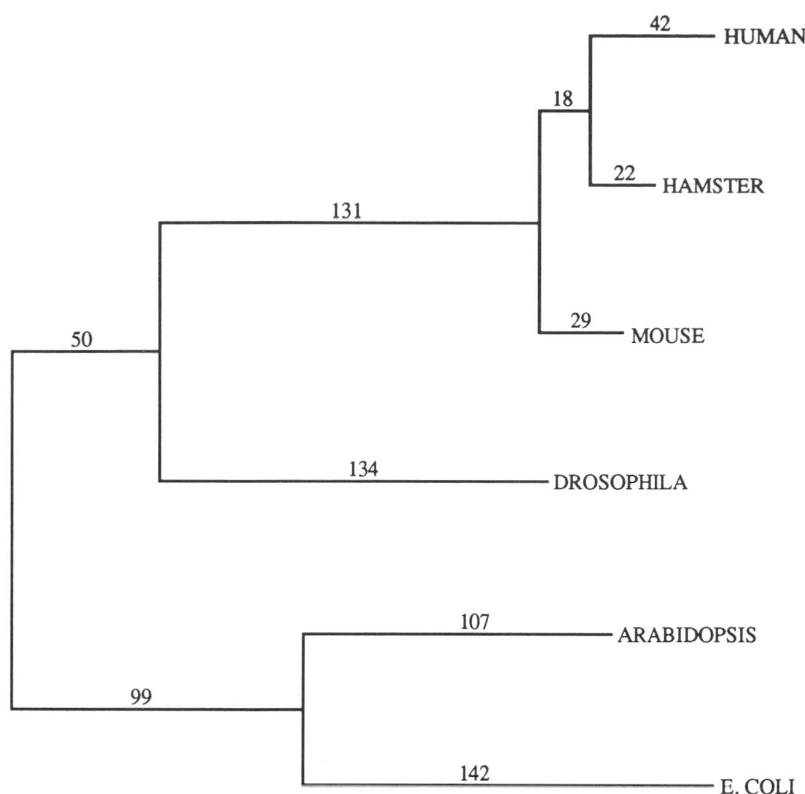


**FIGURE 6.** Molecular phylogeny of six APRT protein sequences. The APRT tree including branch lengths was constructed using TreeAlign.[16] The tree was drawn using TREEDRAW, and the numbers designating the branch lengths were added using a standard Macintosh drawing program.

documentation that accompanies the programs is thorough and extremely helpful. The programs can be implemented on a number of different computers, and Dr. Felsenstein explains how this can be done. The DNAPARS and PROTPARS programs are for inferring parsimonious trees from nucleic acid and protein sequences, respectively. The input file ("infile") for these programs must be properly formatted. Using the aligned sequence file from Hein's TreeAlign, it is possible by editing with a word processor to create a PHYLIP infile. The program CLUSTALV creates a file with the PHYLIP format. Similarly, the Vingron and Argos multiple sequence alignment program produces an infile for PROTPARS.

The computer running time for the PHYLIP programs can be very long with large numbers of lengthy sequences. Therefore, the DNAPARS and PROTPARS runs should be performed as batch jobs. On the other hand, SEQBOOT and CONSENSE, for example, are very fast and can be conducted on-line. With the PHYLIP 3.4 system, it is necessary to rename files because, for some analyses, the output ("outfile") file from one program is the input ('infile') for another program. With a modicum of planning, it is not difficult to run a series of analyses efficiently. PHYLIP 3.4 has tree drawing capabilities which include options for a number of plotting devices.

The *P*hylogenetic *A*nalysis *U*sing *P*arsimony (PAUP) program will create a minimum parsimony tree(s), i.e., a tree(s) with the least number of changes, from nucleic acid sequence datasets with a PCDOS computer equipped with a math coprocessor. In its original form, the input data for the PAUP program must either be passed through the program REDUCSEQ, which is available from Dr. D. C. Swofford (see below) or be prepared by the user. The PAUP package does not multiply align sequences; therefore, the sequence dataset must be aligned prior to using PAUP. The program REDUCSEQ and the PAUP package can be obtained from Dr. D. C. Swofford (Illinois Natural History Survey, 607 Peabody Dr., Champaign, Illinois, 61820, USA) at a modest price. A new version of PAUP is available for MacIntosh™ computers that supports both nucleic acid and protein sequences.

## V. A CAUTIONARY NOTE

Molecular tree-building is not an absolute science. The same dataset of sequences with different algorithms can produce different trees, whereas the same methodology with different datasets from the same taxa can generate noncongruent trees. It is also important to have some indication of the reliability of the tree(s) that a program creates. No single methodology can, with certainty, produce the biologically correct tree. With ten sequences, there are over $2 \times 10^6$ different unrooted bifurcating trees. The algorithms narrow down the possibilities enormously, and measures of reliability (e.g., bootstrapping) help to establish some degree of heuristic confidence in a particular tree. Notwithstanding the adoption of a cautious approach, molecular trees can help determine which sequences of a dataset are closely related and which sequences only share minimal similarity. This type of information, in turn, can focus studies on specific subsets of the original set of sequences. There is also the possibility that unexpected molecular relationships may be revealed by this type of analysis.

## REFERENCES

1. **von Heinje, G.,** *Sequence Analysis in Molecular Biology: Treasure Trove or Trivial Pursuit?,* Academic Press, San Diego, 1987.
2. **Bishop, M. J. and Rawlings, C. J.,** *Nucleic Acid and Protein Sequence Analysis: A Practical Approach,* IRC Press, Oxford, 1987.
3. **Doolittle, R. F.,** Molecular evolution: computer analysis of protein and nucleic acid sequences, *Meth. Enzymol.,* Vol. 183, 1990.
4. **Argos, P., Vingron, M., and Vogt, G.,** Protein sequence comparison: methods and significance, *Protein Eng.,* 4, 375, 1991.
5. **von Heinje, G.,** Computer analysis of DNA and protein sequences, *Eur. J. Biochem.,* 199, 253, 1991.
6. **Fuchs, R., Stoehr, P., Rice, P., Ormond, R., and Cameron, G.,** New services of the EMBL data library, *Nucl. Acids Res.,* 18, 4319, 1990.
7. **Burks, C., Cassidy, M., Cinkosky, M. J., Cumella, K. E., Gilna, P., Hayden, J. E.-D., Keen, G. M., Kelley, T. A., Kelly, M., Kristofferson, D., and Ryals, J.,** GenBank, *Nucl. Acids Res.,* 19 (Suppl. 2221), 1991.
8. **Stoehr, P. and Cameron, G. N.,** The EMBL data library, *Nucl. Acids Res.,* 19 (Suppl. 2227), 1991.
9. **Barker, W. C., George, D. G., Hunt, L. T., and Garavelli, J. S.,** The PIR protein sequence database, *Nucl. Acids Res.,* 19 (Suppl. 2231), 1991.
10. **Bairoch, A. and Boeckmann, B.,** The SWISS-PROT protein sequence data bank, *Nucl. Acids Res.,* 19 (Suppl. 2247), 1991.
11. **Pearson, W. R. and Lipman, D. J.,** Improved tools for biological sequence comparison, *Proc. Natl. Acad. Sci. U.S.A.,* 85, 2444, 1988.
12. **Pearson, W. R.,** Rapid and sensitive sequence comparison with FASTP and FASTA, *Meth. Enzymol.,* 183, 63, 1990.
13. **Altschul, S. F., Gish, W., Miller, W., Myers, E. W., and Lipman, D. J.,** Basic local alignment search tool, *J. Mol. Biol.,* 215, 403, 1990.
14. **Karlin, S. and Altschul, S. F.,** Methods for assessing the statistical significance of molecular sequence features by using general scoring schemes, *Proc. Natl. Acad. Sci. U.S.A.,* 87, 2264, 1990.
15. **Feng, D.-F. and Doolittle, R. F.,** Progressive alignment and phylogenetic tree construction of protein sequences, *Meth. Enzymol.,* 183, 375, 1990.

16. **Hein, J.,** Unified approach to alignment and phylogenies, *Meth. Enzymol.,* 183, 626, 1990.
17. **Vingron, M. and Argos, P.,** A fast and sensitive multiple sequence algorithm, *CABIOS,* 5, 115, 1989.
18. **Schuler, G. D., Altschul, S. F., and Lipman, D. J.,** A workbench for multiple alignment construction and analysis, *Proteins Struct. Funct. Genet.,* 9, 180, 1991.
19. **Lake, J. A.,** A rate-independent technique for analysis of nucleic acid sequences: evolutionary parsimony, *Mol. Biol. Evol.,* 4, 167, 1987.
20. **Saitou, N. and Nei, M.,** The neighbor-joining method: a new method for reconstructing phylogenetic trees, *Mol. Biol. Evol.,* 4, 406, 1987.
21. **De Soete, G.,** A least squares algorithm for fitting additive trees to proximity data, *Psychometrika,* 48, 621, 1983.

# 17

# DNA Mapping in Plants

*Benoit S. Landry*

## I. INTRODUCTION

This chapter describes some of the current approaches and techniques used for RFLP and PCR-based genetic mapping in plants. Detailed description of only the protocols most sensitive to variations will be provided. Other protocols will only be referred to since they are very well explained in the literature and not sensitive to experimental variance.

### A. RFLP-BASED DNA MAPPING

Genetic analysis of plant genomes has been revolutionized in recent years with the use of molecular genetics tools for rapidly developing a large number of genetic markers. In this approach, cloned DNA sequences are used to probe specific regions of a genome for the presence of variations at the DNA sequence level. These variations are detected by restriction endonucleases and revealed by separating DNA fragments according to size by electrophoresis. Polymorphism is seen as differences in the length of genomic DNA fragments homologous to a radioactively labeled cloned DNA sequence. Such variation in fragment length between individuals has been termed restriction fragment length polymorphism (RFLP).[1]

RFLP markers were first used as a tool for genetic analysis in 1974 when a temperature sensitive mutation of adenovirus was associated with a specific RFLP.[2] Other studies demonstrated that DNA differences could be detected directly for small genomes (e.g., mitochondria) or following hybridization with specific sequences for larger genomes.[3,4] The first use of RFLP as a genetic marker in the analysis of disease was in respect to sickle-cell anemia.[5] Botstein et al.[1] later coined the term RFLP and described the theoretical basis of this method and its practical applications for mapping genes associated with disease in humans. Since then, RFLP analysis has been performed for many organisms, including most of the plant crops, to construct detailed genetic linkage maps.[6-8]

### B. PCR-BASED DNA MAPPING

During the accelerated plant genome mapping effort with RFLP markers, a new DNA analysis procedure was developed in 1985; it was called the "Polymerase Chain Reaction/Oligomer Restriction" (PCR/OR) method.[9] This procedure was originally developed for the detection of the sickle-cell mutation in humans.[9] It was rapid and several orders of magnitude more sensitive than standard Southern blotting procedures. In this approach, a known polymorphic restriction site in a DNA sequence is specifically amplified, using a pair of synthetic oligonucleotide primers complementary to alternate strands of the target genomic DNA sequence and DNA polymerase ("PCR" reaction). Double stranded DNA target is melted by raising the temperature, and the reaction mixture is then cooled to allow annealing of the primer to the targeted sequence. The temperature is raised to reach the optimum activity of

0-8493-5164-2/93/$0.00+$.50
© 1993 by CRC Press, Inc.

the polymerase, which then begins synthesis of new strands of DNA on each of the original single stranded DNA molecules. Each temperature cycle doubles the amount of target DNA. After 30 cycles of denaturation, annealing, and polymerization, the theoretical amplification of the original target sequence is in the order of $10^9$.

In the original experiment, solution hybridization of the amplified sequence with an end-labeled complementary probe was followed by digestion of the double-stranded product, with the restriction endonuclease detecting the polymorphism ('OR' reaction). Polyacrylamide gel electrophoresis of the resulting oligomers, followed by autoradiography, revealed the polymorphism. $\beta^a$ (wild type) and $\beta^S$ (sickle-cell anemia) alleles could be distinguished with 1 μg of human genomic DNA in less than 10 h. A small amount (0.5 ng) of low molecular weight, genomic DNA also gave excellent results. This method promised to be generally applicable for other genes and species.

The isolation and commercial availability of a thermostable DNA polymerase greatly improved the original PCR reaction.[10] Since then, more than a thousand scientific publications[11] report the use of the PCR method, and several books describe the methodologies and potential applications.[12,13] It has revolutionized most of the techniques and approaches used to manipulate DNA. The sensitivity of the current technique permits direct visualization on gels of amplified DNA obtained from a single copy sequence using a single cell as starting material.[14]

DNA mapping has not been immune to the sudden changes in analytical techniques that have been generated by PCR. RFLP markers are now being sequenced to identify suitable primers that will allow the amplification of specific polymorphisms in DNA segments of known sequence; these primers are called "Sequence Tag Sites" (STS).[15] More recently, a new type of genetic marker called RAPD (Randomly Amplified Polymorphic DNA) also has evolved from PCR technology.[16] In this approach, a single oligonucleotide primer (typically a 10-mer) is used to direct the amplification instead of the two longer oligonucleotides generally used for targeted PCR reactions.

RAPD and STS markers are promising tools for plant breeders since they do not bear the technical inconveniences of RFLP markers. Only a very small amount of genomic DNA (25 ng) is required, and the analytical process is fast and relatively simple. Radioactive isotopes are not needed to visualize DNA polymorphisms, and only modest laboratory equipment and personal training are necessary. The general dominance of RAPD markers, however, is a disadvantage, but the ease by which saturated genetic maps are constructed with them partially overcomes this problem.[16] STS and possibly RAPD appear to have the potential of being used directly as a selection tool for desirable genotypes in breeding programs.

## II. PROCEDURES

### A. RFLP-BASED DNA MAPPING

The global approach used to construct plant genetic maps using RFLP markers is outlined in Figure 1. In the following section, each component of this process is examined.

#### 1. Preparation of the Probes

##### *a. Cloned Plant DNA Sequences and Plasmid Preparations*

Cloned plant DNA sequences are used as probes to detect polymorphism in specific regions of the genome. They can either be random genomic DNA sequences or cDNA sequences. The specific problems and advantages of different sources of probes have been discussed previously.[17] Cloned DNA sequences from other species can be used, but they should be conserved between species (e.g., ribosomal RNA coding sequences). Less homologous sequences require time-consuming adjustments in the stringency of the hybridization and the washes of the membranes and should be avoided unless necessary for the analyses. cDNA

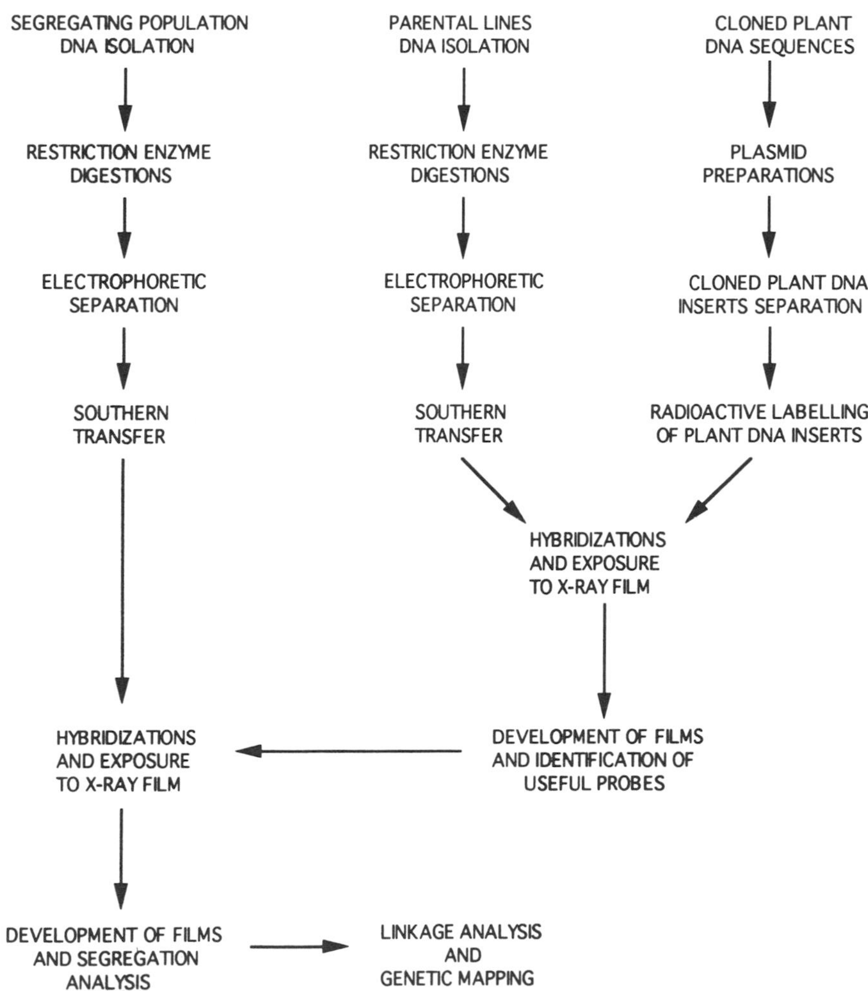


**FIGURE 1.** Experimental steps involved in the construction of a detailed genetic map using RFLP markers.

clones are preferable since they represent expressed DNA sequences and are generally single or low copies. Random genomic DNA clones need to be screened to eliminate highly repetitive sequences; this is a problem for plant species containing many repetitive sequences.[18]

The choice of the vector is critical. Plasmid vectors are preferred. Most of the current plasmid vectors are maintained in high copy number within the bacterial cell (relaxed replication). However, old vectors (e.g., pBR322 series) are under stringent replication within the host bacteria,[19] and it is difficult to obtain sufficient plasmid for probe preparation from a minipreparation; they should be avoided if possible. Lambda and M13-derived vectors are not recommended because of difficulties in isolating their DNA, instability of cloned inserts, and difficulties in storing them, or because of the excessively large size of their inserts.

Numerous protocols related to DNA and cDNA cloning and plasmid preparations are available in the literature, and one of the best sources of well-described procedures is Sambrook et al.[19] We currently use the protocol of Serghini et al. for plasmid preparation.[20] A plasmid miniprep is generally sufficient for several probe labeling reactions.

#### *b. Cloned Plant DNA Insert Separation*

The use of only the plant DNA insert as a probe, instead of the whole plasmid, decreases the level of background hybridization on the membranes. Typically, we digest the plasmid with a restriction endonuclease that cuts the insert from the vector and separates the fragments by gel electrophoresis. Our original procedure, used for several years, was a version of the "freeze-squeeze" protocol[21] modified to make it simple and to handle more than 40 different inserts in a single experiment.[7] Special kits, designed to separate DNA from agarose, are now available commercially, and we prefer using them since they are less time consuming than the freeze-squeeze procedure. The kit we currently use is the Geneclean™ system by Bio101 (LaJolla, CA). Standard protocols are provided by the manufacturer.

#### *c. Radioactive Labeling of Plant DNA Inserts*

Originally, we labeled our DNA probes by "nick-translation" of the inserts in the presence of radioactive nucleotides.[22] We now use commercially available kits based on "random primers" to initiate the nucleotide replacement reaction. An improved version of these random primer kits (QuickPrime™ by Pharmacia, Sweden) uses T7 DNA polymerase, an enzyme that catalyzes a high rate of nucleotide incorporation. The labeling reaction takes only 10 min, and unincorporated nucleotides do not need to be removed from the reaction. Specific activity can reach $1 \times 10^9$ dpm/μg. We routinely label 10 to 40 DNA inserts simultaneously, which will be hybridized to duplicate membranes of either the parental lines or the segregating population. Labeled probes are placed in a boiling water bath for 5 min immediately before hybridization.

### 2. Preparation of Plant DNA for RFLP Analyses

#### *a. DNA Isolations*

Numerous protocols to extract plant DNA have been published in recent years.[23-26] We have tried most of the existing protocols and adopted a basic "CTAB" DNA extraction procedure.[26] This protocol is simple and time efficient, but not all DNA samples prepared this way digest well with every restriction enzyme. To avoid these problems, we modified the "CTAB" protocol by adding a rapid CsCl purification step to the original procedure, by adjusting the buffers to make it suitable for several plant species, and by improving the methodology generally used to perform some steps.

We routinely use this protocol to isolate large amounts of DNA from parental and segregating populations destined to be analyzed with several RFLP markers. The protocol detailed below minimizes the probability of missing data and errors in scoring due to partial digestion that we experienced with other procedures. The step-by-step protocol is as follows:

1. Harvest approximately 50 g of leaf material and freeze immediately at –80°C in a plastic bag. One person can handle 12 different DNA extractions simultaneously.
   For species that accumulate high levels of starch in leaf tissues, it is preferable to place the plants in the dark for 48 h at room temperature prior to collecting the leaves. Young, healthy leaves always yield more DNA. Leaf tissues must be stored at –80°C immediately after harvesting; if stored at –20°C, DNA quality and yield decrease very rapidly.
2. Place opened 250 ml GSA (Sorvall) centrifuge bottles or equivalent on ice and fit a large polypropylene funnel on top. Place one layer of miracloth and then four layers of cheese cloth in the funnel. Wet cheese cloth with 15 ml of extraction buffer.

Funnels (Nalgene, 150 mm diameter) fit snugly inside the neck of the bottle providing a stable support for the sample.

3. Measure approximately 200 ml of ice-cold extraction buffer in a beaker, and pour inside a chilled blender. Take a frozen leaf bag from the freezer, and crush the tissues by squeezing the bag a few times. Do not let the tissues thaw. Discard large vascular tissues (midribs), and drop tissues in the blender. Grind at maximum speed for 20 to 30 sec or until the homogenate has the texture of a sauce.
   Once the frozen tissues are removed from the freezer, they must be processed immediately. The use of a variable rheostat to increase slowly the speed of the blades minimizes splashing and loss of leaf material. This also facilitates rinsing of the blender between samples. The blender is rinsed with distilled water and drained of excess liquid by placing it upside down on paper towels.
4. Immediately pour the homogenate over the cheese cloth into the funnel. Repeat Steps 3 and 4 for two additional samples.
   Processing only three samples simultaneously provides sufficient time to let the homogenate begin filtering through the cheese cloth while not warming up excessively.
5. Hold the corners of the cheese cloth, squeeze content by twisting the ends together onto the miracloth, and discard. Gently squeeze the miracloth, and discard.
6. Centrifuge at 3800 rpm at 4°C for 10 min in a GSA rotor (Sorvall) or equivalent, and continue processing other samples during centrifugation.
7. Discard the supernatant, add 35 ml of extraction buffer to the pellet, and resuspend the pellet.
   Handle the bottles gently when you take them out of the centrifuge to avoid disturbing the pellet. It is best to pour the supernatant carefully in a slow continuous movement since the pellet is very loose. It is easiest to resuspend the pellet with an artist's brush (size #9) by twisting the brush between the thumb and index finger. The pellet contains mostly intact cell nuclei and cell debris.
8. Add 35 ml of lysis buffer and 15 ml of sarkosyl solution. Mix by inverting a few times, and incubate at 60°C for 15 min in a waterbath.
   This step lyses the remaining cells and nuclei. The DNA is now free in the solution and sensitive to shearing if shaken too hard or pipetted too rapidly. The solution should become somewhat viscous at this step depending on the quantity and quality of DNA.
9. Add 100 ml of water-saturated chloroform:isoamyl alcohol (24:1), and invert several times to form an emulsion. Centrifuge at 7000 rpm, 4°C for 10 min.
   If the solution contains a large concentration of DNA, it is more difficult to form the emulsion. The chloroform extracts chlorophyll, some proteins and polysaccharides, and most of the debris. After centrifugation, the upper layer should clarify to a light yellow-green color; the debris, polysaccharides, and proteins accumulate at the interface, and the chloroform should become dark green.
10. Recover the aqueous phase (upper layer) in another GSA bottle with a 25 or 50 ml pipette, and record the volume. Care should be taken not to aspirate debris present at the large interface.
11. Add 0.7 volume of isopropanol to the aqueous phase, rapidly invert the bottle twice, and place it on ice for 30 min.
    This process facilitates the formation of large DNA fibers during precipitation. If the bottle is shaken too hard, DNA fibers will not aggregate sufficiently to be spooled.
12. Swirl the solution in the bottle to aggregate the DNA fibers into one lump. Balance and centrifuge at 8000 rpm, 4°C for 10 min. After 10 min on ice, the bottle can be inverted a few more times to completely mix the solution.

It is often possible to collect the DNA clump without centrifugation using a small plastic pipette to hold the clump by aspiration and transferring it to another tube. This can be done only if a large clump of DNA is formed; we find that a significant amount of DNA often remains in solution and can be lost in this process.

13. Gently pour off the supernatant, and discard. Invert the bottle on paper towel to drain the pellet completely. Dry excess drops of liquid under vacuum, but do not let the pellet dry completely.
14. Add 4.5 ml of CsCl solution, and immediately detach the DNA pellet from the side of the bottle by swirling the bottle of its side. Once this is done, place the bottle at 4°C overnight to let the DNA resuspend gently.
    It is more difficult to resuspend DNA in the presence of a high concentration of CsCl, but we found that it protects DNA from remaining DNAse activity during resuspension and that the overnight incubation is a more gentle procedure, does not require working time, and ensures uniform density of the DNA samples. The presence of RNAse in the CsCl solution degrades most of the RNA and therefore facilitates the recovery of the DNA after the ultracentrifugation.
15. Place the barrel of a 25 ml syringe fitted with a 16 gauge needle on top of a 5.3 ml Quickseal™ (Beckman) ultracentrifugation tube. Using it as a funnel, insert 270 μl of ethidium bromide solution into the tube. Pour in the DNA solution, and complete the volume with the CsCl Solution. Seal tubes, mix the content by rolling the tube on a table, and store it in the dark for no more than one month at room temperature.
    The use of the barrel of a syringe fitted with a needle as a funnel greatly facilitates filling the Quickseal™ tubes.
16. Centrifuge tubes in a vertical rotor (e.g., VTi80 rotor, Beckman) at 75,000 rpm, 15°C for 4 h or at 60,000 rpm, 15°C overnight.
17. Remove tubes from rotor, and visualize the DNA under UV light. Collect the DNA band by puncturing the side of the tube with a 5 ml syringe fitted with a #16 gauge needle. Transfer the DNA into a conical 50 ml disposable centrifuge tube.
18. Add 2.5 ml of 3 *M* sodium acetate buffer (pH = 6.0), and complete to 25 ml with 1X TE buffer. Fill tube with water-saturated isoamyl alcohol, and invert several times to mix phases. Let the phases separate (5 to 10 min), and pour off the organic phase (pink upper phase). Repeat until the aqueous phase has become clear. Pipette off the remaining organic phase after the last extraction.
    The addition of sodium acetate is to prepare for DNA precipitation. The volume is brought up to 25 ml to dilute the CsCl enough so that it will not interfere with the precipitation. Dilution of the DNA also facilitates extraction of the ethidium bromide.
19. Add 25 ml of isopropanol to the aqueous phase, and mix by inverting the tube several times. Centrifuge at 3800 rpm, 4°C for 30 min to collect the DNA precipitate at the bottom of the tube.
    A shorter and faster centrifugation can be used if the tube and the centrifuge can tolerate it.
20. Invert the tube on a paper towel to drain excess liquid, and add 10 ml of 70% ethanol solution to rinse the DNA. Centrifuge at 3800 rpm, 4°C for 10 min. Drain the pellet, and dry slightly under vacuum. Add 400 μl of 1X TE buffer to the pellets, and let resuspend overnight at 4°C. Determine the DNA concentration, and adjust it to either 0.5 μg/μl or 1.0 μg/μl with TE buffer.

The volumes of solutions indicated at the right are sufficient for 12 plant samples, with the exception of the CsCl and stock solutions.

**Extraction Buffer**

| | Concentration | For 2.5 l |
|---|---|---|
| Sorbitol | 50 m*M* | 227.50 g |
| Trizma base | 100 m*M* | 30.25 g |
| $Na_2$-EDTA·$2H_2O$ | 70 m*M* | 65.13 g |
| pH to 8.5 with HCl | | |
| *Just before use, add:* | | |
| Sodium MetaBisulfite | 20 m*M* | 9.5 g |

**Lysis Buffer**

| | | For 450 ml |
|---|---|---|
| 1 *M* Tris-HCl, pH 8 | Stock solution | 90 ml |
| 250 m*M* $Na_2$·EDTA, pH 8 | Stock solution | 126 ml |
| 5 *M* NaCl | Stock solution | 180 ml |
| Distilled $H_20$ | | 54 ml |
| *Just before use add:* | | |
| CTAB* | 55 m*M* | 9 g |

**Sarcosyl Solution**

| | |
|---|---|
| Sodium sarcosyl | 10 g |
| Sterile distilled $H_2O$ | 200 ml |

**CsCl Solution**

| | |
|---|---|
| CsCl | 97 g |
| TE, 1X | 99 ml |
| RNase solution | 1 ml |

**Stock Solutions**

| | | |
|---|---|---|
| NaCl | 5 *M* | 584.40 g/2000 ml |
| $Na_2$-EDTA·$2H_2O$ , pH 8 | 0.25 *M* | 93.05 g/1000 ml |
| Tris-HCl , pH 8 | 1 *M* | 121.10 g/1000 ml |

*Note:* Formulation and method of preparation for TAE, TE, RNAse, sodium acetate, and ethidium bromide solutions are as described elsewhere.[19]

* CTAB = Hexadecyl Trimethyl Ammonium Bromide

### 3. RFLP Analysis

#### a. Restriction Digestion, Electrophoretic Separation, and Southern Transfer

Restriction endonucleases are used as recommended by the manufacturers, except that at least a ten-fold excess of enzyme units is used to digest plant genomic DNA. We routinely scan the parental lines with four restriction enzymes to search for RFLP. They are *Bam*HI, *Eco*RI, *Eco*RV, and *Hin*dIII. During the construction of the RFLP map of lettuce,[27] we tested nine enzymes for their efficiency in detecting RFLP. The above four enzymes were identified as those detecting the highest level of RFLP, and they were the most reliable.[23]

DNA is separated on electrophoresis units specifically designed for the analysis of several samples. We currently use Owl Scientific Plastic units (Model A3; Cambridge, MA). They are very large units designed for RFLP segregation analysis and can accommodate 96 samples when three combs are used. This gives three gels, each approximately 12 cm long. Gels are electrophoresed overnight (45 V for 17 h) using TAE buffer. The electrophoretic conditions for both testing of parental lines with different probes and segregation analysis of polymorphic probes must be the same. The buffer must be changed after every electrophoresis since DNA leaks from the gel into the buffer and re-enters gels that are subsequently run on the same apparatus, thereby increasing background hybridization. Following electrophoretic separation, the gels are mounted for capillary transfer onto positively charged nylon membranes (Hybond-N+, Amersham or ZetaProbe, BioRad) using a standard capillary Southern transfer procedure.[19] Transfer is allowed to proceed overnight. We routinely operate two electrophoresis units (the equivalent of six gels) simultaneously. Commercially available vacuum or electrotransfer units are not as efficient as a capillary transfer since only one or two gels can be handled simultaneously, and this does not leave sufficient time during the day to prepare for the electrophoresis/Southern transfer experiment of the next day. After transfer, membranes are baked at 80°C *in vacuo* for 1 h. Membranes are then placed in plastic bags until needed for hybridization. If only two parental lines are being tested, membranes are cut appropriately to provide duplicates that will allow testing of several probes at the same time. For example, four restriction enzyme digestions of two parental line DNA samples permits the production of nine duplicate membranes with a single electrophoresis unit (eight samples plus two molecular weight markers per set and one empty well between each set). This ensures that all parental and progeny blots are identical.

#### *b. Hybridization*

Hybridizations are carried out at 42°C in the presence of 50% formamide and 10% dextran sulfate with the hybridization solution recommended by the nylon membrane manufacturer. Prehybridization and hybridization are performed in tubes placed in a rotary oven specifically designed for Southern hybridization. We use hybridization ovens from both Robbins (Sunnyvale, CA) and Hybaid (Middlesex, UK). For the Hybaid, we use the twin rotor model since it can accommodate 24 large tubes or 48 small tubes, each containing a different set of membranes being hybridized to a different probe. The use of this system is safer and simpler than the use of plastic bags or boxes.

#### *c. Segregation and Linkage Analysis*

The different strategies to construct segregating populations for genetic mapping are described in Figure 2. A basic principle of this approach is to segregate the highest number of important agronomic traits in the minimum number of crosses. Each additional cross necessitates screening of parental lines for RFLP and the use of a new population for segregation and linkage analyses; this is time-consuming and expensive. It is essential that single plants are used for both male and female parental lines and that a single F1 plant is used to generate the F2 population. Residual heterozygosity and heterogeneity within parental lines will not be tractable and will introduce errors in the map if more than one plant individual is used for male parent, female parent, or F1. This applies even if it is known that the parental lines are highly inbred lines. It is absolutely necessary if it is an outcrossing species.

The choice and size of the segregating populations are based on the purpose of the genetic mapping and the nature of the traits to be analyzed (qualitative or quantitative trait). For these reasons, segregating populations obtained from breeding programs rarely meet these criteria. Plant breeders traditionally are interested in developing efficient selection schemes that will allow rapid progress towards improvement of the numerous characters they have to deal with to produce a superior cultivar. The purpose of gene tagging is not to select a superior genotype. The structure of the segregating populations must generate extreme genotypes, not maximize the probability of identifying a superior individual. For QTL mapping, different crosses are needed to analyze separately each locus involved in a quantitative trait after most QTLs have been identified in a wide cross.

We use 90 F2 plants for general mapping purposes. When a chromosomal region of interest is found, we increase the number of plants for linkage analysis of markers located in this area to obtain a better resolution. Backcross, recombinant inbred, and doubled haploid populations can be used (Figure 2) but the genetic information per plant is lower.[28]

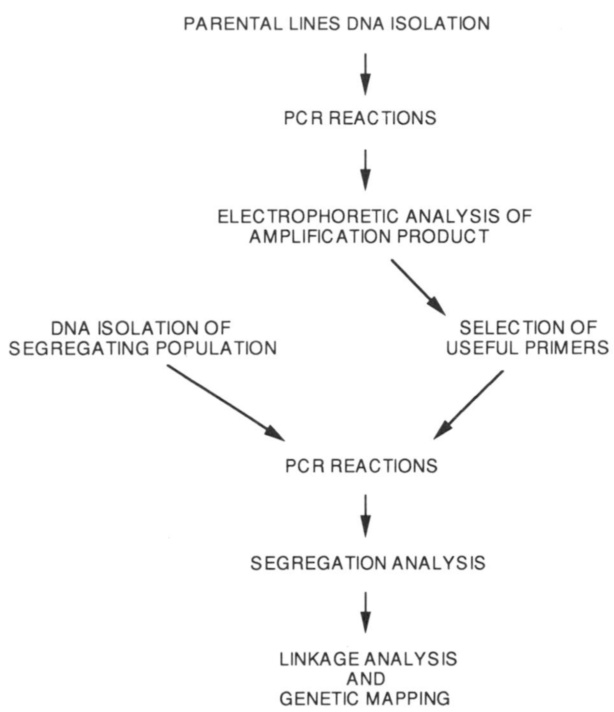


**FIGURE 2.** Experimental steps involved in the construction of a detailed genetic map using STS or RAPD markers.

Electrophoretic conditions to separate DNA of the segregating population should be the same as for DNA of the parental lines. When more than one restriction enzyme detects RFLP with a given probe, it is best to choose the enzyme which displays the clearest polymorphism for segregation analysis. The order of the DNA samples on the gel should be the same for each duplicate and enzyme. This is essential since at least ten probes are tested simultaneously on duplicate sets of membranes. We always record, on the X-ray film, the number of the membrane, the name of the probe, the enzyme used, the date, and the user. During the process of genetic mapping, hundreds of films may be produced and thousands of bands will be scored; errors of identification and scoring will be minimized using this approach.

It is best to use two persons to score X-ray films of the segregating population. Records are entered on a computer by one person while the other one reads the film out loud . Scoring should be done at least twice by a different person. Keep hard copies of the data in at least two different places. When a lane cannot be scored without ambiguity, it is best to consider it as missing data.

One of the best computer programs for linkage analysis is Mapmaker.[29-31] This program is simple to use (especially the Macintosh version) and provides a high degree of precision since a multipoint mapping approach is utilized to determine the linkage relationship between the markers. A modification of Mapmaker (Mapmaker-QTL) has been developed to implement the analysis of loci involved in quantitative traits. The current version of

Mapmaker-QTL runs only on a UNIX-based computer station, which is compatible with A/UX operating system from Apple Computers.

## B. PCR-BASED DNA MAPPING

DNA markers linked to important agronomic traits are rapidly becoming part of plant breeding programs to help in the identification and selection of desirable genotypes in segregating populations.[7,8,17,32-34] However, the general utilization of this approach by breeders is hampered by technical inconveniences of RFLP analysis, the most widely used approach to detect DNA variations. RFLP marker-assisted selection is time-consuming, requires access to a laboratory fully equipped for DNA manipulation, the use of radioactive isotopes, and personnel trained in molecular genetics. These requirements are generally not met by breeding teams. In addition, the RFLP technique necessitates large samples of plant tissue often available only from mature plants. This renders impossible early and nondestructive testing of young seedlings with RFLP markers.

Recently, a new type of genetic marker called RAPD (Random Amplified Polymorphic DNA) has evolved from PCR technology.[16] Although the procedures outlined below are for RAPD analysis, they can also be used for STS-based amplification.

RAPDs are becoming widely used to construct genetic maps of plant species[35] and to tag agronomic traits.[36] However, these genetic markers behave as dominant traits since a single band is either amplified or not amplified at one or more loci. Different alleles at a locus are rarely amplified; quantification of the amplified DNA band is currently not possible due to inherent variation in the efficiency of the amplification reaction between DNA samples. Consequently, dosage differences cannot be used to differentiate homozygotes from heterozygotes for the presence of an amplified band. An additional problem with RAPD markers is the impossibility of predicting whether the same locus will be amplified in different crosses. This generally limits the use of the maps constructed with RAPD markers to the original cross from which they were constructed; a RAPD marker mapping in one region of the genome can map to a completely different region in a different cross. Finally, the amplification reaction is not well understood and sometimes generates spurious results. Careful optimization of the reaction conditions, the utilization of positive and negative controls and testing the reproducibility of the amplification reaction are necessary before drawing conclusions from the results obtained. Nevertheless, this technique is promising and is being improved continuously to make it more reliable for genetic mapping. Figure 3 outlines the approach used to construct plant genetic maps using RAPD markers.

### 1. DNA Isolation from Parental Lines and Segregating Population

Crude DNA extracts from both the parental lines and the segregating population are generally suitable for RAPD analyses. The procedure previously presented (Section II.A.2.1) for DNA isolation from large leaf samples can be scaled down for small amounts of leaf material, and the CsCl purification steps (Steps 14 to 20) can be omitted. The spooled DNA is resuspended in TE-RNAse at the end of Step 13.

An immediate application of the RAPD markers is for plant DNA fingerprinting and seed certification. To apply this new technology, a plant DNA microextraction procedure suitable for RAPD and other PCR-based analyses had to be developed. The procedure needed to be very fast and simple, reproducible, and successful for a broad range of crops so it could be used by a breeding laboratory not equipped for RFLP analysis. It was also important for the procedure to permit harvesting of tissues very early in the life cycle of the plant and to be nonlethal in order to allow recovery of desirable genotypes after RAPD analysis for further testing of other traits. DNA microextraction protocols have been developed recently,[37-41] but none met all the above criteria.

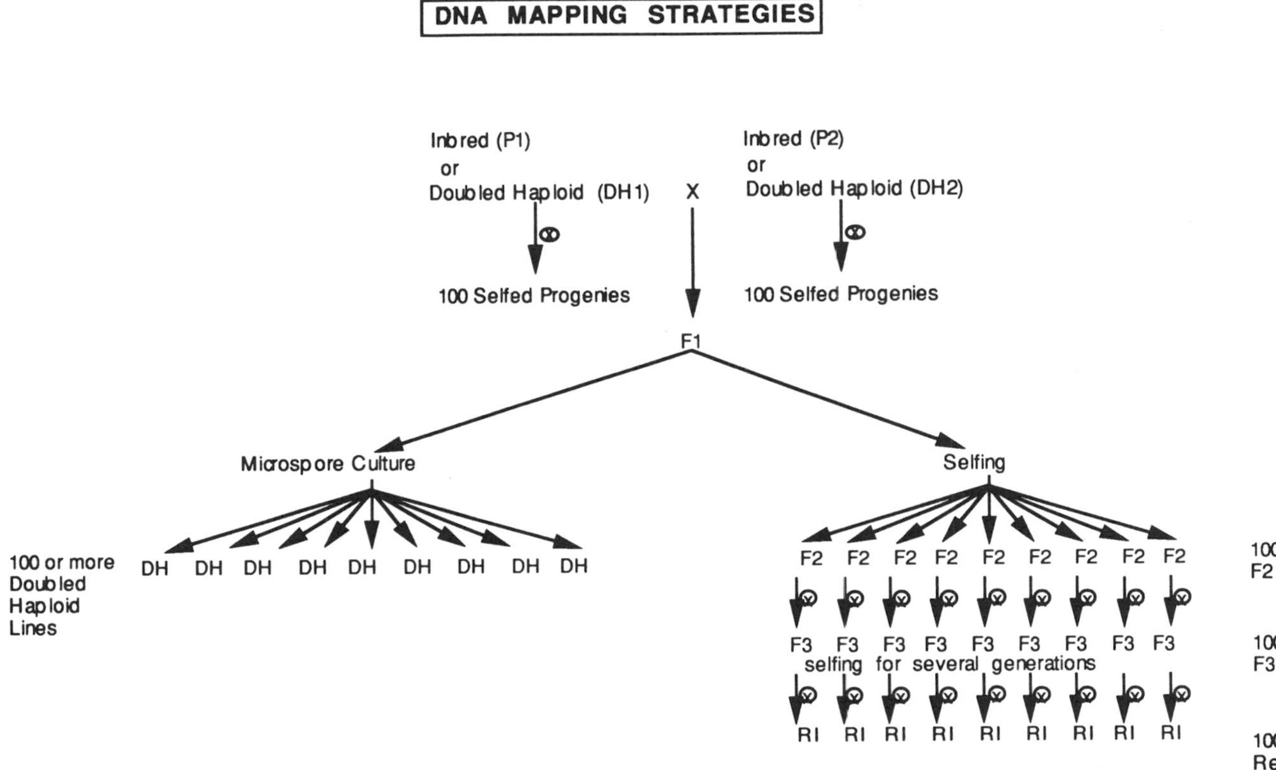


**FIGURE 3.** Some of the most common strategies used for genetic mapping in predominantly selfing species. Single plants are used for P1, P2, and F1. The selfed progenies from P1 and P2 are for DNA isolations and heterozygosity analysis of markers segregating aberrantly. F2 plants and F3 families derived from them are for general mapping purposes and for segregation analysis of Mendelian traits. Doubled haploid populations and recombinant inbred lines are for segregation analysis dominant markers such as RAPD and are ideal for QTL analysis.

The DNA microextraction protocol described below was designed to provide a nondestructive, early DNA diagnostic system to implement DNA marker-assisted selection in plant breeding programs. This procedure has been tested successfully with the following species: *Brassica napus* (canola), *Triticum aestivum* (wheat), *Avena sativa* (oat), *Hordeum vulgare* (barley), *Glycine soja* (soybean), *Zea mays* (corn), *Solanum tuberosum* (potato), and *Lycopersicum esculentum* (tomato).[42] The plants were either grown in a field or a greenhouse. DNA amplification using either RAPD or STS markers was successful with all the tested species.

1. Collect a small leaf disk from a young leaf or a fully developed cotyledon using a common paper punch, and drop the disk directly into a 1.5 ml microcentrifuge tube containing 200 µl of freshly prepared enzyme digestion buffer.
2. Place the opened tube in a vacuum chamber (25 mm Hg), and vacuum infiltrate the solution into the leaf tissues for 30 min. Push the leaf disk into the solution, if necessary, using a disposable pipette tip.
3. Release vacuum, close the tube, and incubate at room temperature in the dark overnight (18 h).
4. Place the tube on ice, and gently tap the side of the tube to help release the protoplasts. Centrifuge at 300 g, 4°C for 5 min to pellet the protoplasts.
   Most of the leaf tissue should have been digested by the enzymes. Disruption of plant tissues by enzymatic digestion of cell walls ensures protection from sample-to-sample contamination and a uniform DNA yield and eliminates the inhibitors of the PCR reaction found when tissues are mechanically disrupted.
5. Pipette off the supernatant, and discard. Add 200 µl of washing buffer, and resuspend the protoplasts by tapping the tube gently. Centrifuge at 300 g, 4°C for 5 min.
6. Resuspend the protoplasts in 160 µl of lysis buffer at room temperature. Add 40 µl of sodium sarkosyl solution to lyse the protoplasts and the nuclei. Invert the tube several times.
7. Pellet the debris by centrifugation at 16,000 g, room temperature for 15 min.
8. Recover the supernatant, add 90 µl of ammonium acetate solution, and 200 µl of isopropanol. Invert the tube several times, and incubate at room temperature for 15 min.
9. Collect the DNA pellet by centrifugation at 16,000 g, room temperature, for 15 min. Pipette off the supernatant, and add 500 µl of 70% ethanol. Invert the tube and centrifuge at 16,000 g, room temperature, for 5 min. Remove the supernatant, and then dry the pellet *in vacuo*.
   Special attention should be paid to the DNA pellet since pellet loss after precipitation is the main reason for unsuccessful amplification.
10. Resuspend the DNA pellet in 50 µl of TE buffer plus RNAse and incubate at 55°C for 10 min to help resuspend the DNA and degrade the RNA. Centrifuge at 16,000 g for 2 min to pellet the debris, and transfer the supernatant to a new tube. Store at –20°C.
    The RNA is digested by adding 10 µg/ml of boiled RNAse A (Sigma, St. Louis) to the resuspended DNA. It is best to degrade the RNA only at the end of the procedure since it helps in precipitating a small amount of DNA. The DNA concentration is best determined by fluorescence in the presence of bisbenzimide (Hoechst dye 33258, Kodak) using a fluorescence spectrophotometer (Model LS5, Perkin-Elmer; excitation: 365 nm; emission: 460 nm). This permits precise determination of the DNA concentration without interference from RNA or protein.

SOLUTIONS FOR DNA MICROEXTRACTION

**Digestion Buffer**

| | Concentration | For 100 ml |
|---|---|---|
| Mannitol | 4% | 4 g |
| Sorbitol | 4% | 4 g |
| $CaCl_2$ | 10 m*M* | 0.11 g |
| MES* | 10 m*M* | 0.21 g |
| pH 6.2 | | |

* MES = 2-[N-Morpholino] ethane-sulfonic acid (potassium salt)

Filter-sterilize the buffer after preparation. The enzymes should be added only before digestion. Since only 200 μl of the digestion buffer is needed per sample, aliquot the amount needed into a tube, then add the enzymes in the following concentrations: Macerase, 1%; Cellulysin, 5%. Dissolve, filter-sterilize, and aliquot 200 μl into 1.5 ml microfuge tubes.

**Washing Buffer**

| | Concentration | For 100 ml |
|---|---|---|
| Trizma base | 50 m*M* | 0.61 g |
| $Na_2 \cdot EDTA \cdot 2H_2O$ | 100 m*M* | 3.72 g |
| Sorbitol | 5.5 % | 5.50 g |
| pH to 8.0 with HCl | | |

**Lysis Buffer**

| | Concentration | For 100 ml |
|---|---|---|
| Trizma base | 50 m*M* | 0.61 g |
| $Na_2 \cdot EDTA \cdot 2H_2O$ | 100 m*M* | 3.72 g |
| pH to 8.0 with HCl | | |

**Sodium Sarkosyl Solution**

| | Concentration | For 10 ml |
|---|---|---|
| Na·Sarkosyl | 5 % | 0.5 g |

This protocol yields 0.25 to 1.00 μg of high molecular weight total DNA and a large quantity of RNA, depending on the species, the age, and the source of tissue (young vs. old tissues; leaf vs. cotyledon). DNA isolated from isolated protoplasts is sufficient to perform a minimum of five, and up to 20, PCR reactions per sample. In a few exceptional cases, we were able to perform at least 500 amplification reactions from a single leaf disk. Amplification products for DNA prepared with this quick procedure are equivalent to those obtained from CsCl-purified DNA. Up to 120 plants can be treated in 2 days, and the procedure lends itself to automation.

## 2. PCR Reaction with RAPD Primers

1. Prepare the PCR reaction buffer, and distribute 23 μl into each 0.5 ml thin wall tube (Cetus/Perkin-Elmer), on ice.
   The PCR reaction buffer contains all of the reagents for the PCR reaction (*Taq* polymerase, RAPD primer, nucleotides, and buffer) except for the plant DNA template. RAPD primers are non-overlapping random 10-mers which can be purchased from Operon Technologies Inc. (Alameda CA), acquired from the University of British

Columbia (Vancouver, B.C.) as part of the UBC RAPD primer synthesis project, or custom synthesized. The *Taq* polymerase should be from Cetus; other commercially available thermostable DNA polymerases are unpredictable.

2. Add 2 µl of plant DNA (or 1/10 volume of the total amount of DNA from a microextraction). Mix well, and then spin down the content for a few seconds at medium speed in a microcentrifuge. Layer the solution with 25 µl of mineral oil; then place the tubes into the temperature cycler.
   Maximum speed is not recommended for centrifugation of thin wall tubes since they can break.
3. Temperature cycling is performed in a Hybaid thermal reactor™ programmed for one cycle as follows:

   30 sec at 94°C (ramping 1.0)
   10 min at 42°C (ramping 2.0)

   Forty-five cycles as follows:

   1 sec at 50°C (ramping 5.0)
   45 sec at 72°C (ramping 1.0)
   5 sec at 94°C (ramping 1.0)
   30 sec at 42°C (ramping 2.0)

   One cycle as follows:

   5 min at 72°C (ramping 1.0)

   The ramping value represents the time in seconds required to increase or decrease the temperature by 1°C. Temperature cycling conditions are specific to each model and brand of thermal cycler apparatus. In addition, each unit must be tested in four replicated experiments where a single large PCR reaction is divided into several tubes placed in each well. This is done to identify wells that systematically produce erratic results.

## SOLUTIONS FOR RAPD AMPLIFICATION

**PCR Reaction Buffer**

| | Final concentration | For 20 reactions |
|---|---|---|
| 10X Salt Buffer | 1X | 50 µl |
| 10X Nucleotide Buffer | 1X | 50 µl |
| RAPD Primer, 20 µ*M* Stock | 0.2 µ*M* | 5 µl |
| *Taq* Polymerase | 0.5 unit/reaction | 2 µl |
| Distilled $H_2O$ | | 353 µl |

Prepare just prior utilization and keep on ice.

**10X Salt Buffer**

| | Final concentration | 10X |
|---|---|---|
| Tris-HCl, pH 8.2 | 10 m*M* | 100 m*M* |
| KCl | 50 m*M* | 500 m*M* |
| $MgCl_2$ | 2 m*M* | 20 m*M* |
| Autoclaved Gelatin | 0.02 % | 0.2 % |

Prepare 100 ml of the 10X buffer in advance, aliquot 1 ml in microfuge tubes, store at 4°C.

**10X Nucleotides Buffer**

| | Final concentration | For 1 ml |
|---|---|---|
| dATP,* 100 m*M* Stock | 2 m*M* | 20 µl |
| dTTP,* 100 m*M* Stock | 2 m*M* | 20 µl |
| dCTP,* 100 m*M* Stock | 2 m*M* | 20 µl |
| dGTP,* 100 m*M* Stock | 2 m*M* | 20 µl |
| Distilled $H_2O$ | | 920 µl |

* dNTPs are from Pharmacia. Aliquot 100 µl in microfuge tubes, then freeze at –20°C. Never dilute the *Taq* Polymerase in distilled $H_2O$.

### 3. Electrophoretic Analysis of Amplified Product and Selection of Useful Primers

RAPD amplification products are analyzed by mixed gel electrophoresis, 3.0% Nusieve:agarose (3:1), (FMC, Rockland ME) for 3 h at 5 Volts/cm in the TAE buffer system and stained with ethidium bromide before UV photography.[19] When resolution is not critical, we use 1.4% agarose gels at a migration of 5 cm (blue dye). Half the RAPD reaction is generally loaded onto the gel.

Useful amplified loci are generally bright bands. They are present in one parental line and completely absent in the other parental line. These are more likely to be stable amplification products. Faint bands generally do not segregate normally in the segregating population and should not be scored.

### 4. Segregation Analyses and Genetic Mapping

Populations most suitable for RAPD mapping are haploid or doubled haploid lines derived from microspores of an F1 plant or recombinant inbred lines derived from repeated selfing of individual plants of an F2 population. Dominance-recessive relationship of alleles does not interfere with the segregation analysis in these populations. Backcross populations can be used, but useful RAPD markers are those that will permit amplification of a DNA fragment only from the recurrent parent. Loss of genetic information occurs when F2 populations are used to map RAPD markers since heterozygotes cannot be separated from homozygotes. PCR amplifications are performed on the DNA samples of the segregating plants under the same conditions as for the parental lines screening. The polymorphic DNA bands are scored for their presence or absence. The time for the electrophoresis can be adjusted for segregation analysis in order to provide the best separation of the polymorphic bands. The technique for scoring and linkage analysis is performed as for RFLP analysis.

## III. CONCLUSION

RFLP and PCR-based DNA markers will likely become integral parts of breeding programs to facilitate selection and to dissect important agronomic traits. In fact, DNA markers have been called "new tools for an old science."[34] They help to increase the efficiency and precision of plant selection. During the next few years, molecular techniques will improve to such an extent that it will be easier to use DNA markers to select a trait than to evaluate the trait itself. The limiting factor in these studies has been, therefore, the lack of knowledge of the genetic basis of most agronomic traits. Quantitative geneticists and breeders must intensify their efforts towards developing a better understanding of the genetic nature underlying agronomic traits using the classical genetic approaches. Although molecular genetics can help the dissection of quantitative traits into Mendelian factors, populations suitable for this precise operation are rarely available. Plant molecular geneticists need the expertise of plant physiologists, pathologists, and breeders to analyze the agronomic traits and to construct and analyze the genetic populations essential for genetic mapping.

## REFERENCES


1. **Botstein, D., White, R. L., Skolnick, M., and Davis, R. W.,** Construction of a genetic linkage map in man using restriction fragment length polymorphisms, *Am. J. Hum. Genet.*, 32, 314, 1980.
2. **Grodzicker, T., Williams, J., Sharp, P., and Sambrook J.,** Physical mapping of temperature-sensitive mutations of adenoviruses, *Cold Spring Harbor Symp. Quant. Biol.*, 39, 439, 1974.
3. **Petes, T. K. and Botstein, D.,** Simple Mendelian inheritance of the reiterated ribosomal DNA of yeast, *Proc. Natl. Acad. Sci. U.S.A.*, 74, 5091, 1977 .
4. **Potter, S., Newbold, J., Hutchison, C., and Edgell, M.,** Specific cleavage analysis of mammalian mitochondrial DNA, *Proc. Natl. Acad. Sci. U.S.A.*, 72, 4496, 1985.
5. **Kan, Y. W. and Dozy, A. M.,** Antenatal diagnosis of sickle-cell anaemia by DNA analysis of amniotic fluid cells, *Lancet*, 2, 910, 1978.
6. **O'Brien, S. J.,** *Genetic Maps: Locus Maps of Complex Genomes*, 5th ed., CSH Laboratory Press, New York, 1990, 1061 pp.
7. **Landry, B. S., Hubert, N., Etoh, T., Harada, J. J., and Lincoln, S. E.,** A genetic map for *Brassica napus* based on restriction fragment length polymorphisms detected with expressed DNA sequences, *Genome*, 34, 543, 1991.
8. **Landry, B. S., Hubert, N., Crete, R., Chiang M., Lincoln, S. E., and Etoh, T.,** A genetic map for *Brassica oleracea* based on RFLP markers detected with expressed DNA sequences and·mapping of resistance genes to race 2 of *Plasmodiophora brassicae* (Wor.), *Genome*, 35, 409, 1992.
9. **Saiki, R. K., Scharf, S., Faloona, F., Mullis, K. B., Horn, G. T., Erlich, H. A., and Arnheim, N.,** Enzymatic amplification of β-globin genomic sequences and restriction site analysis for diagnosis of sickle cell anemia, *Science*, 230, 1350, 1985.
10. **Saiki, R. K., Gelfand, D. H., Stoffel, S., Scharf, R., Higuchi, R., Horn, G. T., Mullis, K. B., and Erlich, H. A.,** Primer-directed enzymatic amplification of DNA with a thermostable DNA polymerase, *Science*, 239, 487, 1988.
11. **Erlich, H. A., Gelfand, D., and Sninsky, J. J.,** Recent advances in the polymerase chain reaction, *Science*, 252, 1643, 1991.
12. **Erlich H. A.,** *PCR Technology; Principles and Applications for DNA Amplification*, 1st ed., Stockton Press, New York, 1989, 246 pp.
13. **Innis, M. A., Gelfand, D. H., Sninsky, J. J., and White, T. J.,** *PCR Protocols; A Guide to Methods and Applications*, 1st ed., Academic Press, San Diego, 1990, 482 pp.
14. **Arnheim, N., Li, H., and Cui, X.,** Genetic mapping by single sperm typing, *Anim. Genet.*, 22, 105, 1991.
15. **Olson, M., Hood, L., Cantor, C., and Botstein, D.,** A common language for physical mapping of the human genome, *Science*, 245, 1434, 1989.
16. **Williams, J. G. K., Kubelik, A. R., Livak, K. J., Rafalski, J. A., and Tingey, S. V.,** DNA polymorphisms amplified by arbitrary primers are useful as genetic markers, *Nucl. Acids Res.*, 18, 6531, 1990.
17. **Landry, B. S. and Michelmore, R. W.,** Methods and applications of restriction fragment length polymorphism analysis to plants, in *Tailoring Genes for Crop Improvement, An Agricultural Perspective*, Bruening, G., Harada, J. J., and Hollaender A., Eds., Plenum Press, New York, 1987, 25.
18. **Landry B. S. and Michelmore R. W.,** Selection of probes for restriction fragment length analysis from plant genomic clones, *Plant Mol. Biol. Rep.*, 3, 174, 1985.
19. **Sambrook, J., Fristch, E. F., and Maniatis, T.,** *Molecular Cloning; A Laboratory Manual*, 2nd ed., CSH Laboratory Press, Cold Spring Harbor, 1989.
20. **Serghini, M. A., Ritzenthaler, C., and Pinek, L.,** A rapid and efficient miniprep for isolation of plasmid DNA, *Nucl. Acids Res.*, 17, 3604, 1989.
21. **Tautz, D. and Renz, M.,** An optimized freeze-squeeze method for the recovery of DNA fragments from agarose gels, *Anal. Biochem.*, 132, 14, 1983.
22. **Rigby, P. W. J., Dieckmann, M., Rhodes, C., and Berg, P.,** Labeling deoxyribonucleic acid to high specific activity in vitro by nick-translation with DNA polymerase I, *J. Mol. Biol.*, 113, 237, 1977.
23. **Landry B. S., Kesseli R. V., Leung H., and Michelmore, R. W.,** Comparison of restriction endonucleases and sources of probes for their efficiency in detecting restriction fragment length polymorphism between four lines of lettuce (*Lactuca sativa* L.), *Theor. Appl. Genet.*, 74, 646, 1987
24. **Draper, J. and Scott, R.,** The isolation of plant nucleic acids, in *Plant Genetic Transformation and Gene Expression*, Draper, J., Scott, R., Armitage, P., and Walden, R., Eds., Blackwell Scientific Publications, Oxford, 1988, 199.
25. **Webb, D. M. and Knapp, S. J.,** DNA extraction from a previously recalcitrant plant genus, *Plant Mol. Biol. Rep.*, 8, 180, 1990.
26. **Bernatzky, R. and Tanksley, S. D.,** Genetics of actin-related sequences in tomato, *Theor. Appl. Genet.*, 72, 314, 1986.

27. **Landry B. S., Kesseli, R. V., Farrara, B., and Michelmore, R. W.,** A genetic map of lettuce (*Lactuca sativa* L.) with restriction fragment length polymorphisms, isozymes, disease resistance genes and morphological markers, *Genetics*, 116, 331, 1987.
28. **Allard, R. W.,** Formulas and tables to facilitate the calculation of recombination values in heredity, *Hilgardia*, 24, 235, 1956.
29. **Lander, E. S. and Botstein, D.,** Mapping Mendelian factors underlying quantitative traits using RFLP linkage maps, *Genetics*, 121, 185, 1989.
30. **Lander, E. S., Green, P., Abrahamson, J., Barlow, A., Daly, M. J., Lincoln, S. E., and Newburg, I.,** Mapmaker: An interactive computer package for constructing primary genetic linkage maps of experimental and natural populations, *Genomics*, 1, 174, 1987.
31. **Lincoln, S. E., Daly, M. J., and Lander, E. S.,** Mapping genes controlling quantitative traits using MAPMAKER/QTL, 2nd ed., Whitehead Institute for Biomedical Research Technical Report, Cambridge, MA., 1990, 205 pp.
32. **Beckmann, J. S. and Soller, M.,** Restriction fragment length polymorphisms and genetic improvement of agricultural species, *Euphytica*, 35, 111, 1986.
33. **Paterson, A. H., Damon, S., Hewitt, J. D., Zamir, D., Rabinowitch, H. D., Lincoln, S. E., Lander, E. S., and Tanksley, S. D.,** Mendelian factors underlying quantitative traits in tomato: comparison across species, generations, and environments, *Genetics*, 127, 181, 1991.
34. **Tanksley, S. D., Young, N. D., Paterson, A. H., and Bonierbale, M. W.,** RFLP mapping in plant breeding: new tools for an old science, *Bio/Technology*, 7, 257–264, 1989.
35. **Tingey, S. V.,** personal communication, 1991.
36. **Michelmore, R. W., Paran, I., and Kesseli, R. V.,** Identification of markers linked to disease resistance genes by bulked segregant analysis: a rapid method to detect markers in specific genomic regions using segregating populations, *Proc. Natl. Acad. Sci. U.S.A.*, 88, 9828, 1991.
37. **Caetano-Anollés, G., Bassam, B. J., and Gresshoff, P. M.,** DNA amplification fingerprinting using very short arbitrary oligonucleotide primers, *Bio/Technology*, 9, 553, 1991.
38. **Clarke, B. C., Moran, L. B., and Appels, R.,** DNA analyses in wheat breeding, *Genome*, 32, 334, 1989.
39. **Irvine, J. M., Oakes, J. V., Shewmaker, C. K., and Crossway, A.,** A rapid screen for the detection of specific DNA sequences in plants, *Gene Anal. Tech. Appl.*, 7, 25, 1990.
40. **Ferre, F. and Garduno, F.,** Preparation of crude extract suitable for amplification of RNA by the polymerase chain reaction, *Nucl. Acids Res.*, 17, 2141, 1989.
41. **Mercier, B., Gaucher, C., Feugeas, O., and Mazurier C.,** Direct PCR from whole blood, without DNA extraction, *Nucl. Acids Res.*, 18, 5908, 1990.
42. **Deragon, J. M. and Landry, B. S.,** RAPD and other PCR-based analyses of plant genomes using DNA extracted from small leaf disks, *PCR Meth. Appl.*, 1, 175, 1992.

# 18

# Random Amplified Polymorphic DNA (RAPD) Analysis

*Kang Fu Yu, Allen Van Deynze, and K. Peter Pauls*

## I. INTRODUCTION

In the past decade, molecular markers have become fundamental tools for plant biologists; they are useful for fingerprinting varieties, establishing phylogenies, tagging desirable genes (to assist their introgression into new varieties), determining similarities among inbreds (to maximize heterosis in hybrids), and mapping plant genomes. Several methods for comparing plants at a molecular level have been developed, including the identification of isozyme, restriction fragment length, or polymerase chain reaction (PCR) markers.

Isozymes were the first molecular markers to be used in plant breeding programs; they are currently being used to screen lines and to tag chromosomes.[1] However, the paucity of isozymes limits their usefulness. Restriction fragment length polymorphisms (RFLPs) are based on differences in fragment lengths obtained by digesting DNA samples with restriction endonucleases. Polymorphisms are due to the presence or absence of restriction sites in the genomes being compared. RFLPs have been used extensively to develop genomic maps,[2,3] establish linkages to traits,[4,5] develop phylogenetic trees,[6,7] and tag chromosomes.[8] These markers have the advantages that they are phenotypically neutral, codominantly inherited, nonspecific to growth stage, and practically limitless in number. However, detection of RFLPs by southern blotting is often laborious, time-consuming, and expensive. In addition, the RFLP procedure usually involves the use of radioactive materials.

The polymerase chain reaction (PCR)[9] results in amplification of specific portions of template DNA that occur between sequences (on the + and – strands) that bind DNA synthesis primers. This cyclic process has three basic steps, including (1) thermal denaturation of DNA, (2) annealing of oligonucleotide primers to the template DNA, and (3) primer extension by a DNA polymerase in the presence of dNTPs. DNA fragments are amplified exponentially for 25 to 45 PCR cycles, separated by electrophoresis, and visualized by ethidium bromide (EthBr) staining.[9,10] PCR was first performed by adding new thermosensitive Klenow fragment (from *Escherichia coli*) to the reaction mixture at the beginning of each cycle. It became a widely used technique in molecular biology after the discovery of a thermally stable DNA polymerase from *Thermus aquaticus*[11] and development of automated thermal cyclers which allowed the DNA amplification to be carried out without intervention.

The advent of automated PCR technology made a new set of markers available to scientists interested in comparing organisms at a molecular level. This chapter describes PCR methods for determining polymorphisms in plants. In particular, the use of random primers to obtain random amplified polymorphic DNA (RAPD, pronounced rapid) markers[12] will be emphasized.

0-8493-5164-2/93/$0.00+$.50
© 1993 by CRC Press, Inc.

## II. POLYMERASE CHAIN REACTION MARKERS

### A. AMPLIFIED SEQUENCE POLYMORPHISMS (ASPS)

ASPs are variations in the length of DNA fragments obtained by PCR from priming sites with known sequences. This method has been used to analyze restriction site polymorphisms, deletion/insertion events, variable number tandem repeats,[13] and base pair substitutions. Since the length of the PCR products for ASPs is generally known, a mixture of primers can be used to allow simultaneous analysis of several polymorphic loci. In this manner, many traits can be screened in a few PCR reactions.[14] The limitation of this technique is that the sequence of the target DNA must be known so that the appropriate primers can be synthesized.

### B. SEMIRANDOM PRIMERS

Several primer strategies have been developed to overcome the requirement for sequence information in PCR analysis. Twenty-one base pair sequences of dinucleotide repeats were used as primers to examine molecular polymorphisms in mice. Eighty-eight percent of the primers produced polymorphisms when they were tested with microsatellite DNA from different inbred strains of mice.[15]

Semi-random primers based on intron splice junction sequences were used to identify polymorphisms in wheat (*Triticum aestivum*), barley (*Hordeum vulgare*), and rye (*Secale cereale*).[16] The 15- to 25-bp primers used in the study generated PCR products when they were used in combinations but not when used alone. Also, due to the nonconserved nature of introns relative to exons, intron-directed primers generated more complex patterns than exon-directed primers. Other options for semi-random primers include conserved gene sequences,[16] poly A signals, translation start signals, and promoter and enhancer sequences.

### C. RANDOM PRIMERS

Random amplified polymorphic DNA (RAPD) markers are obtained by PCR amplification of random DNA segments from single arbitrary primers. Williams et al.[12] were the first to use the procedure on plant samples and suggested it be named RAPD. The arbitrary primers used for the procedure are usually 9 to 10 bp in size; they have a GC content of 50 to 80% and do not contain palindromic sequences. The number of DNA fragments that are amplified is dependent on the primer and the genomic DNA used. A single nucleotide substitution in a primer can result in a complete change in the RAPD pattern; this is an indication of the sensitivity of the system. However, the method is not 100% stringent because much larger numbers of bands are generally observed when bacterial genomes are used as templates than would be predicted. The conditions of the PCR reaction limit the band size to 100 to 3000 bp. Therefore, only DNA fragments within this size range that occur between DNA sequences that are complementary to the primer sequence are amplified. A survey using DNA from 66 $F_2$ soybean plants indicated that RAPD markers are mostly dominant and heritable. Polymorphisms for RAPDs may be due to single base pair changes, deletions of primer sites, insertions that increase the separation of primer sites over the 3000 bp limit, and small insertions/deletions that result in changes in the size of the PCR product. Small insertions or deletions result in codominant inheritance of the marker. Several RAPD markers are usually produced per primer. The frequency of polymorphism ranges from 2.5/primer in *Neurospora* to 0.5/primer in corn. When the PCR products were used as RFLP probes, they mapped at the same location as their PCR amplified markers. Six out of eleven amplified probes tested with soybean DNA identified repetitive DNA sequences.

Like other molecular markers, RAPDs can be used to tag chromosomes and genes, to fingerprint genomes, and to produce genomic maps. The advantages of using these markers are that (1) a universal set of primers can be used for all species, (2) no probe libraries, radioactivity, southern transfers, or primer sequence information are required, (3) only the primer sequence is needed for information transfer, and (4) the process can be automated. The limitation to the use of RAPD markers is that they are dominant. This can be overcome by using more than one closely linked marker.[12]

Welsh et al.[17] used the same technique (but under a different name, arbitrarily primed polymerase chain reaction) to fingerprint human, rice, bacteria, and virus genomes ranging in size from 50 to $3 \times 10^6$ kbp. With this technology, inbred maize lines could be distinguished from each other, and six maize hybrids were shown to have the complement PCR products of their parental inbred lines.[18]

RAPD markers were used to screen different strains of *Leptosphaeria maculans*, which is the causal agent of blackleg disease in crucifers.[19] Differences between avirulent and virulent pathotypes, as well as among isolates within a pathotype, could be distinguished on the basis of specific RAPD markers or by calculating similarity coefficients from the collection of RAPD markers. The results of the RAPD analysis agreed with previous classifications based on RFLPs and cultural assays; they also resolved a question of the ancestry of the fungus (*Leptosphaeria*). The main advantages of the RAPD analysis in this study were deemed to be its sensitivity and speed. Quick DNA preps are available for fungi, and results could be obtained in 1 to 2 days. The investigators suggested that a quick RAPD assay may be used to screen rapeseed seedlots for the presence of the virulent pathotype of the fungus.

Quiros et al.[20] used RAPDs to develop genome-specific primers for the six diploid and amphidiploid Brassica species. Forty-seven primers yielded 65 PCR products with 1 to 15 products per primer. Sixteen, thirty-seven, and twelve RAPD markers were specific for the A, B, and C genomes (as defined by U[21]), respectively. In addition, RAPD markers that were species-specific and specific to accessions within species were identified. This study also tagged six out of the eight possible *B. nigra* chromosomes with RAPDs by using *B. napus-nigra* disomic addition lines.

The efficiency of RAPD markers for tagging genes of interest was illustrated in a study of *Pseudomonas* resistance in near-isogenic-lines (NILs) of tomato.[22] Resistant and susceptible NILs were screened with 144 RAPD primers to yield seven polymorphic markers. Further characterization of four of the markers revealed that two were single copy, one was allelic, and one represented repetitive DNA. The markers mapped within 7.9 c*M* of the disease resistance gene. In this population, RAPD analysis yielded close linkages for 3 of 144 primers tested (2%), and the study was accomplished in 1 month. Previously, five linkages had been established with the same NILs after testing 800 RFLPs (0.5%) over a period of 2 years. Clearly the RAPD analysis was much more efficient than RFLP analysis in this case. In general, the success rate for tagging genes of interest will depend on the number of primers screened, the genome size, and the degree of divergence among the lines.

Plant and animal genomes have been fingerprinted using PCR with short (5 to 21 bp) arbitrary primers that result in low stringency.[23,24] The number of bands ranged from 40 to 100 when separated on polyacrylamide gels. Mixing primers did not yield combinations of the patterns obtained when the primers were used singly, and primers with high GC content yielded more bands. The sensitivity of the method was illustrated by the detection of six polymorphisms among soybean (*Glycine max*) cultivars with a single primer. This was a significant finding as RFLP studies have indicated that soybean has a narrow genetic base and is highly homozygous.

## III. RAPD METHODOLOGY

Efficient use of RAPD markers requires quick DNA extraction, optimum amplification conditions, and appropriate data analysis.

## A. DNA EXTRACTION

Extraction procedures for plant genomic DNA involve disruption of plant cell walls and cell membranes to release the cellular constituents into the extraction buffer, and protection of the DNA from the activity of endogenous nucleases.[25] Many of the extraction methods described in the literature employ either the time-consuming cesium chloride density gradient technique or the cetyltrimethylammonium bromide (CTAB) extraction procedure, which uses hazardous chemicals such as phenol and chloroform.[26-28] Since it is not necessary for the extracted DNA to be high molecular weight in order to produce moderate-sized PCR products,[29] a number of simple and rapid DNA extraction procedures[30,31] are suitable for RAPD analysis. The method presented below is based on a rapid procedure for plant genomic DNA extraction described by Edwards et al.[30] The procedure does not require the use of expensive or hazardous chemicals and uses only milligram quantities of leaf tissue. We have used this procedure for a number of plant species to obtain genomic DNA that can be amplified by random primers. Approximately 4 to 8 h are needed to process 50 to 100 samples, and each sample contains enough DNA for 100 PCRs.

Steps in the procedure:

1. Add 400 µl extraction buffer (Table 1) into a 1.5 ml sterile Eppendorf tube.[a]
2. Use the lid of the Eppendorf tube to pinch out a disk of leaf material into the tube.[b]
3. Homogenize the leaf tissue with a pestel that fits the tube tightly.[c,d]
4. Centrifuge the extracts in a microcentrifuge for 1 min, and transfer 300 µl of the supernatant to a fresh tube.
5. Add one volume (300 µl) of isopropanol, to precipitate the DNA for 5 min, and then centrifuge for 5 min.[e]
6. Discard the supernatent, and air dry the pellet completely (about 30 min on the bench or a few minutes in a SpeedVac concentrator).
7. Add 100 µl of $H_2O$ and allow the DNA to dissolve for 10 min or longer, without agitation.[f]
8. Centrifuge the sample for 1 min, and collect the supernatant. (The supernatant contains a sufficient concentration of DNA so that it can be used directly for RAPD analysis.[g])

*Notes*:

[a] Extraction buffer can also be added after grinding.[30] However, placing the samples directly into the extraction buffer can prevent the leaf tissue from drying out (this is particularly important when many samples are collected or samples are collected from the field) and can protect the DNA from degradation by endogenous nucleases (Figure 1).

[b] For cereals or plants with small leaves or compound leaves, a strip of the leaf or a whole leaf or leaflet can be used. Usually, younger leaves are better than older leaves.

[c] Pestels to fit Eppendorf tubes can be obtained from Mandel Scientific.

[d] A disposable pestel can be used for each sample, or it must be washed thoroughly with distilled water between samples to avoid contamination.

[e] Two volumes of 100% ethanol will also effectively precipitate DNA. A second precipitation or a 70% ethanol wash may improve the results (i.e., repeat Steps 5 to 8).

[f] Recovery of genomic DNA can be improved if it is dissolved at 4°C overnight.

[g] The amount of DNA in the supernatant will vary and can be determined by using a DNA Dipstick (Invitrogen) or by running an agarose gel with standard DNA samples. Spectrophotometric determination of DNA content[32] may not be very accurate because the quick minipreps do not give highly purified DNA samples. The DNA can be stored at 4°C for (at least) 6 months, or longer at –20°C.

**TABLE 1**
**DNA Extraction Buffer**

| Materials | Amount (ml) | Final concentration |
|---|---|---|
| 1 *M* Tris·HCl buffer, pH 7.5 | 10.0 | 200 m*M* |
| 5 *M* NaCl | 2.5 | 250 m*M* |
| 0.5 *M* EDTA | 2.5 | 25 m*M* |
| 10% sodium dodecyl sulfate (SDS) | 2.5 | 0.5 % |
| Water | 32.5 | — |

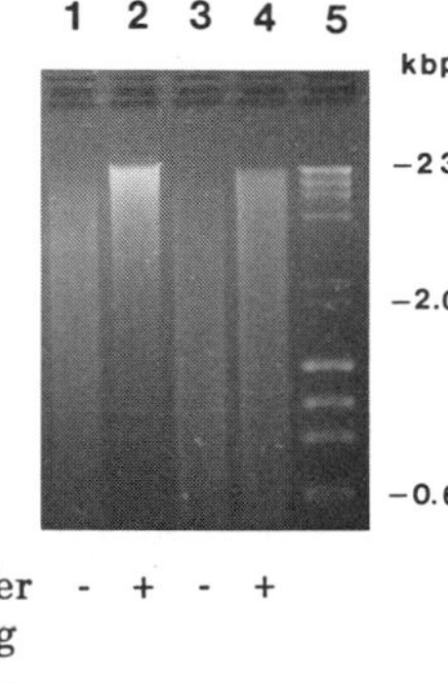


**FIGURE 1.** Comparison of genomic DNA extracted from alfalfa with (Lanes 2 and 4) and without (Lanes 1 and 4) extraction buffer added to the samples before grinding. The time between sampling and grinding was 0 and 3 h for Lanes 3 and 4 and Lanes 1 and 2, respectively.

## B. AMPLIFICATION METHODS

Amplification conditions for RAPD analysis are similar to those used in a normal polymerase chain reaction except that one primer with an arbitrary sequence is used instead of two primers with specific sequences.[12] As a result, amplification in RAPD analysis occurs anywhere in a genome that contains two complementary sequences to the primer that are within the length-limits of the PCR (which is approximately 3kb).

Figures 2 and 3 show the RAPD patterns obtained from a particular DNA sample with different primers and a single primer with different DNA templates, respectively, and indicate that the PCR patterns obtained are dependent on both the template and the specific PCR primer. The range of fragment sizes (from 0.5 to 2.5 kb) and fragment numbers (from 1 to 10) typically found in RAPD patterns is also illustrated in these figures. The standard conditions given below work well for random 10-mers and a wide range of templates, although they may not be optimal for particular combinations. Because there are many factors that can affect the outcome (such as template, primer, dNTP concentrations, as well as PCR program conditions), it is useful to optimize the reaction conditions for each new application.

Steps in the procedure:

1. In a 0.5-ml sterile microfuge tube, mix the materials in the order given in Table 2.[a]
2. Prepare a control reaction without template DNA (replace with water) for each primer used.[b]
3. Overlay the reaction mixture with 50 μl of light mineral oil to prevent evaporation of the sample during repeated cycles of heating and cooling.

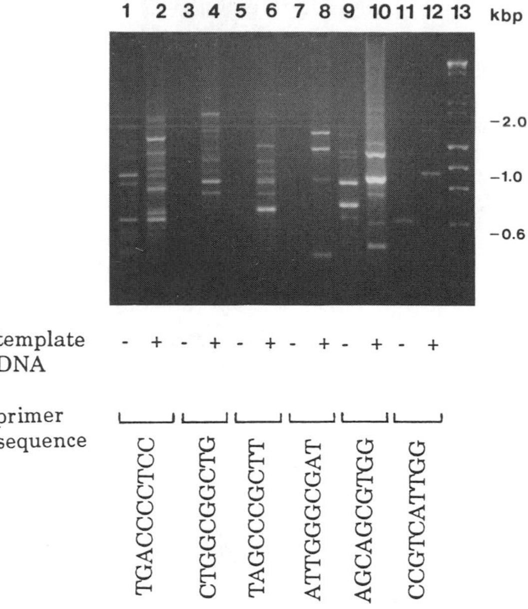

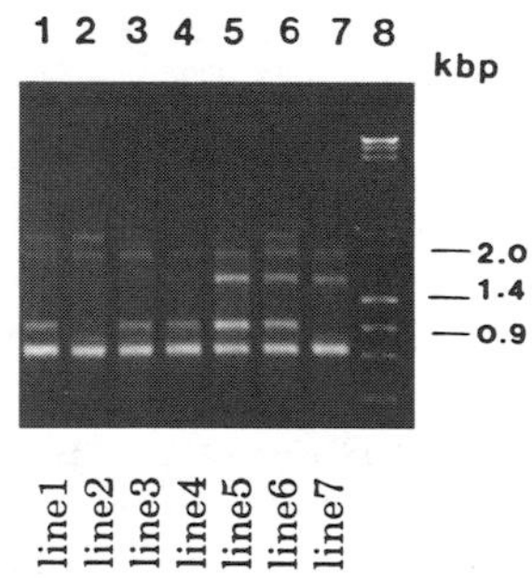


**FIGURE 2.** RAPD patterns from alfalfa genomic DNA. (a) Patterns obtained from the same template DNA and different random primers. Lanes 1, 3, 5, 7, 9, and 11 were controls containing all of the components of the PCR reaction mixture (Table 2) except alfalfa template DNA. (b) RAPD patterns obtained with the same primer plus template DNA from different alfalfa lines.

4. Carry out amplification in an automated thermal cycler programmed for 45 cycles of 1 min at 94°C, 1 min at 36°C, and 2 min at 72°C, using the fastest available transitions between each temperature.[c,d]
5. Amplification products can be stored at 4°C for several days, or at –20°C for much longer periods before separation by gel electrophoresis.

*Notes:*

[a] The quantities of the materials used in a PCR mixture can affect the outcome. In general, template DNA quantities can vary from picogram to nanogram levels depending on the

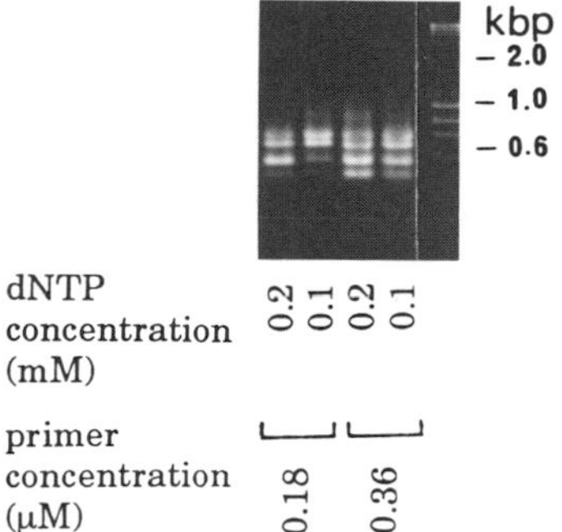


**FIGURE 3.** The effects of primer and dNTP amounts on RAPD patterns produced from bean (*Phaseolus vulgaris*) genomic DNA. Lanes 1 and 2, and 3 and 4 contained 0.18 μ*M* and 0.36 μ*M* of primer, respectively. Lanes 1 and 3, and 2 and 4 contained 0.2 m*M* and 0.1 m*M* of each of the dNTPs, respectively.

**TABLE 2**
**PCR Reaction Mixture**

| Materials | Amount (μl) | Final concentration |
|---|---|---|
| 10X amplification buffer (Table 3) | 2.5 | 10 m*M* Tris·HCl, pH 8.8<br>50 m*M* KCl<br>1.5 m*M* $MgCl_2$<br>0.1% Triton X-100 |
| 0.5 m*M* dATP, 0.5 m*M* dTTP, 0.5 m*M* dCTP, 0.5 m*M* dGTP solution[a] | 5.0 | 0.1 m*M* each dNTP |
| 3 μ*M* primer solution | 3.0 | 0.36 μ*M*[b] |
| Genomic DNA | 2.0 | 25 ng/reaction[c] |
| Taq DNA polymerase (0.75 units/μl) | 2.0 | 1.5 units/reaction[d] |
| Sterile $H_2O$ | 10.5 | — |

[a] The pH of the dNTP solution should be 7.0.

[b] The length of the primer can vary from 5 to 34 nucleotides long. Usually the shorter the primers used, the higher molar concentration of primer needed. The GC content in an oligonucleotide primer should be 40% or greater.[12]

[c] The amount of DNA can vary from pg to ng quantities depending on the genome size used.

[d] The amount of *Taq* DNA polymerase used can vary from 0.5 units/reaction to 2.5 units/reaction. (In our hands the *Taq* polymerase from Promega or BRL work the best.)

genome size. A smear without distinct bands may result if too much template DNA or *Taq* enzyme is added.[12] Figure 3 shows that the RAPD pattern is sensitive to the amounts of dNTPs and *Taq* enzyme that are used. The optima that were determined for dNTP and enzyme levels with the bean DNA were slightly higher than those recommended by the manufacturer of the enzyme (Figure 3).

Control reactions (without template DNA) are important components of PCR analyses because of the sensitivity of the technique.[33] Even minor contaminations of the reagents, pipettors, or tubes that are used can lead to spurious results. A common band in all of the control reactions would be diagnostic of a contamination problem. Some of the primers used in RAPD analysis will give bands without template DNA (Figure 2). This phenomenon has been attributed to primer multimer formation.[12] Most of the fragments seen in the control lanes will be absent when template DNA is included and, therefore, they do not interfere with the analysis. However, those bands which are present in the control and template-containing reactions should not be used for analyses.

**TABLE 3**
**10X Amplification Buffer[a]**

| Materials | Amount (μl) | Final concentration |
|---|---|---|
| 1 *M* Tris, pH 8.8 | 100 | 100 m*M* |
| 5 *M* KCl | 100 | 500 m*M* |
| 1.5 *M* $MgCl_2$ | 10 | 15 m*M* |
| 10% Triton X-100 | 100 | 1% |
| Water | 690 | — |

[a] This is usually provided with the *Taq* DNA polymerase.

c The annealing temperature (*Tm*) can vary according to the length and GC content of the primer as well as the salt concentration in the solution. The following formula has been used to approximate the Tm of the primers:

$$Tm = 81.5 + 16.6(\log M) + 0.41(\%GC) - (500/n) \quad (1)$$

where $n$ = length of primer and $M$ = molarity of the salt in the buffer.[29] To determine the molarity, ignore the contribution of $Mg^{2+}$, and multiply the Tris concentration by 0.67.[29] Annealing works best if done at 2°C above this Tm. This is only an approximation. Therefore, it may be necessary to determine empirically the optimal annealing temperature and time for each primer.

d Each step in the program requires a minimum length of time to be effective, but too long a time unnecessarily prolongs the procedure and may decrease the effectiveness of the *Taq* enzyme. Therefore, it is useful to spend some time testing various PCR programs when developing a new RAPD application. Figure 4 illustrates that large RAPD fragments are particularly sensitive to the primer extension time. Fragments greater than 1 kb were lost in two of the three primers tested with alfalfa DNA when less than 30 sec extension time was used. Figure 5 shows that the GC content of the primer affected its sensitivity to the length of time in the program that was allowed for annealing. RAPD patterns obtained from primers with a high GC content (70% or 80%) were unaffected by a short (5 sec) annealing time, whereas those obtained from primers with 50% or 60% GC content were reduced in intensity even with 30 sec of annealing time.

## C. SEPARATION OF AMPLIFICATION PRODUCTS BY GEL ELECTROPHORESIS

The PCR products can be separated by agarose gel electrophoresis or nondenaturing polyacrylamide gel electrophoresis. The choice of system depends on the size of the DNA fragments to be separated. Agarose gel electrophoresis is relatively simple to perform and separates DNA fragments from 0.5 to 25 kb in size. However, polyacrylamide gel electrophoresis is the method of choice to separate DNA fragments less than 1 kb. The appropriate concentrations of agarose and acrylamide for separating DNA fragments of various sizes are given in Tables 4 and 5, respectively.[34]

Since gel running times are inversely proportional to voltage, doubling the voltage reduces running time by approximately half. However, there is a limit to the voltage that can be applied to a gel. If the voltage is too high, it will generate excessive heat, and this can lead to band diffusion and distortion.

The DNA in the gel can be visualized by staining with ethidium bromide. However, silver staining (which is more complex and time-consuming) gives better resolution, especially of minor fragments separated by polyacrylamide gel electrophoresis.[35]

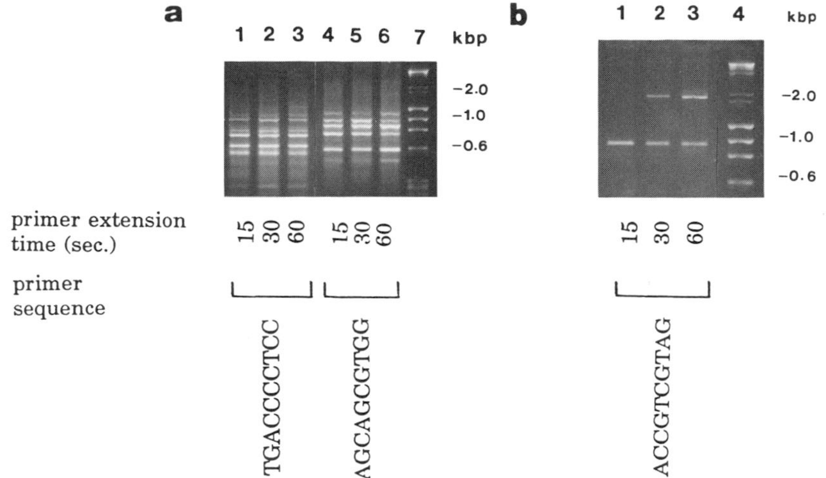


**FIGURE 4.** The effect of primer extension time in the PCR program on RAPD patterns produced from alfalfa genomic DNA. The patterns were obtained with PCR programs consisting of 15 sec at 94°C to denature the DNA, 15 sec at 36°C to allow primers and templates to anneal, and 15 sec at 72°C for primer extension (Lanes a1, a4, and b1) or 30 sec at 72°C (Lanes a2, a5, and b2) or 60 sec at 72°C (Lanes a3, a6, and b3). Patterns in Lanes a1 to a3, Lanes a4 to a6, and Lanes b1 to b3 were obtained with primers TGACCCCTCC, AGCAGCGTGG, and ACCGTCGTAG, respectively.

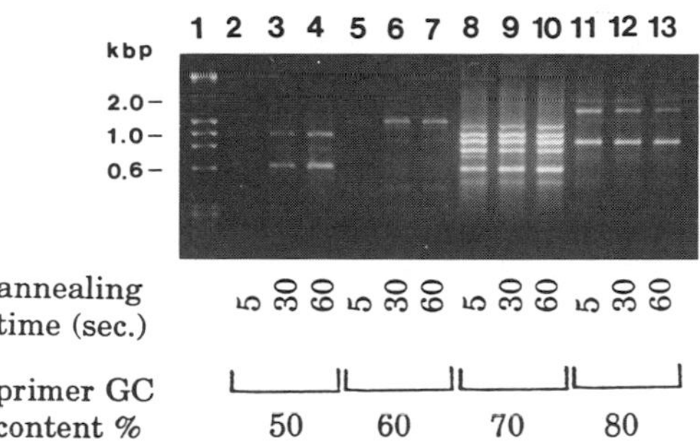


**FIGURE 5.** Effects of primer/template annealing times and primer GC content on RAPD patterns obtained from alfalfa genomic DNA. The patterns were obtained with PCR programs consisting of 5 sec at 94°C to denature the DNA, 5 sec at 36°C (Lanes 2, 5, 8, and 11) or 30 sec at 36°C (Lanes 3, 6, 9, and 12) or 60 sec at 36°C (Lanes 4, 7, 10, and 13) to allow primers and templates and to anneal and 60 sec at 72°C to allow primer extension. Patterns in Lanes 2 to 4, Lanes 5 to 7, Lanes 8 to 10, and Lanes 11 to 13 were obtained with primers TTAGCGGTCT (50% GC), CAAGGGAGGT (60% GC), AGCAGCGTGG (70% GC), and CTGGCGGCTG (80% GC), respectively.

Steps in agarose gel electrophoresis procedure:

1. Prepare an adequate volume of electrophoresis buffer (TAE or TBE; see Table 6) to fill the electrophoresis tank.[a]
2. Add the desired amount of electrophoresis-grade agarose to an appropriate volume of electrophoresis buffer for the gel (see Table 6).
3. Melt the agarose in a microwave oven, and stir to ensure even mixing. (A gel of 1.4% agarose in TAE is commonly used for RAPD analysis.)
4. Cool the melted agarose to 55°C, and pour onto a gel casting platform whose ends have been sealed with stoppers or tape.
5. Insert the comb, and ensure that no bubbles are trapped underneath the teeth.[b]
6. After the gel has hardened, carefully withdraw the comb, and remove the seals from the ends of the platform.

**TABLE 4**
**Appropriate Agarose Concentrations for Separating DNA Fragments of Various Sizes**

| Agarose (%) | Effective range of resolution of linear DNA fragments (Kb) |
|---|---|
| 0.5 | 30–1 |
| 0.7 | 12–0.8 |
| 1.0 | 10–0.5 |
| 1.2 | 7–0.4 |
| 1.5 | 3–0.2 |

**TABLE 5**
**Concentration of Acrylamide Giving Maximum Resolution of DNA Fragments**

| Acrylamide (%) | Size fragments separated (bp) | Migration of bromphenol blue marker (bp) |
|---|---|---|
| 3.5 | 100–1000 | 100 |
| 5.0 | 100–500 | 65 |
| 8.0 | 60–400 | 45 |
| 12.0 | 50–200 | 20 |
| 20.0 | 5–100 | 12 |

Tables 4 and 5 from *Current Protocols in Molecular Biology*, Ausubel et al., Eds. Copyright ©1990. John Wiley & Sons. Reprinted with permission.

7. Place the gel casting platform containing the gel in the electrophoresis tank, and add sufficient electrophoresis buffer to cover the gel to a depth of about 1 mm.
8. Prepare DNA samples by mixing 25 μl of the PCR reaction mixture with 4 μl of the 6X loading buffer (Table 4) in fresh Eppendorf tubes.
9. Load samples into the wells with a micropipette, and take care not to contaminate across wells.
10. Include a lane of molecular markers.
11. Set the voltage to the desired level to begin electrophoresis.
12. When the bromophenol marker dye has migrated almost out of the gel, the separation of the DNA fragments is completed, and the gel can be stained by placing it in water with ethidium bromide (0.5 μg/ml) for 10 min.[c]
13. Visualize the DNA fragments with a UV transilluminator, and record the pattern by photography.

*Notes:*

[a] TBE has a better buffering capacity than TAE and can be reused several times.

[b] We typically use 5 mm thick gels and combs with 1.5 mm × 5.5 mm teeth to give wells of approximately 30 μl.

[c] Ethidium bromide is a mutagen and potential carcinogen. Gloves should be worn, and care should be taken when handling ethidium bromide solution.

**TABLE 6**
**Solutions for Agarose Gel Electrophoresis**

| Materials | Amount | Final concentration |
|---|---|---|
| TAE 50 stock buffer[a] | | |
| Tris base | 242 g | 2.00 *M* |
| Glacial acetic acid | 57.1 ml | 1.00 *M* |
| EDTA | 100 ml 0.5 *M* EDTA (pH 8.0) | 0.05 *M* |
| Water | Bring up to 1 l | |
| TAE 1X buffer[b] working solution | | |
| 50X stock | 40 ml | 0.04 *M* Tris acetate<br>0.001 *M* EDTA |
| Water | 1960 ml | — |
| TBE 5X stock buffer[c] | | |
| Tris base | 54 g | 0.45 *M* |
| Boric acid | 27.5 g | 0.45 *M* |
| EDTA | 20 ml 0.5 *M* EDTA (pH 8.0) | 0.01 *M* |
| Water | Bring up to 1l | |
| TBE 0.5X buffer[b] | | |
| 5X stock | 40 μl | 0.045 *M* Tris-borate<br>0.001 *M* EDTA |
| Water | 1960 ml | — |
| Loading buffer (6X)[d] | | |
| Bromophenol blue | 0.025 g | 0.25% |
| Xylene cyanol | 0.025 g | 0.25% |
| Sucrose | 4.0 g | 40.0% (w/v) |
| Water | 10 ml | — |

[a] Final pH 8.0.
[b] The volume required for one separation using a Bio-Rad electrophoresis unit with a 15 $cm^2$ gel casting platform is 1.5 l running buffer.
[c] Final pH 8.0.
[d] Store at 4°C.

## IV. RAPD ANALYSES

RAPD markers can be used like any other trait to identify or classify genotypes; they segregate in a Mendelian fashion in crosses. Figure 6 shows a RAPD polymorphism between two heterozygous alfalfa plants and its segregation in an $F_1$ population derived from those parents.

### A. BULKING DNA

Considerable time can be saved when analyzing plant populations by bulking DNA samples. With the bulking procedure, one PCR reaction containing a mixture of template DNAs can replace as many as ten PCR reactions containing individual DNA samples. Figure 7 shows that the RAPD pattern from a mixture of alfalfa DNAs (Lane 7) is a superposition of the five RAPD patterns observed in the individual samples (Lanes 2 to 6) that contributed to the mixture. We originally developed the bulking procedure for similarity analyses to compare alfalfa populations that have a large amount of intrapopulation variability caused by the outcrossing nature of this species. It is also be appropriate to bulk individuals in a segregating population according to their phenotypic characteristics when screening for linkages between RAPD markers and plant traits.[36]

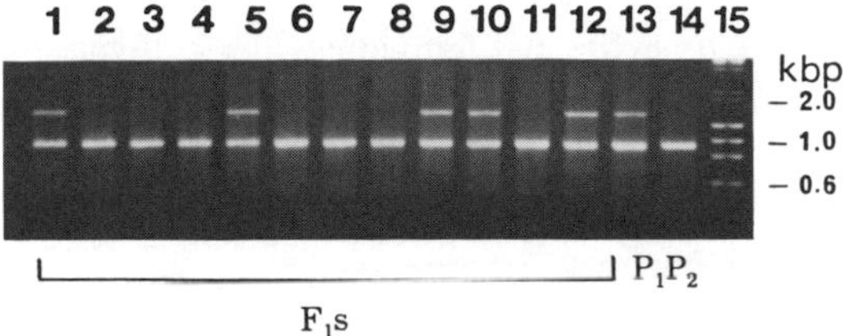


**FIGURE 6.** Segregation of a RAPD polymorphism in alfalfa. Lanes 13 and 14 are RAPD patterns obtained from two heterozygous alfalfa plants; Lanes 1 to 12 are $F_1$ progeny from a cross between the plants in Lanes 13 and 14.

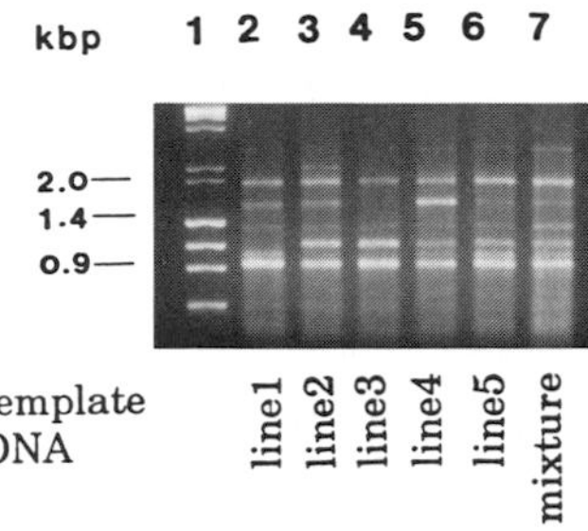


**FIGURE 7.** RAPD patterns from individual and bulked alfalfa samples primed with CTGGCGGCTG. Lanes 2 to 6 are patterns obtained from PCR reactions containing genomic DNA from five alfalfa individuals. The pattern in Lane 7 was obtained from a PCR reaction with the same primer and an equivolume mixture of the DNA samples used for Lanes 2 to 6.

## B. LINKAGES

Linkages between molecular markers and traits of interest are determined by analyzing populations of plants segregating for the trait (or traits). Usually an $F_2$ population is used, but other types of segregating populations such as an $F_1$ population from a cross between heterozygotes, doubled haploid populations, or backcross populations can be used. The simplest analysis for cosegregation of a molecular marker and a trait is a $x^2$ analysis.[2] This analysis may be a simple test of correspondence if the trait is determined by a single dominant gene, or it may be more complex if the trait is determined by multiple and/or recessive genes. In fact, RFLPs linked to quantitative traits such as yield have been determined,[37-39] and this type of analysis should also be possible with RAPD markers once linkage maps consisting of RAPD markers are determined.

Linkage patterns among molecular markers can also be determined. When large enough numbers of markers are available, this information can be used to define linkage maps with programs such as MAPMAKER.[40] RFLP linkage maps for a variety of plant species have been produced, but the use of RAPD markers for this purpose has just begun. Williams et al.[12] showed that RAPD markers and their cognate RFLP probes mapped to the same positions on a soybean linkage map. Furthermore, they suggested that genomic maps consisting of RAPD markers can be obtained more efficiently and with greater density than RFLP-based methods.

## C. IDENTITY/RELATEDNESS

The RAPD patterns can be used to fingerprint plant genotypes or varieties. Previous estimates have indicated that for self-pollinating species the probability of distinguishing 20 inbred lines from each other is 0.99 if 20 RFLP probes are used.[41] Similarly, for an outcrossing species, ten RFLP loci tested against ten individuals per cultivar should suffice to separate one cultivar from another.[42] We have found that between 18 and 70 alfalfa populations can be distinguished from each other on the basis of at least one RAPD marker difference when the RAPD patterns from ten primers are compared (Yu and Pauls, unpublished).

The similarities and differences in the RAPD patterns can be used to calculate genetic distances among individuals or populations. The fraction of shared bands between two samples can be calculated as

$$F = 2X_{1,2}/X_1 + X_2 \tag{2}$$

where $X_{1,2}$ is the number of RAPD fragments with the same molecular weight found in both samples, $X_1$ is the total number of fragments found in one population, and $X_2$ is the total number of fragments found in the other population.[43,44] The genetic distance ($D$) between two samples can be calculated using:

$$D = -\ln(F) \tag{3}$$

Distance values can be used to establish phylogenic trees by the use of various forms of cluster analysis.[45]

## ACKNOWLEDGMENTS

Thanks to Jennifer Kingswell for typing the manuscript and to Kathy Fuchs for providing the data for Figure 3.

Our work was supported by the Ontario Ministry for Food and Agriculture and the Natural Sciences and Engineering Research Council of Canada.

## REFERENCES

1. **Chevre, A. M., This, P., Eber, F., Deschamps, M., Delseny, M., and Quiros, C. F.,** Characterization of disomic addition lines *Brassica napus-Brassica nigra* by isozyme, fatty acids and RFLP markers, *TAG*, 81, 93, 1991.
2. **Helentjaris, T., Slocum, M., Wright, S., Schaefer, A., and Nienhuis, J.,** Construction of genetic linkage maps in maize and tomato using restriction fragment length polymorphisms, *TAG*, 72, 761, 1986.
3. **Landry, B. S., Hubert, N., Etoh, T., Harada, J. J., and Lincoln, S. E.,** A genetic map for *Brassica napus* based on restriction fragment length polymorphisms detected with expressed DNA sequences, *Genome*, 34, 543, 1991.
4. **Osborn, T. C., Alexander, D. C., and Fobes, J. F.,** Identification of restriction fragment length polymorphisms linked to genes controlling soluble solids content in tomato fruit, *TAG*, 73, 350, 1987.
5. **Lee, D., Turner, L., Davies, D. R., and Ellis, T. H. N.,** An RFLP marker for $r_b$ in pea, *TAG*, 75, 362, 1988.
6. **Song, K. M., Osborn, T. C., and Williams, P. H.,** Brassica taxonomy based on nuclear restriction fragment length polymorphisms (RFLPs). 1. Genome evolution of diploid and amphidiploid species, *TAG*, 75, 784, 1988.
7. **Song, K. M., Osborn, T. C., and Williams, P. H.,** Brassica taxonomy based on nuclear restriction fragment length polymorphisms (RFLPs). 2. Preliminary analysis of subspecies within *B. rapa* (syn. *campestris*) and *B. oleracea*, *TAG*, 76, 593, 1988.
8. **McGrath, J. M., Quiros, C. F., Harada, J. J., and Landry, B. S.,** Identification of *Brassica oleracea* monosomic alien chromosome addition lines with molecular markers reveals extensive gene duplication, *MGG*, 223, 198, 1990.

9. **Saiki, R. K., Scharf, S., Faloona, F., Mullis, K. B., Horn, G. T., Erlich, H. A., and Arnheim, N.,** Enzymatic amplification of β-globin genomic sequences and restriction site analysis for diagnosis of sickle cell anemia, *Science*, 230, 1350, 1985.
10. **Mullis, K. B. and Faloona, F. A.,** Specific synthesis of DNA *in vitro* via a polymerase-catalyzed chain reaction, *Meth. Enzymol.*, 155, 335, 1987.
11. **Saiki, R. K., Gelfand, D. H., Stoffel, S., Scharf, S. J., Higuchi, R., Horn, G. T., Mullis, K. B., and Erlich, H. A.,** Primer-directed enzymatic amplification of DNA with a thermostable DNA polymerase, *Science*, 239, 487, 1988.
12. **Williams, J. G., Kubelik, A. R., Livak, K. J., Rafalski, J. A., and Tingey, S. V.,** DNA polymorphisms amplified by arbitrary primers are useful as genetic markers, *Nucl. Acids Res.*, 18, 6531, 1990.
13. **Nakamura, Y., Leppert, M., O'Connel, P., Wolff, R., Holm, T., Culver, M., Martin, C., Fujimoto, E., Hoff, M., Kumlin, E., and White, R.,** Variable number tandem repeat (VNTR) markers for human gene mapping, *Science*, 235, 1616, 1987.
14. **Skolnick, M. H. and Wallace, R. B.,** Simultaneous analysis of multiple polymorphic loci using amplified sequence polymorphisms (ASPs), *Genomics*, 2, 273, 1988.
15. **Love, J. M., Knight, A. M., McAleer, M. A., and Todd, J. A.,** Towards construction of a high resolution map of the mouse genome using PCR-analysed microsatellites, *Nucl. Acids Res.*, 18, 4123, 1990.
16. **Weining, S. and Langridge, P.,** Identification and mapping of polymorphisms in cereals based on the polymerase chain reaction, *TAG*, 82, 209, 1991.
17. **Welsh, J. and McClelland, M.,** Fingerprinting genomes using PCR with arbitrary primers, *Nucl. Acids Res.*, 18, 7213, 1990.
18. **Welsh, J., Honeycutt, R. J., McClelland, M., and Sobral, B. W. S.,** Parentage determination in maize hybrids using the arbitrarily primed polymerase chain reaction (AP-PCR), *TAG*, 82, 473, 1991.
19. **Goodwin, P. H. and Annis, S. L.,** Rapid identification of genetic variation and pathotype of *Leptosphaeria maculans* by random amplified polymorphic DNA assay, *Appl. Environ. Microsc.*, 57, 2482, 1991.
20. **Quiros, C. F., Hu, J., This, P., Chevre, A. M., and Delseny, M.,** Development and chromosomal localization of genome specific markers by polymerase chain reaction in Brassica, *TAG*. accepted for publication.
21. **U, N.,** Genomic analysis in *Brassica* with special reference to the experimental formation of *B. napus* and peculiar mode of fertilization, *Jpn. J. Bot.*, 7, 389, 1935.
22. **Martin, G. B., Williams, J. G. K., and Tanksley, S. D.,** Rapid identification of markers linked to a *Pseudomonas* resistance gene in tomato by using random primers and near-isogenic lines, *PNAS*, 88, 2336, 1991.
23. **Caetano-Anollés, G., Bassam, B. J., and Gresshoff, P. M.,** DNA amplification fingerprinting using very short arbitrary oligonucleotide primers, *Biotech*, 9, 553, 1991.
24. **Caetano-Anollés, G., Bassam, B. J., and Gresshoff, P. M.,** DNA amplification fingerprinting: a strategy for genome analysis, *Plant Mol. Biol. Rep.*, 9, 294, 1991.
25. **Rogers, S. O. and Bendich, A. J.,** Extraction of DNA from plant tissues, in *Plant Molecular Biology Manual*, Gelvin, S. B. and Schilperoort, R. A., Eds., Kluwar Academic Publishers, Dordrecht, 1988, A6:1-10.
26. **Bendich, A. J., Anderson, R. S., and Ward, B. L.,** Plant DNA: long, pure and simple, in *Genome Organization and Expression*, Leaver, C. J., Ed., Plenum Press, New York, 1988, 31.
27. **Murray, H. G. and Thompson, W. F.,** Rapid isolation of higher molecular weight DNA, *Nucl. Acids Res.*, 8, 4321, 1980.
28. **Taylor, B. and Powell, A.,** Isolation of plant DNA and RNA, *Focus*, 4, 4, 1982.
29. **Ausubel, F. M., Brent, R., Kingston, R. E., Moore, D. D., Seidman, J. G., Smith, J. A., and Struhl, K.,** *Current Protocols in Molecular Biology*, John Wiley & Sons, Inc., New York, 1990, chap. 15.
30. **Edwards, K., Johnstone, C., and Thompson, C.,** A simple and rapid method for the preparation of plant genomic DNA for PCR analysis, *Nucl. Acids Res.*, 19, 1349, 1991.
31. **Higuchi, R.,** Simple and rapid preparation of samples for PCR, in *PCR Technology: Principles and Applications for DNA Amplification*, Erlich, H. A., Ed., Stockton Press, New York, 1989, 31.
32. **Maniatis, T., Fritsch, E. F., and Sambrook, J.,** *Molecular Cloning*, Cold Spring Harbor Laboratory Press, Cold Spring Harbor, NY, 1982, 468.
33. **Sambrook, J., Fritsch, E. F., and Maniatis, T.,** *Molecular Cloning*, Cold Spring Harbor Laboratory Press, Cold Spring Harbor, NY, 1989, chap. 14.
34. **Ausubel, F. M., Brent, R., Kingston, R. E., Moore, D. D., Seidman, J. G., Smith, J. A., and Struhl, K.,** *Current Protocols in Molecular Biology*, John Wiley & Sons, Inc., New York, 1990, chap. 2.
35. **Goldman, D. and Merril, C. R.,** Silver staining of DNA in polyacrylamide gels: linearity and effect of fragment size, *Electrophoresis*, 3, 24, 1982.
36. **Michelmore, R. W., Paran, I., and Kesseli, R. V.,** Identification of markers linked to disease-resistance genes by bulked segregant analysis: a rapid method to detect markers in specific genomic regions by using segregating populations. *Proc. Natl. Acad. Sci. U.S.A.*, 88, 9828, 1991.

37. **Paterson, A. H., Lander, E. S., Hewitt, J. D., Peterson, S., Lincoln, S. E., and Tanksley, S. D.,** Resolution of quantitative traits into Mendelian factors by using a complete linkage map of restriction fragment length polymorphisms, *Nature*, 335, 721, 1988.
38. **Tanksley, S. D., Young, N. D., Paterson, A. H., and Bonierbale, M. W.,** RFLP mapping in plant breeding: new tools for an old science, *Biotech*, 7, 257, 1989.
39. **Lander, E. S. and Botstein, D.,** Mapping Mendelian factors underlying quantitative traits using RFLP linkage maps, *Genetics*, 121, 185, 1989.
40. **Lander, E. S., Green, P., Abrahamson, J., Barlow, A., Daly, M. J., Lincoln, S. E., and Newburg, L.,** Mapmaker: an interactive computer package for constructing primary genetic linkage maps of experimental and natural populations, *Genomics*, 1, 174, 1987.
41. **Landry, B. S. and Michelmore, R. W.,** Methods and applications of restriction fragment length polymorphism analysis to plants, in *Tailoring Genes for Crop Improvement*, Brueing, Ed., 25, 1985.
42. **Soller, M. and Beckman, J. S.,** Genetic polymorphism in varietal identification and genetic improvement, *TAG*, 67, 25, 1983.
43. **Nei, M.,** *Molecular Evolutionary Genetics*, Columbia University Press, New York, 1987.
44. **Packer, C., Gilbert, D. A., Pusey, A. E., and O'Brien, S. J.,** A molecular genetic analysis of kinship and co-operation in African lions, *Nature*, 351, 562, 1991.
45. **Swafford, D. L. and Olson, G. J.,** Phylogeny reconstruction, in *Molecular Systematics*, Hillis, D. M. and Moritz, C., Eds., Sinauer Associates, Sunderland, U.S.A., 1990, chap. 11.

# 19

# Detection and Characterization of Plant Pathogens

*Paul H. Goodwin and Annette Nassuth*

## I. INTRODUCTION

Traditionally, the detection and characterization of a plant pathogen has relied on isolation of the microorganism and observation of the symptoms it induces in susceptible host plants. This is a time-consuming process and even then often cannot reliably distinguish between closely related species and strains. Faster and more discriminating detection became possible with the characterization of proteins and nucleic acids from many microbes. Based on this information, antibodies or nucleic acid probes could be chosen that were capable of detecting all members of a species or only one strain, for example a pathogenic versus nonpathogenic strain.

This chapter discusses detection methods that require minimal sample preparation and are therefore suitable for routine testing. Emphasis is placed on features that are unique for the different plant pathogens: viruses, fungi, bacteria, and mollicutes. The discussion includes probe selection, sensitivity of the assay, and, last but not least, sample preparation.

## II. VIRUSES

Assays for viruses should be highly sensitive and accurate as well as relatively inexpensive, adaptable to field conditions, and amenable to large numbers of samples in order to be useful in breeding programs, epidemiological studies, and certification of healthy planting stock.[1] Several assays that fit these criteria have been developed and are discussed below. They are based on the use of antibodies or nucleic acid probes that specifically recognize and bind to the virus coat protein or the viral nucleic acid genome, respectively. This specificity allows screening of crude plant or insect extracts containing a mixture of proteins and nucleic acids. Current practice is to use one of these assays in combination with infectivity tests[1] and/or immunoabsorbent electron microscopy.[2]

Different antibody preparations are available for the detection of many plant viruses. Polyclonal antibodies may detect several isolates, whereas monoclonal antibodies can differentiate between isolates.[3-7] An enzyme directly conjugated to the antibody is usually used to generate the signal indicating that the antibody has bound to the antigen. Easy visual qualitative assessment of the viral antigen is possible by adding a substrate that gives a colored product or induces a color change of an indicator dye. Two widely used enzymes are alkaline phosphatase and horseradish peroxidase.[8] Assays using penicillinase are equally sensitive but less expensive.[9] Other less popular enzymes are β-galactosidase and urease.[10,11] When choosing an enzyme, one must take into consideration endogenous plant or insect enzyme activities which can produce high background reactions.[12]

0-8493-5164-2/93/$0.00+$.50
© 1993 by CRC Press, Inc.

The microplate method of the enzyme-linked immunosorbent assay (ELISA) has been adapted by Clark and Adams[13] for routine testing of plant viruses, and subsequently many variants of this assay have been developed.[8,14-16] It is economical and can be used for quantitative measurement. Usually, an antibody bound to the plate is used to capture the viral antigen, and the same or a different virus-specific antibody, now linked to enzyme, is used to detect the captured antigen (double antibody sandwich (DAS)-ELISA). A disadvantage of this approach is that separate enzyme-antibody conjugates must be prepared for each virus to be tested. In addition, such modified antibodies become more restricted in their capacity to bind to antigens, often causing the reaction to become strain specific.[17] However, this can be counteracted by conjugating the enzyme to a secondary, antiglobulin antibody (indirect double antibody sandwich method) which is commercially available. Another consideration is that the reactivity of antibodies, particularly monoclonal antibodies, may differ greatly depending on the ELISA procedure since adsorption to a surface can alter their binding specificity.[18] Modifications to maximize the sensitivity and to minimize the background signals include the preabsorption of antibodies to healthy host extracts, the use of milk proteins instead of bovine serum albumin as a blocking agent, and the simultaneous incubation of samples and conjugate (cocktail ELISA).[19,20]

The blot immunoassay methods include spotting the crude extract containing the viral antigen onto nitrocellulose or, more expensive but less brittle, nylon membrane. Whatman no. 1 paper has also been used as solid support, but this seems suitable only for viruses which are present in high concentrations, since Whatman paper retains lower quantities of protein than membranes.[21] An advantage of using membranes is that it is possible to store the blot for a prolonged period of time or send it elsewhere for assay. The dot blot immunoassay (DBIA) was found to be eight times more sensitive than ELISA for the detection of tomato spotted wilt virus (TSWV) in extracts from infected *Nicotiana benthamiana* leaves[22] and five times more sensitive for the detection of purified potyviruses.[12] This is likely caused by the deposition of more antigen on the membrane than in the polyvinyl chloride ELISA microplates.[22] The sensitivity of DBIA can be enhanced even further by limiting the area of application and/or by applying a larger volume, and thus more antigen, to a relatively small area with the aid of a dot- or slot-blot apparatus. Routine quantitative measurements are possible with a reflectance densitometer.[23] Antigen can also be applied by pressing a newly cut tissue surface, or squashing an entire structure such as a leaf, flower, or insect, onto the membrane (tissue blot immunoassay). This method requires no sample preparation and is therefore well suited for use in the field for large numbers of samples.[24] Tissues where only a small portion of the cells are infected may give only a weak signal or none at all with ELISA, but test positive on tissue blots when assessed visually or by microscopic observation.[22] Tissue blotting has been used successfully for the detection of a variety of viruses in tissues such as leaves, petioles, stems, and corms. However, not all plants have tissues that lend themselves easily to blotting; e.g., blots can be made from monocot leaves only after removal of the lower epidermis.[25]

Serological methods are not suitable for viroids or for viruses that are poor immunogens or show large serological variation. This and the need for higher sensitivity for the detection of viruses present at lower levels, especially in field-infected plants, prompted the development of detection methods based on nucleic acid probes.[26] The desired level of specificity can be determined through the choice of the probe; different regions of the genome among related viruses are often conserved to different extents. Probes based on conserved genomic sequences have been used successfully to detect several if not all members of a virus group. For example, luteoviruses were detected by probes derived from part of the coat protein gene.[7,27] Isolate-specific cDNA probes can distinguish between biological isolates which appear identical in serological tests using polyclonal antibodies.[28,29] RNA probes were found to be superior to DNA probes because double-stranded DNA probes lose activity by self reassociation and because RNA-RNA hybrids are more stable than DNA-RNA hybrids, thus permitting posthybridization washings under more stringent conditions.[27,30] However, for some tests with RNA probes, an additional wash with a ribonuclease-containing solution was found to be essential to eliminate nonspecific reactions.[27]

[$^{32}$P]-labeled probes are widely used but less desirable for routine tests because of the short half-life of this isotope, the long time needed for exposure, and its potential health hazard. Consequently, nonradioactive probes including biotin-labeled DNA and RNA probes have been developed[26,31] and used succesfully.[32-35] Assays using biotin-labeled probes are faster and are reputed to be equally as sensitive as those using radiolabeled probes.[33,34] An additional advantage is that large quantities of the biotinylated probe can be synthesized once and then stored in aliquots for long periods until needed. Background reactions caused by endogenous biotin present in crude extracts can be minimized by the use of an appropriate extraction buffer.[36]

The dot blot hybridization assay (DBHA) is generally able to detect at least 100X lower amounts of virus than serological tests with polyclonal or monoclonal antibodies.[7,28,30,37-39] For example, the high sensitivity of the DBHA permits detection of virus in individual leafhoppers[40] or whiteflies.[41] For some viruses, such as tomato spotted wilt virus (TSWV), detection levels of DBHA and ELISA are similar. Hybridization is often still preferred because DAS-ELISA gives high background readings with extracts from hosts like pepper.[42]

Virus detection by hybridization is also possible with a tissue blot, but only when the membrane is presoaked with SDS and proteinase K to prevent degradation of viral RNA by plant ribonucleases.[43] This pretreatment is apparently very effective since such blots can be kept for 6 months at ambient temperature without any reduction in their hybridization capacity.[43] The ability to store blots and the fact that plants can be quickly sampled in the field by untrained personnel make the tissue blot hybridization assay a potential tool for large-scale diagnosis. Its sensitivity compares favorably with dot blot hybridizations for plant tissue[43] as well as for large insects such as leafhoppers.[40]

Polymerase chain reaction (PCR)-based assays can detect very low amounts of plant virus nucleic acids[44-48] and viroids[49] and can be more sensitive than ELISA and/or hybridization assays.[44,45,47] As most plant viruses and viroids have an RNA genome, a reverse transcription (RT) step is first required. The amplified DNA product produced by PCR or RT-PCR can be analyzed by gel electrophoresis and/or hybridization. Primers that recognize relatively short homologous sequences present in a wide range of viruses can be used for a general diagnostic assay. Known and suspected members of the potyvirus group were detected by primers based on the core domain of the coat protein and part of the NIb replicase protein.[48] An advantage of PCR is that such primers can detect simultaneously a range of different viruses that normally require several hybridization or serological tests, as has been shown for luteoviruses[46] and geminiviruses.[50] The different virus isolates can subsequently be identified easily by restriction enzyme analysis. The simultaneous addition of several specific primer pairs enables detection of unrelated viruses or viroids in one assay.[51] Identification of amplified products is based on their size and subsequent restriction and/or hybridization analysis. The amplified fragment can also be sequenced and the sequence data used for the design of virus-specific PCR primers.[46] It can be expected that the RT-PCR assay will become common in routine testing of several plant viruses, especially because crude plant extracts can be used, and because reverse transcription and PCR can be performed sequentially in the same tube, thus avoiding pipetting of the sample.[45] While amplification of viral nucleic acid from squash blots would be ideal for screening field samples, the extreme sensitivity of the PCR assay creates problems with contamination from other samples containing viral nucleic acid, and this might be difficult to overcome.

In conclusion, the sensitivity of plant virus detection depends on host, virus, and assay technique. The ELISA assay is often the method of choice when a whole set of antibodies is available and the amount of virus in the sample is at the nanogram range (for method see[8,14-16]). Tissue blot assays have the advantage of being more "field friendly" and can be used to assay for viral antigen or viral nucleic acid (described below). Hybridization generally raises the detection level to the picogram range. PCR-based assays can be even more sensitive if specific primers are used,[45] or more flexible in the range of detection and identification if degenerate primers are used. A relatively simple one-tube protocol[45] (see method below) might be applied for routine laboratory testing.

## A. DETECTION OF PLANT VIRUSES BY TISSUE BLOT IMMUNOASSAY

This procedure is modified from Lin et al.[24] and Hsu and Lawson.[22]

1. Excise a piece of fresh plant tissue using a new razor blade for each piece, or obtain an insect.
2. Squash the sample using a hard object such as a glass rod, or press the newly cut surface with a firm but gentle force onto a dry nylon or nitrocellulose membrane (0.45 μm). The blot can be stored at room temperature until needed.
   Transferred proteins can be visualized by staining with 0.5% Ponceau S in 1% acetic acid for 5 min and destaining in water. Completely destain in water prior to Step 3. This can take overnight for leaf squashes! The Ponceau S solution can be stored at room temperature and reused.
3. Immerse blot in blocking buffer (phosphate-buffered saline solution (PBS, 0.02 *M* $K_2HPO_4$, 0.15 *M* NaCl, pH 7.4) containing 1 to 3% instant nonfat dry milk) for 60 min at room temperature. Rinse in PBS-Tween solution (PBS containing 0.05% Tween-20)
   1 to 2% bovine serum albumin (BSA) can be used instead of milk proteins.
4. Incubate membrane with virus-specific primary antibodies diluted in blocking buffer at room temperature for 1 h, or at 4°C overnight. Wash three times in PBS-Tween solution, 10 to 15 min each time.
   The optimal dilution is usually 1:100 to 1:1000 for polyclonal antibodies and >1:1000 for monoclonal antibodies.[53]
   To use as little valuable primary antibody as possible, one can do the incubation in a sealed plastic pouch. About 1 ml is sufficient for a 5 × 6 cm blot.
   Overnight incubations at room temperature might be required for low titer antisera.
5. Incubate with alkaline phosphatase-labeled anti-immunoglobulin secondary antibodies diluted in blocking buffer at room temparature for 1 h. Wash three times, 10 to 15 min each, in PBS-Tween solution.
   Dilute antibodies as suggested by manufacturer.
6. Incubate in darkness in carbonate buffer (0.1 *M* $NaHCO_3$ and 1 m*M* $MgCl_2$, pH 9.8) containing the substrates nitroblue tetrazolium (NBT, 300 μg/ml) and 5-bromo-4-chloro-3-indolyl phosphate (BCIP, 150 μg/ml) until purple color develops. Stop the reaction by rinsing the membrane in water.
   Store stocks of NBT (50 mg/ml 70% *N,N*-dimethylformamide [DMF]) and BCIP (25 mg/ml DMF), protected from light, at 4°C (up to 3 months).
   Optimal incubation times usually fall between 10 min and 2 h.

*Note:* Use of biotinylated primary antibodies in Step 4 and avidin-enzyme conjugates in Step 5 can improve the sensitivity.[22]

## B. DETECTION OF PLANT VIRUSES BY TISSUE BLOT HYBRIDIZATION ASSAY

This procedure is carried out essentially according to Navot et al.[43]

1. For RNA viruses: saturate the membrane with 0.5% sodium dodecyl sulfate (SDS) and 100 μg/ml proteinase K, and let dry.

2. See Steps 1 and 2 above.
3. Bind nucleic acid to nitrocellulose by baking the membrane for 2 h at 80°C under vacuum, or to a nylon membrane by 2 min exposure to UV light (254 nm). Prehybridization and hybridization conditions are essentially as described for northern hybridizations.[53,54] Optimal stringency conditions will vary with the probe used and must be determined experimentally.

### C. DETECTION OF PLANT VIRUSES OR VIROIDS BY A ONE TUBE PCR AMPLIFICATION PROCEDURE

This procedure is carried out essentially according to Wetzel et al.[45]

1. Put leaf tissue in a plastic bag containing gauze (Bioreba), add sterile water (w/v = 1/4), and grind using a rolling grinder. Collect the plant sap in a 1.5 ml tube, and centrifuge for 10 min in a microfuge at full speed. Dilute the supernatant ten-fold with sterile water.
2. Take 10 μl of the diluted extract, and disrupt virus particles by treatment with 1% Triton X-100 for 10 min at 65°C. Denature target nucleic acid template by incubating for 10 min at room temperature after adding methyl mercury hydroxide to a final concentration of 10 m*M*. Neutralize with 20 m*M* 2-mercaptoethanol for 10 min at room temperature.
3. Add reverse transcription mixture, containing 50 m*M* Tris-HCl, pH 8.3, 50 m*M* KCl, 7.5 m*M* $MgCl_2$, 0.1 U Inhibit ACE (5′,3′ Inc.), 20 μ*M* dNTPs, 0.1 μ*M* primer, and 0.5 U avian myeloblastosis virus reverse transcriptase to the tube to give a final volume of 20 μl. Incubate at 42°C for 45 min.
4. Add 80 μl PCR buffer (10 m*M* Tris-HCl, pH 8.3, 50 m*M* KCl, 1.5 m*M* $MgCl_2$, 0.1% gelatin) containing 1 μ*M* of both primers, 200 μ*M* dNTPs, and 2 U of *Taq* DNA polymerase, and overlay with 100 μl of mineral oil. Amplify for 40 cycles of template denaturation at 92°C (1 min), primer annealing at 62°C (2 min), and DNA synthesis at 72°C (2 min).
5. Analyze reactions by electrophoresis of aliquot, and visualize bands either by ethidium bromide or silver staining.[53,54]
6. Analyze amplified fragments with restriction endonucleases.[53,54]

*Note:* Dilution of reverse transcription mixture results in a decrease of the $MgCl_2$ concentration, avoiding nonspecific hybridizations of primers which can result from high concentrations of $MgCl_2$.

*Note:* Optimal temperatures and time periods for the different steps of the reverse transcriptase and PCR reaction are dependent on the primers and template nucleic acid used and need to be determined experimentally.

## III. FUNGI

To identify plant pathogenic fungi, microscopic spores and spore-bearing structures are examined directly from fungi on infected plant materials or from fungi isolated and cultured on artificial media. In many cases, trained plant pathologists can make rapid presumptive identifications from characteristic disease symptoms. However, distinctive symptoms are not always available, and fungi can be difficult to isolate and identify in culture. Some fungi grow slowly in culture and may take weeks to months to produce the spore-bearing structures needed for identification.[55] It requires considerable skill, experience, and patience to identify many plant pathogenic fungi.

Because of these problems, alternative methods of identification have been sought. Serological methods have been attempted but with limited success, primarily because fungal antigens are complex and not well characterized.[56] Recently, specific DNA probes have begun to be employed to identify plant pathogenic fungi. Because any fungal cell containing DNA can be identified, the lack of spores or spore-bearing structures is unimportant. DNA probes are unaffected by variability in fungal morphological features, and can be used by a wider range of laboratory personnel than those with the specialized expertise required to distinguish minor morphological differences among fungi. Also, slow growing fungi can be more easily identified because target DNA can be extracted from small amounts of mycelium or directly from infected plant material.

Not surprisingly, the limited number of species-specific DNA probes developed so far have been for fungi which are generally difficult to identify quickly. Soil-borne fungi that attack roots and/or crowns cause relatively nonspecific symptoms such as a generalized rot and death of the plant, and many have the identification problems described above when using traditional methods. Examples of soil-borne fungi for which DNA probes have been developed are species of *Phytophthora*,[57,58] *Gaeumannomyces*,[59] *Pythium*,[60] and *Leptosphaeria*.[55] Vascular pathogens also produce nonspecific symptoms, such as foliar chlorosis, wilting, and die-back, and examples of vascular pathogens for which DNA probes have been developed are species of *Phoma*[61,62] and *Verticillium*.[63]

Most DNA probes have been developed to identify fungi by hybridization-based assays such as dot blot hybridization. To select probes, random chromosomal restriction fragments have been cloned.[55,57,58,61] Specificity is selected based on an inability of the cloned fragment to hybridize with DNA from host plants, and DNA from either taxonomically related fungi[57,58] or other fungi associated with the same host.[55,61] The relatively small numbers of clones that need to be screened in order to find a specific DNA probe indicates that there are a number of DNA sequences unique to a fungal species even within a fraction of their genome. In some cases, specificity may be greater than expected. DNA probes specific to *Phytophthora citrophthora* hybridized to all *P. citrophthora* isolates except a collection of isolates from cocoa.[58] Perhaps species-specific DNA probes will prove useful in defining as well as identifying fungal species.

An alternative approach to probe selection is to clone fragments of mitochondrial DNA. The advantage of mitochondrial DNA is that, because of its relatively small size, restriction maps can be developed and analyzed for conserved and nonconserved regions. In *Pythium*, this was done to select DNA probes for certain species.[60] However, some of these probes only hybridized to a subset of isolates which shared the same mitochondrial restriction map, whereas other probes hybridized to more than one *Pythium* species.[60] In the cases of *Gaeumannomyces graminis*[59] and *Peronosclenospora sorghi*,[64] however, cloned mitochondrial fragments were highly specific and only showed little or no homology to other fungi. Although a lack of species specificity may prevent the use of a probe for identification by dot blot hybridization, a nonspecific probe may still be useful in RFLP analysis.

For chromosomal DNA, multiple copy DNA probes have been selected because they have several advantages. Detection of highly repetitive DNA sequences improves both the sensitivity of an assay because the signal is present in multiple copies and its reliability because variation in one copy in the genome would not reduce greatly the total signal observed in a hybridization-based assay. In addition, repetitive DNA has a very high probability of being species-specific. For example, DNA of *Leptosphaeria korrae* which was digested, fractionated, and stained with ethidium bromide, produced several discrete intensely stained bands, and DNA cloned from these bands proved to be specific.[55] One of the cloned sequences had a copy number of 50 to 100 in the *L. korrae* genome.[55] High copy number DNA probes also have been selected by screening clones with labeled DNA of the organism of interest, and then selecting those clones which gave the most intense hybridization signal.[57,58,61] The importance of choosing repetitive DNA as a probe was demonstrated with *Phoma tracheiphila*, which could not be detected in practical applications with probes not containing multiple copy DNA.[61]

The need for high sensitivity of a DNA probe assay is related to its applications in plant pathology. Since root rot and vascular wilt fungi are the most common targets of DNA probes, it is important that a probe be able to detect the organism in roots, stems, leaves, and soil. Hybridization-based assays have been successful in detecting pathogens by extracting DNA from roots of infected tomatoes[65] and turf grass.[55] The use of lignified tissue, such as citrus roots[58] and lemon seedlings,[61] requires that the fungus be allowed to grow for a period of time in culture before DNA detection. Because of their greater sensitivity, PCR-based assays permitted detection of the fungus directly in wheat[66] and lemon[62] tissue, eliminating the need to culture the organism. A PCR-based assay for *Verticillium albo-atrum* also was able to directly detect the pathogen in infected alfalfa stems, leaves, and roots.[63] Low quantities of fungal pathogens in soil probably preclude direct detection by a hybridization-based assay, but such an assay can be used if the fungus is first allowed to grow on culture media or host-bait tissue.[65]

A hybridization-based assay is equivalent to a serological assay (ELISA) in its sensitivity.[55] The most commonly reported limit for unambiguous detection is 1 ng of fungal DNA.[55,57,58,64] To achieve greater sensitivity, a PCR-based assay can be used. DNA primers for PCR can be selected based on sequence data from previously cloned species-specific DNA probes,[62,66] or from sequencing specific regions such as the internal transcribed spacers of ribosomal DNA.[63] For *P. tracheiphila*, only one amplified DNA band of the appropriate size and DNA sequence was obtained with the selected primers, and no amplification products were obtained from DNA of a healthy plant or a related fungus.[62] Only one amplified product also was observed with primers to *V. albo-atrum* and *V. dahliae*.[63] The primers were species-specific, and no amplification was observed using DNA of alfalfa plants inoculated with *V. dahliae* to which it is resistant.[63] For *G. graminis*, however, the primers also amplified DNA from healthy plants and other fungi.[66] These nonspecific amplified products were discernable since they were of a different size than those of *G. graminis*, but they could still give rise to ambiguities and uncertainties in the results. Therefore, nested primers were employed where the products of the first amplification were used in a second round of amplification, with a second pair of primers annealing inside the previously amplified product.[66] This resulted in species-specific amplification with two PCR products, one of which was the predicted size and the other a larger product that may have been amplified from another region of the genome of *G. graminis*.[66] Although DNA sequences from species-specific DNA clones do not guarantee a specific PCR-based assay, PCR-based assays are still sufficiently adaptable to achieve highly specific detection. The benefits in terms of sensitivity are evident for *P. tracheiphila* where a PCR-based assay had a detection limit 0.01 pg of fungal DNA.[62]

## A. DETECTION OF FUNGI IN INFECTED PLANT TISSUE WITH A DNA PROBE

This procedure is based on the method of Tisserat et al.[55]

1. Excise 200 to 400 mg of infected plant tissue.
2. Freeze in 1.5-ml microfuge tube by adding liquid nitrogen.
3. Grind with a steel rod. Construct rod by rounding the tip of a Philips screwdriver with a grinder.
4. Suspend frozen, ground samples in 600 μl 2X CTAB buffer (2X CTAB = 1.4 *M* NaCl, 2% hexadecyltriethylammonium bromide, 1% 2-mercaptoethanol, 10 m*M* Tris-HCl, pH 8.0).
5. Extract twice with chloroform.
6. Precipitate by adding 0.8 vol isopropanol.
7. Resuspend pellet containing DNA in 40 μl TE buffer (TE = 10 m*M* Tris, pH 7.6, 1 m*M* EDTA).

8. Denature DNA at 95°C for 4 min. Transfer 20 µl to a nylon membrane in a slot-blot apparatus. Bake the membrane at 80°C for 2 h.
9. Hybridize with DNA probe as descibed for Southern hybridizations.[53,54]

## IV. BACTERIA

In the field, bacterial diseases of plants can have variable symptoms, and many have long latency periods before symptoms develop. It is also not unusual to see symptoms caused by other abiotic and biotic factors that mimic bacterial diseases. In order to identify plant pathogenic bacteria, the causal agent must be isolated in culture and examined by a series of physiological tests and usually a pathogenicity assay. Because most plant pathogenic bacteria are Gram-negative rods, other characteristics, such as nutritional versatility, enzyme activity, chemical composition of cell components, etc. must also be tested. Pathogenicity assays are important especially to distinguish pathovars which, by definition, can only be differentiated from one another based on host range and symptoms on a particular host. The result is that identification of plant pathogenic bacteria can be very time-consuming, laborious, and sometimes ambiguous and subjective.

Rapid, specific, and sensitive detection of plant pathogenic bacteria is highly important in disease management. Many are transmitted by seed or other planting materials such as tubers and transplants, and the only practical means of control is to plant disease-free materials. To ensure this, certification programs have been developed where fields are surveyed for disease, and planting materials are assayed for bacteria. This entails testing large numbers of samples, and potentially having to detect relatively low numbers of bacteria in seeds and presymptomatic plants. To improve detection, numerous semiselective media have been developed, but physiological tests are still required frequently to confirm the identity of colonies growing on semiselective media.[67] An alternative is serological assays,[56] but monoclonal antibodies may be too strain-specific, and polyclonal antibodies may cross-react with saprophytic and other plant pathogenic bacteria. These problems have been cited frequently by researchers as the rationale for searching for an alternative means of identification such as developing specific DNA probes.[68-72]

Of the current diagnostic DNA probes developed for plant pathogenic bacteria, the vast majority have been designed to detect bacteria known to be transmitted by planting materials. These include the seed-transmitted pathogens of beans, *Pseudomonas syringae* pv. *phaseolicola*[73,74] and *Xanthomonas campestris* pv. *phaseoli*,[75] the seed- and transplant-transmitted pathogens of tomato, *Clavibacter michiganense* subsp. *michiganense*,[70] *P. s. tomato*,[68,76] and *X. c. vesicatoria* (copper resistant strains),[77] and the tuber-transmitted pathogens of potato, *Erwinia carotovora*[72] and *C. m. sepedonicium*.[67,71] Diagnostic DNA probes have also been designed to detect particular genotypes of plant pathogenic bacteria such as tumoregenic strains of *Agrobacterium tumeficiens*,[78] and *Erwinia herbicola* pv. *gypophilae*.[79]

The specificity of diagnostic DNA probes for plant pathogenic bacteria can be at the genus, species, pathovar, or race levels. One approach to developing a diagnostic DNA probe is to base selection on taxonomic comparisons. A specific probe is found by identifying a DNA fragment that is present in the organism of interest but absent in bacteria that are closely related taxonomically. A commonly studied DNA sequence for taxonomic purposes is the 16S ribosomal RNA, and by comparing these sequences between a number of plant pathogenic bacteria, probes specific at the genus level have been identified.[80] However, there were no consistent major sequence differences that were specific at the species or pathovar level. A taxonomic-comparison approach to probe development was successful in screening random cloned DNA fragments to find DNA probes specific to the subspecies, *C. m. sepedomicum*[67,71] and *C. m. michiganense*,[70] and the same approach was also used successfully in developing a specific DNA probe to *Erwina carotovora*, where specificity at the species rather than subspecies level was sought.[72] In contrast, taxonomic comparisons were less successful in selecting specific DNA probes for pathovars of *X. campestris* and *P. syringae*. Random DNA probes were developed both for *X. c. phaseoli* and *P. s. tomato* which were relatively specific but did show some homology to other pathovars of the same species.[75,76] Considering the large number of pathovars of *X. campestris* and *P. syringae* and the uncertainty as to their genetic relatedness, this approach may not provide the most efficient means of obtaining a truly pathovar-specific probe for these bacterial species.

A second approach to developing a specific DNA probe is to use a DNA fragment related to the production of a specific metabolite. The best examples are genes involved in the synthesis of phytotoxins which are produced by one or only a few pathovars. DNA related to phaseolotoxin production specifically detected *P. s. phaseolicola*,[73,74] and DNA involved in coronatine toxin production was used to detect *P. s. tomato*.[68] Tumorigenic strains of *Agrobacterium tumefaciens* and *E. h. gypophilae* could be detected by DNA probes to the T-DNA of the Ti plasmid and auxin biosynthesis genes, respectively.[78,79] In addition to detection of a specific bacterium, this approach can avoid detection of nonpathogenic strains. This is particularly important for bacteria such as *A. tumefaciens* and *E. herbicola*, where nonpathogenic strains can comprise a high proportion of the natural population.

While a probe to genes of specific metabolites may successfully detect pathovars of *P. syringae* known to produce toxins, it may not work for bacteria whose specific pathogenicity factors or other specific metabolites are not well understood. For example, an attempt was made to develop a specific DNA probe by cloning DNA of *E. carotovora* in *Escherichia coli* and screening transformants for growth on a medium which is semiselective for *E. carotovora*.[72] This did not result in a satisfactory probe, and therefore a taxonomic-based approach was used. The result was a DNA probe which identified *E. carotovora* strains with different serological and physiological characteristics, but did not hybridize with DNA from a closely related plant pathogenic bacterium, *E. chrysanthemii*.[72] It can be argued that even if a DNA probe hybridizes to DNA of a limited number of other pathovars of the same species, it is not a serious problem since it is relatively unlikely that those pathovars will be naturally present in nonhost tissue.[75] Therefore, perfect pathovar specificity may not always be essential for practical applications.

Specific DNA probes can also be developed to detect features other than those related to pathogenicity. Probes based on plasmid-encoded genes for copper resistance in *X. c. vesicatoria*[77] and streptomycin resistance in *P. papulans*[81] have been used to monitor the distribution of these genes in the environment. This use of DNA probe technology can be applied to a wide variety of bacterial functions and will be very important in monitoring the distribution of specific genes in nature.

Another use for DNA probes is to distinguish races of bacteria. Avirulence genes would appear to be the ideal candidates to serve as race-specific probes; however, some avirulence genes, such as avrBs2 of *X. c. vesicatoria*, are present in many other *X. campestris* pathovars.[82] Another example of this is the extensive sequence homology between avrB and avrC genes of *P. s. glycinea*.[83] This precludes their use in dot blot hybridizations. On the other hand, a race-specific DNA probe has been reported for *P. solonacearum* race 3.[84]

Detection of plant pathogenic bacteria in disease lesions should be straightforward, considering the relatively large numbers of bacteria present (generally $10^6$ CFU or more). To detect bacteria in plants, the tissue is usually macerated in liquid and incubated to allow the bacteria to diffuse into the liquid. The resulting bacterial suspension can then be spread onto culture medium, as is done in standard bacterial isolations, except that the medium is covered by a nylon or nitrocellulose membrane. After the bacteria grow, the filter is removed, and colony hybridization is conducted. This assay has been used to detect plant pathogenic bacteria in lesion tissue as well as in plant debris, soil extracts, and seed-soak washes.[68,72,74,75] (A seed-soak wash is a liquid such as sterile water, buffer, or semiselective media that is incubated with seeds in order to extract seed-borne bacteria.[85]) Another DNA probe detection assay involves applying an eluted bacterial suspension directly to a membrane in a dot blot manifold.[68,70,72,75,76] The bacteria are then lysed *in situ* and hybridized with a DNA probe.

The sensitivity of dot blot hybridizations for plant pathogenic bacteria varies widely. The reported minimum numbers of cells detectable by a probe range from 200 CFU,[72] $10^3$ CFU,[68,75] $10^5$ CFU,[76] up to $10^6$ CFU[70,71,74] depending on the probe and bacterium. Sensitivity of the colony hybridization procedure is directly comparable to traditional cultural procedures because it requires that colonies of the bacterium of interest remain distinguishable from colonies of other bacteria growing on the membrane. For *E. carotovora*, a single colony was detectable by the DNA probe even when 1000 colonies of soil bacteria were present on the filter.[72] To detect low numbers of *P. s. phaseolicola* in seed-soak washes, the bacteria were first grown on semiselective media, and after 96 h, the colonies were washed off, concentrated, and then spotted onto a nitrocellulose membrane.[74] If there were only a few colonies (1 to 6) of *P. s. phaseolicola* or a large number of saprophytic bacterial colonies (over 50) growing on the media, then the DNA probe assay became unreliable. To obtain reliable results, the putative *P. s. phaseolicola* colonies had to be removed individually, grown on media, and then probed.[74] A similar problem with a dot blot hybridization assay occurred for *P. s. tomato*.[68] This bacterium could only be detected in tomato plants under 5 weeks old, probably because the number of saprophytes increased and pathogen decreased in older infected plants. Another example of this phenomenon was a dot blot hybridization assay for *C. m. sepedonicum*, where the sensitivity of the assay declined when DNA of the target bacterium constituted less than 5% of the total DNA blotted.[69] In general, it appears that DNA of nontarget bacteria can have a considerably negative effect on the sensitivity of a hybridization assay. Therefore, sensitivities of DNA probe assays reported for pure cultures probably greatly overestimate the level of detection actually possible with naturally infected plant tissues.

PCR may be the solution to the problem of DNA probe sensitivity and the presence of nontarget DNA. A PCR-based assay was developed using primers from DNA sequences of the phaseolotoxin gene of *P. s. phaseolicola*.[73] In addition to a high level of sensitivity (1 to 5 CFU per ml of seed-soak wash), the assay was reported not to be affected by the presence of large numbers of nontarget bacteria.[73] PCR-based assays may be critical to achieve detection of plant pathogenic bacteria in soil, presymptomatic plant tissue, and seed-soak washes where bacterial pathogens are present in very low concentrations and saprophytic bacteria are common.

### A. DETECTION OF BACTERIA IN PLANT LESIONS WITH A DNA PROBE

This procedure is based on the method of Cuppels et al.[68]

1. Thoroughly wash plant material with tap water.
2. Excise tissue containing a lesion, approximately 2 $mm^2$, from the plant.
3. Cut into quarters.
4. Place in test tube containing 0.6 ml sterile, distilled water.
5. Incubate at 4°C for 60 min.
6. Filter 0.5 ml of eluate through a nitrocellulose filter previously moistened with 6X SSC, (1X SSC = 0.15 *M* Na Cl, 0.015 *M* sodium citrate) in a dot blot manifold.
7. Apply a vacuum of –65 kPa for no more than 4 min.
8. Lyse the bacteria, and bind the liberated DNA onto the filter as described[54] for *in situ* hybridization of bacterial colonies.
9. Hybridize with DNA probe as described for southern hybridizations.[54]

## V. MOLLICUTES

The importance of improved detection methods is probably greater for the plant pathogenic mollicutes than for any other group of plant pathogenic organisms. This is because only three of the plant pathogenic mollicutes, all helical spiroplasmas, can be cultured.[86] The nonculturable mollicutes are nonhelical and are called mycoplasma-like-organisms (MLOs) because they resemble the true mycoplasmas but cannot be classified adequately without being cultured *in vivo*. The MLOs are, therefore, characterized by the range of susceptible hosts, the specificity of transmission by insect vectors, and the type of symptoms they cause in host plants. MLOs cause decline diseases, which are manifested as a general decline of infected plants, proliferation diseases, which cause a loss of apical dominance and proliferation of plant parts, and phyllody/virescence diseases, which exhibit the production of leaflike green flowers.[86] Unfortunately, the symptoms caused by a MLO can vary depending on plant host, and some symptoms, such as chlorosis and stunting, are common among all diseases caused by MLOs. Considerable confusion has arisen because identification of MLOs is based solely on biological and pathological characteristics.

Alternative assays have been developed to identify MLOs using more reliable and specific traits. Microscopy and DNA-binding stains provide sensitive detection of MLOs but cannot differentiate between them.[86] Serological assays employing monoclonal or polyconal antibodies have been developed and are more reliable and sensitive in detecting MLOs.[86] However, some antibodies have proven to be too specific, whereas others have too much cross-reactivity to be useful in identifying strains and determining their genetic relatedness. Research on serological assays continues, and improvements have been achieved by using partially purified preparations of MLOs as immunogens.[87]

Another approach to MLO detection is to develop an assay employing specific DNA probes. Because they are nonculturable, success in obtaining cloned specific DNA probes for MLOs largely depends on reducing the concentration of host DNA while extracting MLO-infected tissue. Differential centrifugation comparable to that used to isolate plant mitochondria can enrich for MLO DNA because MLOs are similar in size to mitochondria.[88-91] This eliminates much of the host nuclear DNA, which can be reduced even more by treating the preparation with DNase prior to final lysis of the MLO.[91] Further purification of MLO DNA is achieved by CsCl-bisbenzimide equilibrium centrifugation, which results in two well-separated DNA bands. One band is primarily host mitochondrial DNA, and the second lower density band is primarily MLO DNA.[91,92] The MLO DNA is separated because bisbenzimide preferentially binds to stretches of AT, and MLO DNA is AT-rich.[93] The separation of MLO DNA by this method is sufficient so that differential centrifugation may not be necessary.[92] In the case of the western-X MLO, a second lower density DNA band was observed following CsCl-ethidium bromide equilibrium centrifugation, and this band was also enriched in MLO DNA.[89]

A second approach to enriching for MLO DNA involves first isolating phloem sieve tube elements, which are the only plant tissue colonized by MLOs.[86] Midribs of leaves are cut from infected plants, trimmed so that only the vascular tissue remains, and then macerated by enzymes to release sieve tube elements which are removed with forceps.[94-98] This increases the concentration of MLO by more than 30 times compared to that in total leaf midribs,[94] which is important considering the small proportion of infected plant mass composed of MLO cells. Once the seive tube elements are isolated, DNA is extracted for cloning.

Either insect or plant host tissue can be used to obtain MLO DNA for developing probes. Insect hosts are chosen because they contain relatively high numbers of MLOs in their hemolymph and salivary glands.[88,89] However, plant tissues are more commonly used[90-92,94-98] probably because insect vectors can be relatively difficult to raise, and for some diseases caused by MLOs, the insect vector is unknown. Certain plant hosts, such as leaf tip cultures of evening primrose, contain comparatively high numbers of MLOs, which improves the chances of cloning an MLO-specific probe.[91]

The most important feature of a diagnostic MLO DNA probe is that it does not have homology with plant and insect host DNA. Typically, the first step in screening is to hybridize potential probes with DNA from healthy and MLO-infected plants. Later, the MLO-specific probes are hybridized with DNA from several other MLO strains. In addition to determining strain specificity, the amount of cross-hybridization can be used as an estimate of genetic relatedness among MLO strains. The genetic homology among strains has been directly related to the number of probes which cross-hybridize.[89,90,94-97,99] In general, DNA probes for MLOs do not hybridize with DNA of *Spiroplasma* spp.[90,92,96,97,100] and vice versa.[101] Of the DNA probes developed for diagnosis of maize bushy stunt MLO, only 1 out of 14 hybridized with *S. kunkelii* DNA.[88] Among MLO strains, it is common for DNA probes to hybridize with other strains which cause similar symptoms (i.e., decline diseases or phyllody/virescence diseases) but not hybridize with strains which cause different types of symptoms.[89-91] In other cases, the MLO-specific DNA probes range in their degree of strain specificity, and some probes can detect MLOs causing both decline and phyllody/virescence diseases.[94-97] In the case of the apple proliferation MLO, however, all of the MLO-specfic probes were also strain-specific.[92] Cross-hybridization of probes between MLO strains can be an advantage. Detection of chrysanthemum yellows MLO was achieved by using cross-hybridizing probes developed for aster yellows and periwinkle little leaf MLOs.[99] This avoided the need to develop a new diagnostic DNA probe for this strain of MLO.

The most common detection assay for MLOs using a DNA probe is to extract DNA from infected plants or insects and use it in a dot blot hybridization.[88,89,92,94-99,101] An alternative assay for insects involves crushing individual insects on a nitrocellulose membrane and then lysing and probing the cells as per *in situ* hybridization.[88,89,91] For any assay, the sensitivity of detection must be maximized because MLOs rarely reach high populations in their hosts, especially in plants, where they are limited to sieve tube cells.

One means of improving the sensitivity of a hybridization assay for MLOs is to employ extrachromosomal DNA probes. Most MLOs examined contain extrachromosomal DNA[88,94,95,97,99,100] which, when used as a probe, gives stronger hybridization signals than chromosomal DNA probes. For example, a minimum of 0.02 g of plant tissue was required to detect maize bushy stunt MLO with an extrachromosomal DNA probe, whereas 0.3 g of tissue was required for a chromosomal DNA probe.[88] The higher sensitivity is probably due to extrachromosomal DNA existing in multiple copies in MLO cells. A specific DNA probe of *S. citri* was developed from an 8 kb endogenous plasmid which was in sufficiently high copy number to comprise up to 10% of the total cellular DNA.[101] A potential problem with an extrachromosomal DNA probe is that not all isolates may share the same extrachromosomal DNA. This was found for maize bushy stunt MLO, and a mixture of probes was suggested to avoid this difficulty.[88] For other MLOs strains, this is not a problem, and there is evidence of homology between chromosomal and extrachromosomal DNA.[97]

With dot blot hybridization, the sensitivity of MLO detection is reported to be 7 to 15 ng or 15 to 30 ng of MLO DNA depending on the type of host tissue.[92] Comparable sensitivity was recorded for a hybridization assay of clover proliferation and potato witches'-broom MLOs where 2 ng of MLO DNA was the minimum amount detectable.[98] However, with a PCR-based assay for clover proliferation, $10^{-2}$ to $10^{-5}$ ng of MLO DNA was detectable as a visually amplified band in an agarose gel, and if blotted and hybridized with a DNA probe internal to the primers, $10^{-5}$ to $10^{-8}$ ng of MLO DNA was detectable.[102] The sensitivity of this PCR-based assay depended on the primers used for amplification. Sensitivity was depressed in the case of one set of primers probably because of primer-dimer formation which was visible in agarose gels when the amount of target DNA was reduced.[102]

The high AT content of MLO DNA might present a problem for a PCR-based assay. For instance, it is generally recommended that PCR primers have only a 50% AT content.[103] However, the best pair of primers for clover proliferation MLO had AT contents of 74% and 58%, and the amplified DNA fragment had an AT content of 77.5%.[102] Therefore, it appears that the high AT content of MLOs will not prove to be an insurmountable obstacle. Considering the need for maximum sensitivity in detecting MLOs, it appears likely that PCR-based assays will become much more common.

### A. DETECTION OF MLO IN INFECTED PLANTS WITH A DNA PROBE

This procedure is based on the method of Lee and Davis.[94]

1. Grind approximately 0.3 g of midrib or young shoot tissue in liquid $N_2$.
2. Mix and crush the sample with a minipestle in a microcentrifuge tube containing 0.4 ml extraction buffer (0.1 *M* Tris, pH 8.0, 0.5 *M* EDTA, 0.5 *M* NaCl, 0.5% 2-mercaptoethanol, 0.5% SDS).
3. Centrifuge at 2000 rpm for 10 min.
4. Transfer the supernatant to a clean microcentrifuge tube.
5. Centrifuge the remaining pellet at 8000 rpm for 10 min.
6. Remove the supernatant and combine with the first.
7. Heat the supernatant at 65°C for 5 min.
8. Centrifuge at 14,000 rpm for 5 min.
9. Transfer the resulting supernatant to another microcentrifuge tube.
10. Extract with 200 μl phenol + 200 μl chloroform-isoamyl alcohol (24:1).
11. Centrifuge at 14,000 rpm for 5 min.
12. Collect the aqueous phase.*
13. Denature DNA, and transfer to nitrocellulose membrane in a dot blot apparatus. Bake membrane at 80°C for 2 h.
14. Hybridize with DNA probe as described for Southern hybridizations.[53,54]

* For biotinylated probes, nucleic acids are further purified by a second phenol-chloroform extraction, ethanol precipitation, centrifugation, and resuspension of pellets in 6X SSC (50 to 100 μl per pellet).

### B. DETECTION OF MLOs IN INFECTED INSECTS WITH A DNA PROBE

This procedure is performed according to Kirkpatrick et al.[89]

1. Freeze leafhoppers at –20°C.
2. Place on a water-moistened nitrocellulose membrane.
3. Crush the leafhoppers.
4. Place the membrane with the crushed leafhoppers on filter paper moistened with 0.3 *M* NaOH. Incubate 3 min.
5. Transfer the nitrocellulose membrane to a second filter paper also soaked with 0.3 *M* NaOH. Incubate 3 min.
6. Sequentially transfer the membrane to a pair of filter papers soaked with 1 *M* Tris HCl, pH 8.0. Incubate 3 min.
7. Sequentially transfer the membrane to a pair of filter papers soaked with 0.5 *M* Tris HCl, pH 8.0, and 1.5 *M* NaCl. Remove insect debris.
8. Dry nitrocellulose membrane, and bake at 80°C for 2 h.
9. Hybridize with DNA probe as described for Southern hybridizations.[53,54]

## ACKNOWLEDGMENTS

We thank Drs. Cheryl Kuske, Jane Robb, and Raymond Lee for their suggestions regarding the manuscript.

## REFERENCES

1. **Matthews, R. E. F.,** *Plant Virology*, 3rd ed., Academic Press, New York, 1991.
2. **Milne, R. G.,** Immunoelectron-microscopy for virus identification, in *Electron Microscopy of Plant Pathogens*, Mengden, K. and Lesemann, D. E., Eds., Springer-Verlag, New York, 1990.
3. **Bar-Joseph, M., Garnsey, S. M., Gonsalves, D., Moscovitz, M., Purcifull, D. E., Clark, M. F., and Loebenstein, G.,** The use of enzyme-linked immunosorbent assay for the detection of citrus tristeza virus, *Phytopathology*, 69, 190, 1979.
4. **Permar, T. A., Garnsey, S. M., Gumpf, D. J., and Lee, R. F.,** A monoclonal antibody that discriminates strains of citrus tristeza virus, *Phytopathology*, 80, 224, 1990.
5. **Wang, M. and Gonsalves, D.,** ELISA detection of various tomato spotted wilt virus isolates using specific antisera to structural proteins of the virus, *Plant Dis.*, 74, 154, 1990.
6. **Harrison, B. D., Muniyappa, V., Swanson, M. M., Roberts, I. M., and Robinson, D. J.,** Recognition and differentiation of seven whitefly-transmitted geminiviruses from India, and their relationships to African cassava mosaic and Thailand mung bean yellow mosaic viruses, *Ann. Appl. Biol.*, 118, 299, 1991.
7. **Herrbach, E., Lemaire, O., Ziegler-Graff, V., Lot, H., Rabenstein, F., and Bouchery, Y.,** Detection of BMYV and BWYV isolates using monoclonal antibodies and radioactive RNA probes, and relationships among luteoviruses, *Ann. Appl. Biol.*, 118, 127, 1991.
8. **Clark, M. F. and Bar-Joseph, M.,** Enzyme immunosorbent assays in plant virology, *J. Virol. Meth.*, 7, 51, 1984.
9. **Sudarshana, M. R. and Reddy, D. V. R.,** Penicillinase-based enzyme-linked immunosorbent assay for the detection of plant viruses, *J. Virol. Meth.*, 26, 45, 1989.
10. **Neurath, A. R. and Strick, N.,** Enzyme-linked fluorescence immunoassays using β-galactosidase and antibodies covalently bound to polystyrene plates, *J. Virol. Meth.*, 3, 155, 1981.
11. **Chandler, H. M., Cox, J. C., Healey, K., MacGregor, A., Premier, R. R., and Hurrell, J. G. R.,** An investigation of the use of urease-antibody conjugates in enzyme-immuno-assays, *J. Immunol. Meth.*, 53, 187, 1982.
12. **Berger, P. H., Thornbury, D. W., and Pirone, T. P.,** Detection of picogram quantities of potyviruses using a dot blot immunobinding assay, *J. Virol. Meth.*, 12, 31, 1985.
13. **Clark, M. F. and Adams, A. N.,** Characteristics of the microplate method of enzyme-linked immunosorbent assay for the detection of plant viruses, *J. Gen. Virol.*, 34, 475, 1977.
14. **Koenig, R. and Paul, H. L.,** Variants of ELISA in plant virus diagnosis, *J. Virol. Meth.*, 5, 113, 1982.
15. **Van Regenmortel, M. H. V.,** *Serology and Immunochemistry of Plant Viruses*, Academic Press, New York, 1982.
16. **Hill, S. A.,** The ELISA (Enzyme-Linked Immunosorbent Assay) technique for the detection of plant viruses, *Soc. Appl. Bacteriol. Tech. Ser.*, 19, 349, 1984.
17. **Koenig, R.,** ELISA in the study of homologous and heterologous reactions of plant viruses, *J. Gen. Virol.*, 40, 309, 1978.
18. **Dekker, E. L., Porta, C., and Van Regenmortel, M. H. V.,** Limitations of different ELISA procedures for localizing epitopes in viral coat protein subunits, *Arch. Virol.*, 105, 269, 1989.
19. **Zimmerman, D. and Van Regenmortel, M. H. V.,** Spurious cross-reactions between plant viruses and monoclonal antibodies can be overcome by saturating ELISA plates with milk proteins, *Arch. Virol.*, 106, 15, 1989.
20. **Van den Heuvel, J. F. J. M. and Peters, D.,** Improved detection of potato leafroll virus in plant material and in aphids, *Phytopathology*, 79, 963, 1989.
21. **Haber, S. and Knapen, H.,** Filter paper sero-assay (FiPSA): a rapid, sensitive technique for sero-diagnosis of plant viruses, *Can. J. Plant Pathol.*, 11, 109, 1989.
22. **Hsu, H. T. and Lawson, R. H.,** Direct tissue blotting for detection of tomato spotted wilt virus in *Impatiens*, *Plant Dis.*, 75, 292, 1991.
23. **Banttari, E. E. and Goodwin, P. H.,** Detection of potato viruses S,X, and Y by enzyme-linked immunosorbent assay on nitrocellulose membranes (Dot-ELISA), *Plant Dis.*, 69, 202, 1985.

24. **Lin, N. S., Hsu, Y. H., and Hsu, H. T.,** Immunological detection of plant viruses and a mycoplasmalike organism by direct tissue blotting on nitrocellulose membranes, *Phytopathology*, 80, 824, 1990.
25. **Bottacin, A. and Nassuth, A.,** Evaluation of Ontario-grown cereals for susceptibility to wheat streak mosaic virus, *Can. J. Plant Pathol.*, 12, 267, 1990.
26. **Hull, R. and Al-Hakim, A.,** Nucleic acid hybridization in plant virus diagnosis and characterization, *Trends Biotechnol.*, 6, 213, 1988.
27. **Robinson, D. J. and Romero, J.,** Sensitivity and specificity of nucleic acid probes for potato leafroll luteovirus detection, *J. Virol. Meth.*, 34, 209, 1991.
28. **Polston, J. E., Dodds, J. A., and Perring, T. M.,** Nucleic acid probes for detection and strain discrimination of cucurbit geminiviruses, *Phytopathology*, 79, 1123, 1989.
29. **Moseley, J. and Hull, R.,** Comparison of eight isolates of beet yellows virus by filter hybridisation and enzyme-linked immunosorbent assay, *Ann. Appl. Biol.*, 118, 605, 1991.
30. **Varveri, C., Candresse, T., Cugusi, M., Ravelonandro, M., and Dunez, J.,** Use of $^{32}$P labeled transcribed RNA probe for dot hybridization detection of plum pox virus, *Phytopathology*, 78, 1280, 1988.
31. **Forster, A. C., McInnes, J. L., Skingle, D. C., and Symons, R. H.,** Non-radioactive hybridization probes prepared by the chemical labelling of DNA and RNA with a novel reagent, photobiotin, *Nucl. Acids Res.*, 13, 745, 1985.
32. **Habili, N., McInnes, J. L., and Symons, R. H.,** Non-radioactive photobiotin-labelled DNA probes for the routine diagnosis of barley yellow dwarf virus, *J. Virol. Meth.*, 16, 224, 1987.
33. **Rao, A. L. N., Huntley, C. C., Marsh, L. E., and Hall, T. C.,** Analysis of RNA stability and (–) strand content in viral infections using biotinylated RNA probes, *J. Virol. Meth.*, 30, 239, 1990.
34. **Roy, B. P., Abouhaidar, M. G., Sit, T. L., and Alexander, A.,** Construction and use of cloned cDNA biotin and $^{32}$P-labelled probes for the detection of papaya mosaic potexvirus RNA in plants, *Phytopathology*, 78, 1425, 1988.
35. **Roy, B. P., Abouhaidar, M. G., and Alexander, A.,** Biotinylated RNA probes for the detection of potato spindle tuber viroid (PSTV) in plants, *J. Virol. Meth.*, 23, 149, 1989.
36. **Eweida, M., Sit, T. L., and Abouhaidar, M. G.,** Molecular cloning of the genome of carlavirus potato virus S: biotinylated RNA transcripts for virus detection in crude potato extracts, *Ann. Appl. Biol.*, 115, 253, 1989.
37. **Haber, S., Polston, J. E., and Bird, J.,** Use of DNA to diagnose plant diseases caused by single-strand DNA plant viruses, *Can. J. Plant Pathol.*, 9, 156, 1987.
38. **Barbara, D. J., Kawata, E. E., Ueng, P. P., Lister, R. M., and Larkins, B. A.,** Production of cDNA clones from the MAV isolate of barley yellow dwarf virus, *J. Virol. Meth.*, 68, 2417, 1987.
39. **Eweida, M. and Oxelfelt, P.,** Production of cloned cDNA from a Swedish barley yellow dwarf virus isolate, *Ann. Appl. Biol.*, 114, 61, 1989.
40. **Boulton, M. I. and Markham, P. G.,** The use of squash-blotting to detect plant pathogens in insect vectors, in *Developments and Applications in Virus Testing*, Jones, A. C. and Torrance, L., Eds., The Lavenham Press Ltd., Lavenham, 1986, 55.
41. **Polston, J. E., Al-Musa, A., Perring, T. M., and Dodds, J. A.,** Association of nucleic acid of squash leaf curl geminivirus with the whitefly *Bemisia tabaci*, *Phytopathology*, 80, 850, 1990.
42. **Rice, D. J., German, T. L., Mau, R. F. L., and Fujimoto, F. M.,** Dot blot detection of tomato spotted wilt virus RNA in plant and thrips tissues by cDNA clones, *Plant Dis.*, 74, 274, 1990.
43. **Navot, N., Ber, R., and Czosnek, H.,** Rapid detection of tomato yellow leaf curl virus in squashes of plants and insect vectors, *Phytopathology*, 79, 562, 1989.
44. **Vunsh, R., Rosner, A., and Stein, A.,** The use of the polymerase chain reaction (PCR) for the detection of bean yellow mosaic virus in gladiolus, *Ann. Appl. Biol.*, 117, 561, 1990.
45. **Wetzel, T., Candresse, T., Ravelonandro, M., and Dunez, J.,** A polymerase chain reaction assay adapted to plum pox potyvirus detection, *J. Virol. Meth.*, 33, 355, 1991.
46. **Robertson, N. L., French, R., and Gray, S. M.,** Use of group-specific primers and the polymerase chain reaction for the detection and identification of luteoviruses, *J. Gen. Virol.*, 72, 1473, 1991.
47. **Korschineck, I., Himmler, G., Sagl, R., Steinkeller, H., and Katinger, H. W. D.,** A PCR membrane spot assay for the detection of plum pox virus RNA in bark of infected trees, *J. Virol. Meth.*, 31, 139, 1991.
48. **Langeveld, S. A., Dore, J.-M., Memelink, J., Derks, A. F. L. M., van der Vlugt, C. I. M., Asjes, C. J., and Bol, J. F.,** Identification of potyviruses using the polymerase chain reaction with degenerate primers, *J. Gen. Virol.*, 72, 1531, 1991.
49. **Puchta, H. and Sanger, H. L.,** Sequence analysis of minute amounts of viroid RNA using the polymerase chain reaction (PCR), *Arch. Virol.*, 106, 335, 1989.
50. **Rybicki, E. P. and Hughes, F. L.,** Detection and typing of maize streak virus and other distantly related geminiviruses of grasses by polymerase chain reaction amplification of a conserved viral sequence, *J. Gen. Virol.*, 71, 2519, 1990.
51. **Levy, L., Hadidi, A., and Garnsey, S. M.,** Multiplex reverse transcription/polymerase chain reaction for the detection of mixed citrus viroids in a single reaction, *Phytopathology*, 81, 1212, 1991.

52. **Salinovich, O. and Montelaro, R.,** Reversible staining and peptide mapping of proteins transferred to nitrocellulose after separation by SDS-PAGE, *Anal. Biochem.*, 156, 341, 1986.
53. **Ausubel, F. M., Brent, R., Kingston, R. E., Moore, D. D., Seidman, J. G., Smith, J. A., and Struhl, K.,** *Current Protocols in Molecular Biology*, Vol. 2, John Wiley & Sons, New York, 1989.
54. **Sambrook, J., Fritsch, E. F., and Maniatis, T.,** *Molecular Cloning: A Laboratory Manual*, 2nd ed., Cold Spring Harbor Laboratory, Cold Spring Harbor, New York, 1989.
55. **Tisserat, N. A., Hulbert, S. H., and Nus, A.,** Identification of *Leptosphaeria korrae* by cloned DNA probes, *Phytopathology*, 81, 917, 1991.
56. **Miller, S. A. and Martin, R. R.,** Molecular diagnosis of plant disease, *Ann. Rev. Phytopathol.*, 26, 409, 1988.
57. **Goodwin, P. H., Kirkpatrick, B. C., and Duniway, J. M.,** Cloned DNA probes for identification of *Phytophthora parasitica*, *Phytopathology*, 79, 716, 1989.
58. **Goodwin, P. H., Kirkpatrick, B. C., and Duniway, J. M.,** Identification of *Phytophthora citrophthora* with cloned DNA probes, *Appl. Environ. Microbiol.*, 56, 669, 1990.
59. **Henson, J. M.,** DNA probe for identification of take-all fungus, *Gaeumannomyces graminis*, *Appl. Environ. Microbiol.*, 55, 284, 1989.
60. **Martin, F. N.,** Selection of DNA probes useful for isolate identification of two *Pythium* spp., *Phytopathology*, 81, 742, 1991.
61. **Rollo, F., Amici, A., Foresi, F., and di Silvestro, I.,** Construction and characterization of a cloned probe for the detection of *Phoma tracheiphila* in plant tissues, *Appl. Microbiol. Biotechnol.*, 26, 352, 1987.
62. **Rollo, F., Salvi, R., and Torchia, P.,** Highly sensitive and fast detection of *Phoma tracheiphila* by polymerase chain reaction, *Appl. Microbiol. Biotechnol.*, 32, 572, 1990.
63. **Nazar, R. N., Hu, X., Schmidt, J., Culham, D., and Robb, J.,** Potential use of PCR-amplified ribosomal intergenic sequences in the detection and differentiation of verticillium wilt pathogens, *Physiol. Mol. Plant Pathol.*, 39, 1991.
64. **Yao, C. L., Magill, C. W., and Frederiksen, R. A.,** Use of an A-T rich DNA clone for identification and detection of *Peronosclerospora sorghi*, *Appl. Environ. Microbiol.*, 57, 2027, 1991.
65. **Goodwin, P. H., English, J. T., Neher, D. A., Duniway, J. M., and Kirkpatrick, B. C.,** Detection of *Phytophthora parasitica* from soil and host tissue with a species-specific DNA probe, *Phytopathology*, 80, 277, 1990.
66. **Schesser, K., Luder, A., and Henson, J. M.,** Use of polymerase chain reaction to detect the take-all fungus, *Gaeumannomyces graminis*, in infected wheat plants, *Appl. Environ. Microbiol.*, 57, 553, 1991.
67. **Schaad, N. W.,** *Laboratory Guide for Identification of Plant Pathogenic Bacteria*, APS Press, St. Paul, MN, 1988.
68. **Cuppels, D. A., Moore, R. A., and Morris, V. L.,** Construction and use of a nonradioactive DNA hybridization probe for detection of *Pseudomonas syringae* pv. tomato on tomato plants, *Appl. Environ. Microbiol.*, 56, 1743, 1990.
69. **Johansen, I. E., Rasmussen, O. T., and Heide, M.,** Specific identification of *Clavibacter michiganese* subsp. *sepedonicum* by DNA-hybridization probes, *Phytopathology*, 79, 1019, 1989.
70. **Thompson, E., Leary, J. V., and Chun, W., W., C.,** Specific detection of *Clavibacter michiganense* subsp. *michiganense* by a homologous DNA probe, *Phytopathology*, 79, 311, 1989.
71. **Verreault, H., Lafond, M., Asselin, A., Banville, G., and Bellemare, G.,** Characterization of two DNA clones specific for identification of *Corynebacterium sepedomicum*, *Can. J. Microbiol.*, 34, 993, 1988.
72. **Ward, L. J. and De Boer, S. H.,** A DNA probe specific for serologically diverse strains of *Erwinia carotovora*, *Phytopathology*, 80, 665, 1990.
73. **Prossen, D., Hatziloukas, E., Panopoulos, N. J., and Schaad, N. W.,** Direct detection of the halo blight pathogen *Pseudomonas syringae* pv. *phaseolicola* in bean seeds by DNA amplification, *Phytopathology*, 81, 1159, 1991.
74. **Schaad, N. W., Azad, H., Peet, R. C., and Panopoulos, N. J.,** Identification of *Pseudomonas syringae* pv. *phaseolicola* by a DNA hybridization probe, *Phytopathology*, 79, 903, 1989.
75. **Gilbertson, R. L., Maxwell, D. P., Hagedorn, D. J., and Leong, S. A.,** Development and application of a plasmid DNA probe for detection of bacteria causing common bacterial blight of bean, *Phytopathology*, 79, 518, 1989.
76. **Denny, T. P.,** Differentiation of *Pseudomonas syringae* pv. *tomato* from *P. s. syringae* with a DNA hybridization probe, *Phytopathology*, 78, 1186, 1988.
77. **Garde, S. and Bender, C. L.,** DNA probes for detection of copper resistance genes in *Xanthomonas campestris* pv. *vesicatoria*, *Appl. Environ. Microbiol.*, 57, 2435, 1991.
78. **Burr, T. J., Norelli, J. L., Katz, B. H., and Bishop, A. L.,** Use of Ti plasmid DNA probes for determining tumorigenicity of *Agrobacterium* strains, *Appl. Environ. Microbiol.*, 56, 1782, 1990.
79. **Manulis, S., Gafni, Y., Clark, E., Zutra, D., Ophir, Y., and Borash, I.,** Identification of a plasmid DNA probe for detection of strains of *Erwinia herbicola* pathogenic on *Gysophila paniculata*, *Phytopathology*, 81, 54, 1991.

80. **De Parasis, J. and Roth, D. A.,** Nucleic acid probes for identification of genus-specific 16s rRNA sequences, *Phytopathology*, 80, 618, 1990.
81. **Norelli, J. L., Burr, T. J., Lolicero, A. M., Gilbert, M. T., and Katz, B. H.,** Homologous streptomycin resistance gene present among diverse gram-negative bacteria in New York State apples, *Appl. Environ. Microbiol.*, 57, 486, 1991.
82. **Kearney, B. and Staskawicz, B.,** Widespread distribution of *Xanthomonas campestris* avirulence gene avr Bs2, *Nature*, 346, 385, 1990.
83. **Tanaki, S., Dahlbeck, D., Staskawicz, B., and Keen, N. T.,** Characterization and expression of two avirulence genes cloned from *Pseudomonas syringae* pv. *glycinea*, *J. Bacteriol.*, 170, 4846, 1988.
84. **Cook, D. and Sequira, L.,** Isolation and characterization of DNA clones specific for race 3 of *Pseudomonas solonacearum*, *Phytopathology*, 81, 696, 1991.
85. **Saettler, A. W., Schaad, N. W., and Roth, D. A.,** *Detection of Bacteria in Seed*, APS Press, St. Paul, MN, 1989.
86. **Kirkpatrick, B. C.,** Strategies for characterizing plant pathogenic mycoplasma-like organisms and their effects on plants, in *Plant-Microbe Interactions*, Vol. 3, Kosuge, T. and Nester, E. W., Eds., McGraw-Hill, New York, 1989, 241.
87. **Clark, M. F., Morton, A., and Buss, S. L.,** Preparation of mycoplasma immunogens from plants and a comparison of polyclonal and monoclonal antibodies made against primula yellows MLO-associated antigens, *Ann. Appl. Biol.*, 114, 111, 1989.
88. **Davis, M. J., Tsai, J. H., Cox, R. L., McDaniel, L. L., and Harrison, N. A.,** Cloning of chromosomal and extrachromosomal DNA of the mycoplasmalike organism that causes maize bushy stunt disease, *Mol. Plant-Microbe Interactions*, 1, 295, 1988.
89. **Kirkpatrick, B. C., Stenger, D. C., Morris, T. J., and Purcell, A. H.,** Cloning and detection of DNA from nonculturable plant pathogenic mycoplasma-like organism, *Science*, 238, 197, 1987.
90. **Kuske, C. R., Kirkpatrick, B. C., and Seemuller, E.,** Differentiation of virescence MLOs using western aster yellows mycoplasma-like organism chromosomal DNA probes and restriction fragment length polymorphism analysis, *J. Gen. Microbiol.*, 137, 153, 1991.
91. **Sears, B. B., Lim, P. O., Holland, N., Kirkpatrick, B. C., and Klomparens, K. L.,** Isolation and characterization of DNA from a mycoplasmalike organism, *Mol. Plant-Microbe Interactions*, 2, 175, 1989.
92. **Bonnet, F., Saillard, C., Kollar, A., Seemuller, E., and Bove, J. M.,** Detection and differentiation of the mycoplasma-like organism associated with apple proliferation disease using cloned DNA probes, *Mol. Plant-Microbe Interactions*, 3, 438, 1990.
93. **Kollar, A. and Seemuller, E.,** Base composition of mycoplasma-like organisms associated with various plant diseases, *J. Phytopathol.*, 127, 177, 1989.
94. **Lee, I. M. and Davis, R. E.,** Detection and investigation of genetic relatedness among aster yellows and other mycoplasma-like organisms by using cloned DNA and RNA probes, *Mol. Plant-Microbe Interactions*, 1, 303, 1988.
95. **Lee, I. M., Davis, R. E., and De Witt, N. D.,** Nonradioactive screening method for isolation of disease-specific probes to diagnose plant diseases caused by mycoplasma-like organisms, *Appl. Environ. Microbiol.*, 56, 1471, 1990.
96. **Davis, R. E., Lee, I. M., Dally, E. L., Dewitt, N., and Douglas, S. M.,** Cloned nucleic acid hybridization probes in detection and classification of mycoplasma-like organisms (MLOs), *Acta Horticulturae*, 234, 115, 1988.
97. **Davis, R. E., Lee, I. M., Douglas, S. M., and Dally, E. L.,** Molecular cloning and detection of chromosomal and extrachromosomal DNA of the mycoplasma-like organism associated with little leaf disease in Periwinkle (*Catharanthus roseus*), *Phytopathology*, 80, 789, 1990.
98. **Deng, S. and Hiruki, C.,** Genetic relatedness between two nonculturable mycoplasma-like organisms revealed by nucleic acid hybridization and polymerase chain reaction, *Phytopathology*, 81, 1475, 1991.
99. **Bertaccini, A., Davis, R. E., Lee, I. M., Conti, M., Dally, E. L., and Douglas, S. M.,** Detection of chrysanthemum yellows mycoplasma-like organism by dot hybridization, *Plant Dis.*, 74, 40, 1990.
100. **Kuske, C. R., Kirkpatrick, B. C., Davis, M. J., and Seemuller, E.,** DNA hybridization between western aster yellows mycoplasma-like organism plasmids and extrachromosomal DNA from other plant pathogenic mycoplasma-like organism, *Mol. Plant-Microbe Interactions*, 4, 75, 1991.
101. **Nur, I., Bove, J. M., Saillard, C., Rottem, S., Whitcomb, R. M., and Razin, S.,** DNA probes in detection of spiroplasmas and mycoplasma-like organisms in plants and insects, *FEMS Microbiol. Lett.*, 35, 157, 1986.
102. **Deng, S. and Hiruki, C.,** Enhanced detection of a plant pathogenic mycoplasma-like organism by polymerase chain reaction, *Proc. Jpn. Acad.* (Ser. B), 66, 140, 1990.
103. **Saiki, R. K.,** Amplification of genomic DNA, in *PCR Protocols: A Guide to Methods and Applications*, Innis, M. A., Gelfard, D. H., Sninsky, J. J., and White, T. J., Eds., Academic Press, San Diego, 1990, 13.

# 20

# Screening for Inoculant-Quality Strains of Rhizobia

*David H. Hubbell*

## I. INTRODUCTION

Studies of the *Rhizobia*-legume association (RLA) historically have been motivated largely by practical considerations.[1] This unique plant-bacteria association commonly results in "fixation" or reduction of dinitrogen by *Rhizobia* within nodules, which they induce on the roots of compatible host legumes. This activity can provide nitrogen in quantity sufficient to supplement significantly the available soil nitrogen required for growth of the legume. Soil nitrogen is a consumptive element in plant nutrition and must be replenished periodically by chemically fixed nitrogen fertilizer or by biological nitrogen fixation (BNF), as in the *Rhizobia*-legume association. It is the mineral element most often found limiting for crop production. Nitrogen fertilizer is unavailable or prohibitively expensive, especially for subsistence farmers, in many parts of the world.[2] The *Rhizobia*-legume BNF system is, potentially, universally available and economically feasible. It therefore represents a vital technology of inestimable value for increasing and sustaining crop productivity in agricultural soils where soil nitrogen is a real or potential limitation.[3] A critical aspect of the RLA is the fact that it often can be manipulated under nitrogen-limiting field conditions in such a way that BNF, and hence crop production, can be enhanced easily and inexpensively. One management practice by which the RLA is exploited-manipulated is referred to as "inoculation".[1] This is the practice of placing *Rhizobia* on legume seed or in the soil at the time of planting, thus greatly increasing the probability of development of a successful BNF system in the legume crop.

The biotechnology of legume inoculation with *Rhizobia* is multifaceted. The isolation, screening, and final selection of inoculant-quality strains, as discussed here, represent merely the initial stage of the process. Other considerations which must be addressed include choice/identification/development of an appropriate host plant genotype(s), seed acquisition and storage, maintenance and preservation of rhizobial strains, large-scale production of inoculum, inoculant formulation and use, and the logistics of storage, transport, marketing, and distribution. General reviews of mass production and commercialization of microbial inoculants are available.[4-6] Several additional papers are specific in their discussion of production and use of rhizobial inoculants for legumes.[7-12]

The importance of the symbiotic nitrogen-fixing RLA in world agriculture is extensively documented.[13-15] However, despite a hundred years of basic and applied research, the full agronomic potential of this system is still unrealized.[16] The reason is that we do not have sufficient basic understanding of the nature of the *Rhizobia* and their leguminous hosts or of their functioning as a symbiotic system. These are important deficiencies. Plant genotype

0-8493-5164-2/93/$0.00+$.50
© 1993 by CRC Press, Inc.

and rhizobial strain variation are primary considerations in attempting to use the system effectively in production agriculture.[17] Successful management of the system in any given situation depends on a thorough understanding of many plant, bacterial, and environmental factors. Successful legume inoculation is based on optimum strain-genotype pairing under the most conducive and attainable environmental conditions. An early and critical consideration in the formulation of an inoculation methodology is the selection of appropriate rhizobial strains. The characteristics of an inoculant-quality strain are well recognized. It must survive and grow in the soil in competition with other microbes, it must dominate in the process of plant infection-nodulation, and it must be highly effective in nitrogen fixation. The traits which confer these characteristics and, consequently, the means by which these traits might be selected for in the laboratory are poorly known. In essence, what are the tests needed for laboratory screening of potential inoculant-quality strains of rhizobia?

Isolation and purification of rhizobial strains is accomplished easily by traditional microbiological methods. However, there is no satisfactory protocol for the rapid, reliable laboratory screening of these isolates for strains which are superior in their ability to survive in the soil, as well as to infect, nodulate, and fix nitrogen in roots of legumes, i.e., inoculant-quality strains. Absence of such a protocol necessitates the use of laborious field trials of numerous strains to test, at great cost, their performance as inoculants. As a result, the majority of isolates screened as promising by inadequate laboratory procedures are ultimately discarded as symbiotically inferior on the basis of *in vivo* field trials.

The tenets of *Rhizobia*-legume technology are well known and widely practiced.[18] A current laboratory methodology manual is also available in which the treatment of methodology is comprehensive and proceeds logically, in step-wise fashion, from strain isolation and selection to inoculant production and use.[19] Valuable background material on the biology of rhizobia and legumes is also included.[19] A virtue of this manual is that it describes simple, inexpensive methodology which can be used very effectively in minimally equipped microbiological laboratories. A valuable earlier manual is that of Vincent.[20]

There are several authoritative treatments of the theory and practice of selection of strains of *Rhizobia*.[21-23] Halliday emphasizes the importance of screening isolates in order to eliminate inferior strains early, thereby saving much time and effort in further testing of strains which have high potential for poor performance in greenhouse and field trials.[22] We support this approach and advocate an even more rigorous application of screening as standard practice. The purpose of this work is to suggest an approach to developing a protocol for simple, efficient laboratory screening of numerous rhizobial strains. This would be conducted prior to field testing and final selection of inoculant-quality strains.

## II. DEFINITIONS

The following definitions of terms, as used in the present text, are given for purposes of clarity. Nodulating combinations of rhizobia and legumes are referred to as "compatible" or "homologous", whereas non-nodulating combinations are referred to as "incompatible" or "heterologous". Rhizobial strains are referred to as "infective" (nodulating) or "noninfective" (non-nodulating) for a particular legume(s). Infective strains may further be characterized as "effective" (nitrogen-fixing) or "ineffective" (non-nitrogen-fixing). When two or more compatible strains of *Rhizobia* are present in the rhizosphere of a legume, they may be referred to as "dominant" or "aggressive" or, alternatively, as "subordinate" or "submissive" in the sum of their interactions which lead to characteristic differences (ranking) in nodulation of a given host legume. In essence, these terms express relative competitive ability of rhizobial strains, including growth and survival in soil, the infection process, and nodule induction. The term "superior" is used in reference to a strain which has an optimal combination of competitive, infective, and effective traits which qualify it as an inoculant-quality strain. Strains of lesser quality, suboptimal for such traits, are regarded as relatively "inferior".

"Extrinsic" factors are defined here as environmental stress factors which can be demonstrated to affect adversely the occurrence and/or extent of infection-nodulation-nitrogen fixation under specific stress conditions. They influence symbiotic performance quantitatively under reasonable plant growth conditions and may prevent development of the symbiosis in extreme situations.

"Intrinsic" factors are genetically-based rhizobial and legume characteristics which determine the occurrence and extent of development of the symbiosis when environmental factors are favorable. *Rhizobia*-legume combinations which are intrinsically compatible will perform symbiotically to the extent of their genetic potential under optimum conditions of plant growth. This genetic potential varies widely among combinations. The rationale for isolating, screening, and selection is therefore to identify inoculant-quality combinations.

## III. GUIDELINES FOR STRAIN ISOLATION AND LABORATORY SCREENING

The soil, rhizosphere, and rhizoplane environments should not be used as primary sources for isolation of strains. Strains from these environments presumably are adapted to extrinsic factors affecting survival and growth; however, such factors influence but do not directly determine symbiotic properties of infection, nodulation, and nitrogen fixation. These latter properties are selected for during plant passage and are manifested by effective nodules. As a secondary consideration, there is no adequate methodology for accomplishing *ex planta* isolations in an efficient and consistent manner.[24] There are numerous reports of differential and selective media for isolation of specific (marked) strains of introduced *Rhizobia*, but they are not applicable for use in the primary isolation of indigenous *Rhizobia*. Such strains, in any event, would be quite unpredictable in their symbiotic capability since criteria related directly to symbiosis (nodulation and nitrogen fixation) are not selected for in the soil or rhizosphere. Isolates from these latter environments may be highly competitive but of undetermined ability to nodulate and fix nitrogen.

The legume root nodule, as traditionally used, remains the best potential source for isolation of superior strains. A good procedure, when feasible, is to select nodules from a legume growing with unusual vigor under apparently adverse environmental conditions. Strains isolated from such nodules, by virtue of their presence, ostensibly are better adapted to extrinsic factors and therefore dominant in the specific stressed environment. The probability of effectiveness is assessed by the depth of the green color of the foliage, indicating relative adequacy of nitrogen, and the red color of the nodule interior, indicating relative activity of nitrogen fixation. If such plants are not naturally occurring, they may be planted in the soil of concern, under field conditions, where they have the opportunity to "trap" superior native strains of *Rhizobia*.

Strains isolated from nodules have completed all of the requirements for successful establishment of the association and have thus been selected for by the host legume. This does not guarantee that the isolate obtained is superior since double strain occupancy of nodules is well documented.[25] In such cases, an inferior strain may inadvertently be isolated. This might be avoided by testing several single colonies from plating of the crushed nodule suspension.

Conventional wisdom advocates that isolates be obtained from young nodules formed early in the life of the plant, preferably from a seedling. These nodules, usually found in the crown region, presumably are formed by the more aggressive strains present in the mixed population in the soil. As a selection criterion, earliness in nodulation may be important as an indicator of high competitive ability. Subsequent nitrogen fixation should favor early plant growth and survival in nitrogen-limited environments.

However, nodules formed later, on lower parts of the root system, may be preferable for strain isolation. Such strains, presumably more mobile in the rhizosphere, initiate nodules which are deeper in the soil and therefore less susceptible to dehiscence due to onset of adverse environmental conditions (heating, drying) at the soil surface. Conversely, crown nodulation would be preferable under high moisture conditions for the same reason. There is little evidence to support either of these contentions.

Is abundant nodulation a desirable criterion for strain isolation? The evidence is equivocal. Legume root nodules manifest a compensating effect of nodule mass for nodule numbers. On a sparsely nodulated root system, the nodule mass may be increased to compensate for reduced numbers. Total nodule mass and presumably nitrogen-fixing tissue remain the same irrespective of nodule numbers. This would appear to minimize the importance of nodule numbers. Nodule mass may be an inherent and characteristic trait of any given genotype-strain combination. The relative importance of nodule numbers and size has been considered in detail.[21,26]

A rapid qualitative plant growth test to confirm the identity and effectiveness of new presumptive rhizobial isolates should be conducted under optimum conditions prior to screening.[27] A growth pouch assay confirms rhizobial identity and infectiveness by appearance of nodules. Relative effectiveness of the strains can be assessed by the color of plants and interior of nodules and by plant dry weight.

The ultimate objective of obtaining superior strains is of a practical nature. They are intended for use in the commercial production of legume inoculants which have a high probability of performing with maximum symbiotic capability when used in commercial legume production.

Clearly, any strain which is dominant in a given cropping situation will be found to possess the traits which surmount important extrinsic factors present in that environment. These strains can be isolated and their identity and quality confirmed. An alternative is to bypass the arduous process of isolation and screening of local strains and directly field test imported strains which have been selected from presumably similar environments. The assumption is that critical extrinsic factors in the old and new environments are similar, and therefore strains adapted to the original environment will also be adapted to, and therefore compete and dominate in, the new location.[22] Although occasionally successful, this approach often proves unsatisfactory due to the complexity of each situation. Not all of the extrinsic factors influencing dominance at a particular field site can be identified. It is therefore naive to believe that an imported strain, shown to be superior in the original environment, will invariably prove to be superior in its new environment. The environment selects the microorganism, and no two soil environments are identical. They can be expected to differ in many respects, which may or may not be identifiable and thus may or may not be taken into account in the strain selection process. Many unrecognized extrinsic factors or combinations of known extrinsic factors may influence natural selection of strains and therefore ultimately determine the dominance of indigenous strains. Factors commonly recognized as extrinsic are therefore merely the more visible or obvious factors which have been identified. It is for this reason that the time-tested practice of isolating strains from the native rhizobial population, if present, is more successful and the method of choice.[27] They are adapted to the full and heterogeneous range of extrinsic factors of that environment, both recognized and unrecognized. In practice, unfortunately, strains are screened on the basis of their adaptation to only one or a few obvious extrinsic factors.[21] An appreciation for the range of possible extrinsic factors can be obtained from several authoritative reviews.[28-33]

At what point in the passage of rhizobia through the continuum of environments — soil, rhizosphere, rhizoplane, endorhizosphere, nodule — is superiority determined? Is it determined suddenly at a specific point or determined gradually in passage by progressive accumulation of critical traits? At what point in the overall process is strain isolation most likely to result in obtaining a superior (inoculant-quality) strain?

*Rhizobia* must pass through a continuum of five environmental zones on their way to *in vivo* selection for superior nodulating ability. Zones of passage of *Rhizobia* from soil to nodule in the infection-nodulation process may be characterized as follows.

The **soil** selects *Rhizobia* for resistance to extrinsic or environmental stress factors unique to that soil. Initial efforts to isolate inoculant-quality strains should therefore be directed toward isolating indigenous strains since these strains would have the highest probability of being best adapted to that unique combination of extrinsic factors. They are conveniently obtained by isolating from nodules of a "trap" legume of the desired species growing under relevant soil and climatic conditions.

The legume **rhizosphere** is recognized as nonspecifically stimulatory to the growth of rhizobia, whether compatible or incompatible. Ability to survive, multiply, and colonize the legume rhizosphere is not usually correlated with infective ability and is therefore of secondary consideration in screening for dominant strains.[31]

The legume **rhizoplane** (root surface) is the zone where intrinsic plant and bacterial factors critical to the infection process are first expressed. These factors, expressed in the infection stages of (1) recognition-adsorption and (2) root hair cell wall penetration, must be completed in order for infection to proceed, but this does not guarantee that infection will continue to the stage of nodule formation.

The **endorhizosphere** begins at the plant cell wall-membrane juncture. It is the point where entering *Rhizobia* are first able to establish chemical communication with the host cell nucleus. This initial cell-cell contact of interactive intrinsic traits is manifested by reorientation of growth from the root hair tip to the point of penetration, resulting in formation of an infection thread. The manner in which an entering strain communicates with the root hair nucleus is the final level of control determining the initiation of an infection. It ultimately determines the dominance or subordinance of the entering strain.

Any rhizobial strain will be ranked as dominant in a mixture of strains in proportion to the extent of its qualitative and quantitative activity in these five zones.

Progress of studies of screening for inoculant-quality strains is hampered because there is no rapid and effective laboratory methodology for screening strains which are predictably dominant and highly effective under field conditions. There are no good differential or selective media for isolation and identification of rhizobia which are both aggressively infective and highly effective.

The philosophy guiding the process of strain screening varies widely among investigators. Some workers practice selection of strains which are highly specific in their response to certain extrinsic factors, plant genotype, etc. However, these "target traits" are highly subject to variation over time and space. Strains screened using such specific criteria are likely to be very inflexible in terms of their ability to function well consistently in a dynamic environment. Strains should be screened as "broad spectrum" strains having resistance to a wide variety of extrinsic factors. Resistance to individual extrinsic factors should encompass a wide range of values (maximum-minimum) in order to give selected strains the inherent flexibility needed to adapt and emerge consistently as dominant strains in a dynamic environment.

Nodulation (infectiveness) and nitrogen fixation (effectiveness) in legumes are separate characteristics, and both are subject to strain variation.[34] Final selection of strains for use as inoculants must therefore consider their eventual use in either single-strain or multistrain inoculants. There are two schools of thought in relation to the practice of this final selection.[21,35] Those advocates of single-strain inoculants practice selection of strains which are highly specific for a plant genotype-soil environment situation. This finely tuned system leads to excellent inoculation response, but there is minimal flexibility. Changes in plant genotype or soil variables may require a new inoculant strain instead of depending on the previously used inoculant strain, which may be firmly entrenched as a competitor for nodulation but less effective and not easily displaced by a new strain. A second school of thought advocates the use of multi-strain inoculants. In practice, the component strains are selected on the basis of their possession of broad infectiveness and effectiveness traits. They will perform well symbiotically in combination with a variety of plant genotypes under fluctuating plant-growth conditions.

It has been suggested that "engineering" for highly specific dominance traits in *Rhizobia* and legumes may be a promising means of circumventing the problems of selecting for dominant native strains.[35,36] These highly specific genetic alterations are interactive in both *Rhizobia* and legumes and prove inhibitory or stimulatory to nodulation in different situations.[36]

Screening procedures currently are left largely to the whim of individual investigators.[21] A standard protocol for laboratory screening of superior *Rhizobia* from a collection of newly isolated strains is needed. The following is a suggested approach to development of a standardized protocol for laboratory screening of new isolates. This screening procedure would detect subordinate and inferior isolates and provide a rational basis for discarding them; concomitantly, it would detect superior traits selected for under natural conditions. It is based on current knowledge and speculation about the presence and relative importance of extrinsic and intrinsic factors in determining dominant and superior strains.

The trend of evolution has been more for survival, favoring strains capable of adapting to extrinsic factors.[37] Such strains may or may not have the intrinsic factors related to ability to associate highly effectively with legumes. It is therefore not surprising to find that newly isolated strains which perform exceptionally well in survival and nodulation may range from very poor to moderate with regard to nitrogen fixation.[1] The screening would be for strains which are both resistant to extrinsic factors which influence survival and growth in the soil and have intrinsic characteristics which ultimately determine superiority.

## IV. LABORATORY SCREENING TESTS FOR SUPERIOR RHIZOBIA

The following traits of *Rhizobia* are suggested as possible parameters to be included in developing a laboratory screening procedure which will efficiently reduce the number of isolates subjected to final field testing and selection as inoculant-quality strains. They were chosen on the basis that they seem most relevant as diagnostic of strain quality. It must be emphasized that the following approach is merely suggested at present. It has not been evaluated thoroughly in the laboratory.

1. Lectin-mediated recognition-adsorption of *Rhizobia* to root hair cell walls is reported to be an important early event in the control of infection.[38] The occurrence and quantification of adherence of *Rhizobia* to root hairs is determined by microscopy, using the methods of Dazzo et al. (small-seeded legumes),[39] or Rao and Keister (large-seeded legumes).[40] Concomitantly, root hairs can be observed for characteristic deformation ('marked curling') of root hairs, which is typical and diagnostic for compatible *Rhizobia*-legume combinations.[39]
2. Pectinase and cellulase produced by rhizobia are reported to be important in the cell wall penetration phase of root hair infection.[38] Semiquantitative assessment of rhizobial production of these enzymes is accomplished by the methods of Saleh-Rastin et al.[41]
3. Elevated respiration is a metabolic reaction of plants in response to microbial infection.[42] The occurrence and extent of this reaction in legume roots may be a reflection of strain-genotype compatibility. The reaction may be determined by the method of Coleman and Hubbell.[43]

4. Trehalose, an unusual disaccharide reported to enhance microbial resistance to desiccation,[44] accumulates in rhizobia.[45] Trehalose is determined *in vitro* by the gas chromatography method of Streeter.[46] A less cumbersome enzymatic method might be developed utilizing trehalose in combination with glucose oxidase.[47] Trehalose in relation to the RLA is discussed by Mellor.[48] Indirect tests related to survival (resistance to heat and desiccation) are less complex but more time-consuming. They generally involve incubation of rhizobia in soil under appropriate conditions for a specified time(s), followed by enumeration of survivors, as by growth pouch assay.
5. Poly-β-hydroxybutyrate (PHB) is a reserve energy substrate which commonly accumulates in *Rhizobia* and other soil bacteria.[49] Evidence from a variety of sources indicates that it may play an important role in the survival and growth of *Rhizobia* in soil and may also function in nitrogen fixation. PHB is detected visually as the observation of unstained lipid granules in cells of stained smears of *Rhizobia*. PHB is determined *in vitro* by the method of Law and Slepecky.[50]
6. Halotolerance in both *Rhizobia* and legumes is an important characteristic in determining symbiotic performance in saline soils.[51] Production-uptake-accumulation of betaines by *Rhizobia* may enhance salt tolerance.[52] Betaine content is determined quantitatively *in vitro* by the direct methods of amino acid analysis and nuclear magnetic resonance spectroscopy.[53] These analytical methods are expensive and cumbersome; they would not be practical for screening of numerous strains. It may be feasible to develop a more practical Petri dish assay to estimate halotolerance by assessing growth stimulation of *Rhizobia* by betaine on hypersaline and hyposaline media.[54]
7. Intolerance of soil acidity and related problems (Al, Mn toxicity) is a constraint to the establishment-function of many *Rhizobia*-legume associations. Strain tolerance of acid-aluminum stress conditions is conveniently assessed by the method of Ayanaba et al.[55]

Other physiological characteristics of *Rhizobia* may be of value in screening isolates of *Rhizobia*. These traits, variously implicated in establishment of the symbiosis, have not been studied in regard to their value as general screening criteria. They include siderophore production,[56] extracellular polysaccharide,[57] growth temperature,[58] phenolics,[59] antibiotic resistance,[60] and hydrogenase activity.[61] Additional relevant but unrecognized traits undoubtedly exist. All such traits merit careful study and evaluation as potential general screening criteria.

The number, identity, and relative importance of rhizobial traits which influence symbiotic performance under field conditions is unknown. This will certainly vary according to the conditions of each field situation; therefore, the maximum number of traits should be included in a screening protocol in order to maximize the probability that resultant strains will be highly flexible, i.e., will perform as quality inoculants in a variety of dynamic field environments. It is expected that, in the future, additional rhizobial traits will be identified as useful screening criteria, tests will be developed, and the procedures added to the proposed screening protocol to form a standard (but flexible) methodology. The ultimate utility of such a protocol will depend on (1) consistent correlation of chosen traits with quality of strains obtained, (2) simplicity of assay procedures, (3) speed of assay, and (4) economy of procedures.

All screening tests suggested above, as well as tests which may subsequently be proposed, should be conducted on fresh isolates, confirmed superior strains (positive control), and confirmed inferior strains (negative control) and the values compared.

Many rhizobial traits, recognized and unrecognized, may or may not contribute to success of a rhizobial strain in a particular situation. Many of the traits assessed by the preceding protocol may therefore be inconsequential in a specific situation, as for example, when inoculating under field conditions optimum for the symbiotic system. In this case, stringent strain selection using *all* test criteria might 'over-select', i.e., eliminate some isolates which would in fact perform as quality inoculants in that particular situation. However, it can be argued that more comprehensive screening protocols will result in selection of isolates which have a broader base of flexibility in terms of their ability to adapt to future fluctuations in environment.

## V. SUMMARY

The success of legume inoculation is dependent on the quality of the rhizobial strains used. This quality is established and maintained by ongoing programs of strain isolation, screening, and selection.

Current methods for the screening of inoculant quality rhizobial strains are time-consuming and ineffective. As such, they pose a major constraint on the efficient selection of superior strains from large collections of new isolates. This reduces the efficiency of ongoing programs to increase the quality of rhizobial inoculants. It also reduces the incentive to initiate much needed isolate screening programs.

Here, an approach to the development of a standardized but flexible protocol for strain screening which is designed to be accurate, rapid, and simple to perform is outlined. It emphasizes use, whenever possible, of simple procedures and inexpensive equipment, thereby making it practical for use by investigators who must work in modestly supplied and equipped laboratories. All screening tests intended for inclusion in such a protocol should be validated by consistent correlation with symbiotic performance prior to their incorporation in the protocol.

The proposed laboratory screening protocol is intended to encourage strain selection programs by providing a feasible means of rapidly reducing to a minimum the number of fresh rhizobial isolates which merit final testing and strain selection in field trials with desired genotypes. They are "presumptive" inoculant-quality strains. Final confirmation of the quality of potential inoculant strains, whether imported or newly isolated and screened, must be accomplished by rigorous field testing under local conditions prior to large scale use. There is no substitute. Effective standard methods for field testing (*in vivo* screening) and final selection of strains are well described.[19]

## ACKNOWLEDGMENTS

Contribution No. R-02654 from the Florida Agricultural Experiment Station. This work was partially supported by the University of Minnesota Title XII Bean/Cowpea CRSP Project.

## REFERENCES

1. **Burton, J. C.,** *Rhizobium* culture and use, in *Microbial Technology*, Peppler, H. J., Ed., Van Nostrand-Reinhold, New York, 1967, chap. 1.
2. **Graham, P. H. and Hubbell, D. H.,** Soil-plant-rhizobium interaction in tropical agriculture, in *Soil Management in Tropical America*, Bornemisza, E. and Alvarado, A., Soil Science Department, North Carolina State University, Raleigh, 1975, chap. 12.
3. **Graham, P. H.,** Some problems of nodulation and symbiotic nitrogen fixation in (*Phaseolus vulgaris* L.), a review, *Field Crops Res.*, 4, 93, 1981.
4. **Churchill, B. W.,** Mass production of microorganisms for biological control, in *Biological Control of Weeds*, Charudattan, R. and Walker, H. L., Eds., John Wiley & Sons, New York, 1982, chap. 9.
5. **Lisansky, S. G.,** Production and commercialization of pathogens, in *Biological Pest Control — The Glasshouse Experience*, Hussey, N. W. and Scopes, N., Eds., Cornell University Press, Ithaca, NY, 1985, chap. 9.

6. **Walter, J. F. and Paau, A. S.,** Microbial inoculant production and use, in *Soil Microbial Ecology — Applications in Agriculture, Forestry and Environmental Management,* Metting, F. B., Jr., Ed., Marcel Dekker, Inc., New York, 1992, chap. 21.
7. **Date, R. A. and Roughley, R. J.,** Preparation of legume seed inoculants, in *A Treatise on Dinitrogen Fixation, Section IV,* Hardy, R. W. F. and Gibson, A. H., Eds., John Wiley & Sons, New York, 1977, chap. 7.
8. **Thompson, J. A.,** Production and quality control of legume inoculants, in *Methods for Evaluating Biological Nitrogen Fixation,* Bergersen, F. J., Ed., John Wiley & Sons, New York, 1980, chap. 2.
9. **Cleyet-Marel, J. C.,** Seed inoculation and inoculant technology, in *Nitrogen Fixation by Legumes in Mediterranean Agriculture,* Beck, D. P. and Materon, L. A., Eds., Martinus Nijhoff, Dordrecht, Netherlands, 1988, 379.
10. **Roughley, R. J.,** Legume inoculants: their technology and application, in *Nitrogen Fixation by Legumes in Mediterranean Agriculture,* Beck, D. P. and Materon, L. A., Eds., Martinus Nijhoff, Dordrecht, Netherlands, 1988, 259.
11. **Paau, A. S.,** Improvement of *Rhizobium* inoculants, *Appl. Environ. Microbiol.,* 55, 862, 1989.
12. **Burton, J. C.,** *Legume Inoculant Production Manual,* University of Hawaii NifTAL Project, Honolulu, HI, 1992, in press.
13. **Delwiche, C. C.,** Legumes — past, present and future, *BioScience,* 28, 565, 1978.
14. **Vincent, J. M., Whitney, A. S., and Bose, J., Eds.,** *Exploiting the Legume-Rhizobium Symbiosis in Tropical Agriculture,* College of Tropical Agriculture Miscellaneous Publication 145, Department of Agronomy and Soil Science, University of Hawaii, Maui, HI, 1977.
15. **Graham, P. H. and Harris, S. C., Eds.,** *Biological Nitrogen Fixation Technology for Tropical Agriculture,* Centro Internacional de Agricultura Tropical, Cali, Colombia, 1982.
16. **Miller, R. H. and May, S.,** Legume inoculation: successes and failures, in *The Rhizosphere and Plant Growth,* Keister, D. L. and Cregan, P. B., Eds., Kluwer Academic Publishers, Boston, 1991, 123.
17. **Mytton, R. L., Hughes, D. M., and Kahurananga, J.,** Host-*Rhizobium* relationships and their implications for legume breeding, in *Nitrogen Fixation by Legumes in Mediterranean Agriculture,* Beck, D. P. and Materon, L. A., Eds., Martinus Nijhoff, Dordrecht, Netherlands, 1988, 131.
18. **Vincent, J. M., Ed.,** *Nitrogen Fixation in Legumes,* Academic Press, New York, 1982.
19. **Somasegaran, P., Hoben, H., and Halliday, J.,** *The NifTAL Manual for Methods in Legume-Rhizobium Technology,* University of Hawaii NifTAL Project, Honolulu, HI, 1992.
20. **Vincent, J. M.,** *A Practical Manual for the Study of Root-Nodule Bacteria,* International Biological Program Handbook 15, Blackwell Scientific Publications Ltd., Oxford, 1970.
21. **Date, R. A.,** Principles of *Rhizobium* strain selection, in *Symbiotic Nitrogen Fixation in Plants,* Nutman, P. S., Ed., Cambridge University Press, Cambridge, 1976, chap. 12.
22. **Halliday, J.,** Principles of *Rhizobium* strain selection, in *Biological Nitrogen Fixation,* Alexander, M., Ed., Plenum Press, New York, 1984, 155.
23. **Thompson, J. A.,** Selection of *Rhizobium* strains, in *Nitrogen Fixation by Legumes in Mediterranean Agriculture,* Beck, D. P. and Materon, L. A., Eds., Martinus Nijhoff, Dordrecht, Netherlands, 1988, 207.
24. **Fred, E. B., Baldwin, I. L., and McCoy, E.,** *Root Nodule Bacteria and Leguminous Plants,* University of Wisconsin Press, Madison, 1932.
25. **Dart, P. J.,** Legume root nodule initiation and development, in *The Development and Function of Roots,* Torrey, J. G. and Clarkson, D. T., Eds., Academic Press, New York, 1975, chap. 21.
26. **Nutman, P. G.,** Genetics of symbiosis and nitrogen fixation in legumes, *Proc. R. Soc. Lond. B,* 172, 417, 1969.
27. **Kremer, R. J. and Peterson, H. L.,** Isolation, selection and evaluation of *Rhizobium* under controlled conditions, *Commun. Soil Sci. Plant Anal.,* 13, 749, 1982.
28. **Dowling, D. N. and Broughton, W. J.,** Competition for nodulation of legumes, *Annu. Rev. Microbiol.,* 40, 131, 1986.
29. **Bowen, G. D.,** Dysfunction and shortfalls in symbiotic responses, in *Plant Disease, Vol. III,* Horsfall, J. G. and Cowling, E. B., Eds., Academic Press, New York, 1978, chap. 11.
30. **Vincent, J. M.,** Factors controlling the legume-*Rhizobium* symbiosis, in *Nitrogen Fixation,* Vol. II, Newton, W. E. and Orme-Johnson, W. H., Eds., University Park Press, Baltimore, MD, 1980, 103.
31. **Bushby, H. V. A.,** Ecology, in *Nitrogen Fixation, Vol. 2, Rhizobium,* Broughton, W. J., Ed., Clarendon Press, Oxford, 1982, chap. 2.
32. **Vincent, J. M.,** The role of legume, *Rhizobium,* and environment in nitrogen fixation: constraints on symbiotic potential and their removal, in *Nitrogen Fixation by Legumes in Mediterranean Agriculture,* Beck, D. P. and Materon, L. A., Eds., Martinus Nijhoff, Dordrecht, Netherlands, 1988, 275.
33. **Vincent, J. M.,** Ecological aspects of the root-nodule bacteria: competition and survival, in *Nitrogen Fixation by Legumes in Mediterranean Agriculture,* Beck, D. P. and Materon, L. A., Eds., Martinus Nijhoff, Dordrecht, Netherlands, 1988, 165.

34. **Burton, J. C.,** The *Rhizobium*-legume association, in *Microbiology and Soil Fertility*, Gilmour, C. M. and Allen, O. N., Eds., Oregon State University Press, Corvallis, 1964, 107.
35. **Somasegaran, P. and Bohlool, B. B.,** Single-strain versus multistrain inoculation: effect of soil mineral N availability on rhizobial strain effectiveness and competition for nodulation on chick-pea, soybean, and dry bean, *Appl. Environ. Microbiol.*, 56, 3298, 1990.
36. **McCardell, A. J., Cregan, P. B., and Sadowsky, M. J.,** Genetics and improvement of biological nitrogen fixation, in *Soil Microbial Ecology*, Metting, F. B., Ed., 1992, chap. 6.
37. **Bowen, G. D.,** Misconceptions, concepts and approaches in rhizosphere biology, in *Contemporary Microbial Ecology*, Ellwood, D. C., Hedger, J. N., Latham, M. J., Lynch, J. M., and Slater, J. H., Eds., Academic Press, New York, 1980, 283.
38. **Dazzo, F. B. and Hubbell, D. H.,** Control of root hair infection, in *Nitrogen Fixation, Vol. 2*, Broughton, W. J., Ed., Clarendon Press, Oxford, 1982, chap. 9.
39. **Dazzo, F. B., Napoli, C. A., and Hubbell, D. H.,** Adsorption of bacteria to roots as related to host specificity in the *Rhizobium*-clover symbiosis, *Appl. Environ. Microbiol.*, 32, 166, 1976.
40. **Rao, V. and Keister, D. L.,** Infection threads in the root hairs of soybean (*Glycine max*) plants inoculated with *Rhizobium japonicum*, *Protoplasma*, 97, 311, 1978.
41. **Saleh-Rastin, N., Coleman, S., Petersen, M. A., and Hubbell, D. H.,** Improved methods for *in vitro* detection of cell wall modifying enzymes in rhizobia, *Appl. Environ. Microbiol.*, in press.
42. **Wood, R. K. S.,** Alterations in metabolism in diseased plants, in *Physiological Plant Pathology*, Botanical Monographs, Vol. 6, James, W. O. and Burnett, J. H., Eds., Blackwell Scientific Publications, Oxford, 1967, chap. 11.
43. **Coleman, S. E. and Hubbell, D. H.,** Histochemical detection of inoculation-induced respiratory changes in seedling roots of nodulating and non-nodulating soybean genotypes, *J. Histochem. Cytochem.*, 38, 1071, 1990.
44. **Crowe, J. M., Crowe, M. L., and Chapman, D.,** Preservation of membranes in anhydrobiotic organisms. The role of trehalose, *Science*, 223, 701, 1984.
45. **Streeter, J. G.,** Accumulation of α,α-trehalose by *Rhizobium* bacteria and bacteroides, *J. Bacteriol.*, 164, 78, 1985.
46. **Streeter, J. G.,** Carbohydrates in soybean nodules. II. Distribution of compounds in seedlings during the onset of nitrogen fixation, *Plant Physiol.*, 66, 471, 1980.
47. **Elbein, A. D.,** The metabolism of trehalose, in *Advances in Carbohydrate Chemistry and Biochemistry*, Vol. 30, Tipson, R. S. and Horton, D., Eds., Academic Press, New York, 1974, 227.
48. **Mellor, R. B.,** Is trehalose a symbiotic determinant in symbioses between higher plants and microorganisms?, *Symbiosis*, 12, 113, 1992.
49. **Dawes, E. A. and Senior, P. J.,** The role and regulation of energy reserve polymers in microorganisms, in *Advances in Microbial Physiology*, Vol. 10, Rose, A. H. and Tempest, D. W., Eds., Academic Press, New York, 1973, 135.
50. **Law, J. H. and Slepecky, R. A.,** Assay of poly-β-hydroxybutyric acid, *J. Bacteriol.*, 82, 33, 1961.
51. **Velagaleti, R. R. and Marsh, S.,** Influence of host cultivars and *Bradyrhizobium* strains on the growth and symbiotic performance of soybean under salt stress, *Plant Soil*, 119, 133, 1989.
52. **Rudulier, D. L. and Bernard, T.,** Salt tolerance in *Rhizobium*: a possible role for betaines, *FEMS Microbiol. Rev.*, 39, 67, 1986.
53. **Madkour, M. A., Smith, L. T., and Smith, G. M.,** Preferential osmolyte accumulation: a mechanism of osmotic stress adaptation in diazotrophic bacteria, *Appl. Environ. Microbiol.*, 56, 2876, 1990.
54. **Marthi, B. and Lighthart, B.,** Effects of betaine on enumeration of airborne bacteria, *Appl. Environ. Microbiol.*, 56, 1286, 1990.
55. **Ayanaba, A., Asanuma, S., and Munns, D. N.,** An agar plate method for rapid screening of *Rhizobium* for tolerance to acid-aluminum stress, *Soil Sci. Soc. Am. J.*, 47, 256, 1983.
56. **Schwyn, B. and Neilands, J. B.,** Universal chemical assay for the detection and determination of siderophores, *Anal. Biochem.*, 160, 47, 1987.
57. **Puhler, A., Arnold, W., Buenda-Claveria, A., Kapp, D., Keller, M., Niehaus, K., Quandt, J., Roxlau, A., and Weng, W. M.,** The role of the *Rhizobium meliloti* exo-polysaccharides EPS I and EPS II in the infection process of alfalfa nodules, in *Advances in Molecular Genetics of Plant-Microbe Interactions*, Vol. 1, Henneke, H. and Verma, D. P. S., Eds., Kluwer Academic Publishers, Netherlands, 1991, 189.
58. **Munevar, F. and Wollum, A. G., II,** Growth of *Rhizobium japonicum* strains at temperatures above 27°C, *Appl. Environ. Microbiol.*, 42, 272, 1981.
59. **Siqueira, J. O., Nair, M. G., Hammerschmidt, R., and Safir, G. R.,** Significance of phenolic compounds in plant-soil-microbe systems, *Crit. Rev. Plant Sci.*, 10, 63, 1991.
60. **Bromfield, E. S. P. and Jones, D. G.,** The competitive ability and symbiotic effectiveness of doubly labelled antibiotic resistant mutants of *Rhizobium trifolii*, *Ann. Appl. Biol.*, 91, 211, 1979.
61. **Haugland, R. A., Hanus, F. J., Cantrell, M. A., and Evans, H. J.,** Rapid colony screening method for identifying hydrogenase activity in *Rhizobium japonicum*, *Appl. Environ. Microbiol.*, 45, 892, 1983.

# 21

# Isolation and Characterization of Plant Growth-Promoting Rhizobacteria

*Yoav Bashan, Gina Holguin, and Ran Lifshitz*

## I. INTRODUCTION

Biotechnology has opened up new possibilities concerning the application of beneficial bacteria to the soil for the promotion of plant growth and the biological control of soil-borne pathogens.[1-3] Since the large scale release of genetically engineered bacteria to the environment faces a number of regulatory hurdles, the need to isolate and select superior, naturally occurring rhizosphere bacteria continues to be of interest. Apart from rhizobia symbionts, the rhizosphere-associated beneficial bacteria consist of the following genera: (1) *Pseudomonas* and *Bacillus*, which antagonize pathogenic or deleterious microorganisms (biological control) and (2) bacteria that enhance plant growth directly such as *Azospirillum*, *Herbaspirillum*, *Enterobacter*, *Acetobacter*, *Azotobater*, and *Pseudomonas*, as well as many unidentified rhizosphere isolates.

The nutritional and environmental requirements of these bacteria are very diverse, and hence there is no general method that can be used to isolate all species of Plant Growth-Promoting Rhizobacteria (PGPR). Accordingly, a variety of methods have been developed, primarily within the last two decades.

## II. HABITAT

Generally, there are no specific sites where PGPR can be found since all plant roots are associated with numerous species of microorganisms, beneficial as well as pathogenic. However, some strategies have been established:

1. The best place to search for a biological control agent is in the same ecological niche as the target pathogen.[3-4]
2. Wild ancestors of crop plants which have evolved with PGPR can be isolated and used later to inoculate cultivated plants (M. Feldman, unpublished).
3. When a target plant is introduced into the soil, low numbers of native PGPR in the soil can be enriched in the rhizosphere to form large populations for recovery (Y. Bashan, unpublished).
4. Potential PGPR for biocontrol are selected on the basis of their siderophore or antibiotic production *in vitro*.[2]
5. Of the thousands of bacteria isolated randomly from the plant rhizosphere of diverse habitats, many are being selected for growth promotion without regard to taxonomy or the mechanisms involved. This type of approach is common to industrial R&D programs.[2]

0-8493-5164-2/93/$0.00+$.50
© 1993 by CRC Press, Inc.

# III. ISOLATION

Endorhizosphere bacteria (bacteria colonizing the root interior) are distinct from root surface isolates.[5] Thus, different strategies are followed for the isolation of each of these two groups.

## A. ROOT SURFACE BACTERIA

1. To equalize osmotic pressure, roots are first soaked for 10 min in sterile phosphate-buffered saline (PBS) [10 m*M* $K_2PO_4$-$KH_2PO_4$, 0.14 *M* NaCl, pH 7.2], then chopped into small pieces (3 cm), inoculated on the selected enrichment medium or media (described later), and incubated at 25 to 30°C for 1 to 2 weeks until visible growth is apparent.[6]
2. For the isolation of root surface bacteria with root adhesion capacity, the roots must first be washed thoroughly with running tap water, then with sterile distilled water and finally soaked in PBS. The selected enrichment medium is then inoculated with these roots and incubated at 25 to 30°C until visible growth is apparent.
3. Root surface bacteria can be isolated by shaking small pieces of roots for 10 min on a mechanical gyrator shaker in PBS or phosphate buffer-peptone containing, per liter: peptone, 1.0 g; $K_2HPO_4$, 1.21 g; $KH_2PO_4$, 0.34 g.[7] Optional alternative diluents are: PBS plus 0.025% Tween 20, 0.01% Tween 40,[6] or 0.1% tryptic soy broth (TSB).[8-9] Diluted samples are then inoculated on the appropriate medium at the selected temperature until growth is visible.

## B. ENDORIZOSPHERE BACTERIA

The roots are surface-sterilized by soaking in 95% (v/v) ethanol and 0.1% (w/v) acidified $HgCl_2$ for 1 min, respectively, and then washed (a minimum of ten times) with sterile tap water.[10] Alternate disinfection procedures include (1) soaking roots in 70% ethanol for 5 min, in 6.25% sodium hypochlorite for 10 min, followed by several rinses in sterile distilled water[7] or (2) soaking in 10% $H_2O_2$ for 15 sec and then rinsing three times with 0.1 *M* $MgSO_4$.[5]

Plant material is then suspended in 0.05 *M* PBS or 0.1 *M* $MgSO_4$ and ground with a mortar and pestle, or homogenized by a high speed shaft for 1 to 3 min (100 ml of PBS for each 5 g fresh weight of roots). The slurry may be filtered through sterile cotton wool.[11] Diluted or undiluted aliquots are then prepared for inoculation onto an enrichment medium.

An alternative isolation method is to aseptically spread the sterilized roots on nutrient agar supplemented with glycerol (1%) to verify the surface sterility of the roots. Roots are cut lengthwise, placed onto the selected enrichment medium, preferably supplemented with glucose (since glucose-utilizing, acid-producing bacteria seem to dominate in the endorhizosphere), and incubated at the selected temperature until growth is visibly detected.[7] After the incubation period, 0.1 ml of the growth medium is spread onto chosen selective, solid media and incubated again at the same temperature. Isolates must be restreaked several times on the same solid medium until purity is obtained.

## C. DIAZOTROPHIC PLANT GROWTH-PROMOTING RHIZOBACTERIA

### 1. Spermosphere Model for Isolating Diazotrophs from the Rhizosphere

The spermosphere model[12] consists of a seed germinated in the dark which is releasing exudates in a medium free of carbon and nitrogen. This is then inoculated with soil dilutions and incubated under an acetylene atmosphere. The system is appealing for two reasons: (1) the germinating seedling provides the bacteria with the most useful carbon sources they encounter in the soil, thus avoiding bias in the carbon nutrition and (2) the growing seedling consumes any nitrogen made available by the diazotrophs, keeping the medium nitrogen-free and highly selective.

1. 10 g of rhizosphere soil (roots and the soil that adheres to them) is macerated in a mortar and pestle and diluted to 100 ml in PBS within 2 h of sampling. Additional dilutions of the slurry, in PBS, should be made.
2. Seeds are disinfected by any standard method (designated for a particular plant species).
3. Disinfected seeds are placed on the medium surface of 5 ml of semisolid (0.3% agar) N-free, C-free medium in a 35 ml test tube (described below). The spermosphere tubes are kept in the dark at 28°C (or any other desired temperature) for a week. Tubes with a side arm containing 2 ml 1 *N* NaOH, which serves as a $CO_2$ trap, may be used when necessary.
4. When coleoptiles are 1 cm high or more, the spermosphere assemblies are inoculated with 0.5 ml aliquots of soil dilutions obtained from Step #1. Earlier inoculation frequently results in seedling death. A delay in inoculation also allows for the identification of seed contamination from insufficient disinfection.
5. The number of diazotrophs is calculated by the Most Probable Number method (MPN).[13] The estimation is based on the numbers (ten replicate tubes for each dilution) of nitrogenase-positive tubes detected by the acetylene reduction assay.[14]
6. The contents of ten tubes of the highest dilution at which all tubes are ethylene positive, are pooled, homogenized, serially diluted ($10^{-5}$ to $10^{-9}$) and plated on N-free solid medium[15-16] in flat (120 ml or more) serum bottles and incubated under 1% acetylene for 4 to 8 days. (Pooling and homogenization avoid confusing problems arising from the possible transfer of a small piece of root to a high dilution tube.) Colonies which develop should be picked individually, purified on N-free medium, and tested for nitrogen fixation as described above. In some instances, it is necessary to partially hydrolyze the capsular material using 0.5 or 0.1 *N* NaOH to remove any contaminants.

*Note:* This technique does not give a complete survey of nitrogen-fixing bacteria in the rhizosphere but only delineates the numbers and the nature of the most abundant diazotrophs.

### 2. Semisolid Enrichment Cultures of Root Pieces for Isolation of Microaerophilic Diazotrophic Bacteria

Semisolid, N-free media are the key to the isolation of a number of root-associated microaerophilic diazotrophs.[17,18] Techniques using these media are very simple and useful when there is an abundance of diazotrophs associated with roots. Pure cultures can be obtained with only a few purification steps and without much difficulty. This technique was used in the discovery of four *Azospirillum* strains, *Campilobacter nitrofigilis*, *Herbaspirillum seropedicae*, some diazotrophic *Pseudomonas*, *Acetobacter diazotrophicus*, and *Bacillus azotofixans*.

#### a. *Procedure #1*

Intact root pieces (0.5 to 1.0 cm) are placed into semisolid (0.05% agar or less) NFb, OAB, or BL media (described below) and incubated without shaking for 2 to 5 days at 25 to 35°C. (At lower incubation temperatures, the incubation time should be extended.) Following this incubation, a white bacterial pellicle is formed 2 to 10 mm below the surface. An assessment of the ability of the enriched culture to fix nitrogen should be done by the acetylene reduction technique.[14] (Special care should be taken not to disturb the pellicle because this immediately stops nitrogenase activity.) Almost pure cultures can be obtained after one to three subcultures onto the same medium followed by streaking out of 24-hour-old cultures on solid agar plates consisting of the same medium.[18]

#### *b. Procedure #2 (for isolation of the diazotroph Acetobacter associated with sugarcane)*

Roots and stems of sugarcane are washed with tap water, macerated in a blender, and serial dilutions are prepared in a 5% sucrose solution and incubated in semisolid medium (described below). Continue as in Procedure #1 above.[19]

### 3. Nonselective Media for Isolation of Rhizosphere Bacteria in General

The TSA medium described below recovers a wide range of aerobic and facultatively anaerobic gram-negative and gram-positive bacteria.[9] However, it is advisable to try the isolation procedure simultaneously with different isolation media since a nonselective medium will only allow the recovery of a small percentage of the existing PGPR population.

1. 1/10 strength TSBA medium (tryptic soy broth agar) for heterotrophic bacteria contains tryptic soy broth, 3 g and agar, 15 g.[9] A more diverse population of bacteria can be isolated from soil and other environmental samples by using a more diluted (1/100 strength; 0.3 g TSBA) medium.[4]
2. TSA medium contains the following (g/l): meat extract, 3.0; yeast extract, 3.0; peptone from casein, 15.0; peptone from meat, 5.0; lactose, 10.0; sucrose, 10.0; glucose, 1.0; $NH_3^+Fe^{3+}$ citrate, 0.5; NaCl, 5.0; sodium thiosulfate, 0.5; phenol red, 0.024; agar, 12.0; distilled water, 1000 ml; pH 7.4.[20] The elimination of gram-positive bacteria in 1/10 strength TSA is done by adding 2 μg/l of crystal violet[4] or 1.2 g/l of sodium lauroyl sarcosine (SLS) to S1 medium (described in Section III.C.5).

### 4. Media for the Isolation of Different PGPR

The following media can be used for recovery of most common PGPR populations:

1. Selective medium for the isolation of pseudomonads[20] is based on TSA medium supplemented with (μg/ml): basic fuchsin, 9; nitrofurantoin, 10; nalidixic acid, 23; (mg/ml): cycloheximide, 0.9; TTC (triphenyltetrazolium-chloride), 1.4; the TSA base is supplied with basic fuchsin and TTC before autoclaving. Nalidixic acid and nitrofurantoin are sterilized by filtering through 0.45 μm membrane filters and added aseptically to the sterilized TSA medium. Cycloheximide is either sterilized by filtering before addition to the sterile medium, or by adding directly as the nonsterile powder.
2. Coryneforms and other gram-positive bacteria are isolated on D-2 medium[22] in which potassium dichromate (50 mg/l) and cycloheximide (100 mg/l) are incorporated to enhance selectivity, or on methyl-red agar for gram-positive bacteria.[23]
3. King's B medium for pseudomonads contains (g/l): proteose peptone no. 3 (Difco), 20; glycerol, 10 ml; asparagine, 2.25; $K_2HPO_4$, 1.5; $MgSO_4 \cdot 7H_2O$, 1.5; agar, 15.[21] With an incubation temperature of 30°C, the results can be assessed after 48 h.

*Note:* All the above media can be supplemented with benomyl, 20 mg/ml (Benlate, 50%, W.P. Dupont, USA) or Nystatin and actidione (50 mg of each per liter) to reduce fungal growth.[7,9]

### 5. Media for the Isolation of Fluorescent Pseudomonads

The fluorescent pseudomonads (*P. putida* and *P. fluorescens*) are a large group found in the rhizosphere of various crop plants, which can be isolated easily on the following media.

1. Modified King's B medium supplemented with the antibiotics (mg/l): chloramphenicol, 5; cycloheximide, 75; novobiocin, 45; and penicillin G, 75,000 units.[24]
   *Note:* Resistance to the recommended antibiotics is not unique to the pseudomonads.
   *Note:* The original King's B medium is currently accepted as a diagnostic medium for

the detection of fluorescence; however, it is not particularly suitable for the isolation of these bacteria because it is relatively unselective.

2. D4 medium[22] contains (g/l): glycerol, 10.0 ml; sucrose, 10.0; casein hydrolysate, 1 g; $NH_4Cl$, 5.0 g; sodium dodecyl sulfate, 0.6 g (to eliminate nonpseudomonads); $Na_2HPO_4$ (anhydrous), 1.5 g; agar, 15 g. This medium is commonly used for Gram-negative bacteria; however, many other Gram-negative bacteria can grow on it, and the fluorescence cannot be observed.
3. Medium S1[8] contains (g/l): agar, 18; sucrose, 10; glycerol, 10 ml; casamino acids (Difco), 5.0; $NaHCO_3$, 1.0; $MgSO_4 \cdot 7H_2O$, 1.0; $K_2HPO_4$, 2.3; sodium lauroyl sarcosine (SLS), 1.2; and 20 mg of trimethoprim (Sigma), (5-[(3,4,5- trimethoxyphenyl) methyl]-2,4-pyrimidinediamine). The final pH of the medium is 7.4 to 7.6.

*Note:* Trimethoprim is added after the medium has been autoclaved and cooled.
*Note:* This medium has several advantages over other media used for the isolation of fluorescent pseudomonads; it consistently provides high selectivity and good recovery of fluorescent pseudomonads with samples obtained from a variety of habitats. Fluorescence can be observed during the initial isolation.

### 6. Medium for the Spermosphere Model

***Solution A*:** (mg/l): $H_3BO_3$, 750; $ZnSO_4 \cdot 7H_2O$, 550; $CoSO_4 \cdot 7H_2O$, 350; $CuSO_4 \cdot 4H_2O$, 22; $MnCl_2 \cdot 4H_2O$, 10; distilled water 1000 ml.

***Solution B*:** (g/l): $FeSO_4 \cdot 7H_2O$, 0.8; $MgSO_4 \cdot 7H_2O$, 4.0; $Na_2MoO_4 \cdot 2H_2O$, 0.118; $CaCl_2 \cdot 2H_2O$, 4.0; EDTA, 0.8; solution A, 4 ml; distilled water 1000 ml.

***Final basal solution*:** (g/l): $KH_2PO_4$, 1.8; $K_2HPO_4$, 2.7; solution B, 50 ml; Nobel agar, 5; distilled water, 1000 ml; pH adjusted to 6.8 using KOH; sterilization by autoclaving.

The N-free medium used to grow the bacteria from the spermosphere model[15] contains (g/l): yeast extract as starter, 0.1; starch, 5; glucose, 5; mannitol, 5; malic acid, 3.5; plus the basal medium listed above,[25] or the following combined carbon medium.

### 7. Combined Carbon Medium for Isolation of Diazotrophs

The combined carbon medium was designed to incorporate common factors of various N-free media since the basal composition of these media is very similar. Mannitol was included to support the growth of *Azotobacter* sp., biotin, and p-aminobenzoic acid for *Bacillus* sp. Yeast extract was included to supply miscellaneous organic growth factors and may supply 'starter' nitrogen that promotes growth without inhibiting acetylene reduction.

***Solution A:*** $K_2HPO_4$, 0.8 g; $KH_2PO_4$, 0.2 g; NaCl, 0.1 g; $Na_2FeEDTA$, 28 mg; $Na_2MoO_4 \cdot 2H_2O$, 25 mg; yeast extract, 100 mg; mannitol, 5 g; sucrose, 5 g; sodium lactate, 0.5 ml (60%, v/v) distilled water, 900 ml.

***Solution B (g):*** $MgSO_4 \cdot 7H_2O$, 0.2; $CaCl_2$, 0.06; distilled water, 100 ml.

The solutions should be autoclaved separately, cooled, and mixed. To this new solution, filter-sterilized biotin (5 μg/l) and p-aminobenzoic acid (10 μg/l) should be added and the final pH adjusted to 7.0.

### 8. Media for the Isolation of *Azospirillum*

The most commonly used medium for the isolation of *Azospirillum* is semisolid NFb medium.[18,41] Several useful modifications of this medium have been developed and are indicated below.

#### a. NFb Medium (g/l)

DL-malic acid, 5; KOH, 4; $K_2HPO_4$, 0.5; $MgSO_4 \cdot 7H_2O$, 0.2; $CaCl_2$, 0.02; NaCl, 0.1; $FeSO_4 \cdot 7H_2O$, 0.5; (mg/l): $NaMoO_4 \cdot 2H_2O$, 2; $MnSO_4 \cdot H_2O$, 10; 0.5% alcoholic solution (or dissolved in 0.2 *N* KOH) of bromothymol blue, 2 ml; agar, 0.0175 to 0.5%; 1000 ml distilled water, pH 6.8.[18] When this medium is supplemented with low concentrations (0.002 to 0.005%) of yeast extract, it still eliminates the growth of contaminants and permits colony formation under aerobic conditions.[17] In cases where malic acid-KOH inhibits growth, they can be replaced with 10 g/l sodium succinate[26] or 0.5% sucrose. Bromothymol blue can be replaced by 0.001% bromocresol purple.[27]

#### b. OAB Medium

This modification is more suitable for *Azospirillum* growth than for isolation procedures.[28] It is not highly selective for this genus, but it provides increased buffering capacity over the original medium, microelements, a limited amount of $NH_4Cl$ to initiate aerobic growth, and a small amount of yeast extract to shorten the lag phase and aid vigorous growth. It can be used as a liquid, semisolid (0.05% agar), or solid medium. For optimal growth of *Azospirillum* in liquid medium, the culture should be maintained at a constant $pO_2$ of 0.005 to 0.007 atm under an atmosphere of a mixture of $N_2$ and air.

***Solution A***: (g/l) DL-malic acid, 5; NaOH, 3; $MgSO_4 \cdot 7H_2O$, 0.2; $CaCl_2$, 0.02; NaCl, 0.1; $NH_4Cl$, 1; yeast extract, 0.1; $FeCl_3$, 0.01; (mg/l) $NaMoO_4 \cdot 2H_2O$, 2; $MnSO_4$, 2.1; $H_3BO_3$, 2.8; $Cu(NO_3)_2 \cdot 3H_2O$, 0.04; $ZnSO_4 \cdot 7H_2O$, 0.24; 900 ml distilled water.

***Solution B***: (g/l) $K_2HPO_4$, 6; $KH_2PO_4$, 4; 100 ml distilled water.

After autoclaving and cooling, the two solutions should be mixed. The medium pH is 6.8.

*Note:* $FeCl_3$ can be replaced by 10 ml of Fe(III)-EDTA (0.66%, w/v, in water). This medium can also be supplemented with 10 ml of a vitamin solution to enhance its isolation ability for heterotrophic microaerophilic nitrogen-fixing bacteria. The vitamin solution contains: D-biotin (200 mg), calcium pantothenate (40 mg), myoinositol (200 mg), niacinamide (40 mg), p-aminobenzoic acid (20 mg), pyridoxine hydrochloride (40 mg), riboflavin (20 mg), thiamine dichloride (4 mg), in 10 ml and is sterilized by filtration.[29]

### 9. Semiselective Media for *Azospirillum*

#### a. Congo red-NFb

This medium is basically NFb medium supplemented with 15 ml/l medium of 1:400 aqueous solution of Congo-red, autoclaved separately and added just before using.[30]

*Note:* This medium permits the recognition of *Azospirillum* colonies on plates and facilitates the isolation of pure cultures since the colonies appear dark red or scarlet with typical colony characteristics, whereas many soil bacteria do not absorb Congo-red.

#### b. BL and BLCR Media

These two semiselective media[31] are based on OAB medium. BL medium is OAB medium supplemented with (mg/l) streptomycin sulfate, 200; cycloheximide, 250; sodium deoxycholate, 200; and 2,3,5-triphenyltetrazolium chloride, 15. BLCR is BL medium supplemented with an aqueous solution of Congo-red (approximately 1 ml of a 1 mg/ml solution per liter).

*Note:* These media are very suitable for the isolation of *Azospirillum* from the rhizosphere since the colonies are easily recognizable, especially on BLCR medium. However, some strains of *A. brasilense* failed to grow on this medium,[32] and the growth of *Azospirillum* on BLCR medium is significantly slower compared to the original OAB medium (about 10 days incubation time).

### 10. Medium for Isolation of Halophilic Diazotrophs

(g/l): DL-malic acid, 5; KOH, 4.8; $MgSO_4 \cdot 7H_2O$, 0.25; $CaCl_2$, 0.22; NaCl, 1.2; $Na_2SO_4$, 2.4; $NaHCO_3$, 0.5; $K_2SO_4$, 0.17; $Na_2CO_3$, 0.09; Fe (III)-EDTA, 0.077; $K_2HPO_4$, 0.13; yeast extract, 0.02; (mg/l): biotin, 0.1; $Na_2MoO_4 \cdot 2H_2O$, 2; $MnCl_2 \cdot 4H_2O$, 0.2; $H_3BO_3$, 0.2; $CuCl_2 \cdot 2H_2O$, 0.02; $ZnCl_2$, 0.15; agar 2 to 8; 1000 ml distilled water.[29,33] The medium pH is 8.5. Malic acid,

KOH, and agar are dissolved in one half of the total volume and autoclaved. The salt fraction is sterilized by filtration after dissolving the ingredients in one half of the total volume and discarding the precipitate after centrifugation of the medium. Slight modifications in concentrations are also possible.[33]

### 11. Medium for Isolation of Halophilic Rhizosphere Bacteria

Rhizosphere bacteria from xerophytic plants in hypersaline soils (5.0 to 10.7% NaCl) can be isolated by changing the total salt concentration of the medium to 9, 50, 100, 200, and 250 g/l of total salts used.

The basic medium[34] contains 200 g/l of total salt content and is composed of (g/l): NaCl 158.9; $MgCl_2$, 13.8; $MgSO_4$, 20.9; $CaCl_2$, 1.5; KCl, 4.2; $NaHCO_3$, 0.2; NaBr, 0.5.

These basal salt mixtures can be supplemented with one of the three following media, final concentration (g/l): (1) Peptone P (Oxoid), 10. (2) Yeast extract, 10; Proteose Peptone (Difco), 5; glucose, 1. (3) Yeast extract, 5; glucose, 1; with added soil extract. The soil extract is prepared by autoclaving equal volumes of garden soil and the corresponding salt solution; after decanting, the extract is filtered through paper and the other nutrients are added to the filtrate. pH values are adjusted to 7.5 with KOH.

### 12. Medium for the Isolation of Diazotrophic Acetobacter from Sugarcane

This semisolid medium[19] is based on NFb medium, with modifications for this acid tolerant species.

***Solution A*:** (g/l) cane sugar (or sucrose or glucose), 100; $MgSO_4{\cdot}7H_2O$, 0.2; $CaCl_2{\cdot}2H_2O$, 0.02; $FeCl_3{\cdot}6H_2O$, 0.01; agar, 2.2; (mg/l) $Na_2MoO_4{\cdot}2H_2O$, 2; 0.5% solution (dissolved in 0.2 *N* KOH) of bromothymol blue, 5 ml; 900 ml distilled water.

***Solution B***: (g/l) $K_2HPO_4$, 0.2; $KH_2PO_4$, 0.6; 100 ml distilled water.

*Note:* It is advisable to autoclave the two solutions separately and mix them after they are cool. Then, the medium is acidified with acetic acid to pH 4.5. For the purification of isolates, the medium is supplemented with 0.02 g yeast-extract and 15 g agar. The isolation can be improved with the addition of 1% cane juice. The colonies appear dark orange.

### 13. Medium for Isolation of Marine Beneficial Diazotrophs (HGB)

HGB medium is based on OAB medium[28] with several modifications to suit marine bacteria and is practical for diazotrophic vibrios.[70]

***Solution A*** (g/890 ml distilled water): DL-malic acid, 5; NaOH, 3; $MgSO_4{\cdot}7H_2O$, 3; $CaCl_2$, 0.02; NaCl, 20; yeast extract, 0.1.

***Solution B*** (stock solution, g/500 ml distilled water): $FeCl_3$, 0.5; $Na_2MoO_4{\cdot}2H_2O$, 0.1; $MnSO_4$, 0.105; $H_3BO_3$, 0.14; $CuCl_2{\cdot}2H_2O$, 0.0014; $ZnSO_4{\cdot}7H_2O$, 0.012.

***Solution C***: 100 ml of PBS 0.39 *M*, pH 7.6.

Ten ml of Solution B are added to Solution A and autoclaved. After cooling, this solution is mixed with the buffer, Solution C, which should be autoclaved separately.

### 14. Media for the Isolation of *Bacillus*

*Bacillus* spp. are selected by heat-treating dilutions at 100°C for 15 min prior to plating on 1/10 TSA medium.[4]

### 15. Medium for the Isolation of Azotobacter

An efficient N-free medium for the isolation of *Azotobacter*[35] is based on soil extract and on mannitol as carbon source.

***Composition* (g/l):** $KH_2PO_4$, 1.0; $MgSO_4{\cdot}7H_2O$; 0.2; NaCl, 0.2; $FeSO_4{\cdot}7H_2O$, 0.005; soil extract, 100 ml; tap water, 900 ml; agar, 15; and mannitol, 20. The pH is adjusted to 7.6 with NaOH prior to autoclaving. Mannitol and $FeSO_4$ are sterilized separately and added to the rest of the medium when cool. The soil extract is prepared as follows.[36] Non-earth material is discarded from the soil samples, and the soil is pulverized aseptically. Pulverized soil (10g) is shaken in 90 ml of sterile distilled water for 15 min. One milliliter of this suspension is diluted in 9 ml of 0.85% NaCl, and 1.0 ml of this is plated onto the N-free medium. The plates are then incubated at 26°C for 4 to 5 days.

## IV. CHARACTERIZATION

### A. *IN VITRO* TESTS FOR ANTIMICROBIAL ACTIVITY OF RHIZOSPHERE BACTERIA

Selection of PGPR can sometimes be based on bacterial antagonism toward plant pathogens. The major problem in identifying microbial isolates for use as biocontrol agents is the need to screen large numbers of cultures. There are no reports of useful characteristics associated with pathogen repression that could be used to screen isolates from a collection, thereby reducing the numbers that have to be evaluated by bioassay.[9]

#### 1. Agar Assays

(1) The bacterial isolates are inoculated onto the center of a potato dextrose agar plate and incubated at 28°C. Bacterial growth is terminated after 4 days by exposure to chloroform vapors. Spores of the fungal pathogen are suspended in sterile, distilled water and sprayed over the bacterial growth plates. Zones of inhibition are measured after 36 h of growth at 28°C.[37] (2) Spore suspensions of fungal pathogens are obtained by flooding sporulating culture growth on plates with sterile 0.05% Tween 40. Spore suspensions (0.5 ml) are then plated on the same agar medium used to isolate the rhizobacteria and allowed to dry for 2 to 3 h. Rhizobacteria isolates are stabbed in quadrants of agar plates containing the fungal pathogen and incubated (for 7 days, as in the case of *Fusarium oxysporum* at 27°C). Antibiotic production against the pathogen is observed as a zone of growth inhibition around the agar stabs of the rhizobacterial isolates.[6]

*Note:* While many studies have utilized an agar assay to determine pathogen repression by PGPR isolates, this practice is limited by the absence of a plant which can greatly affect the ability of an amended bacterium to survive, colonize, and repress pathogenic fungi.

#### 2. The 'In planta' Assay

The 'in planta' assay is more representative of conditions to which the amended bacteria will be exposed once field trials are initiated.[9] The following is one example of the numerous 'in planta' assays for disease biocontrol.

The fungal pathogen is cultured on potato dextrose agar for 72 h; then one plate is macerated in 100 ml of 0.01 *M* potassium phosphate buffer (pH 6.8). The homogenate is used as inoculum for 4 kg of a sterile mixture of sand and vermiculite (1:1, v/v), amended with calcium carbonate at a rate of 10 g/kg, and packed in sterile glass test tubes (200 mm × 25 mm). The tubes are filled to a depth of 15 cm and moistened with 18 ml of half-strength Hoagland's nutrient solution. Surface-sterilized seeds are placed on the soil surface and inoculated with approximately 50 μl of $10^7$ cfu/ml of the tested PGPR suspension. The seeds are then covered with an additional 2.0 cm layer of infected 'soil mixture', and an additional 3 ml of nutrient solution are added. Reference PGPR strains, as well as untreated seeds planted in pathogen-inoculated and in noninfected mixtures of sand and vermiculite, are used as positive controls. A strain is considered a promising biocontrol agent if it performs as well as one of the reference strains.

### 3. Antagonistic Activity of PGPR Regulated by Siderophores

The majority of fluorescent pseudomonads are siderophore producers. These siderophores efficiently deplete iron from the environment, making it less available to certain competing microorganisms including plant pathogens.[6] The antagonistic activity of pseudomonad PGPRs can be tested by measuring their ability to inhibit the growth of *Erwinia agricola* and *F. oxysporum* on low-iron media such as SR medium and SR-$Fe^{3+}$ (20 μg of $FeCl_3$/ml) media.[24] Presumably, rhizobacteria which are able to inhibit the test microorganism on SR but not on SR-$Fe^{3+}$ produce extracellular iron-chelating siderophores.

## B. CHARACTERIZATION AT THE SPECIES LEVEL

To facilitate species characterization of PGPR, two distinct approaches can be taken: (1) isolating different strains of known species, and (2) attempting to isolate new species of bacteria, which is usually a time-consuming process. In some instances, it is difficult to choose between these two approaches because one does not always know in advance which species will be detected. However, choosing between these two possibilities is important since it can reduce the amount of labor and financial investment in cases where the researcher only needs better strains than he/she already has.

### 1. Primary Screening of New Isolates

This screening should be done according to morphological, physiological, nutritional, and biochemical characteristics in pure culture, with the guiding principle that more tests are better than fewer. Many such tables, lists, and tests have been published. For example, although outdated for several species, the *Bergey's Manual for Systematic Bacteriology*[38] should be consulted, at least for the genus description. The sources for species description data are the following: For *Azospirillum*,[33,39-43] for *Herbaspirillum*,[11] and for *Acetobacter*.[19] As for pseudomonads, many species of *Pseudomonas* isolated from the field are very heterogenous, vaguely defined, and often fail to fit precisely into established taxonomic subdivisions.[45] Members of the genus *Pseudomonas* can be classified into different groups based on (1) phenotypic characteristics,[46] (2) their cultural and biochemical characters,[47,48] (3) rRNA-DNA homology,[49] and (4) composition of ubiquinone and cellular fatty acids with special reference to hydroxy fatty acids.[50-52]

## C. DNA AND RNA HOMOLOGY

To relate a new strain to a known species after completion of the screening steps, DNA and RNA homology tests are reliable and very common tools. These tests are not specific to PGPR. Therefore, any general procedure of DNA and RNA homology can be used to characterize new species of PGPR such as the methods of Johnson[53] used to characterize strains of *Azospirillum*.[54,55] The main limitation of these methods is the fundamental requirement for well-defined reference strains from known culture collections for comparison with newly isolated strains.

## D. PROTEIN PROFILE ANALYSES (FINGERPRINTING)

These powerful methods are based on one- or two-dimensional sodium dodecyl sulfate polyacrylamide gel electrophoresis (SDS-PAGE). Soluble or total proteins of the tested isolates are compared with the corresponding proteins of reference strains, and the comparisons cannot only adequately differentiate reference strains but in many cases are sensitive at the species level. Large numbers of isolates can be characterized and compared in a relatively short period of time. Two-dimensional (2D)-total-protein analysis is suitable for the characterization of even very closely-related strains; strains that produce identical 2D fingerprints are highly related, if not identical.[56,57]These techniques prove useful for patenting procedures.

Each method is divided into two steps; the protein extraction step is crucial.

### 1. Soluble Protein Extraction

Soluble proteins can be obtained by strong sonication of a bacterial suspension.[58,59] The proteins present in the slurry are concentrated by mixing with acetone at a 1:5 (v/v) ratio. After centrifugation at 15,000 × g, the supernatant is discarded and the remaining acetone is evaporated from the pellet under vacuum. The resulting pellet is suspended in 62 m*M* Tris-HCl buffer (pH 6.8) supplemented with 2.3% (w/v) SDS and 5% (v/v) β-mercaptoethanol. These samples are boiled for 3 min prior to electrophoresis. Other soluble protein extraction methods described primarily for *Bradyrhizobium*[60] and *Bacillus*[61] were successfully used for the characterization of *Azospirillum*.[62,63]

### 2. Total Protein Extraction

***Method #1:***[56] Pellets of washed, stationary-phase cells (in phosphate saline buffer, pH 7.2) are suspended in 0.75 ml extraction buffer containing: 0.7 *M* sucrose, 0.5 *M* Tris, 30 m*M* HCl, 50 m*M* EDTA, 0.1 *M* KCl, and 40 m*M* dithiothreitol, and incubated for 15 min at ambient temperature. An equal volume of phenol (saturated with 50 m*M* Tris-HCl, pH 8.0) is then added. The mixture is maintained under continuous shaking. Subsequent to phase separation by centrifugation, the phenol phase is recovered and re-extracted twice with an equal volume of the extraction buffer. Proteins are precipitated from the phenol phase by the addition of 5 volumes of 0.1 *M* ammonium acetate dissolved in methanol and incubation of this mixture for several hours at –20°C. The precipitate is washed twice with cold ammonium acetate solution and finally with cold 80% acetone. The pellet is air-dried and dissolved in 75 μl of lysis buffer consisting of 9.8 *M* urea, 2% (v/v) Nonidet P-40 (LKB), 100 m*M* dithiothreitol, and 2% (v/v) of a mixture of pH 5 to 7 and pH 3.5 to 10 ampholytes (LKB) at a 5:1 ratio. Samples are stored at –80°C.

***Method #2:***[64] Each tested isolate is transferred into one well in each of two 48-well tissue culture plates (Costar) containing 500 μl of TSB (tryptic soy broth). The bacteria are incubated in the wells at 28°C for exactly 48 h. Then, glycerol is added to the fully grown cultures (final concentration, 25% [v/v]). One plate is stored at –70°C, and the other is used for further strain characterization.

The plates containing the fully grown isolates are centrifuged for 30 min at 300 × g (fixed angle) by using a special swinging adaptor. The supernatants are discarded. Each pellet is preincubated for 15 min at 37°C in 10 μl of a 1 mg/ml lysozyme solution. Total cellular proteins are extracted by boiling each pellet at 95°C for 10 min in 50 μl of sample-buffer mixture (2.5% SDS, 0.125 *M* β-mercaptoethanol, 150 m*M* Tris [pH 8.8], 4 m*M* EDTA, 0.75 *M* sucrose, 0.075% bromophenol blue) and sonicating for 10 sec. After the protein solutions have been cooled on ice, 7 μl of a 0.5 *M* iodoacetamide solution is added to each well.

### 3. One-Dimensional Soluble Protein Profile Analysis

Samples are subjected to one-dimensional SDS-PAGE by using 12% resolving and 5% stacking gels.[60,63] A 10 μl sample is loaded on gels containing 50 to 100 μg of protein per well, as estimated by the Bradford procedure.[65] After standard electrophoresis, the gels are stained for 1 h in 0.2% (w/v) Coomassie Brilliant Blue R-250, rinsed overnight in 7% acetic acid, and destained in a solution containing absolute methanol-glacial acetic acid-water (9:2:9, v/v/v). After photographing, the gel is scanned with a gel scanner at the narrowest slit width to give maximum resolution.

The photographs of the fingerprints are compared and sorted in fingerprint types (FPTs), which are numbered sequentially. FPTs are identified by classical biochemical tests in combination with commercial identification kits.[64] This type of characterization can also be based on cell envelope protein patterns as well as by the analysis of lipopolysaccharides.[5]

### 4. 2D Total Protein Profile Analysis

The first dimension of 2D gel electrophoresis, isoelectric focusing (IEF), is carried out on IEF rod gels (8 × 0.1 cm) containing 9.8 *M* urea, 2% (v/v) Nonidet P-40, 6.4% (v/v) ampholytes pH 5 to 7, 1.3% (v/v) ampholytes pH 3.5 to 10, 0.23% *N,N′*-methylenebisacrylamide, and 4% acrylamide. Following prefocusing (15 min at 200 V, 30 min at 300 V, and 60 min at 400 V), 10 µl samples are loaded at the basic end of the gels and overlaid with 10 µl of a solution containing 8 *M* urea, 1% (v/v) ampholytes, 5% (v/v) Nonidet P-40, and 100 m*M* dithiothreitol. The upper (cathode) buffer consists of 20 m*M* NaOH; the lower (anode) buffer consists of 10 m*M* $H_3PO_4$. Focusing to equilibrium is conducted for 20 h at 400 V.

The second dimension (SDS-PAGE) is carried out in slab gels (5% stacking gel, 15% separating gel, 0.1% SDS). The IEF gel rods are extruded directly onto the stacking gels and covered with equilibration buffer (60 m*M* Tris-HCl, pH 6.8, 2% SDS, 100 m*M* dithiothreitol, 10% glycerol, 0.002% bromophenol blue). Low molecular weight range standards should be used. After equilibration for 10 min, gels are run at 15 mA, stained with Coomassie Brilliant Blue R-250, and dehydrated on a slab gel dryer. Results are analyzed from photographs taken of the gels.[56]

## E. STRAIN IDENTIFICATION BY RFLP ANALYSIS

The technique of restriction fragment length polymorphism (RFLP) analysis is particularly useful for specific strain identification.[67] With this procedure, the total genomic DNA is isolated and cleaved with one or a number of endonucleases. The ensuing genomic DNA fragments are size-separated by agarose gel electrophoresis and probed with a cloned DNA fragment. Specificity is determined by both the probe-target sequence and restriction endonuclease digestion patterns. Probe-target sequences are often repetitive, thereby increasing detection sensitivity. Frequently used restriction enzymes have a 6-bp (base pair) recognition site. These two factors, (1) repetitive target sequences and (2) very short enzyme recognition site, essentially eliminate strain variation due to sequence losses or rearrangement. They also generate an RFLP profile that, when compared with a strain's RFLP 'standard' profile, confirms or refutes strain identity.

1. Genomic DNA extraction. Pellets from 20 ml of stationary-phase cells are suspended in 50 m*M* Tris, pH 8.0, containing 20% sucrose, treated with 1 mg/ml lysozyme after the addition of EDTA to a final concentration of 25 m*M*, and lyzed in 0.5% SDS (added as 10%). Following digestion with RNase A (40 µg/ml) and proteinase K (20 µg/ml), DNA is banded in $CsCl_2$ gradients in the presence of ethidium bromide, precipitated with isopropanol, washed with 70% ethanol, and lyophilized. Pellets are suspended in distilled $H_2O$, and DNA concentrations are determined by absorption at 260 nm.
2. Restriction digests and probes. 3 µg of genomic DNA are incubated with 15 units of the appropriate restriction enzyme (e.g., *Eco*RI, *Pst*I or *Pvu*II) in a total volume of 28 µl for 2 h. A plasmid probe (pAM141) is labeled using either the Oligo Labeling Kit (No. 27-9250-01, Pharmacia) or the DNA Labeling and Detection Kit (nonradioactive, No. 1093-625, Boehringer Mannheim). Prehybridization and hybridization are done in 50% formamide, 5% dextran sulfate, 5% blocking agent (Boehringer Mannheim), 5X SSC (1X SSC = 0.15 *M* NaCl plus 0.015 *M* sodium citrate), 0.1% N-lauroylsarcosine, and 0.02% SDS at 42°C. Blots are hybridized for at least 6 h at 42°C and washed at room temperature for 2X 5 min in 2X SSC, 0.1% SDS, and at 68°C for 2X 15 min in 0.1% SSC, 0.1% SDS.
3. Electrophoresis and blots. Restricted samples of genomic DNA are subjected to electrophoresis in 1% agarose[68] and vacuum blotted to GeneScreen Plus (NEN). Size markers are produced by cutting phage lambda DNA (BRL) with both *Hind*III and *Bgl*II.

### F. STRAIN IDENTIFICATION BY GAS CHROMATOGRAPHIC ANALYSIS

The method of identifying bacterial strains by gas chromatographic analysis[69] of cellular fatty acids is very useful and very accurate and can identify strains at the species level. However, an individual researcher needs to have easy access to a major bacterial culture collection. In addition, manual analysis of gas chromatograph patterns is extremely laborious, especially when there is no primary clue as to the identification of the genus in question. Computer software, which screens aerobic and clinical bacterial libraries, is commercially available. However, unless a laboratory is set up to do this on a routine basis, commercial identification services using this technique are recommended.

## V. STORAGE

Isolates can be stored for short periods on 1/10 TSBA slants (OAB[28] or the combined carbon medium[16] for diazotrophs) at 2°C. For prolonged periods, isolates can be stored in 30% glycerol solutions at –15 and –70°C.[37] For indefinite storage, isolates should be lyophilized to dryness by any conventional lyophilization technique.

## ACKNOWLEDGMENTS

Y. B. participated in this paper for the memory of the late Mr. Avner Bashan from Israel. We thank Mr. Roy Bowers for constructive English corrections.

## REFERENCES

1. **Bashan, Y. and Levanony, H.,** Current status of *Azospirillum* inoculation technology: *Azospirillum* as a challenge for agriculture, *Can. J. Microbiol.,* 36, 591, 1990.
2. **Kloepper, J. W., Lifshitz, R., and Schroth, M. N.,** *Pseudomonas* inoculants to benefit plant production, in *ISI Atlas of Science: Animal and Plant Sciences,* 60, 1988.
3. **Kloepper, J. W., Lifshitz, R., and Zablotowicz, R. M.,** Free-living bacterial inocula for enhancing crop productivity, *Trends Biotechnol.,* 7, 39, 1989.
4. **Tipping, E. M., Onofriechuk, E. E., Zablotowicz, J. W., Kloepper, J. W., and Lifshitz, R.,** Screening of bacteria isolated from peat for biocontrol of *Pythium ultimum,* in *Proceedings Symposium Peat and Peatlands,* Vol. II, Overend, R. P. and Juglum, J. K., Eds., Canadian Society of Peat and Peatlands, St. Foy, Quebec, 1989.
5. **Van Peer, R., Punte, H. L., De Weger, L. A., and Schippers, B.,** Characterization of root surface and endorhizosphere Pseudomonads in relation to their colonization of roots, *Appl. Environ. Microbiol.,* 56, 2462, 1990.
6. **Kremer, R. J., Begonia, M. F., Stanley, L., and Lanham, E. T.,** Characterization of rhizobacteria associated with weed seedlings, *Appl. Environ. Microbiol.,* 56, 1649, 1990.
7. **Lalande, R., Bissonnette, N., Coutlée, D., and Antoun, H.,** Identification of rhizobacteria from maize and determination of their plant-growth promoting potential, *Plant Soil,* 115, 7, 1989.
8. **Gould, W. D., Hagedorn, C., Bardinelli, T. R., and Zablotowicz, R. M.,** New selective media for enumeration and recovery of fluorescent pseudomonads from various habitats, *Appl. Environ. Microbiol.,* 49, 28, 1985.
9. **Hagedorn, C., Gould, W. D., and Bardinelli, T. R.,** Rhizobacteria of cotton and their repression of seedling disease pathogens, *Appl. Environ. Microbiol.,* 55, 2793, 1989.
10. **De Freitas, J. R. and Germida, J. J.,** Plant growth promoting rhizobacteria for winter wheat, *Can. J. Microbiol.,* 36, 265, 1990.
11. **Lindberg, T. and Granhall, U.,** Isolation and characterization of dinitrogen-fixing bacteria from the rhizosphere of temperate cereals and forage grasses, *Appl. Environ. Microbiol.,* 48, 683, 1984.
12. **Thomas-Bauzon, D., Weinhard, P., Villecourt, P., and Balandreau, J.,** The spermosphere model. I. Its use in growing, counting, and isolating $N_2$-fixing bacteria from the rhizosphere of rice, *Can. J. Microbiol.,* 28, 922, 1982.

13. **Postgate, J. R.,** Viable counts and viability, in *Methods in Microbiology,* Vol. 1, Norris, F. N. and Ribbons, D. W., Eds., Academic Press, New York, 1969, 611.
14. **Hardy, R. W., Burns, R. C., and Holsten, R. D.,** Application of the acetylene-ethylene assay for measurement of nitrogen-fixation, *Soil Biol. Biochem.,* 5, 47, 1973.
15. **Omar, A. M. N., Richard, C., Weinhard, P., and Balandreau, J.,** Using the spermosphere model technique to describe the dominant nitrogen-fixing microflora associated with wetland rice in two Egyptian soils, *Biol. Fertil. Soils,* 7, 158, 1989.
16. **Rennie, R. J.,** A single medium for the isolation of acetylene-reducing (dinitrogen-fixing) bacteria from soils, *Can. J. Microbiol.,* 27, 8, 1981.
17. **Döbereiner, J.,** Isolation and identification of root associated diazotrophs, *Plant Soil,* 110, 207, 1988.
18. **Döbereiner, J. and Day, J. M.,** Associative symbioses in tropical grasses: characterization of microorganisms and dinitrogen-fixing sites, in *Proc. 1st Int. Symp. Nitrogen Fixation,* Newton, W. E. and Nyman, C. J., Eds., Washington State University Press, Pullman, U.S.A., 1976, 518.
19. **Cavalcante, V. A. and Döbereiner, J.,** A new acid-tolerant nitrogen-fixing bacterium associated with sugarcane, *Plant Soil,* 108, 23, 1988.
20. **Stolp, H. and Gadkari, D.,** Nonpathogenic members of the Genus *Pseudomonas,* in *The Prokaryotes,* Vol. I, 2nd ed., Starr, M. P., Stolp, H., Trüper, H. G., Balows, A., and Schlegel, H. G., Eds., Springer-Verlag, New York, 1986, 722.
21. **King, E. D., Ward, M. K., and Raney, D. E.,** Two simple media for the demonstration of pyocyanin and fluorescin, *J. Lab. Clin. Med.,* 44, 301, 1954.
22. **Kado, C. I. and Heskett, M. G.,** Selective media for isolation of *Agrobacterium, Corynebacterium, Erwinia, Pseudomonas,* and *Xanthomonas, Phytopathology,* 60, 969, 1970.
23. **Hagedorn, C. and Holt, J. G.,** Ecology of soil arthrobacters in Clarion-Webster toposequences of Iowa, *Appl. Microbiol.,* 29, 211, 1975.
24. **Sands, D. C. and Rovira, A. D.,** Isolation of fluorescent pseudomonads with a selective medium, *Appl. Microbiol.,* 20, 513, 1970.
25. **Berge, O., Heulin, T., and Balandreau, J.,** Diversity of diazotroph populations in the rhizosphere of maize (*Zea mays* L.) growing on different French soils, *Biol. Fertil. Soils,* 11, 210, 1991.
26. **Tyler, M. E., Milam, J. R., Smith, R. L., Schank, S. C., and Zuberer, D. A.,** Isolation of *Azospirillum* from diverse geographic regions, *Can. J. Microbiol.,* 25, 693, 1979.
27. **New, P. B. and Kennedy, I. R.,** Regional distribution and pH sensitivity of *Azospirillum* associated with wheat roots in eastern Australia, *Microb. Ecol.,* 17, 299, 1989.
28. **Okon, Y., Albrecht, S. L., and Burris, R. H.,** Methods for growing *Spirillum lipoferum* and for counting it in pure culture and in association with plants, *Appl. Environ. Microbiol.,* 33, 85, 1977.
29. **Reinhold, B., Hurek, T., Niemann, E.-G., and Fendrik, I.,** Close association of *Azospirillum* and diazotrophic rods with different root zones of Kallar grass, *Appl. Environ. Microbiol.,* 52, 520, 1986.
30. **Rodríguez Cáceres, E. A.,** Improved medium for isolation of *Azospirillum* spp., *Appl. Environ. Microbiol.,* 44, 990, 1982.
31. **Bashan, Y. and Levanony, H.,** An improved selection technique and medium for the isolation and enumeration of *Azospirillum brasilense, Can. J. Microbiol.,* 31, 947, 1985.
32. **Horemans, S., Demarsin, S., Neuray, J., and Vlassak, K.,** Suitability of the BLCR medium for isolating *Azospirillum brasilense, Can. J. Microbiol.,* 33, 806, 1987.
33. **Reinhold, B., Hurek, T., Fendrik, I., Pot, B., Gillis, M., Kersters, K., Thielemans, S., and de Ley, J.,** *Azospirillum halopraeferens* sp. nov., a nitrogen fixing organism associated with roots of Kallar grass (*Leptochloa fusca* (L.) Kunth), *Int. J. Syst. Bacteriol.,* 37, 43, 1987.
34. **Quesada, E., Ventosa, A., Rodríguez-Valera, F., and Ramos-Cormenzana, A.,** Types and properties of some bacteria isolated from hypersaline soils, *J. Appl. Bacteriol.,* 53, 155, 1982.
35. **Kole, M. M., Page, W. J., and Altosaar, I.,** Distribution of *Azotobacter* in Eastern Canadian soils and in association with plant rhizospheres, *Can. J. Microbiol.,* 34, 815, 1988.
36. **Page, W. J.,** Sodium-dependent growth of *Azotobacter chroococcum, Appl. Environ. Microbiol.,* 51, 510, 1986.
37. **Juhnke, M. E., Mathre, D. E., and Sands, D. C.,** Identification and characterization of rhizosphere-competent bacteria of wheat, *Appl. Environ. Microbiol.,* 53, 2793, 1987.
38. *Bergey's Manual for Systematic Bacteriology,* Williams & Wilkins, U.S.A.
39. **Bashan, Y., Singh, M., and Levanony, Y.,** Contribution of *Azospirillum brasilense* Cd to growth of tomato seedlings is not through nitrogen fixation, *Can. J. Bot.,* 67, 2429, 1989.
40. **Bashan, Y., Mitiku, G., Whitmoyer, R. E., and Levanony, H.,** Evidence that fibrillar anchoring is essential for *Azospirillum brasilense* Cd attachment to sand, *Plant Soil,* 132, 73, 1991.
41. **Döbereiner, J., Marriel, I. E., and Nery, M.,** Ecological distribution of *Spirillum lipoferum* Beijerinck, *Can. J. Microbiol.,* 22, 1464, 1976.

42. **Magalhães, F. M. M. and Döbereiner, J.,** [Occurrence of *Azospirillum amazonense* in some amazonian ecosystems], *Rev. Microbiol., São Paulo,* 15, 246, 1984 (in Portugese).
43. **Tarrand, J. J., Krieg, N. R., and Döbereiner, J.,** A taxonomic study of the *Spirillum lipoferum* group, with descriptions of a new genus, *Azospirillum* gen. nov. and two species, *Azospirillum lipoferum* (Beijerinck) comb. nov. and *Azospirillum brasilense* sp. nov., *Can. J. Microbiol.,* 24, 967, 1978.
44. **Baldani, J. I., Baldani, V. L. D., Seldin, L., and Döbereiner, J.,** Characterization of *Herbaspirillum seropedicae* gen. nov., sp. nov., a root-associated nitrogen-fixing bacterium, *Int. J. Syst. Bacteriol.,* 36, 86, 1986.
45. **Lifshitz, R., Kloepper, J. W., Scher, F. M., Tipping, E. M., and Laliberté, M.,** Nitrogen-fixing pseudomonads isolated from roots of plants grown in the Canadian high arctic, *Appl. Environ. Microbiol.,* 51, 251, 1986.
46. **Stanier, R. Y., Palleroni, N. J., and Doudoroff, M.,** The aerobic pseudomonads: a taxonomic study, *J. Gen. Microbiol.,* 43, 159, 1966.
47. **Hugh, R. and Gilardi, G. L.,** *Pseudomonas,* in *Manual of Clinical Microbiology,* 2nd ed., American Society for Microbiology, Washington, D.C., 250, 1974.
48. **Holding, A. J. and Collee, J. G.,** Routine biochemical tests, in *Methods in Microbiology,* Vol. 6A, Norris J. R. and Ribbons, D. W., Eds., Academic Press, New York, 1971, 1.
49. **Palleroni, N., Kunisawa, J. R., and Contopoulou, R.,** Nucleic acid homologies in the genus *Pseudomonas, Int. J. Syst. Bacteriol.,* 23, 333, 1973.
50. **Ikimoto, S., Kuraishi, H., Komagata, K., Azuma, M., Suto, T., and Murooka, H.,** Cellular fatty acid composition in *Pseudomonas* species, *J. Gen. Appl. Microbiol.,* 24, 199, 1978.
51. **Oyaizu, H. and Komagata, K.,** Grouping of *Pseudomonas* species on the basis of cellular fatty acid composition and their quinone system with special reference to the existence of 3-hydroxy fatty acids, *J. Gen. Appl. Microbiol.,* 29, 17, 1983.
52. **Watanabe, I., Rolando S. O., Ladha, J. K., Katayama-Fujimura, Y., and Kuraishi, H.,** A new nitrogen-fixing pseudomonad: *Pseudomonas diazotrophicus* sp. nov. isolated from the root of wetland rice, *Can. J. Microbiol.,* 33, 670, 1987.
53. **Johnson, J. L.,** Genetic characterization, in *Manual of Methods for General for General Bacteriology,* Gerhardt, P., Murray, R. G. E., Costilow, R. N., Nester, E. W., Wood, W. A., Krieg, N. R., and Phillips, G. B., Eds., American Society for Microbiology, Washington, D.C., 1981, 450.
54. **Falk, E. C., Döbereiner, J., Johnson, J. L., and Krieg, N. R.,** Deoxyribonucleic acid homology of Azospirillum amazonense Magalhães et al. 1984 and emendation of the description of the genus *Azospirillum, Int. J. Syst. Bacteriol.,* 35, 117, 1985.
55. **Falk, E. C., Johnson, J. L., Baldani, V. L. D., Döbereiner, J., and Kreig, N. R.,** Deoxyribonucleic and ribonucleic acid homology studies of the genera *Azospirillum* and *Conglomeromonas, Int. J. Syst. Bacteriol.,* 36, 80, 1986.
56. **De Mot, R. and Vanderleyden, J.,** Application of two-dimensional protein analysis for strain fingerprinting and mutant analysis of *Azospirillum* species, *Can. J. Microbiol.,* 35, 960, 1989.
57. **Lambert, B., Meire, P., Joos, H., Lens, P., and Swings, J.,** Fast-growing, aerobic, heterotrophic bacteria from the rhizosphere of young sugar beet plants, *Appl. Environ. Microbiol.,* 56, 3375, 1990.
58. **Bally, R., Givaudan, A., Bernillon, J., Heulin, T., Balandreau, J., and Bardin, R.,** Numerical taxonomic study of three $N_2$-fixing yellow-pigmented bacteria related to *Pseudomonas paucimobilis, Can. J. Microbiol.,* 36, 850, 1990.
59. **Kersters, K. and De Ley, J.,** Identification and grouping of bacteria by numerical analysis of their electrophoretic protein patterns, *J. Gen. Microbiol.,* 87, 333, 1975.
60. **Kamicker, B. J. and Brill, W. J.,** Identification of *Bradyrhizobium japonicum* nodule isolates from Wisconsin soybean farms, *Appl. Environ. Microbiol.,* 51, 487, 1986.
61. **Shivakumar, A. G., Gundling, G. J., Benson, T. A., Casuto, D., Miller, M. F., and Spear, B. B.,** Vegetative expression of the delta-endotoxin genes of *Bacillus thuringiensis,* sub-species Kurstaki in *Bacillus subtilis, J. Bacteriol.,* 166, 194, 1986.
62. **Bilal, R., Rasul, G., Qureshi, J. A., and Malik, K. A.,** Characterization of *Azospirillum* and related diazotrophs associated with roots of plants growing in saline soils, *World J. Microbiol. Biotechnol.,* 6, 46, 1990.
63. **Sundaram, S., Arunakumari, A., and Klucas, R. V.,** Characterization of azospirilla isolated from seeds and roots of turf grass, *Can. J. Microbiol.,* 34, 212, 1988.
64. **Lambert, B., Leyns, F., Van Rooyen, F., Gosselé, F., Papon, Y., and Swings, J.,** Rhizobacteria of maize and their antifungal activities, *Appl. Environ. Microbiol.,* 53, 1866, 1987.
65. **Bradford, M.,** A rapid and sensitive method for the quantitation of microgram quantities of protein utilizing the principle of protein-dye binding, *Anal. Biochem.,* 72, 248.
66. **O'Farrell, P. H.,** High resolution two-dimensional electrophoresis of proteins, *J. Biol. Chem.,* 250, 4007, 1975.

67. **Brown, G., Khan, Z., and Lifshitz, R.,** Plant growth promoting rhizobacteria: strain identification by restriction fragment length polymorphisms, *Can. J. Microbiol.*, 36, 242, 1990.
68. **Maniatis, T., Fritsch, E. F., and Sambrook, J.,** Molecular cloning: a laboratory manual, Cold Spring Harbor Laboratory, Cold Spring Harbor, NY, 1982.
69. **Sasser, M.,** Identification of bacteria through fatty acid analysis, in *Methods in Phytobacteriology*, Klement, Z., Rudolph, K., and Sands, D. C., Eds., Akademiai Kiado, Budapest, 1990, 199.
70. **Holguin, G., Guzman, M. A., and Bashan, Y.,** Two new nitrogen-fixing bacteria from the rhizosphere of mangrove trees: their isolation, identification and in vitro interaction with rhizosphere *Staphylococcus* sp., *FEMS Microbiol. Ecol.*, 101, 207, 1992.

# Index

## A



## B

## C

## D

## E

## H



## I

## K



## L



## M

## N



## O

## Q



## R

## S

## T

## U



## V



## W

## X